International Conference on Energy, Power and Mechanical Engineering (EPME2019)

IOP Conference Series: Materials Science and Engineering
Volume 793

Guangzhou, China
20 - 22 December 2019

ISBN: 978-1-7138-3505-9
ISSN: 1757-8981

Printed from e-media with permission by:

Curran Associates, Inc.
57 Morehouse Lane
Red Hook, NY 12571

Some format issues inherent in the e-media version may also appear in this print version.

Institute of Physics
Dirac House, Temple Back
Bristol BS1 6BE UK

Phone: 44 1 17 929 7481
Fax: 44 1 17 920 0979

techtracking@iop.org

Additional copies of this publication are available from:

Curran Associates, Inc.
57 Morehouse Lane
Red Hook, NY 12571 USA
Phone: 845-758-0400
Fax: 845-758-2633
Email: curran@proceedings.com
Web: www.proceedings.com

TABLE OF CONTENTS

IOP Conf. Series: Materials Science and Engineering **793** (2020) 011001 doi:10.1088/1757-899X/793/1/011001

Preface

As one of the seminars of 2019 International Academic Exchange Conference on Science and Technology Innovation, the International Conference on Energy, Power and Mechanical Engineering (EPME 2019) was successfully held in Guangzhou, China on December 20-22, 2019. Thanks to all participants for their enthusiastic attendance, we all had a productive conference and enjoyed three wonderful days in Guangzhou.

We are honored to invite Prof. David Bassir from French University of Technology as the conference Chairman. During the conference, Chairman David Bassir first had a opening speech to welcome all scholars and attendees, then he also gave a keynote speech on his research on drones, aka unmanned aerial vehicles (UAVs). Followed by Prof. Bassir's wonderful talk, we had Dr. Dae-Eun Kim from Yonsei University to present his insightful speech on machine friction and tribology. Finally, we had Prof. Longsheng Lu from School of Mechanical and Automotive Engineering, South China University of Technology to perform his speech on mechanism engineering. Heated Q&A was held after the keynote speeches. Even during the tea break, scholars continued to discuss these topics, forming a lively and strong academic vibe on site.

During the conference, we received numerous papers from scholars of different universities and institutes, then we selected some excellent paper and compiled this proceedings of EPME 2019 after the rigid review process. These papers cover a variety of topics on energy & power engineering, mechatronics, vehicle engineering, intelligent manufacturing, aerospace engineering, mechanical engineering, electronic equipment and application, and other related areas. We guaranteed all paper have met the requirements of international publication and are ready to be published.

At last, we would to like to express our sincere gratitude to the conference Chairman, the distinguished keynote speakers and all the participants attending the conference, who made this conference a great success. We hope this conference and the proceedings of it could serve as a good reference for those who work in the field of Energy, Power and Mechanical Engineering. We would like to see you next year. Thank you.

The Committee of EPME 2019

IOP Conf. Series: Materials Science and Engineering **793** (2020) 011001 doi:10.1088/1757-899X/793/1/011001

Committee member

Chairman of the Seminar

David Bassir, Mathematical Center & Applications (CMLA), CNRS -UMR 8536 –ENS Cachan French University of Technolog

Speaker at the seminar

David Bassir, Mathematical Center & Applications (CMLA), CNRS -UMR 8536 –ENS Cachan French University of Technolog

Dae-Eun Kim, Department of Mechanical Engineering, Yonsei University

Longsheng LU, South China University of Technology Deputy Dea

Academic Committee

Ebrahim Babaei, University of Tabriz

Qibai Huang, Huazhong University of Science and Technology

Li Guo, Hunan University

Tian Soon Lee, Multimedia University

FengLu, Shenyang Jianzhu University

Guangyi Ma, Dalian University of Technology

Lyudmila Mihaylova, University of Sheffield

John Mo, RMIT University

M.Shahidul Islam, Universiti Malaysia Sarawak

Anil L. Wanare, Savitribai Phule Pune University

Xuewen Wang, Taiyuan University of Technology

Chunbo Xiu, Tianjin Polytechnic University

Liudong Xing, University of Massachusetts

Asher Yahalom, Ariel University

Chunbo Zhu, Harbin Institute of Technology

Youqun Zhao, Nanjing University of Aeronautics and Astronautics

IOP Conf. Series: Materials Science and Engineering **793** (2020) 011002 doi:10.1088/1757-899X/793/1/011002

Peer review statement

All papers published in this volume of *IOP Conference Series: Materials Science and Engineering* have been peer reviewed through processes administered by the proceedings Editors. Reviews were conducted by expert referees to the professional and scientific standards expected of a proceedings journal published by IOP Publishing.

IOP Conf. Series: Materials Science and Engineering **793** (2020) 012001 doi:10.1088/1757-899X/793/1/012001

Discussion on the recycling ecological chain and commercial model of decommissioned batteries

Juanjuan Yin [1*], Mingyu Pan [2], Zhen Chen [2], Jing Li [1]

[1] Sichuan Energy Internet Research Institute, Tsinghua University, Chengdu 610213, China

2 State Grid Electric Power Research Institute of China, Beijing100075, China

1446131529@qq.com

Abstract. When the battery capacity is reduced to less than 80%, it is no longer suitable for electric vehicles, if the decommissioned batteries directly scrapped treatment, it will cause a great waste of resources, and even cause environmental pollution. The reuse mode of lithium battery can be divided into two modes: Cascade utilization and recovery. This dissertation introduces these two methods, summarizes the current situation of power battery reuse, draws on the successful experience domestic and foreign, and combines the actual situation of domestic enterprises to improve the reuse ecosystem and business model of decommissioned batteries. The paper puts forward the business model of 'Internet + Recycling Network + specialization processing', and constructs the closed-loop network of retired power battery recovery, which integrates government supervision, enterprise operation and consumer participation.

1. Introduction

In recent years, with the rapid development of electric vehicle industry in China, power batteries as the core component have become a research hotspot in the industry. With excellent characteristics and mature technologies, lithium batteries have snatched a huge share of commercial market in its application on new-energy vehicles[1]. The sales of new-energy vehicles and shipment of lithium batteries are shown in Figure 1. According to the battery industry development target put forward in the 'Action Plan for Promoting Development of Automobile Power Battery Industry' issued in March 2017, the total capacity of the power battery industry will exceed 100 GWh by 2020. The large-scale production, sales and utilization of power batteries will definitely result in large-scale scrapping and recycling. When the capacity of a power battery attenuates to below 80% of that when it is initially used, it will fail to meet the standards for use by electric vehicles. This requires recycling and reuse of decommissioned power batteries[2]. Estimation of decommissioning of lithium power battery is shown in Figure 2.

Under the situation that the recycling market has not yet fully developed, an enormous quantity of decommissioned batteries will bring about a series of sever problems. On the one hand, the gap in the recycling market makes it impossible to stack decommissioned batteries in fixed places. Randomly throwing or stacking of batteries tends to cause cracking of the battery shells in worse environmental conditions. This will expose the anode and cathode and the electrolyte in the outside world, causing chemical reaction and resulting in such environmental pollution as heavy metal pollution and high alkali concentration of soil[3,4]. On the other hand, decommissioned power batteries still have 70~80% capacity. The remaining capacity and power can be used to satisfy requirements lower than that in the new-energy vehicles. Direct scrapping will be a huge waste of the remaining value of power batteries. If their value of a full life cycle can be utilized, its cost-efficiency can be effectively improved[5]. Moreover, since the battery-driven system forms 30%~45% of the total cost of electric vehicles and power batteries 75~85% of the total cost of the battery-driven system, the lack of a recycling system for decommissioned batteries is not conducive to lowering the cost and sales price of new-energy vehicles and will seriously hinder the development of the new-energy vehicle industry. Therefore, cascade utilization and recycling of batteries will be an effective solution to the problem related to decommissioned batteries and the technologies involved therein will play a key role in forming a closed loop of new-energy vehicle industrial chains[6].

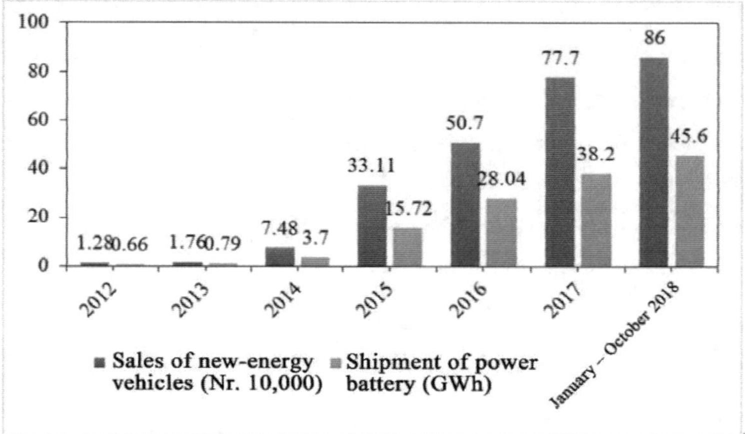

Figure 1. Sales of New-energy Vehicles and Shipment of Power Batteries

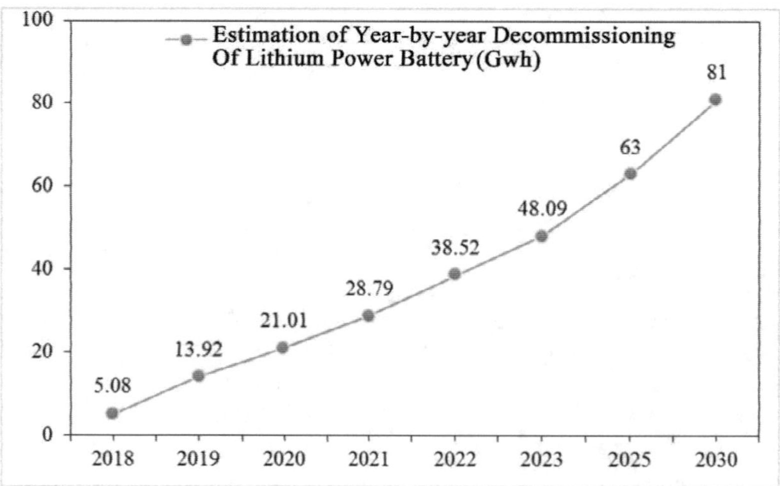

Figure 2. Estimation of Decommissioning of Lithium Power Battery

At present, China has unveiled various policies guiding the establishment of a market for recycling of decommissioned lithium batteries. The 'Technical Policy for Recycling Power Batteries of Electric Vehicles' (2015 Revision) acts as a guidance for enterprises in reasonably carrying out design, production and recycling of power batteries for electric vehicles for the purpose of building a power battery recycling system with both upstream and downstream enterprises involved. In the same year, the Ministry of Industry and Information Technology released the 'Requirements of the Industry Standards for the Comprehensive Utilization' which puts forward requirements for scaled, standard, professional and comprehensive use of power batteries. The Extended Producer Responsibility System defines power battery producers as the responsible bodies for power battery recycling. According to industrial analysis, the lithium battery recycling market is reckoned to reach a scale of RMB 15.6 billions by 2020 and the compound growth rate from 2018 to 2020 will be 40.84%. In this dissertation, regarding decommissioned batteries, and issues of how to take up a highly valuable market and how to actively and correctly respond to national policies, the basis for dividing the recycling system and ecosphere is explored, and a new commercial model for reference by industry is proposed.

2. Ecosystem for Recycling of Decommissioned Lithium Battery

In the current stage, it is urgent to systematize and commercialize the recycling ecosystem. Major problems faced by the green battery recycling industry of China include: 1) Difficulty in battery recycling; It is estimated that 2018 will witness a peak period of power battery decommissioning. However, there is yet no specialized battery recycling network in China and no mandatory or incentive policy requiring or encouraging customers to spontaneously return decommissioned batteries. The scattered and extensive way of recycling will undoubtedly cause environmental pollution and even disorder of the industry. 2) Certain risks to be borne by enterprises; According to national laws and regulations, only enterprises with the permit of handling dangerous waste are allowed to carry out recycling and disposal activities. With no sound market and difficulty in negotiating benefit distribution, enterprises of certain scale and meeting the conditions are not willing to engage in the lithium battery recycling business. Worse still, some small enterprises directly scrap the batteries recycled or use them as power banks, which not only disturbs the market order, but also brings environmental and safety hazards. In this section, the current industrial ecosystem is analyzed based on cascade utilization and disassembling for recycling of batteries, putting forward suggestions on standard industrial layout.

2.1. Cascade Utilization

Cascade utilization generally refers to the process in which a battery with capacity attenuates to below 80% after being used by new-energy vehicles is used again in such places as energy storage stations of commercial residential buildings and telecom base stations, and then scrapped and recycled after its capacity becomes too low. Power batteries on the market are mostly made of lithium iron phosphate (LFP) and ternary compounds. According to comparison of these two kinds of batteries discharging under 1C discharge rate, with the increase of the number of cycles, the capacity of the ternary battery attenuates rapidly while that of the lithium iron phosphate battery still remains at above 50% after 1,000 times [7]. Therefore, LFP batteries are more suitable for cascade utilization.

For cascade utilization considering battery safety and capacity, power batteries, according to their multilevel structures, can be used in different scenarios in the form of battery packs, modules, cells and so on. Currently, such scenarios for cascade utilization include telecom base stations, battery charging (swapping) stations, household energy storage, portable power sources and so on. Among them, base stations promise a huge market and are the most suitable application scenarios. China Tower Corporation Limited is a model in terms of the scale of cascade utilization. It has made more than 3,000 base stations models for cascade utilization in 12 provinces of China. These base stations have been in good operation so far with profits constantly increasing.

2.2. Disassembling for Recycling

Disassembling for recycling is a way of turning power batteries into resources to realize high-efficiency cyclic utilization of battery materials and valuable metals. Waste batteries with low safety and having gone through cascade utilization shall be disassembled for recycling, to keep valuable resources. In this case, the scarcity of upstream raw materials and the price fluctuation can be alleviated in a certain degree. Power batteries are disassembled into shells, anode and cathode , diaphragms and electrolyte after such pretreatment procedures as discharging, disassembling, crashing and separation, and are then subject to different recycling procedures to get valuable materials. Battery recycling processes generally include the dry process, wet process and biological process. The dry process can be further divided into physical separation and high-temperature pyrolysis while the wet process includes acid - alkali leaching, organic solvent extraction and ion exchange[8]. The recycling methods adopted by technically leading domestic recycling enterprises are shown in Table 1.

Table 1. Resource Recycling Methods of Technically Leading Domestic Recycling Enterprises

Enterprise	Recycling Method	Main Product
GEM Co., Ltd	Dry and wet processes	Spherical cobalt powder
Guangdong Brunp Recycling Technology Co., Ltd.	Dry and wet processes	Battery-level cobaltosic oxide and Ni-Co lithium manganate
Ganfeng Lithium	Dry and wet processes	Lithium carbonate and battery-level lithium chloride

2.3. Industrial Layout Analysis

Two principles need to be followed during building of the recycling ecosystem according to the mode of utilization of waste batteries: ① Since batteries are dangerous, safety insurance is required during transportation for recycling and treatment to avoid battery damage. National standards, codes and safety rules shall be observed; ② Since cascade utilization is mostly for wind power, PV power and energy storage applications, research on geological factors is required during selection of the site for battery's industrial ecosphere with priority given to areas where the recycling industry is mature and cascade utilization is feasible. In addition, the principle of 'consumption nearby' also needs to be considered. Upstream and downstream enterprises in the battery industry are closely related. The upstream industry involves battery material manufacturers and battery manufacturers that shall consider the actual needs of downstream new-energy vehicle manufacturers. The downstream industry involves vehicle manufacturers, automobile dealers and battery recycling enterprises that shall also share the responsibility of manufacturers in tackling battery pollution. On July 26, 2018, the Ministry of Industry and Information Technology released a list of enterprises qualified for recycling of waste power batteries. Among them are Huayou Cobalt, Highpower International, GEM, Brunp Recycling Technology and Guanghua Sci-tech. From various perspectives, battery recycling industrial layout and building of reuse ecosphere mainly focus on the Yangtze River delta, Pearl River Delta and surrounding cities of Beijing where relevant industries are rapidly developing. In such areas, it is easy to realize regional integration, carry out battery recycling and reuse, and alleviate the government supervision pressure. Meanwhile, with large number of registered electric vehicles and various recycling methods, emphasis shall be placed on developing such cities into green recycling cities to form a complete industrial chain comprising battery's key material production, manufacturing equipment, sales channels and recycling and reuse with battery technological innovation as the core, so as to create an ecosphere for cluster industrial development.

3. Commercial Mode

Relevant policies related to battery reuse have been released in recent years, constantly driving market and commercial mode adjustment. With experience at home and abroad and according to the Extended Producer Responsibility System, a new commercial mode of 'Internet + recycling network + specialized disposal' is put forward. Under this mode, the responsibility body is made clear and a reuse network driven by a profit chain with clear division of work is built and the Internet is used as a tool for optimization to form an efficient and standard management mode.

3.1. Recycling Network

The recycling network is key to the full-life-cycle ecosphere, and is also the basis for realizing healthy industrial cycles from battery decommissioning from new-energy vehicles, to battery material recycling and battery re-production. Based on different orientations, the power battery recycling network is divided into two types: producer-oriented recycling network and recycling network oriented at a professional third party.

Huayou Cobalt, Ganfeng Lithium and other battery material enterprises possess rich mineral resources and advanced production and recycling technologies, and boast recycling industry layout both at home and abroad. Domestic power battery and vehicle manufacturers such as CATL and BYD with key battery manufacturing technologies and channel advantages realize low-cost recycling and reuse in energy-storage, low-speed vehicle and other cascade utilization scenarios, forming a producer-oriented recycling network.

Domestic professional third-party recycling enterprises are mostly engaged in traditional battery recycling or comprehensive reutilization of relevant waste resources. Through market expanding, they have entered the lithium battery recycling field and accumulated certain operating experience. Nevertheless, a large quantity of manpower and material resources are still needed to establish new recycling networks and logistics channels. GEM, Ganzhou Highpower Technology and Brunp Recycling Technology as the third-party recycling enterprises have been deeply rooted in the recycling field for many years, making them the first batch of model enterprises for recycling of waste batteries. With relatively mature traditional battery and lithium battery recycling technologies and qualification in treatment of dangerous wastes, they have built a business-to-business battery full-life-cycle industrial chain. Through cooperation with the middle and down streams and expansion of the recycling network based on regional division by the enterprises themselves, they can make up for the shortage of recycling channels and alleviate the pressure from building a huge recycling network. For example, through cooperation between GEM and BYD, a production line for processing waste battery materials has been built with the annual quantity of cobalt resources recycled being more than 4,000 t. A recycling network has been built by Ganzhou Highpower Technology and BAIC BJEV together, dedicating to recycling, disposal and reuse of battery renewable resources.

In terms of setting of recycling stations, convenience for customers shall be taken as a starting point in the design of recycling station layout to solve the difficulty in vehicle's waste battery handling. In addition to fixed recycling points at the dealers, streets and bus companies, online booking service for recycling may be provided for large-capacity batteries not easy to recycle. Customers upload battery numbers and pictures to provide battery information for the recycling suppliers, for regular arrangement of personnel for battery recycling door to door. Considering cascade utilization mostly happens at base stations and communities, special recycling points are required, to satisfy recycling needs and realize online and offline collaborative operation.

3.2. Recycling System

To promote healthy circulation of battery recycling ecosphere, identification and coding systems as well as the deposit refund system may be put into full use. The National Development and Reform Commission, Ministry of Industry and Information Technology, Ministry of Ecology and Environment, Ministry of Commerce and General Administration of Quality Supervision, Inspection and Quarantine co-issued the 'Technical Policy for Recycling Power Batteries of Electric Vehicles (2015)' which clearly specifies establishment of the power battery coding system to form a waste battery traceability system. When handling relevant expenses related to waste battery recycling and disposal, battery manufacturers need to stick unified company information labels and permanent codes on their products to facilitate battery recycling and defining of responsible bodies. The battery coding system enables the government to better supervise the market, making waste battery recycling a necessity and punishing irresponsible enterprises.

Moreover, to arouse the initiative of customers in battery recycling and prevent environmental problems caused by random throwing of lithium batteries, customers may be required to pay a deposit when buying new batteries or products with batteries in a unified manner and they can get a refund of the deposit when returning the used batteries. The deposit refund system can promote battery recycling. For reference, the amount of deposit may be 1%~2% of the vehicle price. In addition, vehicle enterprises shall take responsibility for battery-caused pollution, and shall make certain payment to the foundations established by the government or a third-party organization as industrial subsidy. When the third-party enterprises finish recycling and disposal, they may get certain amount of subsidy for battery disposal [9].

3.3. Commercial Mode of 'Internet + Recycling Network + Specialized Disposal'

Under the new trend of 'Internet +', the combination of 'recycling network + specialized disposal' enables flowing of massive data. The battery databases can be established through identification and coding, for real-time tracking of the battery status; by virtue of big data, the third-party enterprises can constantly improve the battery recycling efficiency; by retrieving battery disposal information, the government can easily supervise the recycling market. Meanwhile, customers can get access to relevant battery information via APP or web pages, quickly and conveniently estimating the value of waste batteries, thus quickly placing orders or applying for battery recycling through the Internet [10]. A closed-loop battery recycling network formed based on the mode of 'Internet + recycling network + specialized disposal' is shown in Figure 3.

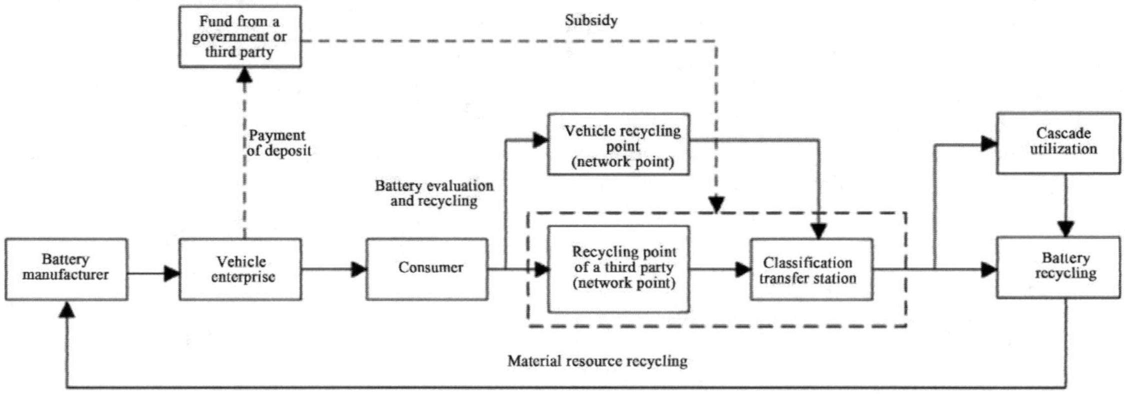

Figure 3. Recycling Network Based on the Mode of 'Internet + Recycling Network + Specialized Disposal'

4. Summary

The recycling industry of decommissioned battery in China is at the stage of development, with an urgent need of sophisticated recycling enterprises to dispose a large scale of waste batteries. Therefore, a series of relevant policies from Chinese government are required to promote an orderly market development. Such two principles as 'Safety and Consumption Nearby' shall be adhered for establishment of a recycling ecosystem. Since battery-related industries are mostly distributed in Beijing and the Yangtze River delta, layout of the ecosphere in such areas is promotive for building industrial parks of battery recycling, thus promoting development of the power battery recycling industry. Under the trend of 'Internet +' and based on the extended producer responsibility system, the closed-loop commercial mode of 'Internet + recycling network + specialized disposal' is put forward with recycling network, recycling system and fund flowing mode as the bases. Therefore, government supervision, enterprise running and customer participation are integrated, with coordinated development under the precondition of respecting the interests of all parties.

Acknowledgements

This work was supported by the State Grid Corporation Science and Technology Project of China (Grant No. 52020118000G).

References

[1] Ren D , Feng X , Han X , et al. Research progress on the life cycle safety evolution of lithium ion batteries . *Energy Storage Science and Technology*, 2018, **7 (06)**: 9-18.
[2] Neubauer J, Pesaran A. PHEV/EV li-lon battery second-use project. Denver, USA: National Renewable Energy Laboratory, 2010
[3] Pei Huang. The large-scale recycling of lithium batteries has become an industry

problem .*Energy Research and Utilization*, 2018 **(05)**: 14-15.

[4] Gao G , He X , Li Y , et al..Recycling Status of Power Lithium Ion Batteries for Waste Vehicles .*Environmental Engineering*, 2017,**35 (10)**: 135-140.

[5] Tao Z , Jia X .Research on Business Model of China's Power Battery Recycling Industry .*Auto Industry Research*, 2018 (10): 33-42.

[6] Liu W , Gao H .The urgency of China's power lithium battery recycling and countermeasures .*Industrial Economic Forum*, 2017,**04 (06)**: 26-31.

[7] Wang Y , Li Y , Zhang N , Zhang H .Comparison between power lithium iron phosphate battery and ternary battery .*Power Technology*, 2015,**39 (09)**: 1824-1825.

[8] Jiang H , Chen Y , Liu H .International Lithium-ion Battery Recycling Technology Route and Enterprise Overview .*New Materials Industry*, 2018 **(03)**: 26-30.

[9] Wu J, Wang X, Chen R, and Jia T. Discussion on the current status of waste battery recycling and the recycling model under the new situation .*Environment and Sustainable Development*, 2018,**43 (01):** 83-85..

[10] Tao Z , Jia X .Research on Business Model of China's Power Battery Recycling Industry .*Auto Industry Research*, 2018 **(10):** 33-42..

IOP Conf. Series: Materials Science and Engineering **793** (2020) 012002 doi:10.1088/1757-899X/793/1/012002

Research on electromagnetic radiation interference of high voltage pulse device

Xu Yuheng, Zhang Xiangjin, Wang Ruibo and Yue Lianyong

School of Mechanical Engineering, Nanjing University of Science and Technology, No. 200 Xiaolingwei Street, Xuanwu District,Nanjing,China

Xuyuheng1994@163.com

Abstract. In view of the strong electrostatic field interference caused by the miniaturization of high voltage pulse capacitor circuit, the protection method of electronic equipment under the strong electrostatic field interference of high voltage pulse device is proposed. Based on the background of miniaturization of all electronic security system, a high-voltage capacitor pulse discharge model is established. The surrounding radiation of high-voltage capacitor instantaneous discharge is simulated by using the COMSOL software. The electromagnetic radiation interference of high-voltage capacitor to the system is analyzed, and the anti-radiation interference measures are proposed based on the electromagnetic shielding mechanism. It provides a basis for the actual circuit design and structure layout, and provides a reference for the stability and safety of the electronic security system and further miniaturization development.

1. Introduction

The high voltage pulse discharge device is a system which uses the very insensitive characteristics of the exploding foil detonator to realize the safety control and initiation of initiating devices. In the process of the impact plate detonator, because of the large current generated by the internal metal bridge foil during the discharge of the high-voltage pulse capacitor, the electric explosion will drive the flyer to impact the insensitive explosive column and detonate it. In this process, the metal bridge foil needs to get enough energy to produce plasma in a short time, which requires that the output pulse current of high-voltage pulse power device can meet the requirements in amplitude and time characteristics[1]. With the electromagnetic environment becoming more and more complex, electromagnetic interference has become the main factor affecting the work of electronic equipment. Because of the characteristics of miniaturized ignition control system, it is very sensitive to electromagnetic interference,high-voltage switch and single-chip microcomputer and other components are easy to be damaged. The transient magnetic field strength is determined by di / dt[2]. According to the experimental results in the references, when the rise time is 10 ns, the pulse electric field with the peak value of kilovolt per meter can change the working state of the MCU and other controllers. In the high voltage pulse discharge device, the peak value of the transient current pulse of the pulse charging circuit is about 100A, the rising time of the pulse is about 10 μs, and the intensity of the induced electric field is far lower than the critical value that can cause damage to the MCU, so the boosting process will lead to the damage of the single chip microcomputer[3]. It is necessary to analyze the electrostatic field generated in the discharge process of the high-voltage converter, and design the shielding measures.

IOP Conf. Series: Materials Science and Engineering **793** (2020) 012002 doi:10.1088/1757-899X/793/1/012002

2. Electromagnetic radiation interference principle of high voltage pulse device

2.1. electromagnetic interference theory of high voltage capacitor

Generally, for initiation of impact plate detonators, the high-voltage pulse capacitor needs to reserve about 1300V high-frequency transformer for pulse boosting, convert the low voltage into high voltage and charge the capacitor through the high-voltage silicon stack until the detonation signal arrives, and the instantaneous discharge of the capacitor triggers the shock plate detonator. When the high-voltage converter is working, the capacitor will not discharge immediately after the charging reaches the predetermined target[4]. Therefore, the capacitor needs to maintain a high voltage of 1300V. The electrostatic field produced by the full charge of energy storage capacitor often makes the surrounding single chip microcomputer and other devices work abnormally or even damage[5].

The expression of electric field intensity of point charge q at any point b in space is as follows:

$$\vec{E} = \frac{q}{4\pi\varepsilon_0}\frac{(r_a - r_b)}{(r_a - r_b)^3} = \frac{q}{4\pi\varepsilon_0 R^2}\vec{a}_R \tag{1}$$

$\vec{a}_R$ is the unit vector from point b to point a.

The electric field intensity generated by the charge system at a certain point in space is equal to the vector sum of the electric field intensity generated at that point when each point of the charge system exists alone. Therefore, the electric field intensity of n-points charge to point "O" in s-plane is as follows:

$$\vec{E} = \sum_{i=1}^{n}\frac{q}{4\pi\varepsilon_0}\frac{(\vec{r_O} - \vec{r_i})}{(\vec{r_O} - \vec{r_i})^3} \tag{2}$$

$\vec{r_i}$ is the distance vector of point charge q_i to O.

The voltage of o point generated by strong electrostatic field is:

$$U = Ed \tag{3}$$

U is the potential difference, where d is the linear distance along the electric field line between two points. According to the principle of formula (1) and (2), when the energy storage capacitor of the electronic safety system is fully charged, a strong electrostatic field will be generated in the surrounding space.

The experimental results show that the strong static electric field is caused by the capacitance coupling between the devices with and without points. As shown in Figure 1, the voltage at both ends of the device is equal to the sum of the working voltage and the electrostatic field strength. The higher the voltage of the live device or the closer it is to the device, the more obvious the interference is[6].

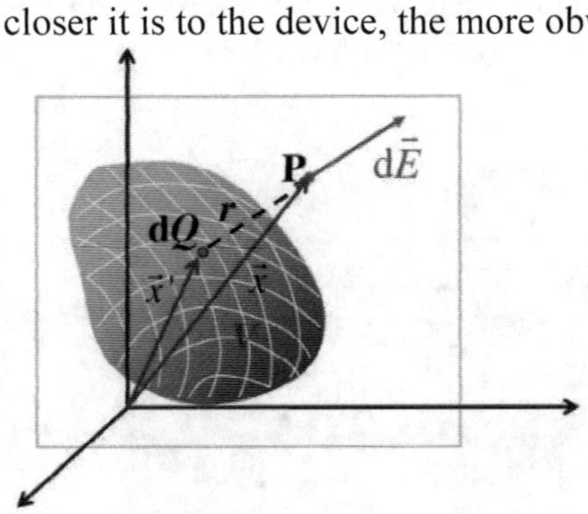

Figure 1. Schematic diagram of electric field intensity excited by space distribution point charge.

2.2. electromagnetic shielding principle

It can be seen from the above results that when the superposition voltage of the electrostatic field coupling at both ends of the device is too large and exceeds the withstand threshold of the device, the working state of the device will be disturbed. Therefore, to reduce the coupling voltage of electrostatic field as much as possible is an important condition to ensure the normal operation of MCU and other devices in the strong electrostatic field environment. When the device is placed near the side of the high-voltage plate, it will bear high voltage and cause damage to the device. Therefore, electromagnetic shielding measures should be taken. Shielding body can absorb energy, reflect energy and counteract energy for external interference electromagnetic wave and internal electromagnetic wave from conductor, cable, component, circuit or system, so shielding body has the function of reducing interference. When the frequency of electromagnetic interference is high, the eddy current produced in low resistivity metal materials is used to counteract the external electromagnetic wave, so as to achieve the shielding effect[7].When the frequency of the interference electromagnetic wave is low, the material with high permeability should be used, so that the magnetic field line is limited in the inner part of the shield to prevent diffusion to the shielding space. In the process of electromagnetic shielding, there are three main ways to attenuate the magnetic field: (1)Reflection loss of shield surfaceR_e; (2) Absorption loss through shield A; (3) Multiple reflection loss in shielding body B.

Among them, the shielding effectiveness is quantitatively evaluated by SE, $SE = R_e + A + B$[8].

$$\begin{cases} R_e = 134.68 + 10\lg(\sigma_r f r^2/\mu_r)dB \\ A = 131h\sqrt{f\mu_r\sigma_r}dB \\ B = 10\lg\begin{cases} 1 + \exp(60.32h\sqrt{f\mu_r\sigma_r})dB \\ -2\exp(30.16h\sqrt{f\mu_r\sigma_r})\cos(30.16h\sqrt{f\mu_r\sigma_r})dB \end{cases} \end{cases} \quad (4)$$

h is the thickness of the shield, σ_r is the relative conductivity, μ_r is the relative permeability, f is the frequency, and r is the distance from the shield to the radiation source.

Table 1 shows the electrical properties of some metal materials.

Table 1 Electrical properties of metallic materials

Material	$\sigma_r/(s\cdot m^{-1})$	μ_r	f/Hz	r/m
1045	5×10^6	150	10^3	10^{-3}
Iron	1×10^7	500	10^3	10^{-3}
Copper	5×10^7	1	10^3	10^{-3}
Aluminum	3×10^7	1	10^3	10^{-3}

3. Electromagnetic radiation modeling and simulation

The high voltage capacitor selected in this paper is multilayer chip ceramic capacitor (MLCC). Its main structure mainly includes three parts: ceramic medium, metal inner electrode and metal outer electrode. The multilayer chip ceramic capacitor is a multilayer composite structure, which is simply said to be a union of several simple parallel plate capacitors[9].

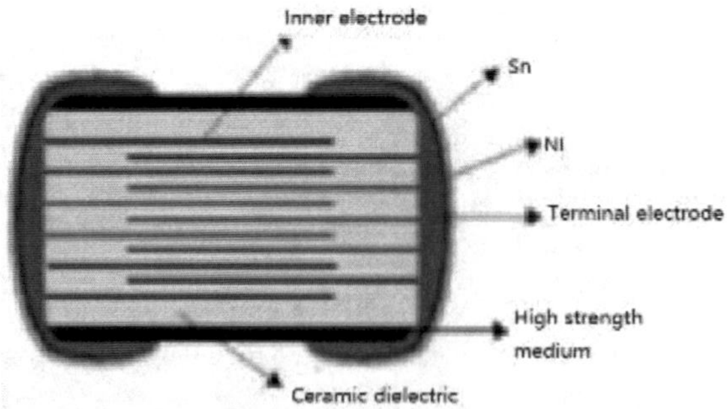

Figure 2. MLCC internal structure diagram.

From the internal structure, it can be seen that MLCC capacitors are formed by alternately laminating and pressing multiple layers of ceramic dielectric films, which is a parallel form of multiple capacitors. The voltage drop between each layer of the film is small and non-polar. Therefore, in order to simplify the model structure when modeling, the influence of the strong magnetic field generated by the high voltage of the two-layer plates on the internal structure can be ignored. In this section, the MLCC capacitor actually used is equivalent to a two-layer planar structure for modeling and analysis.

Establish two planar plates in the space. The material of the planar plates is silver-palladium alloy, and the intermediate medium is ceramic. The two plates form a capacitor, and an air field is added around the capacitor. One of the plane plates is grounded, and the other plane plate voltage is 1300V. The model grid is shown in the Figure 3.

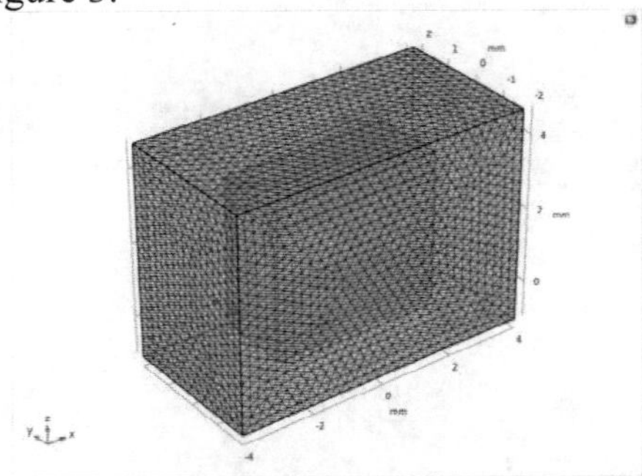

Figure 3. Mesh map of storage capacitors.

The calculation results are as follows. In order to facilitate analysis and comparison, considering that the shield model has a symmetrical type, take the central section and line segment a (from point (0,0,3.75) to point (0.0.6.5)) for observation.

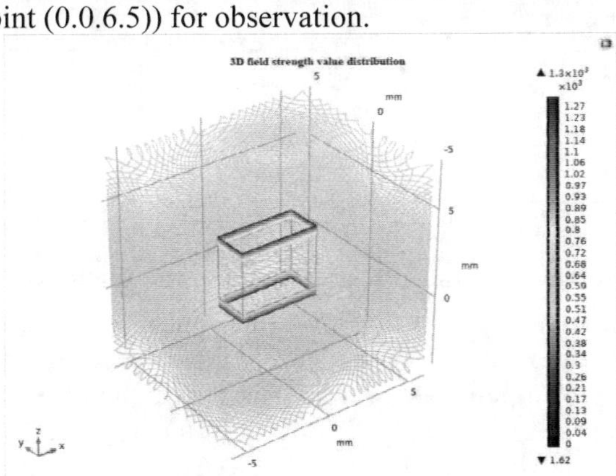

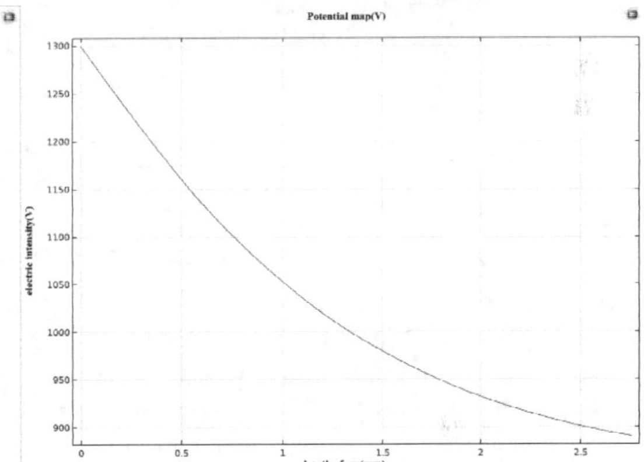

(a) 3D field strength value distribution **(b) $\vec{a}$ direction field strength value distribution.**

Figure 4. Contour map of strong electrostatic field.

Figure 4 (a) is the spatial distribution map of the strong electrostatic field contour, Figure 4 (b) is the electrostatic field distribution map in the a⃗ direction, and the abscissa is the distance from the surface of the high voltage plate, as can be seen from the figure, The maximum electric field strength is in the direction of the axis of symmetry of the capacitor team, and the farther the electric field contour line diffuses outward, the smaller the potential value is, and the potential value is about 900V to 2.5mm from the high-voltage plate. It can be seen from the figure that when the semiconductor device is arranged along the axis of the high-voltage end, the potential difference to which it is subjected is the largest.

4. Electric field shielding protection

Because this design is under the constraint of small volume, high-voltage capacitors cannot be independently arranged, and interference is easy to crosstalk through the cable between the boards. Also, the requirement of minimum volume and easy processing under the constraint of small volume, so consider using a thin-walled metal cylindrical shield, as shown in the figure. 6. Because the other side of the high-voltage capacitor is arranged as a pyrotechnic device, the device protection on the same side of the PCB is mainly considered in this design. The shield is a thin-walled cylindrical structure with a wall thickness of 0.5mm, and the material is defined as 1045 steel.

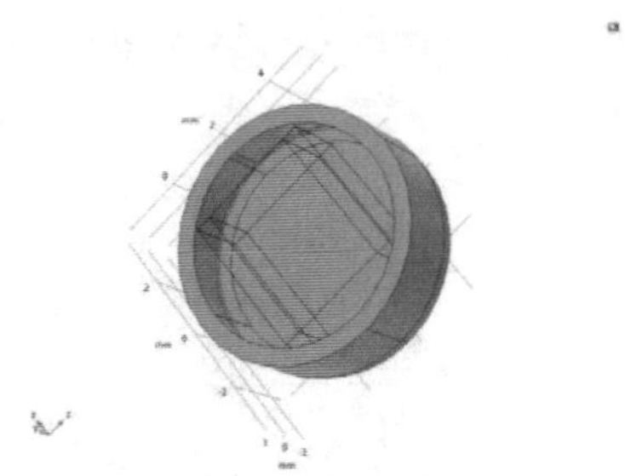

Figure 5. Shield Model.

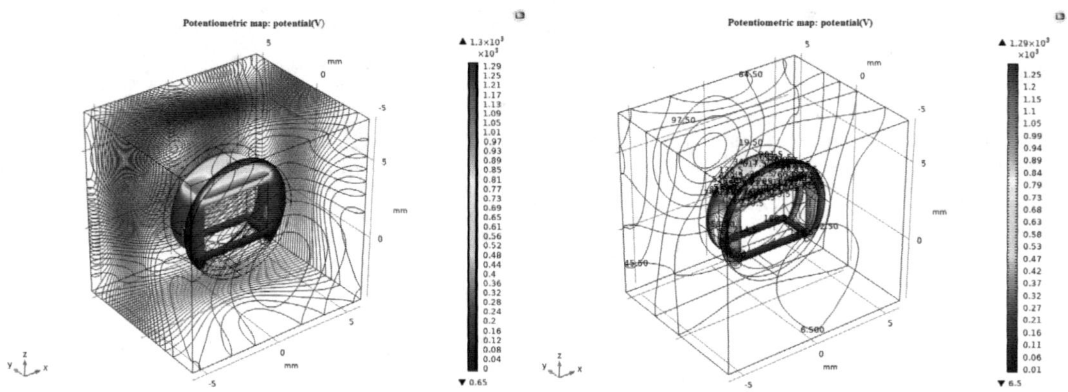

Figure 6. Contour value distribution of the strong electrostatic field of the shield.

Add 1300V potential value to the high voltage plate, and ground the low voltage plate and the shield. It can be seen from the figure that the high-voltage region is concentrated inside the cavity of the shield and the surface of the inner cavity, and the electric field rapidly decays after passing through the cavity. The electric field attenuation is greatest along the negative direction of the Y-axis, that is, the direction in which the PCB is stacked. In order to visually show the change of the electric field, the spatial distribution of the electric field is analyzed from the horizontal and vertical tangent planes.

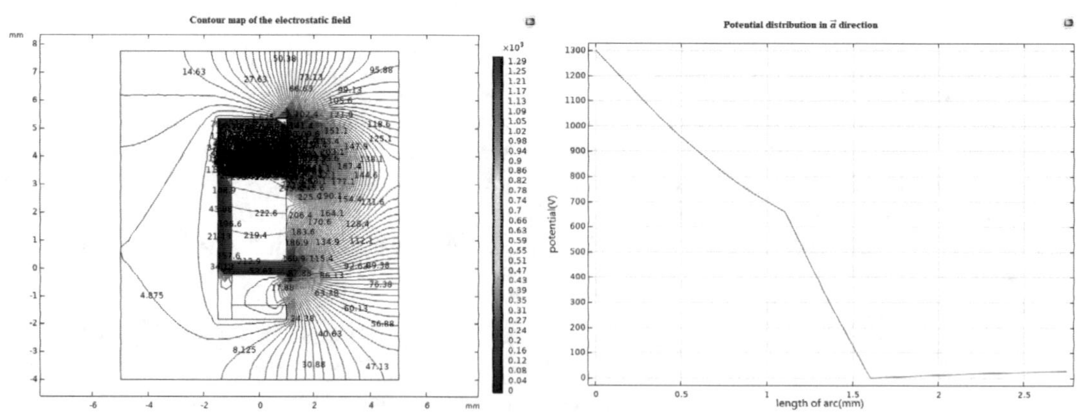

(a) A vertical plane tangential contour map of the electrostatic field (b) Potential distribution in $\vec{a}$ direction

Figure 7. Contour distribution of electrostatic field in vertical tangent plane.

The above figure (a) is the contour map of the electrostatic field in the vertical tangent plane (XOZ plane). It can be seen that The shield has the best effect on the left space in the strong electrostatic field diagram because the shield is unidirectionally arranged. The field strength has been reduced to less than 5V at the center of the left 2mm away from the shield. Figure 2 is the distribution of the electric field strength in the direction of $\boldsymbol{a}$ .When the top of the plate passes through the cavity upward, the electric field linearly decays and the amplitude is small. When passing through the shield case, due to the absorption effect of the metal, the electric field decays at a faster rate, When the upper surface of the shell decays to zero, the field strength gradually increases after the shell exits due to the diffusion of the electric field on the opposite side. The more the electric field is directly to the right, the stronger the field strength is, the maximum is about 50V. The change of the electric field at the bottom is the same as that of the top, but the field strength is lower than the top due to the direction of the low-voltage plate. The field strength directly below is about 8.125V.

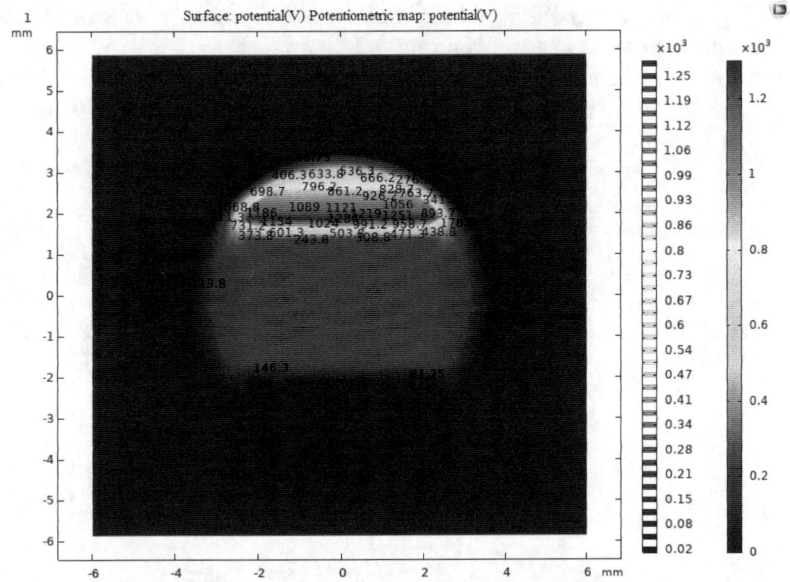

Figure 8. Contour distribution of the strong electrostatic field on the horizontal tangent plane.

The figure above is the contour map of the strong electrostatic field on the plane horizontal tangent plane (XOY). It can be seen from the figure that the strong electric field is concentrated in the high voltage plate and the cavity of the shield.

From the above results, it can be seen that the copper shield can effectively shield the strong electrostatic field radiation generated when the high-voltage capacitor is discharged, and it can ensure that the single-chip computer used is not disturbed by the excessive coupling voltage in the strong electrostatic field.

5. Conclusion

This paper presents a method of single-chip microcomputer protection under the strong electrostatic field interference of a miniaturized pulse discharge device. The electrostatic field strength distribution and potential gradient in the three-dimensional space around the high-voltage energy storage capacitor of the electronic safety system were obtained through simulation calculations, and the structure of the shield was given based on the single-chip microcomputer's withstanding coupling voltage threshold. The simulation method used in this paper is generalizable. It can modify the model, loading voltage, boundary conditions, and material properties to meet the calculation of different electrostatic fields, and use COMSOL post-processor to extract various electrostatic field results for analysis. Due to the limitation of simulation conditions, the modeling and simulation process in this paper ignores the effects of circuit boards and other device materials on the electrostatic field in space, which inevitably differs from the actual situation. In the next,it is necessary to continue studying the electromagnetic radiation interference of high voltage pulse device.

References

[1] Yang Lin 2015 *Research on the technology of full electronic safe detonation of warhead*.Harbin

Institute of Technology. Harbin p 28

[2] Qin Xin,Zhu Peng,Xu Cong,Yang Zhi,Zhang Qiu and Shen Ruiqi 2019 *Discharge characteristics of high voltage capacitor based on MOS controlled thyristor Energetic* materials. **27** Nanjing p 418

[3] Zhang Baoyue,Kang Xingguo and Zhen Song 2015 *Protection method of single chip microcomputer under strong electrostatic interference in electronic safety system Journal of Detection & Control*. **37** Xi An pp 28-30

[4] Tang Linlin 2014 *Analysis shielding and utilization of strong magnetic field environment of electromagnetic orbit shell fuze*. Nanjing p 24

[5] Cao Ronggang,Zou Jun and Yuan Jiansheng 2009 *Measurement and analysis of electromagnetic field radiated by pulse power supply, High power laser and particle beams*. **9** Beijing p 1428

[6] Feng Yongzhu,Song Zongpu 2000 *Segmented pressure measurement method for dielectric loss of electrical equipment under strong electric field interference Shandong Electric Power Technology*. **1** Jinan p 3

[7] He Yong,Zhang Xiangjin 2019 *Research on Electromagnetic Radiation Interference of Miniaturized Pulsed Laser Fuze Launching System Applied Laser*. **39** Shanghai p 151

[8] Feng Yin,Zhang He and Zhang Xiangjin 2011 *Electromagnetic interference of pulsed laser fuze transmitting and receiving module High power laser and particle beams*. **23** Mianyang p 250

[9] Fei Zhenlin 2009 *Study on the Composition and Process of High Temperature MLCC Ceramic Materials*. University of Electronic Science and technology Chengdu p1

IOP Conf. Series: Materials Science and Engineering **793** (2020) 012003 doi:10.1088/1757-899X/793/1/012003

Research on Calibration Technology of Five-Degrees-of-Freedom Tandem Robots

Zhe LI[1], Zhijing Zhang[1*], He Yao[2]

Department of Mechanical Engineering, Beijing Institute of Technology, Beijing 100081, China;

*E-mail: zhzhj@bit.edu.cn

Abstract. In the robot feedback control system of visual feedback, the calibration of the hand-eye system is very important, because it directly affects the accuracy of the robot. Aiming at the small operating space and compact structure of the micro-assembly robot system, a fixed-vision hand-eye calibration method is proposed. The method first obtains the coordinates of 15 spatial points in the robot base coordinate system, and then uses the camera pose measurement principle to obtain the coordinates of the corresponding spatial points in the camera coordinate system, and then establishes the calibration equations. Based on the least squares method, the coordinate transformation matrix between the hand-eye systems is calibrated. The experimental results show that the constructed hand-eye system coordinate transformation matrix is effective and reliable. The calibration accuracy and effectiveness of the method are verified by relevant grabbing experiments.

1. Introduction

Most of the existing robot systems use the vision system as the only or important means of target pose detection [1-3]. Due to the structural size limitations of micro-assembly robots and the strict precision requirements in precision assembly operations, the calibration methods of hand-eye systems commonly used in macro-operated robots is not useful. For example, in the calibration of the hand-eye system of the macro-operated robot, the binocular vision [4] and the structured light vision sensor scheme [5] are usually used to acquire the pose information of the object, and the three-dimensional coordinates of the operating object in the camera coordinate system are determined. However, in a micro-assembly robot system, it is very difficult to install two micro-cameras due to the size constraints of the assembly space structure. For other calibration methods of the hand-eye system, in recent years, many scholars have conducted a lot of research, and the literature proposed a calibration method based on binocular vision [6]. By changing the position of the robot end tool, the target point is measured in combination with the binocular vision system, and the coordinate conversion calculation is performed to solve the TCP coordinate system [7]. However, due to space constraints in the field of micro-assembly, this method is difficult to implement; the literature proposed to introduce stereoscopic 3D reconstruction technology into the hand-eye model of single-camera structured light scanning system, and proposed the calibration algorithm of phase matching technology [8]. The calibration of the vision system is realized, but the accuracy of the projector is high, and it is difficult to meet the requirements in actual operation [9].

At present, the calibration technique of the hand-eye system of micro-assembly robots is still immature, and many theoretical methods are in constant research and exploration. In this paper, the fixed visual hand-eye micro-assembly system is studied and its calibration method is studied. The method overcomes the problems in the calibration of the binocular camera and the error of the movement of the

IOP Conf. Series: Materials Science and Engineering **793** (2020) 012003 doi:10.1088/1757-899X/793/1/012003

stage, and the calibration of the hand-eye system of the robot can be completed by a simple operation of a single camera.

2. Camera imaging model

The pinhole camera model has four coordinate systems: world coordinate system, camera coordinate system, image physical coordinate system, and image pixel coordinate system. For the two-dimensional plane, the coordinates of the world coordinate system of the real world space point are $P_w(X_w, Y_w)$, the corresponding camera coordinate system coordinates are $P_c(X_c, Y_c)$, the coordinates of the corresponding image physical coordinate system are $P'(X', Y')$, and the coordinates of the corresponding image pixel coordinate system are $p(u, v)$, according to the pinhole Camera model.

World coordinate system to camera imaging model coordinate system: In addition to the rotation matrix R between the two coordinate systems, there is also a translation matrix t. The relationship is as shown in equation (2-1):

$$\begin{bmatrix} X_c \\ Y_c \\ Z_c \end{bmatrix} = R \begin{bmatrix} X_w \\ Y_w \\ Z_w \end{bmatrix} + t \tag{2-1}$$

The relationship between the camera coordinate system and image physical coordinate system is as shown in Figure 1 and equation (2-2)

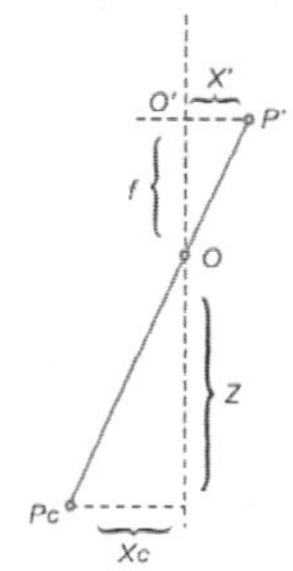

Figure 1 similar triangles

$$\frac{Z_c}{f} = \frac{X_c}{X'} = \frac{Y_c}{Y'} \tag{2-2}$$

Image physical coordinate system to image pixel coordinate system. As shown in Figure , the relationship is as shown in equation (2-3):

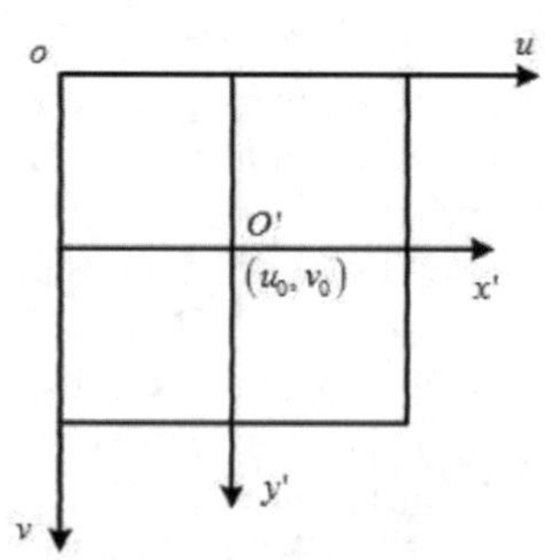

Figure 2 Image physical coordinate system and image pixel coordinate system

$$\begin{cases} u = \dfrac{x'}{dx} + u_0 \\ v = \dfrac{y'}{dy} + v_0 \end{cases} \tag{2-3}$$

After deformation, it can be transformed into a matrix form, as shown in equation (2-4):

$$\begin{bmatrix} u \\ v \\ 1 \end{bmatrix} = \begin{bmatrix} \dfrac{1}{dx} & 0 & u_0 \\ 0 & \dfrac{1}{dx} & v_0 \\ 0 & 0 & 1 \end{bmatrix} \begin{bmatrix} x' \\ y' \\ 1 \end{bmatrix} \tag{2-4}$$

dx represents the actual physical size of each pixel along the x-axis; dy represents the actual physical size of each pixel along the y-axis. (u_0, v_0) indicates the coordinates of the optical center corresponding to the image pixel coordinate system.

The relationship between the world coordinate system and the image pixel coordinate system is obtained by the equation (2-1)、(2-2)、(2-3)、(2-4), as shown in the equation (2-5):

$$Z_c \begin{bmatrix} u \\ v \\ 1 \end{bmatrix} = K(RPw + t) = K\left(R \begin{bmatrix} X_w \\ Y_w \\ Z_w \end{bmatrix} + t \right) \tag{2-5}$$

K is the internal parameter matrix of the camera, and R and t are the external parameters of the camera.

3. Fixed visual hand-eye calibration algorithm

Hand-eye calibration refers to the transformation matrix between the camera coordinate system and the robot coordinate system. There are two main forms of hand-eye system installation: one is the Eye-in-Hand system (the camera is mounted directly at the end of the robot); the other is the Eye-to-Hand system (the camera is fixedly mounted and the position does not change with the movement of the robot). Due to the compact structure of the micro-assembly system, the mechanical arm load is small. Therefore, we chose to use a fixed installation of the camera.

For the calibration of the Eye-to-Hand system, the homogeneous coordinate transformation relationship between the world coordinate system and the image pixel coordinate system is as shown in equation (3-1):

$$[x', y', 1] = [x, y, 1] \cdot \begin{bmatrix} m & n & 0 \\ o & p & 0 \\ t_1 & t_2 & 1 \end{bmatrix} \tag{3-1}$$

$[x' \quad y' \quad 1]$ is the coordinate of the target point in the image pixel coordinate system, $[x \quad y \quad 1]$ is the coordinate of the target point in the world coordinate system, $\begin{bmatrix} m & n & 0 \\ o & p & 0 \\ t_1 & t_2 & 1 \end{bmatrix}$ is a 3×3 orthogonal matrix rotated by the world coordinate system relative to the image pixel coordinate system.

Therefore, the relationship can be obtained as (3-2):

$$\begin{aligned} x' &= x \cdot m + y \cdot o + t_1 \\ y' &= x \cdot n + y \cdot p + t_2 \end{aligned} \tag{3-2}$$

An expression obtained by amplifying the above expression (3-3)

IOP Conf. Series: Materials Science and Engineering **793** (2020) 012003 doi:10.1088/1757-899X/793/1/012003

$$\begin{aligned} x' &= x\times m + 0\times n + y\times o + 0\times p + 1\times t_1 + 0\times t_2 \\ y' &= 0\times m + x\times n + 0\times o + y\times p + 0\times t_1 + 1\times t_2 \end{aligned} \tag{3-3}$$

The matrix representation is as shown in equation (3-4)

$$\begin{bmatrix} x' \\ y' \end{bmatrix} = \begin{bmatrix} x & 0 & y & 0 & 1 & 0 \\ 0 & x & 0 & y & 0 & 1 \end{bmatrix} \cdot \begin{bmatrix} m \\ n \\ o \\ p \\ t_1 \\ t_2 \end{bmatrix} \tag{3-4}$$

Amplify the matrix to obtain the equation (3-5)

$$\begin{bmatrix} x_1' \\ y_1' \\ x_2' \\ y_2' \\ \vdots \\ x_n' \\ x_n' \end{bmatrix} = \begin{bmatrix} x_1 & 0 & y_1 & 0 & 1 & 0 \\ 0 & x_1 & 0 & y_1 & 0 & 1 \\ x_2 & 0 & y_2 & 0 & 1 & 0 \\ 0 & x_2 & 0 & y_2 & 1 & 0 \\ & & \cdots & & & \\ x_n & 0 & y_n & 0 & 1 & 0 \\ 0 & x_n & 0 & y_n & 0 & 1 \end{bmatrix} \cdot \begin{bmatrix} m \\ n \\ o \\ p \\ t_1 \\ t_2 \end{bmatrix} \tag{3-5}$$

Finishing the equation (3-6):

$$\begin{bmatrix} x_1' \\ x_2' \\ x_n' \\ \cdots \\ y_1' \\ y_2' \\ y_n' \end{bmatrix} = \begin{bmatrix} x_1 & 0 & y_1 & 0 & 1 & 0 \\ x_2 & 0 & y_2 & 0 & 1 & 0 \\ x_n & 0 & y_n & 0 & 1 & 0 \\ & & \cdots & & & \\ 0 & x_1 & 0 & y_1 & 0 & 1 \\ 0 & x_2 & 0 & y_2 & 0 & 1 \\ 0 & x_n & 0 & y_n & 0 & 1 \end{bmatrix} \cdot \begin{bmatrix} m \\ n \\ o \\ p \\ t_1 \\ t_2 \end{bmatrix} \tag{3-6}$$

Find the least squares solution with MATLAB

$$\begin{bmatrix} m \\ n \\ o \\ p \\ t_1 \\ t_2 \end{bmatrix} = \begin{bmatrix} x_1 & 0 & y_1 & 0 & 1 & 0 \\ x_2 & 0 & y_2 & 0 & 1 & 0 \\ x_n & 0 & y_n & 0 & 1 & 0 \\ & & \cdots & & & \\ 0 & x_1 & 0 & y_1 & 0 & 1 \\ 0 & x_2 & 0 & y_2 & 0 & 1 \\ 0 & x_n & 0 & y_n & 0 & 1 \end{bmatrix}^{-1} \cdot \begin{bmatrix} x_1' \\ x_2' \\ x_n' \\ \cdots \\ y_1' \\ y_2' \\ y_n' \end{bmatrix} \tag{3-7}$$

4. Experimental design

The robotic hand-eye system calibration experiment device includes: auto focus camera, five-degree-of-freedom tandem robot.

The experimental steps are as follows: First, a hexahedral base is installed at the end of the robot, a fixture is mounted on the base-a three-jaw chuck, and then a standard cylinder is clamped with a three-jaw chuck, and one end of the standard cylinder is dyed with a red dye. And put a white paper on the tray position.

The control robot points a red dot on the white paper, records the coordinates in the robot coordinate system, and then moves to a different position and repeats 10 times. After completion, the camera takes

IOP Conf. Series: Materials Science and Engineering **793** (2020) 012003 doi:10.1088/1757-899X/793/1/012003

a picture and the resulting experimental image is shown in Figure 3.

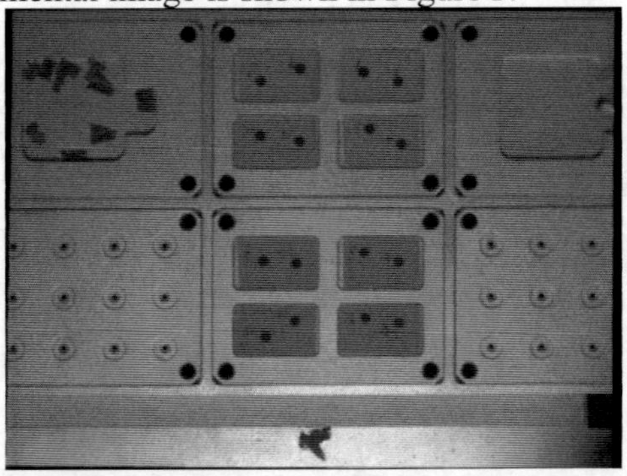

Figure 3 Point the picture in the camera

The coordinates of the center of each point in the graph are extracted according to the Hough transform. The extracted target points are in the world coordinate system, and the coordinates in the image pixel coordinate system are shown as Table 1.

Table 1 World coordinate system, original coordinates in the image pixel coordinate system

Serial number	Coordinates in the world coordinate system	Coordinates in the Image pixel coordinate system
1	-13.0167, 295.2323	1635.8636, 414.0444
2	-0.9312, 295.2159	1846.4604, 411.9381
3	-9.7094, 269.3443	1678.0402, 836.6050
4	11.3305, 261.219	1654.8481, 978.2444
5	-8.5305, 217.2812	1706.5456, 727.8610
6	-6.6071, 195.0013	1729.6037, 108.1503
7	3.6713, 190.171	1915.6967, 91.7871
8	33.1817, 300.9095	2407.0381, 1.488
9	44.3605, 298.2609	2610.1980, 2.5299
10	35.5652, 278.8685	2421.6405, 7.2489
11	44.9946, 275.0712	2590.3613, 5.7392
12	34.6878, 234.7774	2333.6303, 01.8452
13	47.0987, 230.1313	2561.1963, 83.6319
14	38.1771, 209.0045	2369.8713, 46.3625
15	48.3348, 204.6028	2557.4551, 22.7496

The resulting transformation matrix $R=\begin{bmatrix} 0.0629 & 0.0117 & 0 \\ -0.0023 & 0.0584 & 0 \\ -118.3451 & 298.4647 & 1 \end{bmatrix}$

The error $\varepsilon = [x', y', 1]\cdot R^{-1} - [x, y, 1]$, $[x' \;\; y' \;\; 1]$ is the coordinate of the target point in the image pixel coordinate system, which is the coordinate of the target point in the world coordinate system, and R is the 3×3 orthogonal matrix of the world coordinate system rotated relative to the image pixel coordinate system. Furthermore, the error of the position coordinates of each point based on the current rotation transformation matrix robot in the world coordinate system is [0.3681 0.1377 0.1987 0.1840 0.3733 0.2858 0.0907 0.0793 0.2482 0.1238 0.1929 0.2420 0.0389 0.3243 0.1451], the maximum error is 0.3681 mm.

5. Conclusion

In this paper, a calibration method for the hand-eye system of a micro-assembly robot is proposed, and the transformation matrix between the micro-assembly hand-eye systems is obtained. The validity and accuracy of the calibration results were verified by the prototype system. Compared with other calibration methods, the calibration process of the method is simple and convenient. Besides, it doesn't require complicated auxiliary equipment and has good practicability.

References

[1] Ji ZHU Z, JI Q. Novel eye gaze tracking techniques under natural head movement [J]. IEEE Transactions on Bio-medical Engineering,2007,54(12) : 2246-2260.

[2] Xu X, Zhu D, Zhang H, et al. TCP-based calibration in robot-assisted belt grinding of aero-engine blades using scanner measurements [J]. The International Journal of Advanced Manufacturing Technology, 2017, 90(1-4):635-647

[3] PRIBANIC T, CIFREK M, PEHAREC S. Light plane position determination for the purpose of structured light scanning [C]. IEEE ISIE 2005, Dubrovnik, Croatia, USA, 2005: 1315-1320.

[4] MUIS A, OHNISHI K. Eye-to-hand approach on eye-in-hand configuration within real-time visual serving [J]. IEEE/ASME Trans on Robotics,2005,10(4) : 404-410.

[5] MEDIONIG, KANG S B. Emerging topics in computer vision [M]. New Jersey, USA: Pearson Prentice Hall, 2005: 215-245.

[6] JIMENEZ R, CERES R, PONS J L A survey of computer vision methods for locating fruit on trees [J]. Transactions of the ASAE, 2003, 43(6) : 1911-1920.

[7] Roth Z, Mooring B, Ravani B. An overview of robot calibration [J]. IEEE Journal on Robotics and Automation,1987, 3(5): 377-385

[8] He Q X, You Z, Kong X D. Positioning Error Compensation Method of Industrial Robot Based on Closed-loop Feedback of Position and Orientation[J]. China Mechanical Engineering, 2016, 27(07): 872-876 (in Chinese)

[9] Tan J, Xi N, Wang Y. A singularity-free motion control algorithm for robot manipulators—a hybrid system approach [J]. Automatica, 2004, 40(7): 1239-1245

IOP Conf. Series: Materials Science and Engineering **793** (2020) 012004 doi:10.1088/1757-899X/793/1/012004

Research on welding deformation of side beam of freight train bogie

Liyana[1], Suliang[1], Wangwen[2]

[1] College of locomotive and vehicle engineering, dalian jiaotong university, liaoning 116028;

[2] CRRC QIQIHAR ROLLING STOCK CO.,LTD.Dalian R&D Center

*Corresponding author's e-mail: su1994liang@163.com

Abstract. The welding deformation directly affects the production quality and service life of the bogie, so controlling welding deformation is an important engineering problem in the production process of the bogie. In this paper,the side beam of freight train bogie is taken as the research object, the heat source of the local joint mode is checked based on the thermal elastic-plastic theory and the plastic strains are extracted to establish the inherent strain database with the welding simulation software.Then, the welding deformation of the whole side beam is carried out by using the inherent strain method based on thermal elastic-plastic theory. By quantitative analysis of the displacement deformation, the results show that simulation numerical results are similar to factory experimental data.At the same time the rationality and reliability of the established inherent strain database are verified. It is also proved that the inherent strain method based on thermal elastic-plastic theory is a fast and effective way to predict the welding deformation, which provides technical support for the prediction and control of the welding deformation of the whole frame of the fast freight train.

1. Introduction

Welding is commonly used in mechanical processing and it is an extremely important processing method in the manufacturing process of railway vehicle bogie. Because the residual deformation will inevitably occur in the welding process, the residual deformation will directly affect the production quality and service life of the bogie. Therefore, at the present stage of railway development in China, the research on welding deformation of bogie has extremely important practical engineering significance[1].

With the rapid development of computer technology in recent years, numerical simulation of welding deformation has also played a great role in deformation research. A lot of research results have been obtained in welding deformation simulation of locomotive and vehicle.The thermal elastic-plastic finite element method is used to simulate the welding process by recording every moment of the welding process, so as to obtain the results of the temperature field, stress field and strain field at every moment. For example, Sun chuanzheng from southwest jiaotong university simulated the whole welding process of the bogie frame with this method and obtained the welding deformation of the bogie under the actual production conditions[2]. Due to the influence of computer capacity, the application of three-dimensional thermal elastic-plastic finite element method to simulate welding deformation of large components will be limited, but the inherent strain method can solve this problem well. When the size and distribution of the inherent strain are known, it is applied to the action area as an initial value. The welding deformation of large components can be calculated by elastic calculation. For example, Zhu Ping from Tianjin University used this method to predict the welding deformation of Bogie of Mumbai Metro[3].

In this paper, under the actual production process conditions, a three-dimensional geometric model of the bogie side beam is built and the typical local models are extracted according to the joint form, geometric parameters and process parameters. Then based on thermal elastic-plastic theory, the inherent strain database is established by extracting the local plastic strains on joints. At last, the deformation of the whole model is calculated with the mapped joint database,which provide some references for actual welding of bogie side beam of express freight train.

2. Introduction of Bogie Side Beam Model

2.1 Bogie Side Beam

Bogie is an important load-bearing component of locomotive and vehicle, which plays the role of supporting the body. The welding deformation of side beam directly affects the overall production quality of bogie, so the welding variation of side beam is controlled.

Side beam is composed of upper (16 mm) and lower cover plate (20 mm), web plate (14 mm), spring seat and sleeve. The base material of side beam is Q345A.The geometric model of side beam is shown in Fig.1.

2.2 Local Joint Model

According to the different types of joints, thickness of plates and process parameters, two joints are adopted: double-sided and single-sided welded T-joint, shown in Table 1 and Fig.2.

Table 1. Description of Joints

No	Type	Location	Dimension	Layers
1	double-sided welded T-joint	the middle of the lower cover plate and the web	200 mm × 94 mm × 116 mm	7
2	single-sided welded T-joint	the middle of the upper cover plate and the web,	200 mm × 80 mm × 120 mm	5

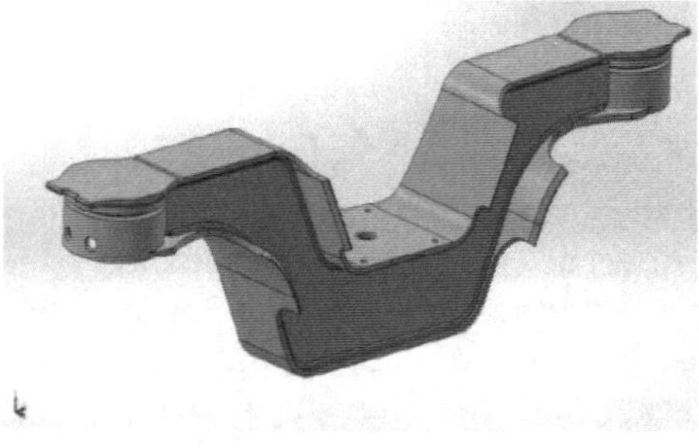

Figure 1. Side Beam Geometry Model

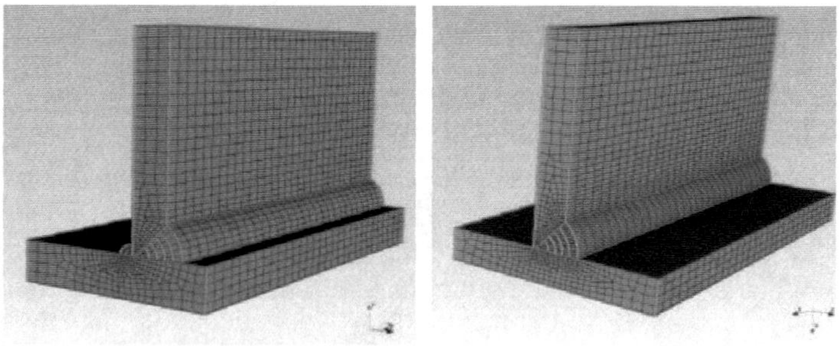

(a) Double-sided welded T-joint (b) Single-sided welded T-joint

Figure 2. Finite element model of T-joint

2.3 Welding Scheme

The welding deformation caused by six main welds of side beam between webs and upper, lower cover plates is mainly considered. The distribution of six main welds is shown in Fig.3, and the number of layers per weld is shown in Fig.4. The detailed welding process description is shown in Table 2.

IOP Conf. Series: Materials Science and Engineering **793** (2020) 012004 doi:10.1088/1757-899X/793/1/012004

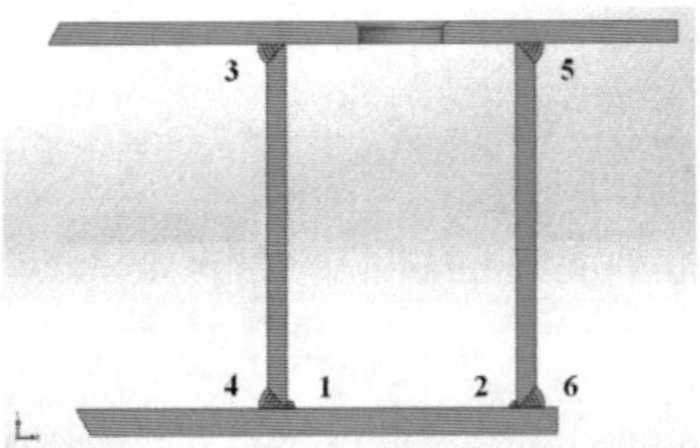

Figure 3. Sequence Number of Six Welds

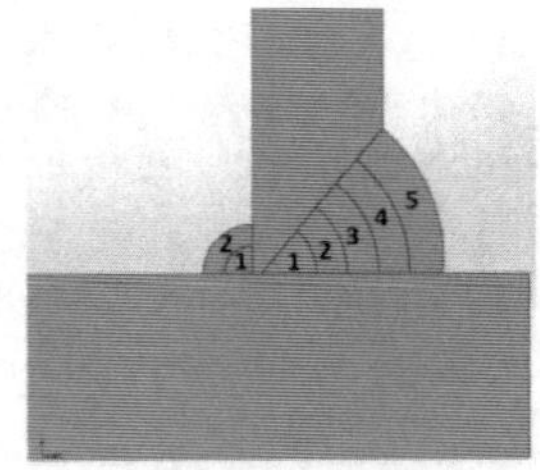

Figure 4 .Number of layers per weld

Table 2. Description of Welding Process

Welding sequence	Explain	Remarks
1	1-1、2-1	1-1 is the first layer of No. 1 weld, 2-2 is the second layer of No. 2 weld, and so on.
2	2-2、1-2	
3	3-1、4-1、5-1、6-1	
4	3-2、4-2、5-2、6-2	
5	3-3、4-3、5-3、6-3	
6	3-4、4-4、5-4、6-4	
7	3-5、4-5、5-5、6-5	

3. Calculation of heat source and plastic strain for Local Model

The accuracy of heat source checking directly affects the accuracy of inherent strain database. The welding parameters of local model heat source checking are shown in Table 3.

Table 3. Welding parameters

Groove form	Groove angle	Groove depth	Blunt edge	welding method	Welding wire
V-groove	Slope angle 50 degrees	14mm	0.5mm-11mm	gmaw	SM-70(ER50-6)Ø1.2
Protective gas	Weld material	Base material	Thermal convection coefficient	ambient temperature	Interlayer temperature
20%CO2+80% Ar	Q345A	Q345A	20W/(m²·k)	20°C	<100°C

Fig.5 and Fig.6 are respectively the strain results of local plastic strain of welded T-joints.

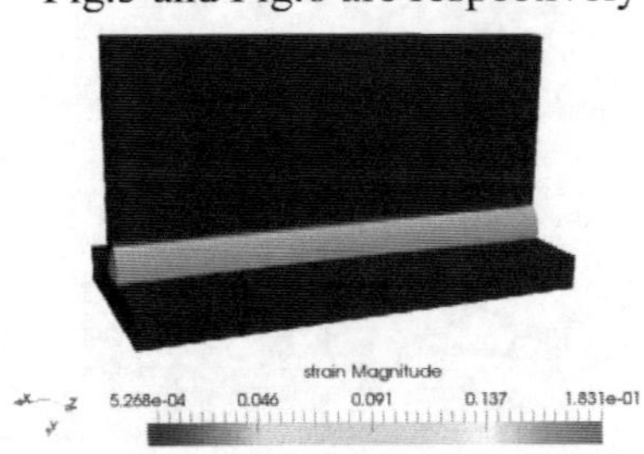

Figure 5 .Strain results of single-sided welded T-joint

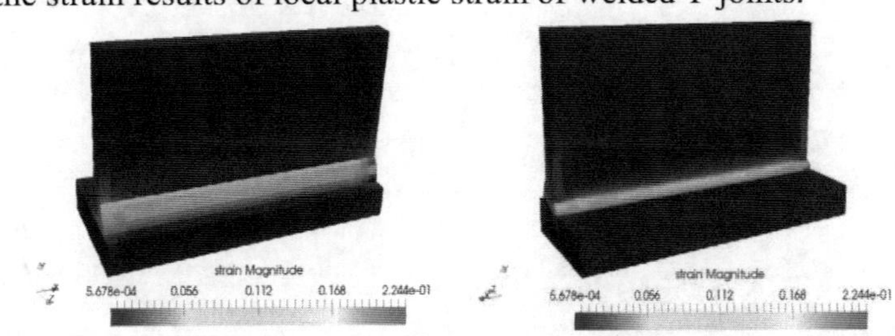

Figure 6 .Strain results of double-sided welded T-joint

4. Calculation and Result Analysis

4.1 Extraction and Establishment of Inherent Strain

IOP Conf. Series: Materials Science and Engineering **793** (2020) 012004 doi:10.1088/1757-899X/793/1/012004

The extraction radius of local plastic strain for each layer of weld can be determined by the range of strain results for each layer of local joint. According to the law of strain distribution, the maximum linear distance which can include the inherent strain region is selected. Inherent strain database is shown in Table 4.

Table 4. Inherent strain database

Weld Sequence Number	Strain direction					
	XX（X Direction）		YY（Y Direction）		ZZ（Z Direction）	
	MIN	MAX	MIN	MAX	MIN	MAX
1-1	-0.00666696	0.024652	-0.0425451	0.0209707	-0.0174792	0.0447186
1-2	-0.00999595	0.0340229	-0.0593649	0.032815	-0.0277338	0.064911
3-1	-0.00790944	0.0296367	-0.0724171	0.0217609	-0.0173627	0.0691026
3-2	-0.00832088	0.028951	-0.071391	0.0231193	-0.0189444	0.0679748
3-3	-0.00974087	0.02933	-0.0730706	0.0234046	-0.0194767	0.0685246
3-4	-0.0130494	0.0304866	-0.0493287	0.0221104	-0.0207003	0.0345855
3-5	-0.016489	0.0359886	-0.0332113	0.032365	-0.0249184	0.0291924
4-1	-0.0105564	0.0340391	-0.0745641	0.0271177	-0.0220328	0.0814115
4-2	-0.0120663	0.0255148	-0.0725146	0.011822	-0.0127627	0.0807358
4-3	-0.0108391	0.0213225	-0.043764	0.010728	-0.0106305	0.0436322
4-4	-0.0103737	0.0207295	-0.0346696	0.00751138	-0.00752363	0.0213824
4-5	-0.0106442	0.0213972	-0.0323672	0.0114897	-0.0142981	0.0235924

4.2 Analysis of the calculation results

Using the plastic strain database value of each layer of local welded joint as the mapping source, map each layer of weld corresponding to the overall model, and an elastic calculation is carried out.

The central points of the two webs along the X direction of 800mm, 680mm, 480mm, 380mm, 180mm, - 180mm, - 380mm, - 480mm, - 680mm, - 800mm are selected as the measuring points (as shown in Fig. 7). The comprehensive displacement and other deformations in three directions of the 11points are quantitatively analyzed,as shown in Fig. 8.

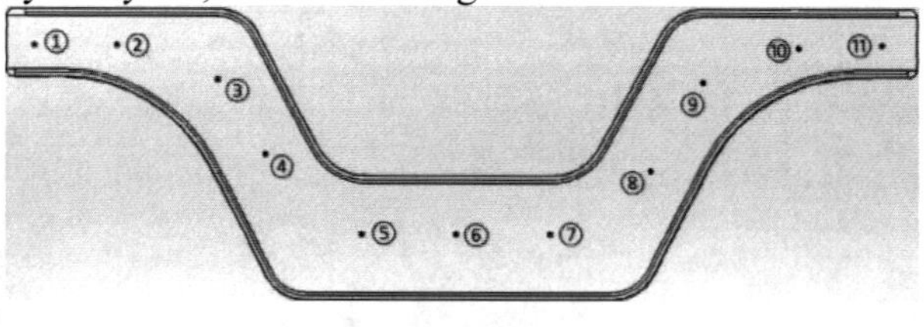

Figure 7. The location of Measuring Points

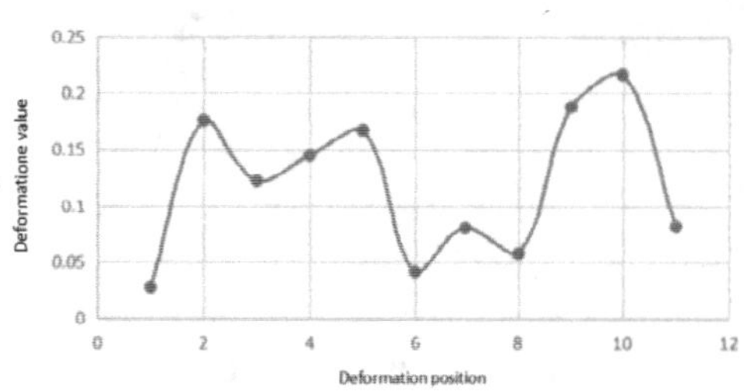

（a）Longitudinal deformation (X)

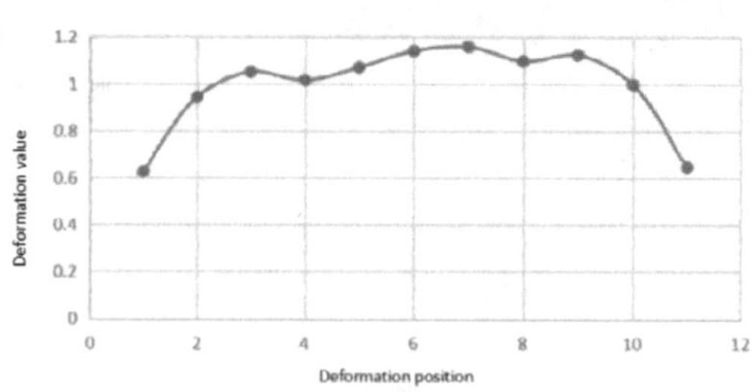

（b）Vertical deformation(Y)

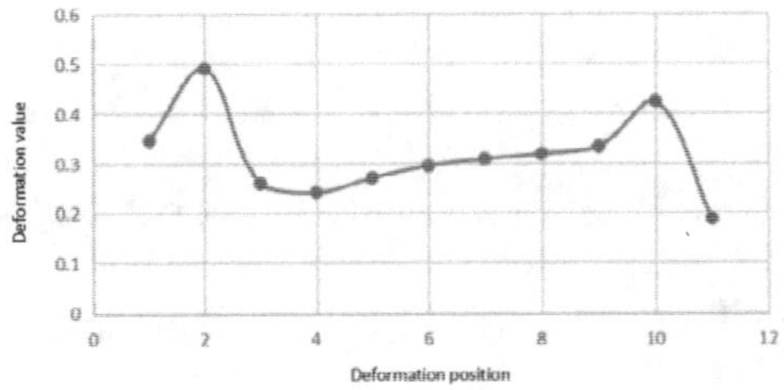

(c) Transverse deformation (Z)

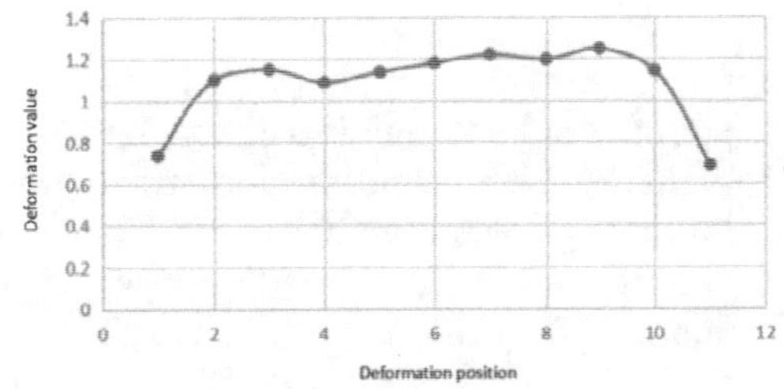

(d) Comprehensive deformation

Figure 8 .Deformations in various directions

According to the analysis of the figure, the maximum deformation position of the side beam web in the longitudinal direction (X) is at the position of the slope of the curved surface at both ends, and the maximum value is 0.22mm.The deformation trend of the vertical (Y) is convex, which means the deformation of both ends is small and the deformation of the middle is large, and the maximum value is 1.16mm.The maximum deformation position in the transverse direction (Z) is in the middle position, and the maximum deformation amount is 0.49mm. The comprehensive deformation trend is the same as the vertical deformation , and the maximum deformation is 1.24mm.The vertical deformation of the side beam web obtained from the factory statistics is about 1.10mm-1.30mm, and the longitudinal and lateral deformations are very small. The experimental data are similar to the calculated simulation values.

5. Conclusion

Based on thermal elastic-plastic theory,the heat source of the local joint is checked and the plastic strain is extracted to establish the inherent strain database. Then using the inherent strain method, the elastic deformation of the whole model is calculated. The results of welding deformation are compared with the statistical data of the factory. It is found that the simulation results are similar to the statistical data and the error is within a reasonable range. The following conclusions can be given:

(1) Reasonable checking of heat source for local joints of side beams. The 5 layers of welds (21607 elements) for single-side welded T-joints need 8 hours for thermal elastic-plastic calculation. And the 7 layers of welds (27504 elements) for double-side welded T-joints need 12 hours for calculation. But It only takes 20 minutes to calculate the elastic deformation for the 24 layers of welds (718501 elements) of the overall side beam model. So it can be seen that the inherent strain method based on thermal elastic-plastic theory is an economical and applicable method, which can be used to calculate the welding deformation of large structures.

(2) According to the analysis of welding deformation data, the vertical deformation of the side beam is the largest and the deformation trend is convex.The maximum deformation is 1.16mm, and the error between simulation and experiment is 5.45%-10.76%.

(3) The rationality and validity of the inherent strain database established by extracting three main directions of plastic strain are verified by comparing simulation results with experiments.

Acknowledgments

This work was supported by"Research on Multi-domain Simulation Technology of Express Freight Train under multi-load Conditions" (2017YFB1201302－11) , the National Key Research and Development Program of China and "Welding Deformation Prediction and Control of Key parts of High-speed Trains", the Major Project from the Institute of Railway Transportation Equipment Technology of Liaoning Province.

IOP Conf. Series: Materials Science and Engineering **793** (2020) 012004 doi:10.1088/1757-899X/793/1/012004

Reference:

[1] Sun Kai. Research on Numerical Simulation of welding of truck bogie sub-frame [D]. Southwest Jiaotong University, 2011.

[2] Sun Chuanzheng. Welding numerical simulation of new train bogie frame [D]. Southwest Jiaotong University, 2012.

[3] Zhu Ping. Research on prediction of bogie welding deformation based on inherent strain method [D]. Tianjin University, 2012.

IOP Conf. Series: Materials Science and Engineering **793** (2020) 012005 doi:10.1088/1757-899X/793/1/012005

Research on water vapor measurement and removal methods for the spacecraft sealed cabin

Yue Wu[1], Qingling Guo[1,2], Weifeng Yuan[1], Zheng Xie[1], Xi Wei [1], Jing Wang [1]

[1] Beijing Institute of Spacecraft Environment Engineering, Beijing 100094, China

[2] Corresponding author's e-mail address: 1159497408@qq.com

Abstract. In order to develop a water vapor measurement and dewatering device suitable for ground test of spacecraft sealed cabin. A moisture measuring device suitable for the sealed cabin was designed. The existing moisture content measuring sensors were equipped with auxiliary devices to protect the electrical circuit of the sensor, and then the sensor was calibrated and analyzed. After calibration, it can meet the requirement of moisture measurement in vacuum environment. A small chamber was built to study water change rule in the pumping process. In the water removal system, the cabin was connected to the vacuum pump system through a two-way sealing flange, which can complete the water removal outside the vacuum chamber by pumping. When the pressure of the sealed cabin reached 100 Pa, high purity nitrogen gas was added to 97 kPa, then pumped in again. The results show that the water removal method can ensure that the dew point temperature of the sealed chamber is lower than 0°C during the vacuum thermal test. After protected, the dew point sensor can measure moisture at low pressure. The water removal system can effectively remove the moisture in the sealed cabin, thus ensuring the completion of the vacuum thermal test.

1. Introduction

During the vacuum thermal environment test of the spacecraft sealed cabin, a large amount of water vapor was found in the cabin with the process of the test.

These vapors come from two aspects: on the one hand, it is due to the water vapor generated during the training of astronauts in the cabin; on the other hand, it is due to the water vapor brought by the operators during the installation and debugging of the equipment in the cabin [1].

Excessive humidity of water vapor will affect the working status of various electronic instruments and equipment sensitive to moisture content. In order to ensure the successful completion of all testing in the thermal test of the sealed cabin, it is necessary to accurately measure the moisture content of gases in the sealed cabin of spacecraft and control the dew point temperature of gases in the sealed cabin below 0°C[2].

Because of the large volume and large gas amount of the sealed cabin, the requirement of dehumidification is very high. In addition, the sealed cabin is located in the vacuum container during the vacuum thermal test, the existing water vapor measurement and dewatering methods can not meet the dewatering requirements of the sealed cabin.

Aiming at the dehumidification problem of spacecraft sealed cabin, a series of studies were carried out, and a moisture measurement and dehumidification system for sealed cabin was designed to ensure the successful completion of the thermal test of sealed cabin.

2. Water vapor measurement method

According to the source of water in the sealed cabin[3], the following conclusions are drawn by analyzing the characteristics of water in the sealed cabin: a) the measured object is air (gas in sealed cabin); b) the gas pressure in the sealed cabin was negative pressure (50kPa) during the measurement; c) the air humidity in the sealed cabin is relatively high, and the relative humidity may reach more than 90%; d) the temperature of the gas in the sealed cabin is between 18°C ~20°C during the determination. It can be considered that the temperature of the tested gas sample is stable; e) most of the water molecules are adsorbed on the surface of materials and equipment by van der Waals force. The structure of the adsorbed surface molecules does not change much. Therefore, the chemical properties of the adsorbed water molecules and the adsorbed surface remain unchanged, which belong to physical adsorption.

According to the different measuring principles, the measuring methods of water vapor can be divided into gravimetric method, electrolysis method, dew point method and resistance-capacitance method[4]. Analyzing the merits and demerits of various measuring methods, and combining with the characteristics of moisture content in the sealed cabin, this paper select the resistance-capacitance method to measure the moisture in the sealed cabin. The resistance-capacitance method uses metal aluminum as the measuring probe, and forms an oxide film on the surface of metal aluminum by electrochemical method. Then a thin layer of metal is deposited on the film. Thus, the metal aluminum of the probe core and the metal film on the surface of the probe constitute a capacitor, and the porous alumina film acts as the dielectric layer of the capacitor. When measuring, the gas is fully contacted with the probe, and the moisture in the gas is adsorbed by the porous alumina film, which results in the change of the impedance of the capacitor. The variation of capacitor impedance is related to vapor concentration, and the moisture content of gas is measured. The DMT152 type measuring sensor is selected in this paper.

Considering the use of sensors in vacuum and cryogenic environment, an auxiliary device is designed as shown in Figure 1. The sensor is built in an auxiliary device. The measuring port is connected with the flange of the auxiliary device, and the measuring signal is drawn out through a sealed plug. The auxiliary device is in a sealed state of atmospheric pressure, which can protect the electrical circuit in the sensor, thus ensuring the normal use of the sensor.

Figure 1. Humidity sensor support device

3. Calibration method of moisture measurement sensor

Usually the calibration pressure of the sensor is atmospheric pressure. It is necessary to measure the moisture content in the low pressure environment this time. In order to verify the measurement performance of DMT152 sensor at low pressure, calibration tests were carried out. Its process is as follows: 1) place DMT152 sensor and SH-2001 standard dew point meter in the same position in vacuum container; 2) measure water content under atmospheric pressure; 3) Turn on the vacuum pump and measure the moisture content under 0.5 atmospheric pressure; 4) repressure to atmospheric environment and measure moisture content under atmospheric pressure. The results of standard dew point meter and DMT152 sensor are compared as shown in Table 1.

IOP Conf. Series: Materials Science and Engineering **793** (2020) 012005 doi:10.1088/1757-899X/793/1/012005

Table 1 Dew point test record table

Serial number	Standard dew point meter（°C）	DMT152（°C）	Pressure value（kPa）	Time	Error（°C）
1	6.4	7.90	100.38	18：34	1.5
2	0.4	0.68	56.26	19：28	0.28
3	5.4	6.49	100.00	19：37	1.09

As can be seen from Table 1, all the errors are within the range of 2°C, which meets the technical requirements of the project.

4. Study on dewatering technology of sealed cabin

4.1 Variation rule of water in the process of vacuum pumping

In order to remove the moisture in the sealed cabin, first of all, the change rule of water in the process of vacuum pumping should be studied. A test system is established as shown in Figure 2. It includes vacuum vessel, vacuum acquisition system, platinum resistance temperature measurement system, vacuum measurement system and water pressure measurement equipment. The effective volume of the vacuum vessel is $0.2m^3$. The physical adsorption of water on the material surface is simulated by spraying water on the surface of the degreased gauze. Use stainless steel plate to simulate the cabin material, and the gauze with water is placed on the stainless steel plate. The gauze with added water is placed on the stainless steel plate. The gauze size is 270 x 400 mm and the amount of added water is 30g. A platinum resistance temperature sensor is arranged on the surface of gauze and the bottom of stainless steel plate. The temperature of water evaporation surface and material bottom are measured respectively. The water vapor partial pressure is measured by dew point meter. Open the vacuum pump and vacuumize the container until the vacuum degree reaches the limit (below 100Pa). Measure the change of partial pressure of water vapor in the container, as shown in Figure 3.

The experimental results show the variation rule of water in the sealed cabin during vacuum pumping process. With the process of pumping, the partial pressure of water vapor decreases continuously. One part of the liquid water changes to water vapor and the other to ice. Without external heat source, the sublimation rate of ice is very small, and there is no method relying solely on vacuum pumping. The method of extracting all water from the container is consistent with previous studies[5]. Therefore, the method of applying external heat source is adopted to provide the sublimation latent heat needed and accelerate the extraction of water vapor.

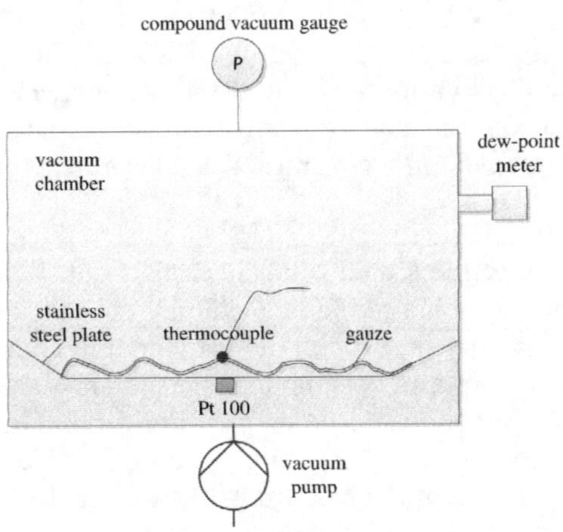

Figure 2. Schematic diagram of the small container test system

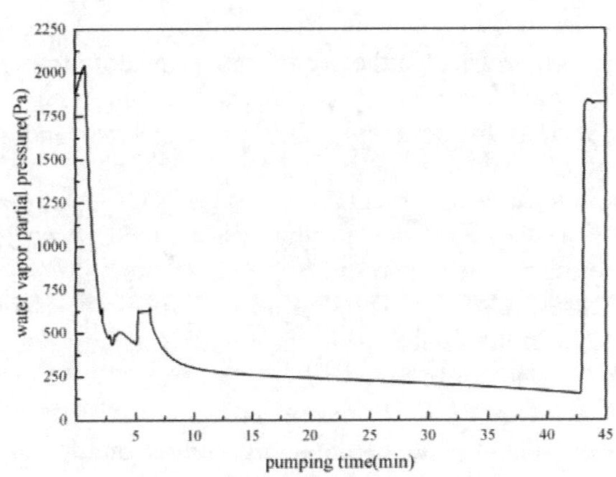

Figure 3. Variation of water vapor partial pressure during pumping

IOP Conf. Series: Materials Science and Engineering **793** (2020) 012005 doi:10.1088/1757-899X/793/1/012005

4.2 Dewatering system of spacecraft sealed cabin

Based on the above pumping and dewatering tests, the dewatering system of sealed cabin for vacuum thermal test is designed, including vacuum pump set, cold trap, sealing flange and vacuum pipeline, as shown in Figure 4. Because the sealed cabin is in the space environment simulation container during the vacuum thermal test, its dewatering system can only be outside the container. The metal bellows are used to connect the sealed cabin with the oil-free vacuum pumping unit outside the container through the sealing flange, and the water vapor in the sealed cabin is removed by vacuum pumping. A cold trap is placed in front of the vacuum pump set to absorb water vapor in the pipeline.

The flange connection in the space environment simulator needs both vacuum pumping and inflation. Besides, the temperature of heat sink in the container is very low and the sealing effect is difficult to guarantee, so the flange connection should be minimized in the process of connecting the pipeline. The structure of the sealing flange is shown in Figure 5. The radial and axial two-way seals are adopted to seal the flange. This structure can ensure the double-way seals without interfering with each other and play the role of multi-channel seals. Meanwhile, insulation measures are installed to the flange to ensure that it is used in room temperature environment. In addition, a heating belt is installed on the outside of the pipeline and a multi-layer heat insulation component is wrapped to heat the pipeline and the gas so as to avoid condensation of water vapor on the inner wall of the pipeline.

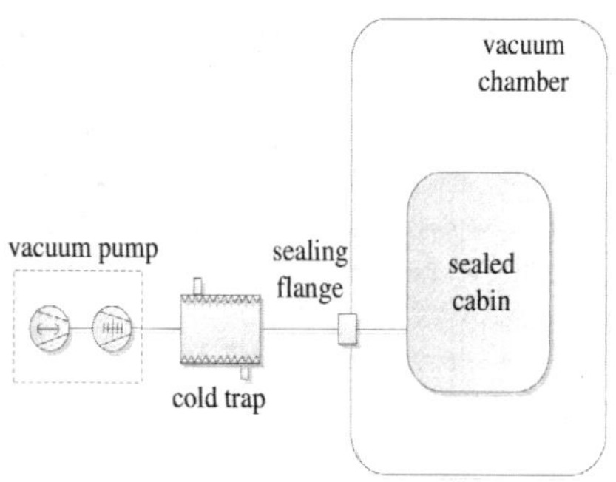

Figure 4. Sealed cabin dewatering system

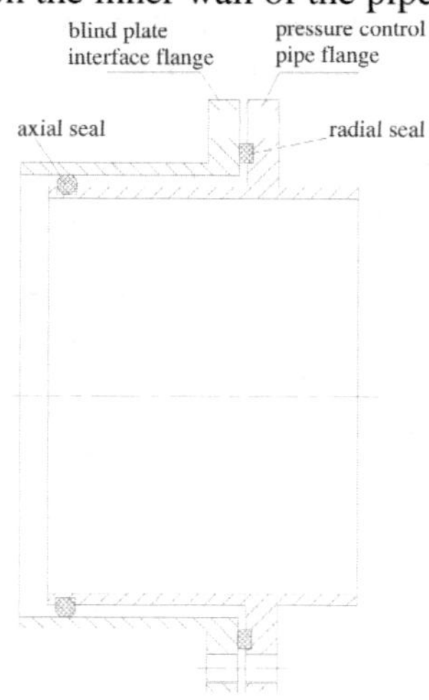

Figure 5. Structure of the sealing flange

The cold trap is installed on the pipeline outside the container to collect moisture in the sealed cabin. It is cooled by liquid nitrogen and the inner gallbladder structure is used. In addition to collecting water in the sealed cabin, the cold trap can also collect various organic compound molecules with high boiling point in the sealed cabin, which is of great significance for purifying the sealed cabin, especially for pollutant composition analysis and material selection in the manned sealed cabin.

4.3 Dewatering process and results of sealed cabin

Using the above dewatering system, the dewatering process of the sealed cabin is carried out during the thermal test, and the dew point temperature in the cabin is monitored in real time by the moisture measuring sensor. The dewatering process is as follows: From 10:30 to 12:30, the vacuum unit is used to pump the sealed cabin through the sealing flange. The suction lasts for 2 hours, and the pressure in the sealed cabin drops to 100 Pa; from 12:30 to 14:30, the sealed cabin holds pressure for 2 hours; from 14:30 to 17:20, the sealed cabin is filled with pure nitrogen to 96 kPa; from 17:50 to 18:47, the sealed cabin is pumped again to reduce the dew point temperature of the gas; from 18:58 to 20:15, the sealed cabin is re-pressurized to 97 kPa by using the air passing through the liquid nitrogen cold trap. So far, the dewatering test of the sealed cabin is completed. The dew point temperature curve of the gas in the dewatering process of the sealed cabin is shown in Figure 6.

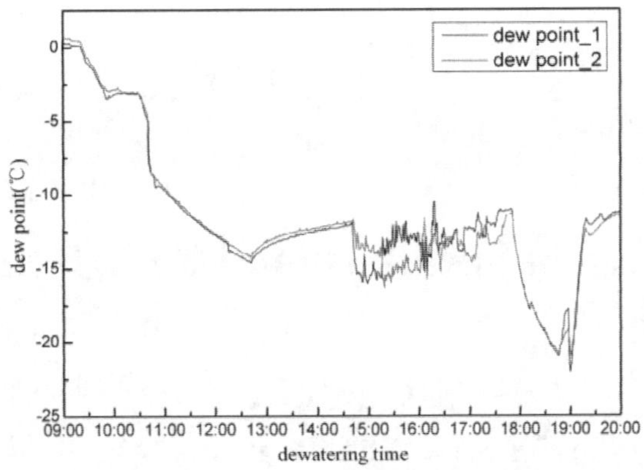

Figure 6. Dew point curve in the process of dewatering

As can be seen from Figure 8, compared with the first pumping and dewatering, the dew point temperature decreases faster, and the dew point temperature obtained is lower in the second pumping after adding high purity nitrogen and the dewatering efficiency is improved significantly. During the whole thermal balance test period, the dew point temperature of the gas in the spacecraft sealed cabin is always below 0°C, which meets the test requirements. Therefore, the dewatering system and the dewatering method established in this paper are feasible for dewatering the sealed cabin.

5. Conclusion

Based on the problem of humidity control and the physical characteristics of water vapor in spacecraft sealed cabin, this paper presents a humidity measurement method and dewatering method suitable for spacecraft sealed cabin.

The existing sensors are improved and calibrated to meet the water content test requirements of spacecraft sealed cabin in low pressure environment.

According to the variation rule of water in the process of vacuum pumping, the dewatering technology of sealed cabin was established and the dewatering system of sealed cabin was developed.

The sealed cabin is dewatered prior to the thermal test so as to the dew point temperature of the gas in the spacecraft sealed cabin is always below 0°C during the whole thermal balance test period, which avoids the influence of moisture on the operation of electronic equipment.

This method has an ideal dewatering effect, which guarantees the safety of electronic equipment in spacecraft during thermal test and ensures the smooth process of thermal test. It provides a new thought for the dehumidification of large spacecraft sealed cabin.

References

[1] FAN Yufeng, LIU Bingqing, HUANG Jiarong, MAN Guanglong. The Research of the Air Thermodynamic Process with the Pressure Relief and Recovery in Airlock Cabin[J]. Journal of Astronautics, 2013 34(2): 293-298

[2] LIU Dongxiao, MAN Guanglong, CAO Jianfen, HAN Haiying, FAN Yufeng, Temperature-humidity independent control method for manned spacecraft sealed-cabin and its experimental validation [J] Spacecraft Environment Engineering, 2014 31 (4): 386-391

[3] ZHU Guangchen, JIA Shijin. The ground verification of spacecraft EVA functions[J]. Manned Spaceflight, 2009, 15(3): 48-53

[4] WANG Zhaohui, LI Zhida, Humidity Measurement and Calibration Method[J]. Metrology & Measurement Technology, 2006 26(z1): 96-101

[5] ZHANG Shiwei, ZHANG Zhijun, et al.. Thermal dynamical study of vacuum freezing of liquids[J]. Chinese Journal of Vacuum Science and Technology 2009, 29(6): 619-623

IOP Conf. Series: Materials Science and Engineering **793** (2020) 012006 doi:10.1088/1757-899X/793/1/012006

Design and Numerical Simulation of a High Speed Centrifugal Compressor

Qingyang Yu[1], Peng Wu[2] , Yelin Deng[1]

[1] School of Rail Transportation, Soochow University, Suzhou, 215006, China

[2]School of Mechanical And Electrical Engineering, Soochow University, Suzhou, 215006, China

E-mail: yelin.deng@suda.edu.cn

Abstract. The issue of energy consumption has attracted worldwide attention, therefore the performance requirements of rotating machinery are getting higher and higher in modern commercial run. Active magnetic bearings have the advantages of high speed and low friction, so they are widely used in high-speed rotating machines such as high-speed centrifugal compressors. A stage of impeller and diffuser was designed. Numerical simulations of the flow field were performed using CFX. The pressure and velocity field in the model stage (impeller and diffuser) were obtained and analyzed. The results show that the model stage has uniform velocity field distribution and regular flow field, and the performance parameters can well meet the industrial requirements.

1. Introduction

In modern commercial run, due to the mechanical friction of the traditional motor that supports the rotor with mechanical bearings, the existing compressors are limited in speed. The compressors that use magnetic levitation bearings have extremely low frictional resistance at high or super high speed. And they can maintain stable performance under low temperature or high temperature [1-3].The impeller is the core component of a centrifugal compressor. How to achieve energy saving and high efficiency is the main content of research on centrifugal compressor [4-5].The current study designs a high-speed compressor. Through the simulation and analysis of the airflow inside the centrifugal compressor, the complicated curved surface and airflow flow characteristics of the impeller during its rotation are understood in detail. The current paper is organized in three parts. First, a centrifugal compressor is designed by the CF turbo. Second, the 3D flow field of the stage is simulated utilizing CFD software. Third, the flow characteristics of the compressor is discussed, with particular emphasis on velocity field of impeller and diffuser.

2. Three-dimensional model design

Considering the actual manufacturing process, the designed impeller blades are Ruled surface blades in the current paper, which are used especially to enable flank milling for manufacturing. The mean surface is generated by spatial movement of a straight line. To reduce the flow loss, the NACA6509 blades are used, and they are made of titanium alloy. The three-dimensional model of CF turbo impeller and diffuser is shown in Figure 1.

Figure 1. Three-dimensional model of CF turbo impeller and diffuser

The current paper designs a high-speed centrifugal compressor, with considering the material and key size structure and speed of the impeller and diffuser. The ternary flow impeller has twisted blades, whose shapes are determined according to the flow requirements. Therefore, the ternary impeller has stable and efficient characteristics. The paper uses a unshroud impeller with splitter blades. And the specific design parameters are presented in Table 1.

Table 1. Impeller parameters

Impeller parameters	Value
Rated pressure ratio	3
Rated speed (r / min)	40000
Inlet temperature (K)	293
Total inlet pressure (Pa)	101325
Impeller inlet diameter (mm)	138
Impeller inlet hub diameter(mm)	28
Number of main blades	9
Number of shunt blades	9
Impeller outlet diameter (mm)	223
Impeller exit installation angle(°)	38
Specific power	3.9

3. Theoretical basis of numerical simulation

3.1. Governing Equations

(1) Mass Conservation Equation

$$\frac{\partial \rho}{\partial t} + \nabla \cdot (\rho \vec{U}) = 0 \tag{1}$$

(2)Momentum Conservation Equation

$$\frac{\partial (\rho \vec{U})}{\partial t} + \nabla \cdot (\rho \vec{U}\vec{U}) = -\nabla \mathrm{p} + \nabla \tau + \vec{S}_M \tag{2}$$

$$\tau = \left[\mu \nabla \cdot \vec{U} + (\nabla \cdot \vec{U})^T - \frac{2}{3}\delta \vec{U}\right] \tag{3}$$

Among them, $\vec{S}_M$ is the momentum source term, tau is stress tensor.

(3)Energy Conservation Equation

$$\frac{\partial (\rho T)}{\partial t} + div(\rho \mu \mathrm{T}) = div\left(\frac{\lambda}{C_p}\mathrm{gradT}\right) + S_h + \phi \tag{4}$$

3.2. Turbulence Model

Turbulence is a continuous pulsation of the fluid velocity, temperature, pressure, and other parameters that change with time and space. Its essence is a three-dimensional unsteady vortex motion.

(1) SST Model

IOP Conf. Series: Materials Science and Engineering **793** (2020) 012006 doi:10.1088/1757-899X/793/1/012006

The SST model is derived from changes in the standard k-ε model. SST k-ε is a hybrid model, which retains the advantages of the model near the wall and uses the model far away from the wall. The specific formula is as follows,

$$v_t = \frac{a_1 k}{\max\,(a_1\omega; \Omega F_2)} \tag{5}$$

$$\frac{D\rho k}{Dt} = \frac{\partial}{\partial x_i}\left[(\mu + \sigma_k\mu_t)\frac{\partial k}{\partial x_j}\right] + \tau_{ij}\frac{\partial u_i}{\partial x_i} - \beta^*\rho\omega k \tag{6}$$

$$\frac{D\rho\omega}{Dt} = \frac{\partial}{\partial x_j}\left[(\mu + \sigma_\omega\mu_t)\frac{\partial \omega}{\partial x_j}\right] + \frac{\gamma}{v_t}\tau_{ij}\frac{\partial u_i}{\partial x_j} - \beta\rho\omega^2 + 2(-F_1)\rho\sigma_{\omega^2}\frac{1}{\omega}\frac{\partial k}{\partial x_j}\frac{\partial \omega}{\partial x_j} \tag{7}$$

Among them, Ω is the vorticity, F_1andF_2 are mixed functions, which are functions of the distance from the point to the wall surface.

4. Numerical simulation

4.1. Calculation settings

1. Mesh: The mesh is a unstructured mesh. The leading and trailing edges of the impeller blades and diffuser blades have been refined. The total number of grid points is 13.04 million. The mesh is shown in Figure 2.

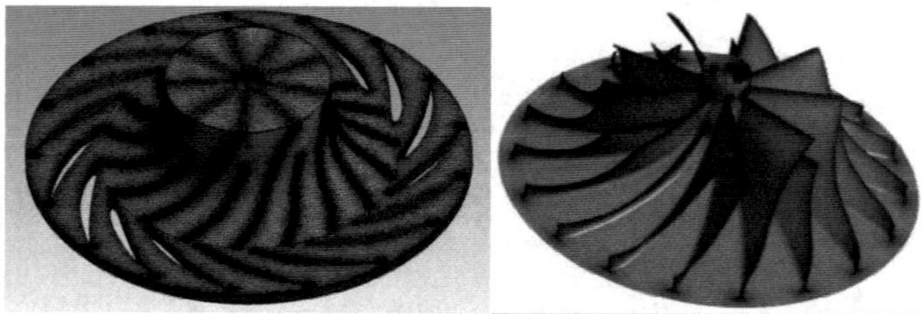

Figure 2. Model grid effect diagram

2. Boundary conditions: Adiabatic, non-slip boundaries are used on the wall.

3. Computational setup: The dynamic and static interface settings of the impeller and diffuser are set using Frozen Rotor,and the convergence criterion is 1E05.

5. Compressor flow characteristics analysis

5.1. Analysis of pressure field characteristics

Figure 3 represents that the overall pressure distribution in the impeller is relatively uniform. Along the direction of fluid flow, the static pressure on the pressure surface of the impeller increases uniformly, and the pressure at the turning of the suction surface at 1/2 of the diffuser blade drops suddenly. When the pressure reaches the end of the diffuser blade, the pressure rises sharply. This phenomenon of a sharp decrease in pressure increases with the generation of shock waves, which will lead to the loss of kinetic energy. From the pressure nephogram, it is indicated that along the direction of fluid flow, the fluid pressure inside the volute is slightly lower than the fluid pressure outside, and the pressure is lower near the volute, but the overall distribution is regular.

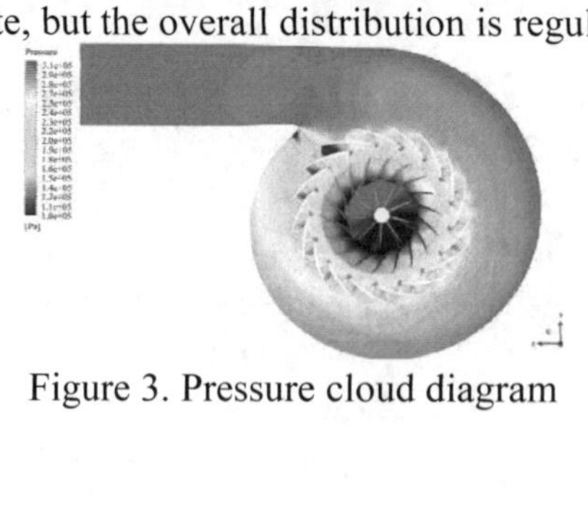

Figure 3. Pressure cloud diagram

IOP Conf. Series: Materials Science and Engineering **793** (2020) 012006 doi:10.1088/1757-899X/793/1/012006

5.2. Analysis of flow field characteristics

The impeller streamline diagram and compressor streamline diagram are shown in Figures 4 and 5, respectively. The reason why that there are some irregular flow lines at the impeller blade tip clearance in figure 4 is the leakage and flow of the tip clearance, but it indicated that the tip leakage loss are little in the impeller. The flow velocity in the diffuser increases sharply at the corresponding sudden pressure changeand secondary flow occurs at the trailing edge of individual blades,which is due to the influence of the reverse pressure gradient. The effect of the diffuser blade at different positions on the flow field at the outlet is large, but the flow separation is small, and the overall flow is stable.

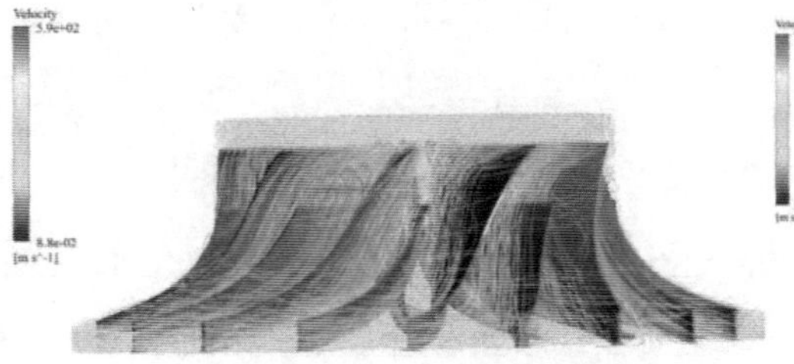

Figure 4. Impeller streamline diagram

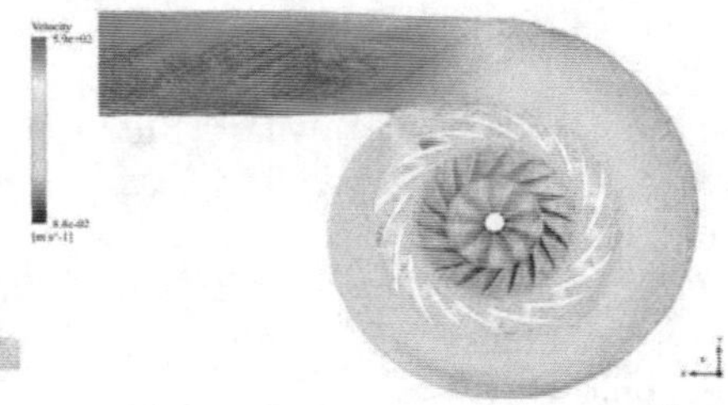

Figure 5. Compressor streamline diagram

5.3. Analysis of velocity field characteristics

The speed vector diagram of the impeller is shown in Figure 6. The flow angle of attack at the leading of the blade and the front mounting angle of the blade has a low coefficient of variation. Due to the high speed, it can be seen from the figure6 that the high-speed section is mainly concentrated on the inside of the main blade, and the speed is significantly reduced after the splitter blade, and the speed at the tip is higher than the surrounding speed.

The speed vector diagram of the diffuser is shown in Figure 7. The deviation of the flow angle of attack at the leading edge of the diffuser blade is basically zero, which causes very little flow loss. The stability of the flow at the outlet of the diffuser greatly affects the performance of the compressor. The distribution of the field relatively intuitively reflects the stability of the flow at the outlet of the diffuser. It can be seen from the figure that the flow at the diffuser outlet is evenly distributed, and there is basically no separation in the flow channel to minimize loss.

Figure 6. Impeller speed vector

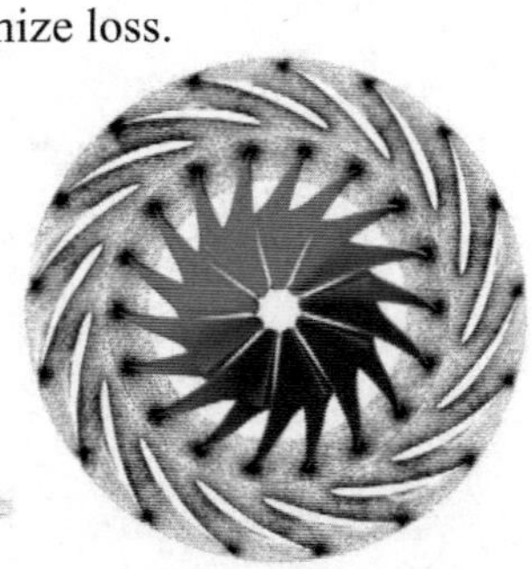

Figure 7. Vector of diffuser speed

6. Conclusion

The numerical results show that the actual pressure ratio of the compressor meets the design requirements. Through numerical simulation of the impeller, it can be seen that the flow path of the impeller is relatively narrow, the working distance of the impeller is long, the flow separation is limited, and the efficiency is high. An unavoidable leakage occurs at the top of the impeller, resulting in secondary flow. However, there is small loss of flow due to the low intensity of secondary flow. In addition, this impeller has advantages of air-distribution uniformity and reasonable pressure distribution. Through the simulation analysis of the diffuser, the deviation of the flow angle of attack at the leading edge of the diffuser blade is basically zero, and the air flow at the blade trailing edge tends to be stable, which reduces the flow loss. And the design is reasonable.

High-speed centrifugal compressors with ternary flow theory for blade design and excellent magnetic suspension bearings are bound to have very important application value and broad application prospects in practical industrial production.

References

[1] Tian Li. Research on the Fault Tolerant Control System of Active Magnetic Bearing [D]. Huazhong University of Science and Technology, 2019.

[2] Jing Tian. Structural Optimization Analysis of high Speed Axial Magnetic Bearings[D]. Shangdong University, 2019.

[3] Reza Safaeian,Hossein Heydari. Study effect of the back irons in passive magnetic bearings with alternating radially magnetized rings by using 2D analytical method[J]. Compel,2019,38(6).

[4] Ziliang Li,Ge Han,Xingen Lu,Enliang Huang,Shengfeng Zhao. Improving the operating range using a centrifugal compressor with a tandem impeller[J]. Aerospace Science and Technology,2019.

[5] Evandro L.L. Pereira,Cesar J. Deschamps. Numerical analysis and correlations for radial and tangential leakage of gas in scroll compressors[J]. International Journal of Refrigeration,2019.

Force Analysis of Fork Type Connecting Structure Pin Roll of Aircraft Vertical Tail in Comprehensive State

Tianle Xing[1]

[1]School of Aeronautic Science and Engineering, Beihang University, Beijing, 100191, China

*TianleXing@hotmail.com

Abstract. Aiming at the stress problem of pin roll connection at the vertical tail of aircraft, the thesis establishes a simplified simulation model of vertical tail. From the mechanical perspective, it takes into account the gravity, assembly force of interference fit and the influence of air resistance on the fork force. Compared with the yield stress of titanium alloy material used in the actual fork, it is concluded that actual stress is much smaller than the yield stress of the material, which verifies the rationality of pin roll design.

1.Introduction

The empennage is a stabilizing and maneuvering device installed at the rear of the aircraft. The vertical tail is the empennage in the vertical direction, which is composed of a fixed vertical stabilizer and a movable rudder. The vertical stabilizer plays the role of directional stability, while the rudder plays the role of directional manipulation.

Aerodynamic and mass loads act on the vertical empennage, among which the aerodynamic load consists of a balancing load, maneuver load, and side force increment under the gusts during a turbulent air flow. The fork connection, also known as the clamping groove connection or lug connection, is a structural form that is widely used in the empennage connection. The fork structure generally consists of lugs, rod pieces and pin rolls. The fork connection structure is simply constructed and easily reassembled, with clear force transferring. However, it is not capable of bearing excessive wing plane load. In this thesis, a simplified vertical tail model is established, in which both sides are connected with the rear fuselage by three fork structures, and titanium alloy (Ti-6Al-4V) is selected as the material. The gravity and interference assembly stress are respectively obtained via mechanical calculation, and the AnsysWorkbench finite element simulation is used to analyze the influence of air resistance on fork stress. The rationality of pin roll design is verified by synthesizing the above factors.

2. The Effect of Gravity on Pin Roll

There are three connection points between the vertical tail and the fuselage bulkhead, which are statically indeterminate structures. It is impossible to directly solve the component force of gravity on six (three on each side) pin rolls.

As is measured by Solidworks, the center of gravity position for the vertical tail model is located at the place of (3972.09,2204.97, -202.10), with the very front end of the vertical tail as the origin. The coordinates of six connection points can also be directly measured. Considering one side, the vertical tail is a symmetric model, whose center of gravity position falls after the position of three pin rolls.

IOP Conf. Series: Materials Science and Engineering **793** (2020) 012007 doi:10.1088/1757-899X/793/1/012007

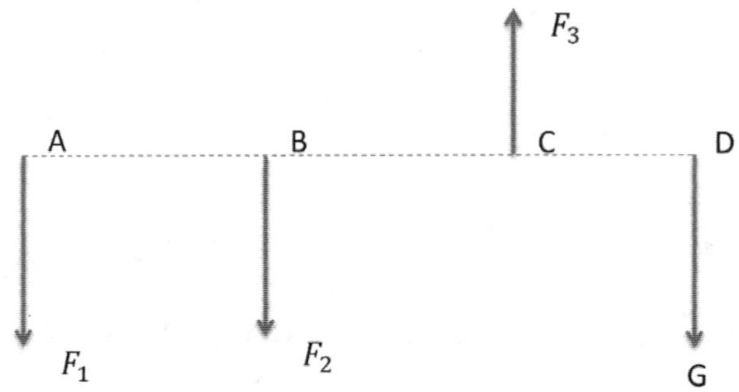

Figure 1.Gravity component analysis of vertical tail

The four points in Figure 1 are schematic diagrams where pin roll fixed point and action of gravity point are projected on the same line. The direction of component force at point C is upward. If only point B exists, the direction of component force at point B is downward. Adding a point A will not change the direction of component force at point B, which increases the stability of the structure, whereas the component force at point A is downward. The equilibrium equation can be obtained as:

- $F_1 + F_2 + G = F_3$
- $F_1x_1 \times F_2x_2 \times Gx_4 - F_3x_3 = 0$

This equation cannot be solved directly, and there are no other conditions. However, it can be figured out that point C is under the biggest force among the three points of application. When there are only points B and C, the effects of gravity are all borne by points B and C, and the component force of point B reaches the ultimate maximum. In this case, the equilibrium equation can be solved:

- $F_2 + G = F_3$
- $F_2x_2 + Gx_4 - F_3x_3 = 0$

Where: $G = 1500kg. x_2 = 1579.44mm, x_3 = 2169.90mm, x_1 = 1005.35mm.$

The maximum of the solution is: $F_{2_{max}} = 4578.27N\ , F_{3_{max}} = 6078.27N$

Under the action of gravity, the pin is in direct contact with the fork hole, and the local stress on the contact surface becomes the extrusion stress. When the extrusion stress is too large, significant plastic deformation will occur in the local area of contact between the hole and pin, thus affecting the normal fit. The thickness of the lug is δ, and the diameter of the pin or hole is d. According to the test and analysis results, the maximum extrusion stress is:

- $\sigma_{max} \approx \frac{F_{3max}}{\delta d} = \frac{6078.27N}{15mm \times 50mm} = 8.104 \times 10^6 Pa$

3.Calculation of Interference Fit Stress

The vertical tail is equipped with the hole-axis assembly, of which the basic dimension is 30mm. The fit tolerance of pin roll is $\varnothing 30t6\binom{+0.054}{+0.041}$, while that of the assembled hole is $\varnothing 30H7\binom{+0.021}{0}$. [1] Its maximum magnitude of interference is $Y_{max} = EI - es = 0.054mm$.

By adopting the Harold method for calculating interference fit stress, the hole-axis fit is simplified to an interference fit model with two equal-length components. As is demonstrated in the following figure (Figure 2), the common interference amount after the assembly of two components is: $\delta_1 = a_2 - b_1 = Y_{max} = 54\mu m$. [2]

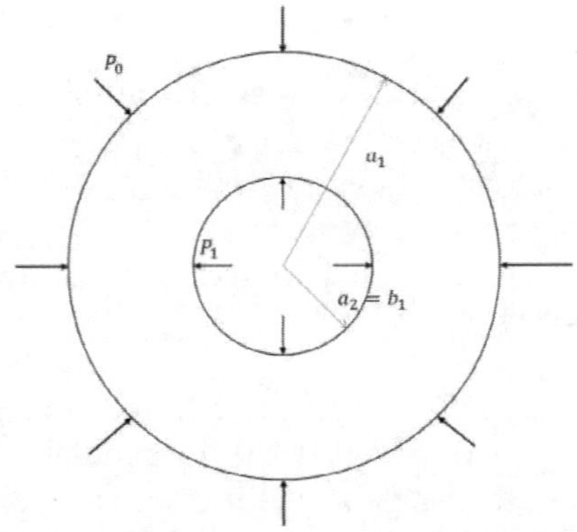

Figure 2. Interference fit model of two equal-length components

The equation applicable to the interference fit of two components is: $\delta_1 = -p_0 k_{3,1} + p_1(k_{4,1} + k_{1,2}) - p_2 k_{2,2}$, where $p_0 = 0$, $p_2 = 0$, conversion factor $k_{4,1} = \frac{b_1}{E}\left(\frac{a_1^2+b_1^2}{a_1^2-b_1^2} - \mu_1\right)$, $k_{1,2} = \frac{a_2}{E}\left(\frac{a_2^2+b_2^2}{a_2^2-b_2^2} - \mu_1\right)$. Among them, titanium alloy (Ti-6AL-4V) is selected for the vertical tail assembly structure: elastic modulus $E = 1.078 \times 10^{11}\text{Pa}$, Poisson's ratio $\mu = 0.31$. Size of outer ring component $a_1 = 23.34\text{mm}$, $b_1 = 15\text{mm}$, size of solid shaft $a_2 = 15\text{mm}$, $b_2 = 0\text{mm}$

Substitute into the formula for calculation:

- $k_{4,1} = \frac{b_1}{E}\left(\frac{a_1^2+b_1^2}{a_1^2-b_1^2} - \mu_1\right)$

$$= \frac{15 \times 10^{-3}}{1.078 \times 10^{11}}\left(\frac{(23.34 \times 10^{-3})^2 + (15 \times 10^{-3})^2}{(23.34 \times 10^{-3})^2 - (15 \times 10^{-3})^2} - 0.31\right) = 2.918 \times 10^{-13}$$

- $k_{1,2} = \frac{a_2}{E}\left(\frac{a_2^2+b_2^2}{a_2^2-b_2^2} - \mu_1\right)$

$$= \frac{15 \times 10^{-3}}{1.078 \times 10^{11}}\left(\frac{(15 \times 10^{-3})^2 + 0}{(15 \times 10^{-3})^2 - 0} - 0.31\right)$$

$$= 9.601 \times 10^{-14}$$

The stress of interference fit is: $P_1 = \frac{\delta_1}{k_{4,1}+k_{1,2}} = \frac{54\times10^{-6}}{2.918\times10^{-13}+9.601\times10^{-14}} = 1.392 \times 10^8 Pa$

4.Estimation of Interference Fit Assembly Force

For vertical tail, the hole-axis assembly is adopted, of which the basic dimension is 30mm. The fit tolerance of pin roll is $\varnothing 30\text{t6}\left(^{+0.054}_{+0.041}\right)$, while that of the assembled hole is $\varnothing 30\text{H7}\left(^{+0.021}_{0}\right)$. Its maximum magnitude of interference is $Y_{max} = EI - es = 0.054\text{mm}$.

The contact stress P_1 of the fitting surface is generated due to the magnitude of interference, which can be calculated according to the referencing formula[2]:

- $P_1 = \frac{E\delta}{4r_2} \times \frac{r_3^2 - r_2^2}{r_3^2} = \frac{E\delta}{4r_2} \times \frac{\left(\frac{r_3}{r_2}\right)^2 - 1}{\left(\frac{r_3}{r_2}\right)^2}$

Set $K = \frac{\left(\frac{r_3}{r_2}\right)^2 - 1}{\left(\frac{r_3}{r_2}\right)^2}$. Take $K = 1$. According to the maximum magnitude of interference Y_{max}, the estimated value of contact pressure is:

IOP Conf. Series: Materials Science and Engineering **793** (2020) 012007 doi:10.1088/1757-899X/793/1/012007

- $P_1 = \frac{E}{4r_2} \times Y_{max}$

Titanium alloy (TI-6AL-4V) is selected for the vertical tail assembly structure: elastic modulus $E = 1.078 \times 10^{11}Pa$, radius of contact surface $r_2 = 30mm$.

Substitute it into the formula for estimation:

- $P_1 = 4.851 \times 10^7 N/m^2$.

The interference fit stress is proportional to the pressed area of fitting surface and friction coefficient. Its formula is as follows:

- $P = 2P_1 \pi r_2 lf$

L is the length of fitting surface, while f is the static friction coefficient.

Aircraft assembly is usually carried out below 20 °C, and the temperature during service does not exceed 50 °C. Referring to the figure, the estimated friction coefficient of titanium alloy (TC4) is $f = 0.6$, and the length of fitting surface is $l = 40mm$.

Substitute it into the formula for estimation:

- $P = 2.195 \times 10^5 N$

This estimated value is larger than the exact value. Consequently, a reasonable assembly structure can be designed, and the size of components can be determined, so as to smoothly realize the hole-axis assembly.

5.The Impact of Air Resistance on Pin Roll

The workbench is used to establish the model of vertical tail under the vertical side force (Figure 3). The analysis results are mainly influenced by fluid analysis, while structural analysis plays a small role in the deformation results.[3] The unidirectional fluid-solid coupling analysis is adopted.[4] Unidirectional fluid-solid coupling analysis refers to one-way data transmission at the interface. The CFD analysis results are transferred to the solid structure analysis, but no solid structure analysis results are transmitted to the fluid analysis. Among which, fluid analysis is performed in CFX, whereas structural analysis is conducted in Transient Structural. Because it is a unidirectional fluid-solid coupling analysis, only the extracted fluid calculation results are required to be applied to the solid calculation.

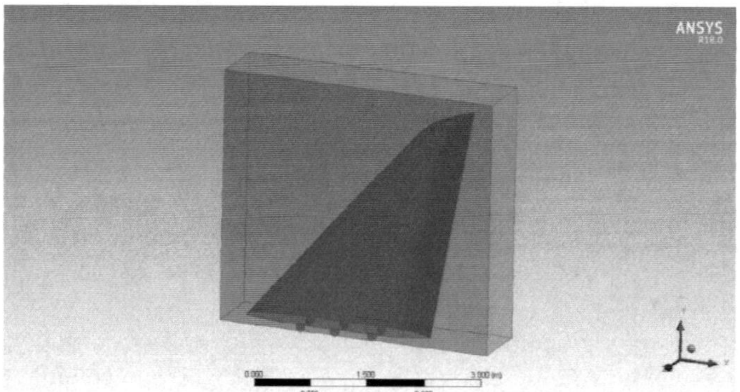

Figure 3. Internal surface model of lug hole

In the fluid analysis, the wind speed is set as 900m/s, and the fixed constraint of vertical tail is the internal surface model of the lug hole. As is shown in Figure 4:

IOP Conf. Series: Materials Science and Engineering **793** (2020) 012007 doi:10.1088/1757-899X/793/1/012007

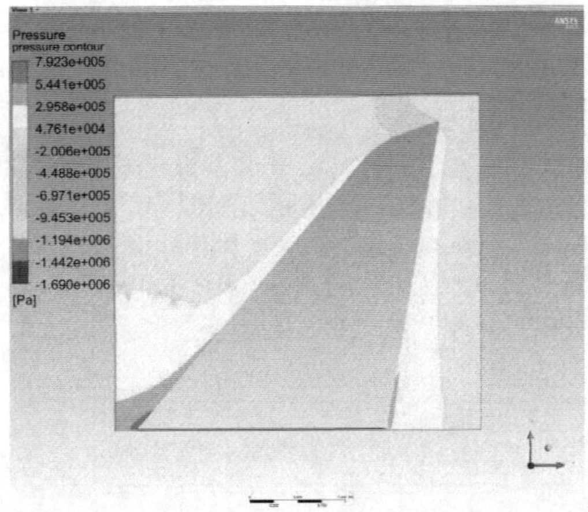

Figure 4. Nephogram of fluid analysis pressure in CFX

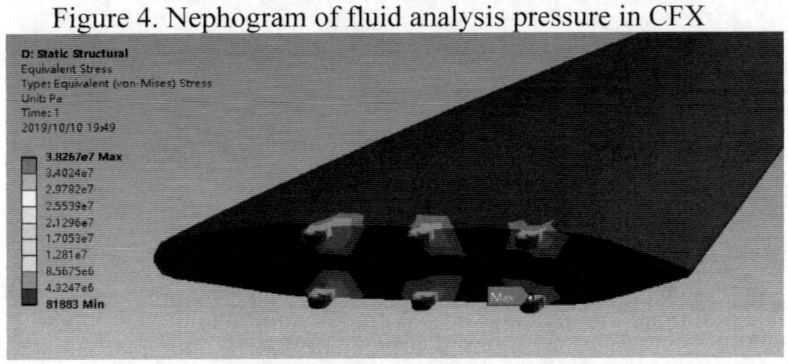

Figure 5. Stress concentration point at the pin

It can be seen from Figure 5 that when subjected to the side force, stress concentration point is at the pin, and the pin end of the tail bears the most force. The maximum stress is 3.83×10^7Pa.

6.Conclusion

In this paper, the force of the pin shaft connecting the vertical tail and the rear fuselage of the aircraft is studied, and the allowable stress of the pin shaft is compared with the actual stress.

This paper analyzes the actual working condition of the pin shaft, and finds three aspects that affect the force on the pin shaft: the gravity of the vertical tail, the assembly force of interference fit and the air resistance of the vertical tail during flight. The gravity of the vertical tail and the assembly stress are analyzed by numerical calculation method, and the influence of vertical tail air resistance on the pin shaft is analyzed by finite element analysis method.

After the above analysis, the pin junction end at vertical tail is the most stressed. Counting the influence of gravity, interference assembly force and flight resistance, the maximum force is:

- $\sigma_m = \sigma_G + \sigma_A + \sigma_f = 8.104\text{MPa} + 139.2\text{MPa} + 38.3\text{MPa} = 185.604\text{Mpa}$

Taking into account the high reliability requirements of aircraft parts, the safety factor of pin design is selected as 3.5 according to experiences, and the maximum force after considering the safety factor is:

- $\sigma_m = 3.5 \times 185.604\text{Mpa} = 649.614\text{MPa}$

The yield limit of the titanium alloy (Ti-6Al-4V) used at the pin is 827MPa, and the applied stress exceeds its yield stress. Therefore, the pin roll design is reasonable, and it can meet the requirements of the complex environment of flight.

IOP Conf. Series: Materials Science and Engineering **793** (2020) 012007 doi:10.1088/1757-899X/793/1/012007

References

[1] Zheng, YR.(1992)Estimation of the Assembly Force of Interference Fits. Machinery Manufacturing Engineer., 07: 27-28.
[2] Jiang,JC.(1991)Stress Calculation of Interference Fit. Mechanical Science and Technology for Aerospace Engineering., 05:16-19.
[3] Zhou,LF.(2011) Introduction to the Method of Unidirectional Fluid-solid Coupling in Computer Simulation Analysis. Science & Technology and Enterprise., 15: 58-58.
[4] Song,XG Cai, L., Zhang, H.(2012) ANSYS Fluid-solid Coupling Analysis and Engineering Projects . China Water & Power Press,Beijing.

IOP Conf. Series: Materials Science and Engineering **793** (2020) 012008 doi:10.1088/1757-899X/793/1/012008

Target Detection and Tracking System Based on FPGA

Xiuli Yu[1], Bingqi Li[1], Mingshuai Dong[1*] and Weimin Yin[1]

[1]Automation school, Beijing University of Posts and Telecommunications, Beijing, China

E-mail:dongmingshuai@bupt.edu.cn

Abstract. The parallel processing mode of Field Programmable Gate Array (FPGA) and the ability to perform multiple tasks repeatedly meet the real-time and flexible requirements of target detection and tracking. The FPGA of Cyclone IV of ALTERA company is adopted as the hardware processor for the system platform to filter and preprocess color conversion of the collected image. The edge detection method of Sobel operator is adopted to realize the convolution and binarization processing of the third-order matrix of pixel points and the operator template, and the inter-frame difference algorithm is applied to the moving target. At the same time, the tracking algorithm based on centroid feature matching can realize the real-time tracking of target position. The system uses the characteristics of high real-time, low power consumption, parallel processing and strong expansibility of the FPGA, transplants machine vision algorithm into the FPGA, reduces the delay of image processing, and completes the hardware implementation of the target detection and tracking algorithm. Finally, the experiment verifies the real-time performance and effectiveness of the algorithm.

1. Introduction

With the development of integrated circuits, the application of integrated image processing chips, FPGA and Digital Signal Processing (DSP) processors become more and more extensive. Among them, the design cost of special integrated image processing chips is relatively high, and the flexibility, universality and portability of the chip are unsatisfactory. The calculation speed of DSP processors is very fast, which is suitable for completing the calculation of complex algorithms. But, the limitations of hardware optimization prevent the system from being widely used. On the contrary, because of its special design structure, FPGA can process data in pipelined and parallel ways[1][2]. It has the characteristics of reconfiguration, which enables it to complete various situations, and has better compatibility and flexibility. Therefore, in the work of target detection and tracking which requires high computation speed and strict real-time performance, the advantage of FPGA is obvious[3]. Using FPGA processor instead of computer to process target detection and tracking can make the integrated system have the characteristics of small size, low cost, low energy consumption, fast speed, high flexibility, good real-time performance and good compatibility. Therefore, the research on image processing and target tracking based on FPGA has certain economic interests and broad application prospects.

This paper presents a target detection and tracking system based on FPGA. In this system, an image acquisition and display platform is built with FPGA as the hardware processor. The collected image data is stored in the memory. On this basis, moving objects in the scene are detected, and the features of moving objects are extracted by image segmentation. Finally, the tracking effect of the target is achieved by controlling the rotation of the platform and the movement of the car.

IOP Conf. Series: Materials Science and Engineering **793** (2020) 012008 doi:10.1088/1757-899X/793/1/012008

2. Overview of Overall System Functions

The system uses the FPGA as the processor, and designs the functional modules of the system, including image acquisition module, image preprocessing module, SDRAM data storage module, image algorithm processing module, VGA display driver module, motion control module, configuration module and mobile car based on STM32 controller, as shown in figure 1.

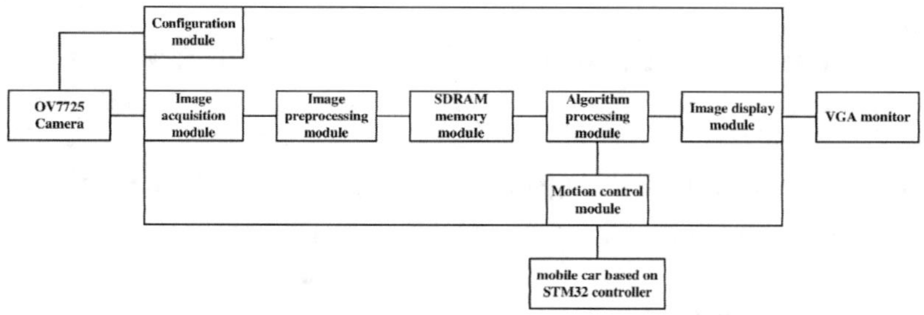

Figure 1 Architecture of System Platform Moduls

The system platform uses the pipeline image processing architecture of parallel processing to reduce the delay of image processing. The dual-core architecture processor of FPGA + ARM is adopted. The parallel algorithm is implemented by FPGA. ARM can undertake some tasks of serial data processing and some control tasks. CMOS monocular industrial camera OV7725 is used for image acquisition. SDRAM memory for data storage and VGA display with 640*480 resolution and 60Hz refresh rate for image display. And the common interface and transmission protocol of visual system are used, such as USB, CAMERALINK, serial port, LVDS interface, I2C interface and so on. The communication between each functional module depends on the instantiation of the functional module and signal reference of the top-level file. Finally, the whole hardware platform is built.

2.1. VGA Display Driver Module

This module is used to drive VGA display. According to the protocol analysis of VGA, combined with the development mode and working principle of the FPGA, the VGA driver is decomposed into the structure shown in figure 2.

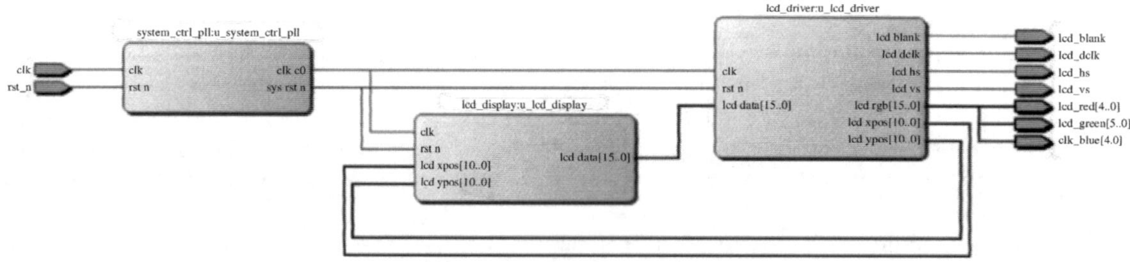

Figure 2 VGA driver structure partition

The system_ctrl_pll module belongs to the global clock management module, which is responsible for the input and output of the whole system's global clock; the lcd_display module belongs to the image display module, which receives the data information of the pixels, outputs the displayed data, and finally completes the image production task; the lcd_driver module belongs to the VGA controller. The driver module is responsible for receiving display data, outputting RGB display data, and timing control and display control.

2.2. RAM Driver Module

SDRAM chip is selected to design the RAM drive module to store the collected image data, and the ping-pong operation of the memory is used to solve the frame interleaving problem. The principle is shown in figure 3.

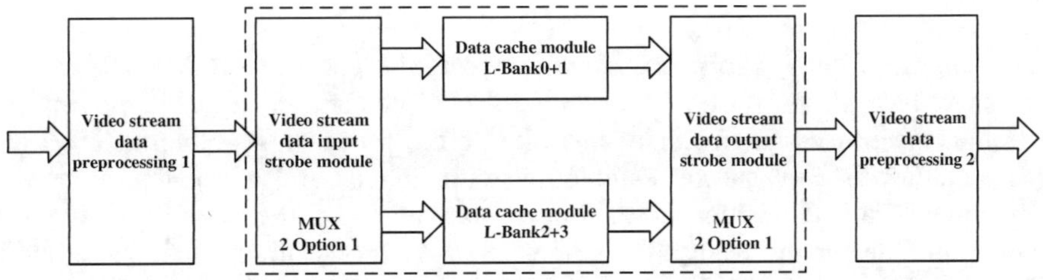

Figure 3 The schematic diagram of table tennis operation principle

In the Sdram_Control_2Port file, two FIFO memories are instantiated. Video stream data is input. Two consecutive frames of images are not written into the same BANK through the gated module, and then the image data stored in the two FIFOs are output sequentially through the output gated module, that is, 0+1B is achieved through the buffer. The incomplete *ping-pang* operation of ANK and 2+3 Bank data. After testing, the function works well.

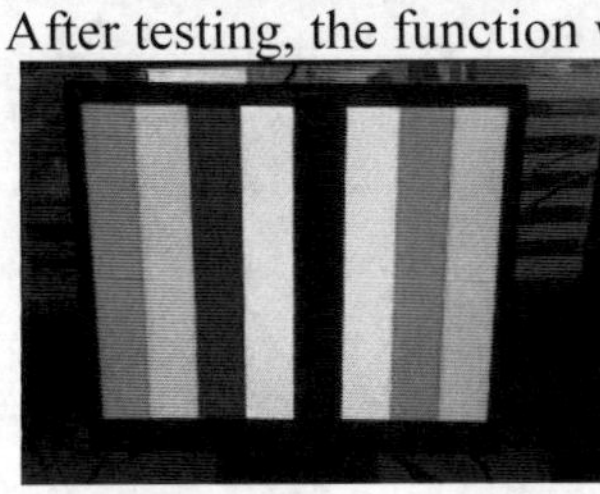

Figure 4 Table tennis operation effect (vertical stripe)

Figure 5 Table tennis operation effect (horizontal stripe)

2.3. Camera Driver Module

This module is used to drive the camera to collect image data, capture the light source data through the photosensitive sensor, and then convert the captured analog signal into digital signal through AD conversion processing for subsequent modules to calculate and store.

The board selected in this article has reserved the interface of the CMOS camera module, and the interface follows the interface definition of all official CMOS modules. The camera model selected OmniVision's OV7725 Sensor as the image acquisition source[4].

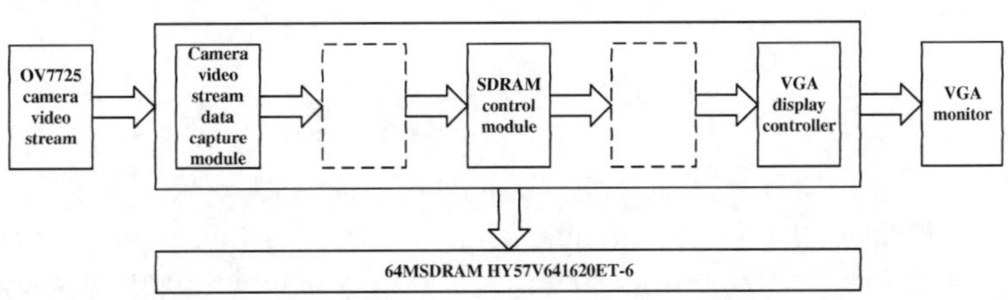

Figure 6 Structure flow chart of video acquisition and display system based on OV7725 camera

3. Hardware Implementation of Vision Algorithms

When designing a complete vision system, image processing algorithms should be divided into three categories: bottom, middle and high-level algorithms according to their different functions and tasks. The following figure is the flow chart of the system algorithm processing.

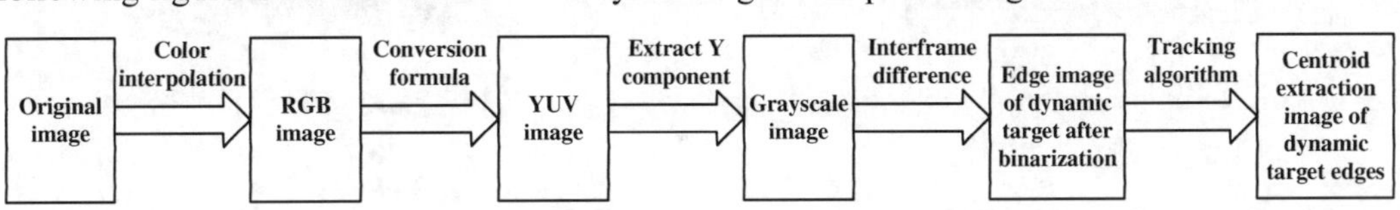

Figure.7 Flow chart of system algorithm

3.1. Image Preprocessing Algorithms

Image preprocessing algorithm is to process the image at the pixel level, and does not segment and recognize the image. Its main purpose is to process the complex and changeable moving image locally or globally, so that the processed image is suitable for the processing of higher-level algorithm. Its main processing method is convolution calculation with images and templates, so as to change the structure of the pixel data. Considering the logic realization of the algorithm, the preprocessing algorithm is very suitable for the front-end devices whi//ch can be directly connected with the image sensor, just like the FPGA we used. In the preprocessing part, the original image is transformed into RGB image, RGB image into YCbCr image, YCbCr image into gray image, median filter and Sobel edge detection are completed.

3.1.1.Convert Original Image into Gray Image. In the process of image acquisition by camera, the output data of OV7725 sensor is Raw data (Raw RGB), which is the original image file. It includes image data acquired from scanners, digital cameras and so on. After color interpolation, it becomes RGB image. Then the RGB image is transformed into YCbCr image by the conversion formula between RGB and YCbCr, as shown in (1).

$$\begin{cases} Y = 0.299R + 0.587G + 0.114B \\ Cb = 0.568(B - Y) + 128 = -0.172R - 0.339G + 0.511B + 128 \\ Cr = 0.713(R - Y) + 128 = 0.511R - 0.428G - 0.083B + 128 \end{cases} \tag{1}$$

Because VHDL language is needed for hardware transplantation, but VHDL can not perform floating-point operations, so the equivalent transformation of formula is carried out by expanding the value by 256 times and then moving the value to the right by 8 Bit bit, as shown in formula (2):

$$\begin{cases} Y = ((77R + 150G + 29B) \gg 8) \\ Cb = ((-43R - 85G + 128B) \gg 8) + 128 \\ Cr = ((128R - 107G - 21B) \gg 8) + 128 \end{cases} \tag{2}$$

Then the algorithm formula hardware is transplanted to the FPGA, and the formula is combined by the adder, shift register and multiplier in the FPGA. If the above formula is used directly, when the simultaneous expression is less than 0, the operation will be wrong. Therefore, before transplanting the hardware of the algorithm, the operation will be split and transformed. First, 128 is extracted into brackets, and the formula (3):

$$\begin{cases} Y = (77R + 150G + 29B) \gg 8 \\ Cb = (-43R - 85G + 128B + 32768) \gg 8 \\ Cr = (128R - 107G - 21B + 32768) \gg 8 \end{cases} \tag{3}$$

Because R, G and B are all positive numbers, the value of Y must be greater than 0, and the calculated Cb and Cr must be greater than 0, so they can be calculated directly without spillover.

After the above processing, the gray image is obtained, and the original image is compared with the gray image to obtain figure 8 and figure 9:

Figure 8 Original image

Figure 9 Gray level image

3.1.2. Median Filtering. Noise exists in the gray-scale image, so filtering can not only remove the solitary noise, but also preserve the edge characteristics of the image and avoid the obvious blurring of

IOP Conf. Series: Materials Science and Engineering **793** (2020) 012008 doi:10.1088/1757-899X/793/1/012008

the image.In this paper, we use the fast sorting method as shown in figure 10 to get the median of the pixels. The steps are as follows:

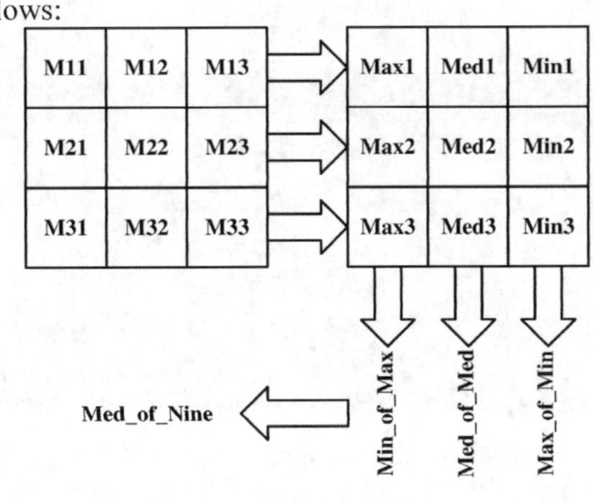

Figure 10 Quick sorting method for median principle diagram

The RTL principle of the above process is shown in figure 11. Input 9 pixels and output target median pixels:

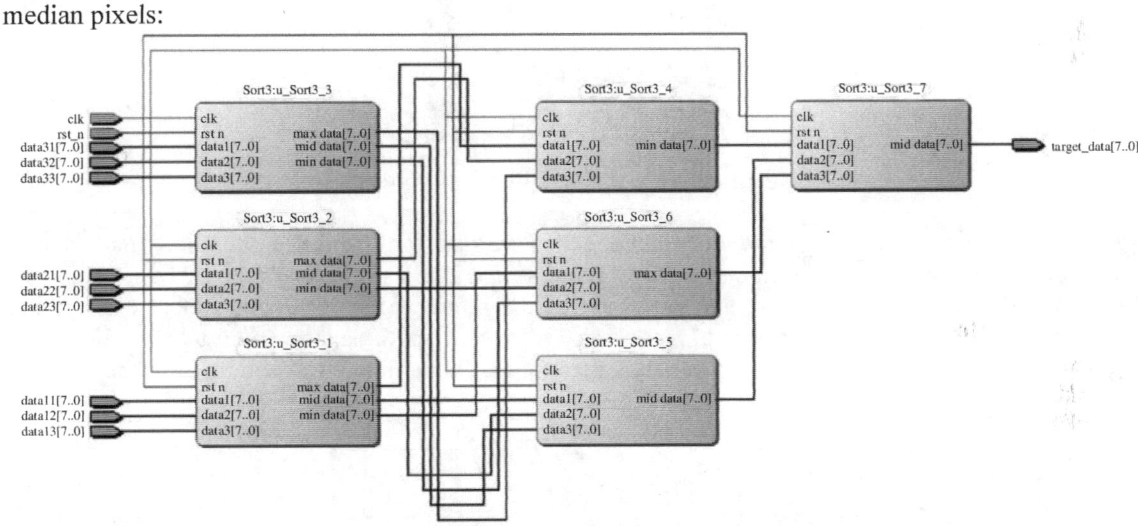

Figure 11 Median RTL schematic diagram for quick sorting

Through the simulation test of Modelsim, figure 12 is obtained.

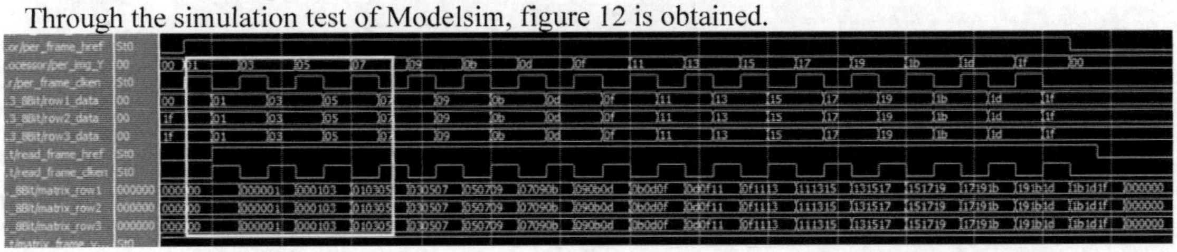

Figure 12 Generates third-order pixel matrix Modelsim simulation

In the yellow box above, you can clearly see that the data is input into the shift register and generate a third-order matrix to verify the correctness of the method. The simulation test of median filter is also carried out by Modelsim, and the simulation figure is shown in figure 13.

IOP Conf. Series: Materials Science and Engineering **793** (2020) 012008 doi:10.1088/1757-899X/793/1/012008

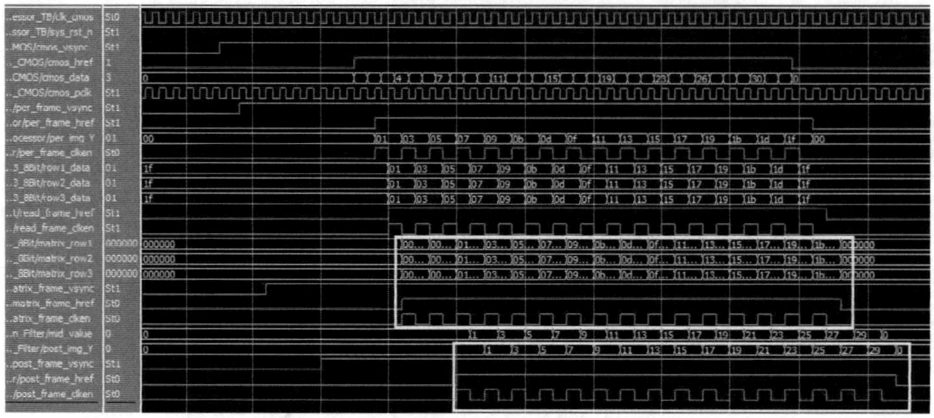

Figure 13 Modelsim simulation waveform of median filtering

The image after median filtering is shown in figure 14. The image is relatively dark because the maximum value is discarded.

Figure 14 Image processed by median filtering

3.1.3. Sobel Edge Detection Algorithms. Sobel operator is a discrete difference operator, which is used to calculate gradient approximation of image brightness function. Using Sobel operator at any pixel of the image, the corresponding gray gradient value of the point will be obtained. Sobel convolution operator contains a series of 3x3 matrices, which refer to transverse and longitudinal calculators respectively. By plane convolution of these two operators with the image to be processed, the gradient values of the pixels in the X and Y directions of the image can be obtained respectively[5].

Sobel edge detection algorithm can be carried out according to the following steps in figure 15:

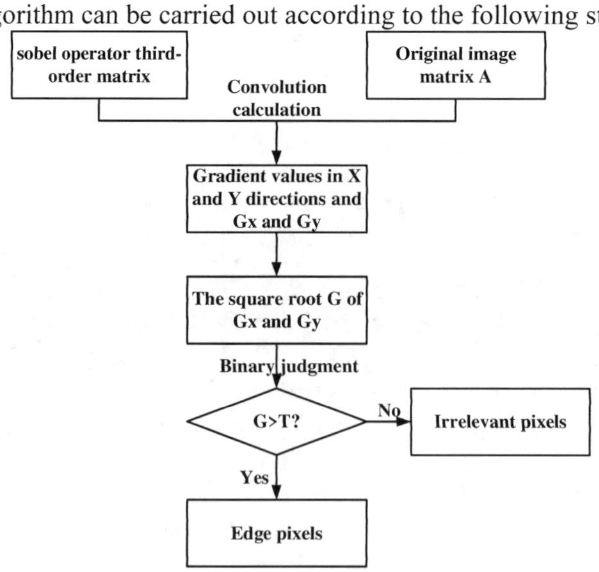

Figure 15 Shift RAM shift register IP core module

Then, we use Modelsim to simulate and test the correctness of Sobel edge detection algorithm, as shown in figure 16.

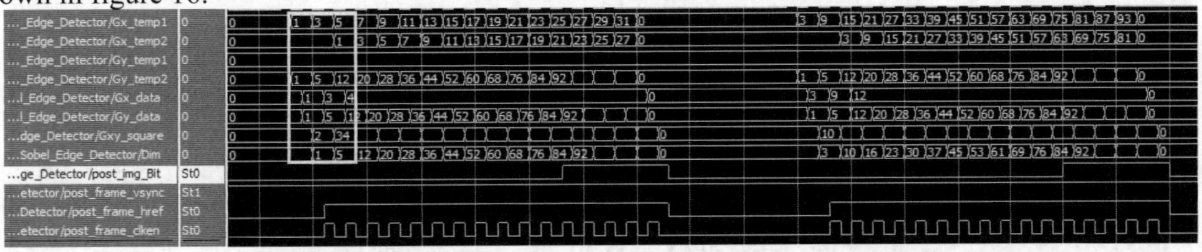

Figure 16 Modelsim simulation of Sobel edge detection method

The logic circuit representation of the calculation formula of the Sobel edge detection algorithm above is shown in figure 17. According to the graph, the logic relationship of the hardware circuit of the algorithm in the FPGA can be clearly seen.

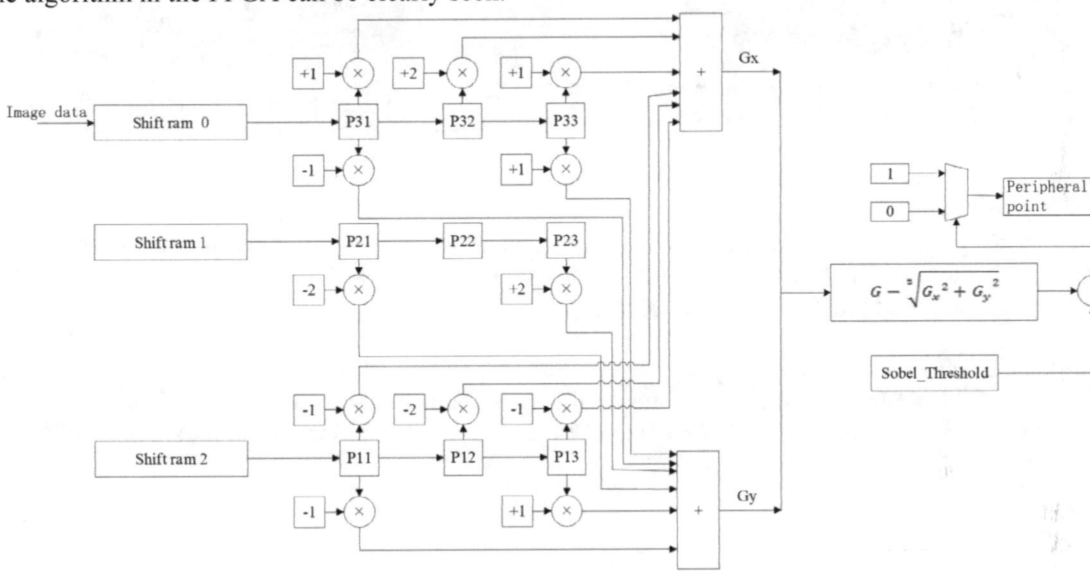

Figure 17 Logic circuit design of Sobel edge detection algorithm

The image processed according to Sobel edge detection algorithm is shown in figure 18.

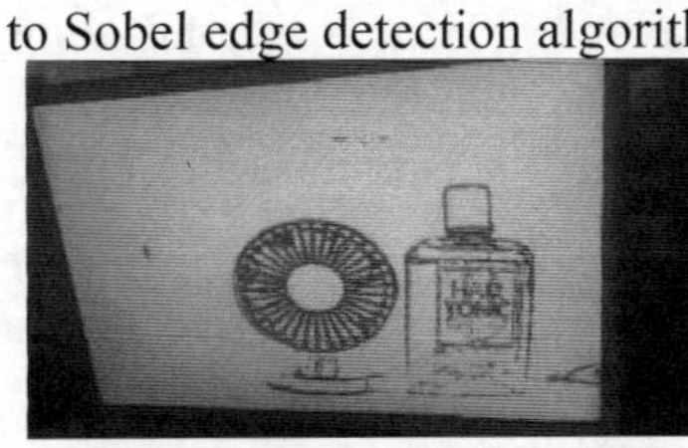

Figure 18 Sobel Edge Detection Processed Image

3.2. Target Detection Algorithms

In this paper, we use the inter-frame difference algorithm to extract moving objects. The main idea is to analyze the images of different frames in succession in the video images with moving objects. The different parts of the image are the regions where the moving objects are located. By doing the difference operation on the images of different frames in succession, we can extract the moving objects. By establishing a mathematical model, the pixel values of two consecutive adjacent frames are calculated differentially and then binarized to extract moving objects in the image. If the video frame image at any time t is expressed as f (x, y, t), then the video frame image at time t + 1 is expressed as f (x, y, T + 1). The calculation formula is shown in formula (4):

$$f_d(x,y,t,t+1) = \begin{cases} 1, & |f(x,y,t) - f(x,y,t+1)| > T \\ 0, & |f(x,y,t) - f(x,y,t+1)| \le T \end{cases} \quad (4)$$

In formula (4), T is a pre-set differential threshold. By comparing the results of the differential operation with the threshold, the binary processing is carried out. The region with 0 pixel value is identified as the background region, and the region with 1 pixel value is identified as the moving target region. That is to say, the threshold judgement is to identify the pixels with relatively large changes in the two frames as the target pixels, and the smaller changes as the background pixels without changes.

Figure 19 Background white wall when the frame difference method is used to process the static background

Figure 20 Moving fan after the frame difference method is used to process the moving fan

3.3. Target Tracking Algorithms

Target tracking algorithm is the process of segmenting and locating the determined moving target from the collected continuous multi-frame images, and finally obtaining the complete trajectory of the moving target or predicting the location of the target[6][7][8].

Because the edge feature extraction method is used in the target detection algorithm, the tracking algorithm based on feature matching is adopted. In the previous stage, the edge information of the target extracted from the image is processed by the inter-frame difference method to track the target. That is to say, the centroid position is calculated by the edge information of the moving target[9]. The formula is as follows:

$$\bar{x} = \frac{\sum \bar{x}_i A_i}{\sum A_i} \quad (5)$$

$$\bar{y} = \frac{\sum \bar{y}_i A_i}{\sum A_i} \quad (6)$$

Among them, $\bar{x}_i$ is the abscissa of the pixels, $\bar{y}_i$ is the ordinate of the pixels, and $\bar{A}_i$ is the pixel of the pixels. The centroid coordinates of $(\bar{x}, \bar{y})$ type can be obtained by calculation. The image is processed by the inter-frame difference method and the moving object center extraction algorithm under the square area around the core (the area is increased for observation), as shown in figure 21.

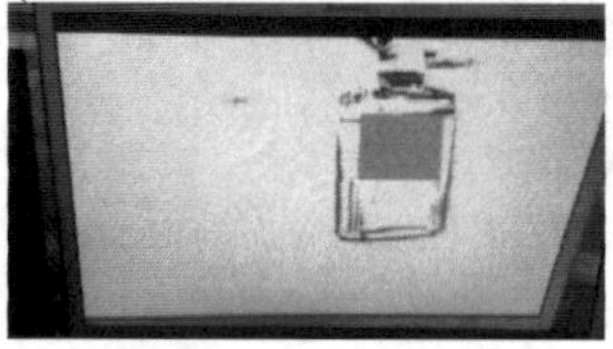

Figure 21 Interframe Difference Method for Moving Target Centroid

3.4. Target Tracking System Experiments

The object detection and tracking system can detect moving objects and mark their centroid. After comparing with the coordinates of the image center, the direction of centroid movement can be judged and the car can track the centroid. The implementation process of the algorithm is to determine the position of the centroid coordinate of the moving object relative to the central coordinate of the VGA display screen, and control the signal 1 of left rotation or 0 of right rotation sent by the FPGA to

STM32. The signal is transmitted to the STM32 controller of the car through the serial port line for motion control.

In the motion control algorithm program of STM32, the signal value received by STM32 is judged. When it is 1, the left-wheel motor of the car is controlled to be stationary, the right-wheel rotates at a certain speed, and the rotation time is preset to achieve the effect of turning the car to the left. When it is 0, the right-wheel motor of the car is controlled to be stationary and the left-wheel rotates at a certain speed. Rotate and preset the rotation time to achieve the effect of turning the car to the right. The flow chart of the tracking experiment is shown in figure 22.

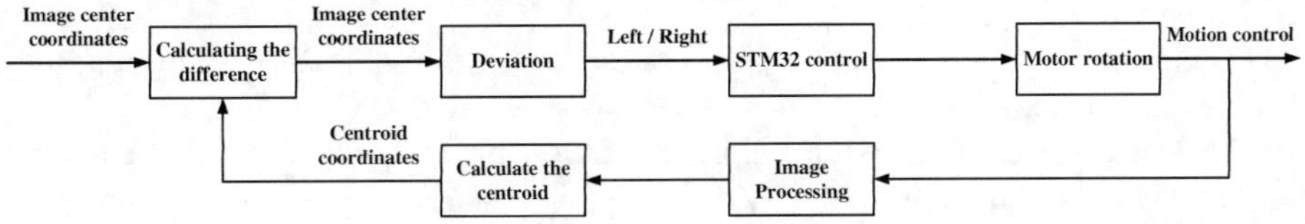

Figure 22 Flow chart of target centroid tracking experiment

Experiments show that the car can complete the expected rotation and move with the moving target. figure 23, 24 and 25 are the results of the target detection and tracking system controlling the car to track the moving target.

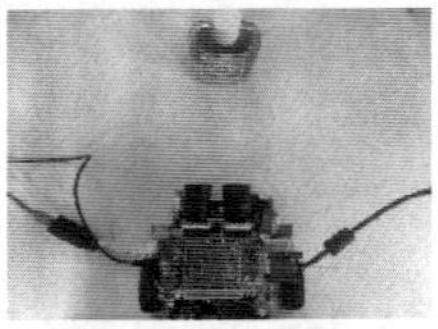

Figure 23 Tracking Starting Point.

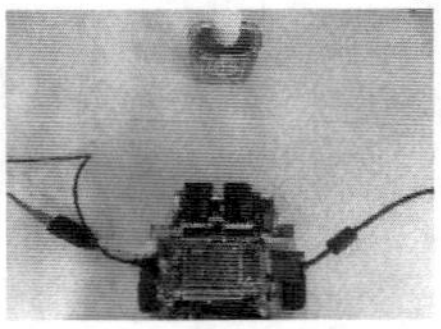

Figure 24 Tracking Left Rotation.

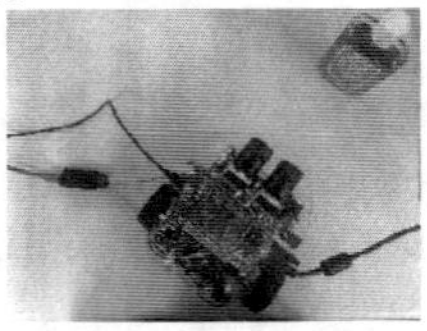

Figure 25 Tracking Right Rotation.

4. Conclusion

The system platform built in this paper uses two FIFOs to design *ping-pang* storage, which solves the frame interleaving phenomenon when using a single SDRAM, and is verified by the color bar experiment. The formulas of the inter-frame difference algorithm are realized and transplanted by program, and the binary images of objects in static and moving states are obtained. The start-up program of the inter-frame difference algorithm is designed, and the control system switches between gray image and the image after the inter-frame difference processing. Selecting the edge characteristics of the moving object for tracking algorithm based on feature matching, that is, extracting the edge information of the moving object, calculating the centroid position of the moving object, and finally controlling the car to track the centroid position in real time. The centroid effect map of moving object is obtained by experiment, and the tracking of object by car is also realized.

The system platform uses the FPGA as the core of the hardware processor of image processing, which improves the speed and real-time of image processing, reduces the volume and energy consumption, and meets the requirements of the vision processing system towards the direction of miniaturization, integration, real-time and intelligent development.

References

[1] Wang Yandong. Design and implementation of stepper motor cooperative motion control system based on embedded real-time system [D]. Hebei University of Technology, 2016.

[2] Zhang Jin, Liu Haiyan, Wu Shihai. Application of FPGA in Moving Object Detection and Tracking [J]. Electronic Measurement Technology, 2010, 33 (12):64-67.
[3] Chen Quanjin. Research on hardware implementation of dynamic image detection and tracking algorithm based on video image [D]. University of Electronic Science and Technology of China, 2012.
[4] Zeng Yongxi, Huang Meihong, Chen Weihao. Video surveillance system based on FPGA + OV7725 [J]. Fujian Computer, 2015, 31 (09):8-9+39.
[5] Gonzalez R C, Woods R E. Digital Image Processing[M]. Addison-Wesley Longman Publishing Co., Inc., 2001.
[6] Xiao Dan. Intelligent video analysis system based on motion detection and tracking [D]. Tianjin University, 2012.
[7] Xiong Bei. Design of Bayer image color restoration based on FPGA [D]. Chengdu: Chengdu University of Technology, 2015.
[8] Li Tong. Research on multi-target tracking technology under intelligent video surveillance [D]. Hefei: University of Science and Technology of China, 2013.
[9] Jain R, Nagel H H. On the analysis of accumulative difference pictures from image sequences of real world scenes. [J]. IEEE Transactions on Pattern Analysis & Machine Intelligence, 1979, 1(2):206-214.

Research on Fault Diagnosis Method of Gearbox Based on SA and BP-AdaBoost

Yangyang Zhang[1], Yunxian Jia[1*], Weiyi Wu[1] Xiaobo Su[1] and Dong Liang[1]

[1]Army Engineering University, Shijiazhuang, Hebei, 050003, China

[*]Corresponding author's e-mail: 2108007979@qq.com

Abstract. In order to solve the problems of difficult selection of state feature parameters and poor accuracy of single BP Neural Network in gearbox fault diagnosis, a gearbox fault diagnosis method based on SA and BP-AdaBoost is proposed. Taking the typical fault of a gearbox as the research object, the vibration signal of the typical working state of the gearbox is collected through the preset fault experiment. The time domain statistical parameters with high sensitivity is selected as the feature vectors by the Sensitivity Analysis method. Then these state feature vectors are input into BP-AdaBoost model for training and testing, and their results are compared with those of BPNN model. The results show that the proposed method can quickly and effectively diagnose the fault of gearboxe, and is better than BPNN model.

1. Introduction

Gearbox is an important part of vehicle transmission system, mainly composed of gears and bearings. With the extension of operation time, the performance of gearbox gradually deteriorates, and it is one of the mechanical devices with high failure rate because it often works continuously under the harsh environment such as heavy load and high speed[1]. At the same time, due to the important role of gearbox in the transmission device, its health condition will affect the working state of the whole vehicle chassis system. In case of failure, if not handled in time, it will lead to serious consequences[2]. Therefore, the study of gearbox health monitoring and early fault diagnosis is of great significance for the prevention of serious vehicle faults[3].

For the fault diagnosis of modern mechanical equipment, artificial Neural Network is simple, operable and intelligent. Among them, BP Neural Network is the most widely used one[4,5]. Liang Chao et al.[6] used BP Neural Network as classifier for rotor fault diagnosis, Zhang Yingsong et al.[7] established a fault diagnosis model based on BP Neural Network and applied it to the fault diagnosis model of turbofan engine. There are also scholars who use improved BP Neural Network for fault diagnosis of gearbox or gearbox. In reference[8], a fault diagnosis method based on Chaos Quantum Particle Swarm Optimization BP Neural Network is proposed to identify the fault mode of gearbox of wind turbine. In reference[9], the BP Neural Network is improved by using dynamic learning rate, and the gearbox fault diagnosis network model is established. However, the learning algorithm of BP network has a slow convergence speed, and it is easy to fall into the local optimal value, resulting in the low accuracy of diagnosis. The AdaBoost algorithm is introduced into BP Neural Network model to build BP-AdaBoost fault diagnosis model, which can effectively improve the diagnosis accuracy of the original single BP Neural Network.

Based on the above research, this paper proposes a gearbox fault diagnosis method based on sensitivity analysis (SA) and BP-AdaBoost. There are many time-domain statistical parameters of vibration signal, so sa is used to select the time-domain statistical parameters with high sensitivity as the

IOP Conf. Series: Materials Science and Engineering **793** (2020) 012009 doi:10.1088/1757-899X/793/1/012009

state eigenvector and reduce the dimension of the state eigenvector. Then these eigenvectors are input into BP-AdaBoost model for training and testing, which can quickly get accurate diagnosis results.

2. SA and BP-AdaBoost

2.1. Sensitivity Analysis

In the process of time-domain analysis, considering that there are many time-domain statistical parameters of vibration signal, many time-domain indexes will greatly increase the calculation time, and some unnecessary parameters will also form interference, which may increase the error. Therefore, it is very important to select the appropriate characteristic parameters to form the fault characteristic vector, which requires that after extracting the time-domain characteristic parameters, these parameters should be further studied Through sensitivity analysis, the characteristic parameters which can effectively represent the state of gearbox are selected. The calculation formula of sensitivity analysis is:

$$u = \left| \frac{X_i - X}{X} \right| \tag{1}$$

Where u is the sensitivity, X_i is the parameter to be analyzed, and X is the reference parameter. Usually, the time-domain characteristic parameters of normal state are taken as reference parameters, and the time-domain characteristic parameters of fault state are taken as parameters to be analyzed.

2.2. BP-AdaBoost

The main idea of BP-AdaBoost algorithm is: first, each sample is initialized to equal weight, and then iterates n times with weak classifier. After each iteration, the weight is updated according to the classification result. For the failed samples, a larger weight is given, and more attention is given to the next iteration. Weak classifiers get a sequence of classification functions through multiple iterations. Each classification function will give a certain weight according to the classification error, and finally combine multiple weak classifiers to form a strong classifier with small and stable classification error. The algorithm flow of BP-AdaBoost is shown in Figure 1, and its main steps are as follows.

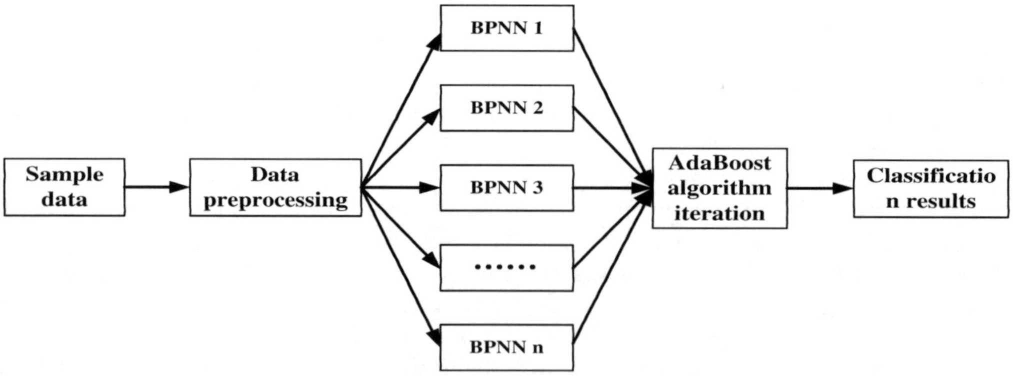

Figure 1. Algorithm flow

Step 1: data selection and network initialization. Select m group training samples from the sample space and initialize the sample weight $D_t(i) = 1/m$. According to the dimension of sample input and output, the structure of neural network is determined, and the weight and threshold of BP Neural Network are initialized.

Step 2: weak classifier prediction. When the first t weak classifier is trained, the BP Neural Network is trained with training data and the output of the training result is predicted. The prediction sequence $g(t)$ and the sum of prediction errors e_t are obtained

$$e_t = \sum_i D_t(i) \quad i = 1, 2, \cdots, m(g(t) \neq y) \tag{2}$$

Where $g(t)$ and y are predicted classification results and expected classification results respectively.

Step 3: calculate the weight a_t of the prediction sequence according to the prediction error e_t. The formula is

$$a_t = \frac{1}{2}\ln\left(\frac{1-e_t}{e_t}\right) \tag{3}$$

Step 4: update the training sample weight. Adjust the weight of the next round of training samples according to the weight a_t of the prediction sequence, and the adjustment formula is

$$D_{t+1}(i) = \frac{D_t(i)}{B_t} \cdot \exp\left[-a_t y_i g_t(x_i)\right] \quad i = 1, 2, \cdots, m \tag{4}$$

Where B_t is the normalization factor, the purpose is to keep the weight proportion unchanged and the sum of the weights equal to 1.

Step 5: synthesize a strong classifier. After T training rounds, we get group weak classification function $f(g_t, a_t)$ and group strong classification function $h(x)$.

$$h(x) = \mathrm{sign}\left[\sum_{t=1}^{T} a_t \cdot f(g_t, a_t)\right] \tag{5}$$

3. Experimental verification

In order to verify the effectiveness of the proposed method, a group of experiments are set up to verify the real experimental data. The vibration analysis and fault diagnosis experiment platform system of QPZZ-Ⅱ rotating machinery is adopted in the experiment. The gearbox used in the experiment is mainly composed of a pair of meshing gears. According to the common failure modes of gearbox, this experiment simulates four operation states of gearbox, namely pitting, wear, tooth breakage and normal state. During the experiment, the acceleration monitoring data in the Y direction of the load side of the gearbox input shaft are collected by the acceleration sensor, the sampling frequency is 5120 Hz, and 150 groups of sample data are collected for each state.

In order to extract fault information more comprehensively, this paper selects 14 commonly used time-domain statistical parameters, which are: mean value, root mean square value, square root amplitude, absolute mean amplitude, mean square value, maximum value, minimum value, peak-peak value, waveform index, peak value index, pulse index, margin index, skew index and kurtosis index.

The above characteristic parameters are numerous, which will inevitably produce interference or overlap, and the 14 dimensional data sample will increase the calculation amount, so the effective solution is to carry out sensitivity analysis. Sensitivity analysis is conducted for 14 characteristic parameters of each state, and the analysis results are shown in Figure 2.

It can be seen from the above figure that the first, fifth and 13th characteristic parameters are highly sensitive, and there are great differences between different fault modes, so the recognition ability of different states of gearbox is better. Therefore, three characteristic parameters, namely mean value (P1), mean square value (P2) and skew index (P3), are selected as the characteristic parameters of gearbox state. The state characteristic parameters of 150 groups of sample data in each state are calculated to obtain 150 state eigenvectors in each state. Some state eigenvectors are shown in Table 1.

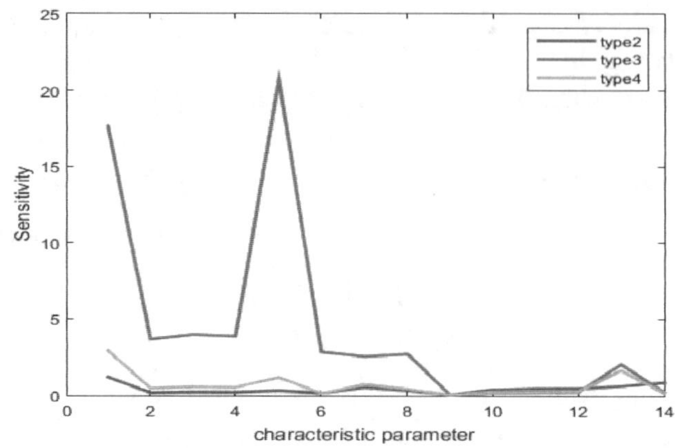

Figure 2. Sensitivity analysis of characteristic parameters

Table 1. State eigenvector

	P1	P2	P3
	0.0486	19.5797	-0.0803
Type 1	0.0308	17.3548	-0.0769
	0.0280	18.4701	-0.0687
	0.0327	11.3584	-0.1667
Type 2	0.0425	12.5395	0.0692
	0.0395	13.0963	-0.1683
	0.1104	383.1858	-0.0336
Type 3	0.0486	392.4404	0.0070
	0.0890	409.6765	0.0050
	0.0659	40.6511	-0.1370
Type 4	0.0153	37.7756	-0.0523
	-0.0038	36.5522	-0.1858

The BP-AdaBoost fault diagnosis model is established, and in the 160 groups of state eigenvectors of each state, 100 groups are randomly selected as training samples, and the rest 50 groups are selected as test samples. Input the training samples into the model in turn to train them. After the training, input the test samples to get the gearbox fault diagnosis results, as shown in Table 2.

Table 2 Test results

	Type 1	Type 2	Type 3	Type 4	Total accuracy rate
Test samples	50	50	50	50	
Accuracy quantity	49	48	47	50	97%
Accuracy rate	100%	96.67%	95%	100%	

From the test results, it can be seen that the method proposed in this paper can accurately identify the operation state of the gearbox, and the total accuracy reaches 97%. In order to further verify the superiority of the proposed method, three groups of comparative experiments are carried out below, and the experimental results are summarized as shown in Table 3.

Table 3 Comparison of test results

	BPNN	BP-AdaBoost	SA & BP-AdaBoost

Total number of test samples	200	200	200
Accuracy quantity	173	186	194
Accuracy rate	86.5%	93%	97%
Running time	1.713 s	3.436 s	2.097 s

Obviously, AdaBoost integrated learning method can effectively enhance the fault diagnosis ability of BPNN, and the accuracy of BP-AdaBoost fault diagnosis model is higher than BPNN. SA method can significantly reduce the dimension of data, reduce the calculation time, and improve the accuracy.

4. Conclusions

This paper presents a fault diagnosis method of gearbox based on SA and BP-AdaBoost. The results of experiment and simulation show that Sensitivity Analysis (SA) method can extract the feature parameters with high sensitivity and reduce the dimension of data. The state feature vector composed of three feature parameters can represent the key and useful fault information. BP-AdaBoost method can use the state feature vector to diagnose the fault mode of gearbox quickly and accurately, which has certain value in engineering practice reference significance and application value.

Acknowledgments

This work was supported by National Natural Science Foundation of China (71871220),

References

[1] Duan, L.X., Xie, J.Y., Wang, K., et al. (2017) Gearbox fault diagnosis based on auxiliary data sets under different working conditions, 36 (10): 104-108 + 116.
[2] Jiang, X.X., Li, S.M., Zhou, D.W., et al. (2017) Time frequency ridge fusion method for fault identification of planetary gearbox under different operating conditions in time. Journal of vibration engineering, 30 (01): 127-134.
[3] He, Z.J., Chen, J, Wang, T.Y., et al. (2012) Theory and application of mechanical fault diagnosis. Higher Education Press, Beijing.
[4] Hu, J., Yan, F.W.. (2011) Research on the method of gasoline engine misfire diagnosis based on BP Neural Network. Automotive engineering, 33 (02): 101-105.
[5] Kong, X.C., Zhao, W.Z., Wang, C.Y. (2017) Joint estimation of SOC of lithium battery based on bp-ekf algorithm. Automotive engineering, 39 (06): 648-652.
[6] Liang, C., Lu, P., Gao, N., et al. (2018) LPP based rotor vibration fault feature extraction method. Journal of vibration engineering, 31 (03): 539-544.
[7] Zhang, Y.S, Wu, C., Tang, H.L, et al. (2017) Decision making method for maintenance of a turbofan engine based on fault detection. Journal of Aeronautical power, 32 (01): 82-88.
[8] Cheng, J.T., Ai, L., Duan, Z.M., et al. (2017) Application of cqpso-bp algorithm in gearbox fault diagnosis of wind turbine. Journal of solar energy, 38 (08): 2112-2116.
[9] Mo, Y.M., Qi, F., Zhang, J. (2017) Research on Gearbox Fault Diagnosis Based on dynamic learning rate and BP Neural Network. Mechanical design and manufacturing, 11: 49-52.

IOP Conf. Series: Materials Science and Engineering **793** (2020) 012010 doi:10.1088/1757-899X/793/1/012010

Research on Vehicle Exterior Fire Suppression Techniques

Gaoshou Deng, Runing Lin, Xiaoxiang Shi, Wen Zhang, Hui Chen

China North Vehicle Institute of Technology, 100072, Beijing, China

gaoshoudeng@163.com

Abstract. This paper studies automatic vehicle exterior fire detection and suppression techniques. The main difficulties include: 1) complex and changeable environment outside the vehicle; 2) low sensitivity of the fire detector; 3) quick dissipation of fire extinguishing agent in the open space outside the vehicle; 4) extinguishing agent nozzle blocking due to dust and sand. This paper selects a special ultra-fine dry powder fire extinguishing agent for automobiles to design the protective nozzle and to study the automatic detection controller with the functions of condition monitoring and fault diagnosis, with an aim to improve fire suppression reliability. Based on the virtual and real vehicle tests, it is verified that the fire extinguishing effect of the designed system meets the design requirements.

1. Introduction

The local fire outside the vehicle tends to spread rapidly to the whole vehicle. If it is not extinguished in time, the personnel and vehicle safety will be in great danger. It might be too late to put out the vehicle exterior fire using the portable fire extinguisher, and sometimes the person may fail to get out of the car due to the fierce fire. Therefore, the research on efficient and reliable automatic vehicle exterior fire suppression techniques is of great significance for personnel and vehicle safety.

The current research on vehicle fire suppression techniques mainly focuses on interior firefighting. Exterior firefighting techniques, however, are relatively difficult to study due to the unfavorable factors including 1) complex and changeable environment outside the vehicle; 2) low sensitivity of the fire detector; 3) quick dissipation of fire extinguishing agent in the open space outside the vehicle; 4) extinguishing agent nozzle blocking due to dust and sand.

The research on vehicle exterior fire suppression techniques mainly covers the selection of fire extinguishing agent, automatic fire detection controller, protective nozzle design, fire extinguishing system layout inside vehicle, and verification tests.

2. Selection of fire extinguishing agent

At present, available fire extinguishing agents include inert gas, fine water mist, halogenated alkanes, aerosols and ultra-fine particles. Their characteristics are analyzed as follows.

- Inert gas

One of the representative inert gases for fire suppression is IG-541, whose fire extinguishing mechanism is suffocation. Its main disadvantages are 1) low fire suppression efficiency; 2) high concentration requirement for fire extinguishing; therefore, it is demanding in terms of sealing and strength of the container; 3) large size, making it unsuitable for vehicle exterior fire suppression.

- Fine water mist

The particle size of the water mist is generally between 20 and 120 μm, and its fire extinguishing mechanism is cooling and suffocation. Its fire suppression ability is affected by the size of the water droplets and the working pressure of the nozzle (higher working pressure of the nozzle, stronger ability to generate droplets). Therefore, the equipment and technology that can turn water into ultra-fine water mist is very important. However, it is not suitable for extinguishing vehicle exterior fire.

- Halogenated alkanes

Haloalkane fire extinguishing agents mainly include heptafluoropropane, trifluoroiodomethane (code: F23), etc. Heptafluoropropane, featuring non-conductivity, low toxicity, and zero ozone depletion potential (ODP), is efficient and fast in fire suppression. However, its GWP reaches 2900 and it will remain in the atmosphere for up to 37 years, making it unsuitable for long-term use.

The trifluoroiodomethane concentration for fire extinguishing is 3.9%. When CF3I is thermally decomposed, it will generate I ions, strongly inhibiting the combustion reaction. However, it will generate toxic gases under high temperature and its synthesis is costly, so it is not used widely. Haloalkane fire extinguishing agents are not suitable for extinguishing vehicle exterior fire.

- Aerosols

Aerosols are formed by dispersing ultra-fine solid or liquid particles in a gaseous medium. This type of fire extinguishing agent can form a fire aerosol substance. The fire extinguishing mechanism of the hot aerosol includes chemical suppression, decalescence and cooling, and oxygen concentration reduction; therefore it has high fire extinguishing efficiency. However, it also has some disadvantages: ① the flame is sprayed outward during combustion, which may cause a secondary fire; ② the surface temperature of the equipment is too high; ③ the thermal stability is poor; ④ it is difficult to suppress class A deep-seated fire. Therefore, it has limitations in application and is not suitable for extinguishing vehicle exterior fire.

- Ultra-fine particles

Ultra-fine particles can effectively suppress combustion, with multiple fire extinguishing mechanisms, such as endothermic cooling, gas-phase chemical suppression, solid-phase chemical suppression, and reduction of oxygen concentration. In addition, ultra-fine particles are small in size, usually 3~5 μm, which is larger than the specific surface area. So, they have a high frequency of interaction with the flame and a strong ability to capture free radicals, therefore a high fire extinguishing efficiency.

The advantages of ultra-fine dry powder fire extinguishing agent include 1) zero ozone depletion potential (ODP); 2) zero GWP; 3) non-toxicity and no harm; 4) non-corrosion to protective materials; 5) non-irritation to human body; 6) high fire extinguishing efficiency.

Automobile-specific ultra-fine dry powder fire extinguishing agent belongs to the type of ultra-fine particle fire extinguishing agent, and it can adapt to harsh external conditions such as high temperature, low temperature, high humidity, and vibration and long-term vehicle operation. It is suitable to be used not only in the relatively enclosed total submerged fire extinguishing, but also in open local submerged fire extinguishing as well as in the open space fire suppression. Its efficiency is 40 times that of the water-based extinguishing agent, 6 to 10 times that of the ordinary ABC dry power extinguishing agent, 2 to 3 times that of the halon extinguishing agent, and 10 to 12 times that of the heptafluoropropane fire extinguishing agent. It can be used to extinguish A, B, C, E fires on the vehicle as well as fire outside the vehicle.

3. Automatic fire detection controller technique

At present, automatic fire detectors used in cars, such as heat-sensitive wires and optical detectors, are susceptible to the outside interference, leading to lower detector sensitivity and the generation of false signals. By researching highly reliable detectors based on state monitoring and fault diagnosis technologies, we can improve the sensitivity and reliability of fire detection. Based on the high temperature characteristics of the flame, the temperature for fusing the special insulation layer between the two metal wires and sending out the signal is generally set as 85~150 °C. This detection method is highly reliable and can not be easily affected by environmental factors.

However, the failure of the detector itself will affect the reliability. Therefore, it is necessary to improve detector reliability by monitoring and diagnosing detector failures, such as damage of the insulation layer, disconnection of the detector, damage of the metal protective layer, and grounding of the metal wire. The control circuit between the automatic controller and the fire extinguisher is also prone to failure. The control circuit can be used to realize the functions of circuit self-diagnosis and fault alarm and to determine whether there is short circuit or open circuit, thereby improving the reliability of the automatic fire suppression system. The working principle of automatic fire control technique is shown in Figure 1.

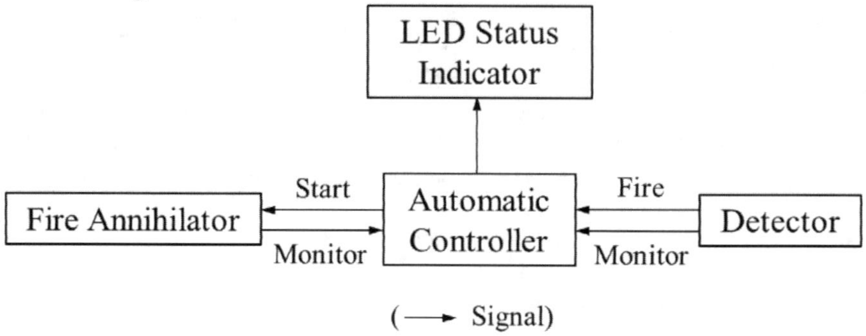

Figure 1. Working principle of automatic fire control technique

4. Protective nozzle

Disk-shaped protective nozzle are used to extinguish fires on circular or roughly circular planes and rectangular or roughly rectangular planes, and the nozzle is located in the middle of the plane. The disc-shaped protective nozzle is composed of a nozzle, a protective cover and a guide rod. The protective cover, located at the upper part of the nozzle, is made of non-metal or light metal. The guide rod is fixed on the top surface of the nozzle through a screw connection. When not in use, the protective cover shields the orifices.

During fire suppression, the fire extinguishing agent in the pipeline pushes the protective cover upward, thus exposing the orifices, ensuring the normal and smooth discharge of the fire extinguishing agent. The advantages of the design include 1) the protective cover can effectively prevent the orifices from being blocked by mud, dust, gravel, rain and snow, etc., and keep the inside of the orifices and the pipeline clean and unblocked, so as to ensure the reliable operation of orifices; 2) when the orifices are not shielded, the lower flange of the protective cover can also help to ensure the right spraying direction of the fire extinguishing agent, thus improving the fire extinguishing efficiency. The disk-shaped protective nozzle is shown in Figure 2.

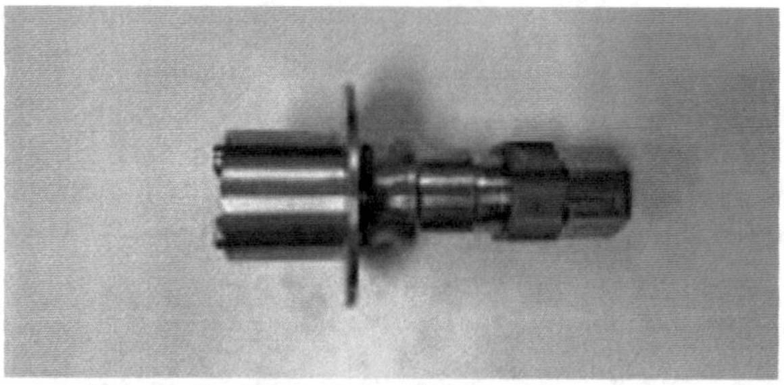

Figure 2. Disc-shaped protective nozzle

5. Layout of fire extinguishing system

The vehicle exterior fire extinguishing system is composed of the fire detector, automatic controller, fire extinguisher, pipeline and nozzle. When a fire occurs, the detector detects the burning flame and sends a signal; the automatic controller activates the fire extinguisher; the fire extinguishing agent is

IOP Conf. Series: Materials Science and Engineering **793** (2020) 012010 doi:10.1088/1757-899X/793/1/012010

transported along the pipeline to the nozzle and is sprayed out instantly, covering the flame root, and extinguishing the flame at one time. Considering the limited number of fire extinguishers that can be put in the vehicle, the possibility of scattered ignition points and simultaneous ignition of multiple parts as well as the pipeline length and characteristics of pressure distribution, it is necessary to reasonably allocate the limited fire extinguishing bottles on the vehicle.

The vehicle exterior fire extinguishing system can be developed either for newly developed vehicles or for equipped ones. The layout design of the system shall be conduct based on the close negotiation with the vehicle designer or user, and it shall not interfere with other vehicle components. The computer-aided design can help to arrange the fire extinguishing system on the overall design drawing of the vehicle, setting the positions of the nozzle, the fire extinguishing bottle, and the pipeline. The overall layout of the vehicle exterior fire extinguishing system is shown in Figure 3.

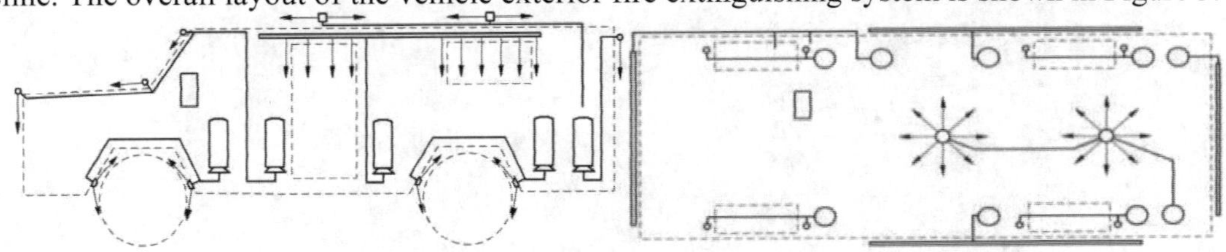

Figure 3. Layout of vehicle exterior fire extinguishing system

6. Conclusion

Tests are needed to verify whether the vehicle exterior fire extinguishing system is reasonable and whether its fire extinguishing effect meets the design requirements. Usually, a combination of virtual test and real vehicle test is adopted, and the real vehicle test is performed only after the virtual test is passed. The advantages of adopting virtual test include 1) save test costs; 2) reduce test risks; 3) shorten the development cycle; 4) increase success rates. The effect of the virtual test is shown in Figure 4, and the effect of the real car test is shown in Figure 5.

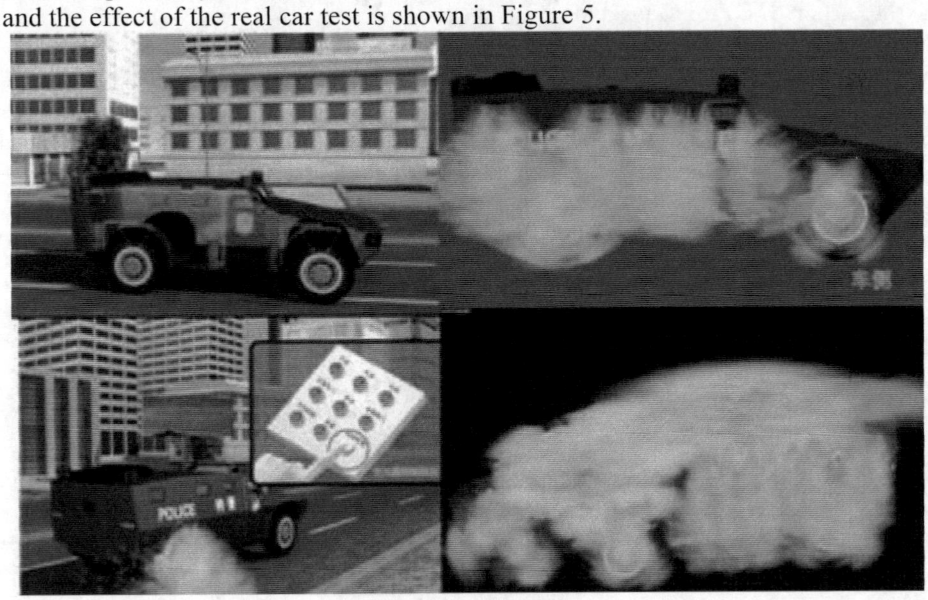

Figure 4. Virtual test

Figure 5. Real vehicle test

References

[1] Shi Xiuzhi. On cold aerosol fire extinguishing techniques [J]. Fire technology and product information. 2000 (10): 27-28.

[2] Zhu Rong, Tang Zhuhua.A Preliminary Study on the Configuration and Use of Fire Extinguishers [J]. Fire Management Research, 2005 (4): 483-485.

[3] Du Jianhua, Zhang Rencheng, Ding Huan.The current status and development trend of automobile fire detection research [J]. Fire Science and Technology, 2012, 31 (4): 436-439.

[4] Chen Tao, Zhao Lizeng, Fu Xuecheng. Research on automatic fire extinguishing technology based on ultra-fine dry powder in initial electrical fire [J]. China Safety Science Journal, 2015.

IOP Conf. Series: Materials Science and Engineering **793** (2020) 012011 doi:10.1088/1757-899X/793/1/012011

Research on Improvement Methods of Automotive NVH Performance

LIU Xing

(Sichuan Aerospace Vocational College，Chengdu 610100，China)

Abstract: The performance of automotive NVH will directly affect the production and sales of automobiles. This article summarizes the generation and control of the main NVH problems in automobiles, and discusses the methods to improve the NVH performance of the vehicle from three aspects: free damping treatment of the body structure, analysis of the contribution of body panels, and analysis of key hard-point forces of the vehicle steering system.

1. Introduction

With the development of automotive technology, the key technology gaps among engine manufacturers, chassis, and body of various vehicle manufacturers have narrowed. The performance of automotive NVH (Noise, Vibration, and Harshness) will directly affect the production and sales of complete vehicles. During the driving of a car, the main sources of vibration noise include: vibration noise generated by engine excitation, vibration noise generated by the interaction of automobile tires and the road surface, and vibration noise generated by friction between the air flow and the vehicle body. After the various vibration noises are superimposed, they are transmitted to the entire body, causing the vibration of the body structure, radiating the noise to the interior of the vehicle, and making the passengers in the vehicle feel uncomfortable.

2. Basic Theory of Vehicle Vibration Performance and Acoustics

2.1 Car Vibration Performance

The vibration performance of a car is mainly evaluated by its ride comfort characteristics. During the driving process of the car, vibration and impact will occur, which will affect the comfort of the driver and passengers. Due to the uneven road surface, the generated incentives will be input to the car, and then passed to the people in the car through the tire system, suspension system, and seat system. Under different driving conditions, different human accelerations are obtained, and the ride of a car is characterized by the response of the human body to vibrations.

The human body's response to vibration depends on the frequency and intensity of the vibration, and also has a direct relationship with the psychological and physical factors of the human body. There are also some differences in the sensitivity of different people to vibration. At present, the evaluation of ride comfort mainly includes two parts, namely subjective evaluation method and objective evaluation method. The subjective evaluation method mainly considers human factors, and realizes the comprehensive evaluation of vehicles based on actual statistical methods. The objective evaluation method mainly relates to vehicle vibration isolation performance and mechanical vibration physical parameters, and can fully display actual evaluation indicators.

2.2 Basic Theory of Acoustics

Studies have shown that the hearing frequency of the human ear is 20 Hz to 2000 Hz, and the sensitivity to different frequencies is not the same. Sound can form physical phenomena, including sound pressure and sound power. Sound pressure is the change that occurs when atmospheric pressure is disturbed by sound waves. Sound power indicates the strength of the sound source and determines the speed and amplitude of sound propagation.

Due to the wide range of sound pressure energy, for convenient representation, logarithmic coordinates are often used instead of absolute coordinates. At the same time, from the perspective of sound reception, the subjective "loudness" is not directly proportional to the absolute value of sound pressure, but approximately proportional to the logarithm of sound pressure. Therefore, considering these two reasons, logarithmic coordinates are commonly used in acoustics to measure sound pressure, expressed in terms of sound pressure level, and the unit is dB (decibel).

The effective value of the sound pressure at the point to be measured is p_e, and the sound pressure level L_p is defined as:

$$L_p = 20\lg\frac{p_e}{p_{ref}} \tag{1}$$

In the formula, p_{ref} is the reference sound pressure, and is taken as $2\times10^{-5}\text{Pa}$. This value is the audible valve sound pressure of normal human ear to 1KHz sound.

According to the principle of energy superposition, the average value L_{pm} of the sound pressure level can be obtained:

$$L_{pm} = 10\lg(\frac{1}{m}\sum_{i=1}^{m}10^{0.1L_{pi}}) \tag{2}$$

Where $L_{pi}(i = 1,2,\ldots,m)$ is the sound pressure level measured at the i measurement point, and m is the total number of measurement points.

3. Generation and Control of Automotive Major NVH Problems

3.1 Source of Car Noise

Automotive noise mainly comes from engine noise, wind noise, and structural vibration radiation noise. The noise radiated by these noise sources will be introduced into the vehicle through air and solids. In the process of solid-borne noise, that is, the vibration caused by the excitation of the engine and the road surface can be transmitted to the vehicle body through the floor structure, thereby causing continuous vibration of the vehicle body, and then the vibration radiation noise of the body plate wall is transmitted to the vehicle, resulting in the interior of the vehicle noise. In the process of airborne noise, the noise mainly comes from automobile engine noise, tire noise, fluid noise caused by body and air friction, and propagates into the car through the body gap to form interior noise. In general, high-frequency noise above 800Hz is mainly airborne noise, low-frequency noise from 5 Hz to 300Hz is mainly caused by body vibration, and mid-to-high frequency noise from 30 Hz to 300Hz is mainly caused by body wall vibration.

3.2 Evaluation of Interior Noise
In order to reduce the harm to people caused by noise, many countries in the world have implemented relevant standards for the control of automobile internal noise. In the course of research, the American Society of Automotive Engineers conducted statistics on a large amount of data, and proposed that the noise value inside the cab of a truck cannot exceed 88dB, and also pointed out the maximum value of noise sound pressure level.

Table 1 Maximum Internal Noise of SAE Recommended Trucks in the United States

Octave Center Frequency (Hz)	63	125	250	500	1000
Sound Pressure Level (dB)	103	97.5	92	86.5	81

Due to the different structures between vehicles, different driving conditions have different effects on vehicle noise, resulting in large differences in noise levels in cars. In order to ensure that the intelligibility of people's voice communication in the cab is not affected, the noise value in the car cannot exceed 70dB.

3.3 Control Method of Automobile Vibration Noise
The first is vibration isolation treatment, adding elastic damping elements such as rubber pads to the vibration transmission path of the car can reduce the transmission of vibration sources to the vibration body and reduce the vibration noise of the interior structure. Both the engine suspension damping and the suspension system shock absorption damping can reduce the body vibration and reduce the interior noise. The second is sound insulation and noise reduction. For the noise of the engine and the outside of the car, appropriate sound insulation materials are selected for sound insulation operations to achieve effective control of noise. The last is the application of sound-absorbing materials. Commonly used sound-absorbing materials are asbestos, slag wool, and felt, all of which are porous materials. The more micropores and gaps inside the material, the smaller the gap, the better the sound absorption effect.

4. Automotive NVH Performance Improvement Method

4.1 Body Structure Free Damping Treatment
If the noise in the car is caused by the vibration of the body structure, and the frequency is concentrated below 200Hz, a method of damping the body structure can be adopted. The additional damping material is locally coated on the floor, and the modal strain capacity analysis of the entire bottom plate is performed, and the modal strain energy of the bottom plate within 200 Hz is superimposed on each other. Perform free damping treatment to keep the thickness of the damping layer at about 2mm. By coating the body structure with a damping material, the low-frequency noise in the vehicle can be effectively reduced.

4.2 Analysis of Body Panel Contribution
During the acoustic-solid coupling finite element calculation, the calculation of the structural

modal factors and the calculation of the plate acoustic contribution found that the structural modal and body plates contributed the most to the interior noise. At the same time, the vehicle body is divided into areas such as the roof, front floor, left front door, and right rear door to display the peak noise data of the vehicle in different areas. By comparison, it is found that the tailgate and the roof cover have a greater contribution to the peak of interior noise. By effectively improving the body structure, the interior noise can be reduced and the interior noise design standard can be achieved.

4.3 Force Analysis on Key Hard Points of Vehicle Steering System

4.3.1 Steering System Model Building

During the establishment of the rack-and-pinion steering system model, the main research contents include the steering wheel, steering shaft, steering gear, steering transmission shaft, and lateral stabilizers that affect steering accuracy. The steering wheel is connected to the auxiliary frame through a rotating pair; the steering shaft is connected to the auxiliary frame through a cylindrical pair; the two ends of the steering transmission shaft are connected to the knuckle arm and the gear shaft respectively; the steering rack and the steering gear adopt a translation pair connect. In order to improve the stiffness of the suspension roll angle, adjust the ratio of the stiffness of the front and rear suspensions to avoid large lateral inclination or lateral vibration of the vehicle body at high speeds, a lateral stabilizer bar needs to be installed in the front suspension. For the establishment of the transverse stabilizer bar model, it mainly involves two left and right symmetrical parts, and a rotating pair needs to be applied at the point of symmetry to ensure that the two parts can rotate relative to each other about their axial directions.

4.3.2 Force Analysis of Key Hard Points in Steering System

Compare actual measurement data with effective simulation of lateral stabilizers, establish a rigid-flexible multi-body dynamics model of the steering system, simulate straight-line driving on a B-grade road at 60km / h, and obtain the hard points of the connection points between the steering system and the body Force changes. A force spectrum analysis is performed on each hard point of the hard point force curve, and the actual force situation of the hard point under static conditions is calculated to obtain a static force curve where each hard point deviates from the equilibrium position. By analyzing the hard-point static force curve, the structure optimization of the steering system can obviously improve the NVH performance of the entire vehicle.

5. Result Analysis

Based on the free damping treatment of the body structure, the change of the body panel structure, and the optimization of the vehicle steering system structure, a multi-body dynamics model of the entire vehicle system was established, and the vibration of both sides of the wheels in the same direction was used to re-perform the vehicle vibration and do a comprehensive calculation analysis about noise performance. The calculation results show that the vehicle body reference point vibration speed and noise sound pressure level are significantly reduced, and the NVH performance of the vehicle is significantly improved.

References

[1] Liu Xing, Yan Bing, Zhang Shengjie, Tang Qin. Experimental research on the 36th order

noise source of vehicle generators [J] .Weite Motor, 2016 (44): 35-36 + 40.
[2] Zhang Yu, Wang Mengxiang. Exploration on the Construction of Automobile NVH Technology Course in Higher Vocational Colleges [J]. Automobile Maintenance and Repair, 2019 (14): 50-51.
[3] Li Zhongkui, Wu Kaifeng, Yuan Liang, Xu Zebin, Gong Kan. Automotive body structure design based on NVH performance [J]. Automotive Engineers, 2019 (06): 24-28 + 54.
[4] Xu Jinli, Luo Shijun, Jia Bing, Chen Duanxun. Research on the Influence of Additional Bending Moment on Universal Vehicle's NVH Performance [J]. Mechanical Transmission, 2019, 43 (04): 37-43.
[5] Li Haibing. Comprehensive Evaluation Method and Application of Automobile NVH [J]. Environmental Technology, 2018, 36 (06): 70-76.
[6] Xu Bing, Wang Guochun. Research on Optimization Design of Platform Body Based on NVH Performance [J]. Enterprise Technology and Development, 2018 (09): 33-37.
[7] Liang Huiyao, Feng Huanhong, Ye Fuyu. Improved NVH performance of the new damping structure of air-conditioning compressors [J]. Bus Technology and Research, 2018, 40 (03): 51-53.

IOP Conf. Series: Materials Science and Engineering **793** (2020) 012012 doi:10.1088/1757-899X/793/1/012012

Analysis of manhole structure Safety of draft tubes in hydropower plant

Wenbo Li[1], Jun Zhang[1], Lanlan Liu[2], Xiaokai Li[3], Jun Wang[1], Hengyi Zhou[1], Yi Xie[1]

[1]Hunan Electric Power Research &Test Institute, Changsha, Hunan, 41007, China

[2]State Grid Hunan Transmission Maintenance, Changsha, Hunan, 41007, China

[3]State Grid Hunan Hydropower company, Changsha, Hunan, 41007, China

[*]Corresponding author's e-mail:liwb@alum.imr.ac.cn

Abstract. The fault of the manhole on draft tube in a hydropower plant was investigated and analyzed. The results showed that the main reason for the cracking and leakage of the manhole on draft tube was that: the four corners of the manhole were not rounded, so fatigue cracks appeared under the effect of long-term alternating loads, which caused the leakage at the four corners of the manhole on draft tube during unit operation.

1. Introduction

The draft tube of a hydropower station refers to a pipeline structure that connects the outlet of a turbine runner with the tailwater to make full use of part of the hydropower [1]. The form and size of the draft tube, which has a significant impact on the efficiency of the turbine, are proposed by the turbine manufacturer, [2]. During the maintenance of the unit, maintainers need to enter the draft tube for testing. Therefore, a manhole door needs to be installed on the draft tube to facilitate the access of the maintainers, which involves the opening structure design of manhole on draft tube and the manhole reinforcement, so as to ensure that the manhole strength meets the requirements [3].

2. Detection and analysis

During a certain unit maintenance, the manhole at tailwater was inspected, and it was found that there was cavitation and weld seam cracking on the manhole reinforcement liner, and cracks in the four corners of the manhole at tailwater.

The manhole at tailwater of this unit was a square hole, as shown in Fig. 1. During the inspection, a severe cavitation area with a length of 150mm was found on the surface of the tailwater manhole liner; After the original cavitation site was repaired by welding, the cavitation of different degrees began to appear around the repaired area, as shown in Fig. 2.The fillet welds connecting the liner and the draft tube were detected, one of which had serious cavitation and developed into a crack with a crack length of about 50mm, and the reinforcing plate and the draft tube cracked.

Fig. 1 External appearance of the tailwater manhole

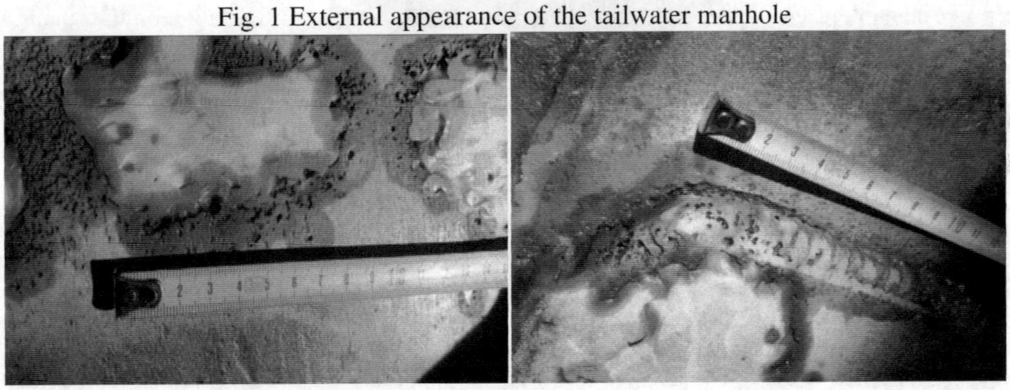

Fig. 2 Cavitation and cracking of tailwater manholes

After removing the manhole reinforcement liner, the tailwater manhole was detected, focusing on the four corners of the tailwater manhole. Cracks were detected in three of the four right angles, one of which was a crack in the lower left corner of the tailwater manhole, with a length of about 30mm; There was a crack in the lower right corner of the manhole with a length of about 130mm (Fig. 3). The tailwater manholes of the same type of unit were inspected and similar defects were found. The four corners of the tailwater manholes had different degrees of cracks which caused different degrees of water leakage (Fig. 4).

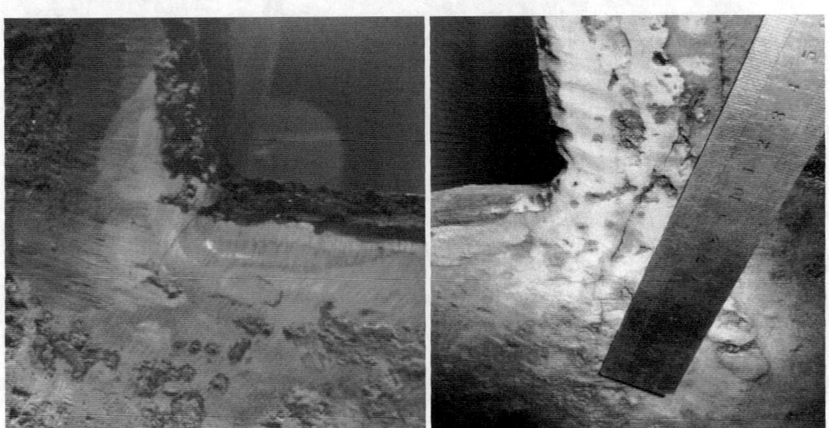

Fig. 3 Cracks on the tailwater manhole

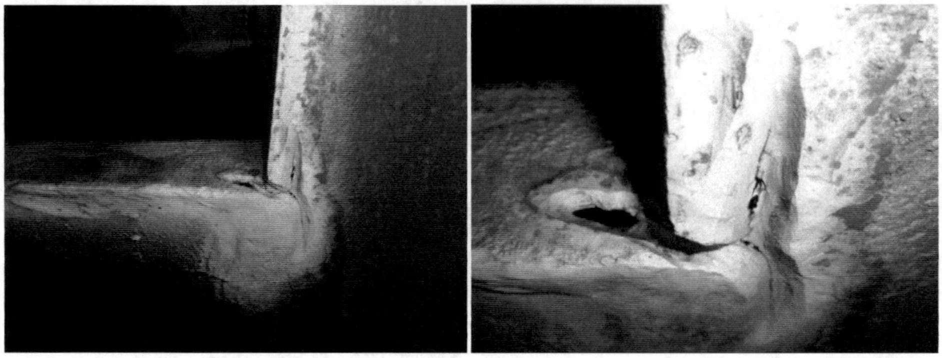

Fig. 4 Cracks on the four corners of other tailwater manholes

Troubleshooting revealed that the cracks were mainly concentrated in the four corners of the manhole. The tailwater manhole was a square hole, and there was obvious stress concentration on the square hole at right angles. According to the clause 4.2.1.17 of GB/T 15468-2006 "Basic Technical Conditions of Hydraulic Turbines", when the square manhole door is used for the draft tube, the four corners should be rounded. The four corners of the tailwater manhole were right-angled, not rounded, so they did not meet the standard requirements. Under long-term cavitation and welding repair cycles, the four corners of the tailwater manholes were rough and uneven, easy to cause local stress concentration.

When the operation conditions of the unit were poor, such as when the unit was being operated in the vibration zone, the draft tube of the unit would vibrate obviously. The draft tube, including the tailwater manhole, was under the alternating load. The unit was put into operation in 1987 and has been in operation for more than 30 years. The alternating stress had been acting on the tailwater manhole for many years. Because the four angles of the manhole were right angles, the stress concentration was obvious compared with other parts, and long-term alternating load effects led to fatigue cracks there.

The stress situation of the tailwater manhole was simulated and analyzed. The element solid185 was used to divide the finite element mesh, with a set internal pressure load Pe=0.2392MPa. The constraint condition was that the top surface around the tailgate was fixed. The hydraulic load surface is as follows:

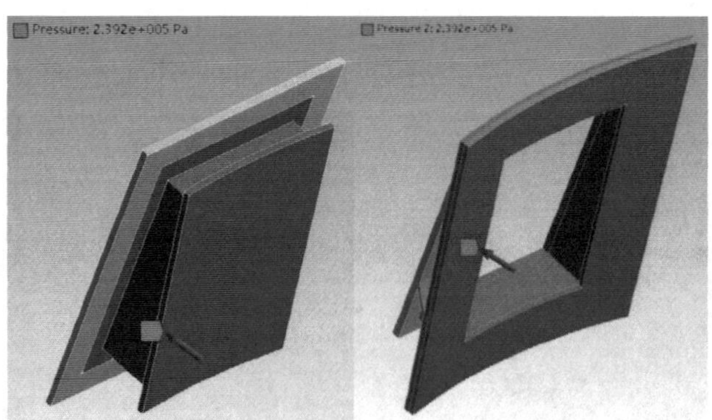

Fig. 5 Hydraulic load applied surface of the tailgate

By calculation, the maximum deformation of the structure was 0.4mm, appearing in the middle of the inner panel of the door cover, where the middle of the inner panel was empty, so the deformation was relatively large under the effect of water pressure. The maximum stress was at the contact between the support plate of the manhole seat and the reinforcement liner of the draft cone tube. The results are shown in Fig. 6 and Table 1.

Table 1 Calculation results of stress on the tailgate

Thickness	Equivalent stress	Allowable stress（GB/T 15468-2006）	Pm（ASME）	Pl+Pb+Q（ASME）

Manhole cover	33.5	93.75	150	450
Manhole base	80.5	93.75	150	450
Manhole base (Rounded)	14.1	93.75	150	450

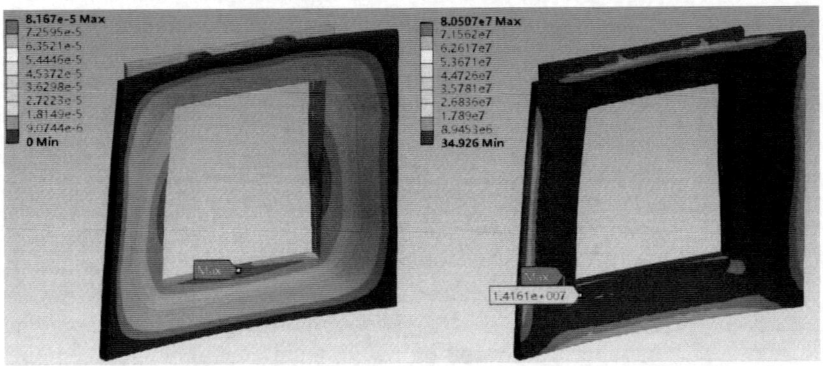

(a) Deformation of the manhole base（b）Stress on the manhole base

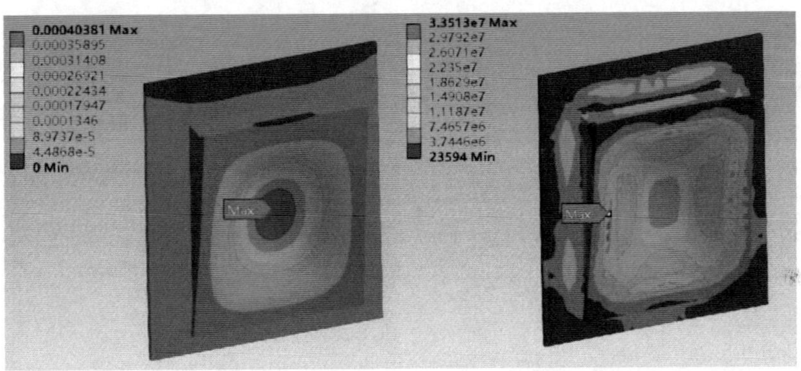

（c）Deformation of the manhole cover（d）Stress on the manhole cover

Fig. 6 Finite element calculation results of manhole cover and base of the draft tube

The calculation results showed that the stresses of the entrance door and the entrance seat of the tailwater spinal canal met the requirements of GB / T15468-2006 and ASME. At the right angle of the entrance seat, the stress is small and the safety is high. The rounded corners will further reduce the stress, but the impact on the safety of the entrance door is small [4,5].

3. Conclusions

In conclusion, the tailwater manholes met the standard requirements in terms of structural design, but because the tailwater manholes had been in service for more than 30 years, the draft tube was severely cavitated, the tube wall was locally thinned, and the strength was insufficient. After repeated repair welding due to cavitation, the four corners were not rounded and the surface was rough, which made the local stress concentration obvious and induced crack initiation and propagation.

References

[1] Access Manhole Reinforcement Design for A Penstock, WANG Hong-qiang, YUNNAN WATER POWER,32(3) 70-74

[2] Boundary effect analysis and stress calculation of manhole for steel penstocks in Qiapuqihai hydropower station, TIE Han, xibei water power, 2006(1),30-32

[3] Stress analysis of penstock manhole of Luosheng Hydropower Station, Yang Qiang, Xu Jiang, Feng Jun, Chen Zhangmiao, hubei water power, 2007(5), 8-12
[4] Structural Design of Deeply Embedded Penstock of A Hydropower Project, Zhang Liping, Zhu Yingru, xibei water power 2017(5),36-38.
[5] Design of Penstock for Huangdeng Hydropower Station, Ang Haihong, Guan Zhongrui, Deng Yongjun, water power,2019 (6), 26-30

Analysis of water jetting happened in volute manhole of the hydropower plants

Wenbo Li[1], Lanlan Liu[2], Xiang Feng[1], Jian Xiao[1], Fei Qi[1], Hengyi Zhou[1], Jun Zhang[1]

[1]Hunan Electric Power Research &Test Institute, Changsha, Hunan, 41007, China

[2]State Grid Hunan Transmission Maintenance, Changsha, Hunan, 41007, China

[3]State Grid Hunan hydropower company, Changsha, Hunan, 41007, China

[*]Corresponding author's e-mail:liwb@alum.imr.ac.cn

Abstract. The water injection in the volute manhole of a hydropower plant was investigated and analyzed. The analysis results showed that the main reasons for the cracking and water jetting happened in the volute manhole were that, firstly, the opening of the volute manhole was not reasonable, the reinforcement structure was damaged, and the reinforcement strength was insufficient. Secondly, the welding quality was poor, which caused the weld defect to become a cracking source and expand into a crack.

1. Introduction

The overflowing metal components of hydropower plants guide the water flow and transform the hydraulic force into mechanical force. Once the overflowing metal components are damaged, serious accidents will occur such as unit damage and water flooding of the plant. Among them, large-scale overflowing pipelines mainly include pressure steel pipes and volutes. In order to facilitate maintainers to enter the interior of the overflowing components for maintenance, pressure steel pipes and draft tubes are usually provided with manhole doors [1]. Opening holes in the overflowing pipelines involves the structural design of manholes and the reinforcement of manholes, so as to ensure that the strength of manholes meets the requirements and prevent the occurrence of insufficient strength which will lead to the accidents such as the cracking of the overflowing pipelines and water flooding of the plant [2-4].

2. Detection and analysis

During the operation of a unit, a large amount of water was jetting near the manhole door of the pressure steel pipe (Fig. 1). The operator discovered the situation in time through the industrial television system, and immediately took emergency measures such as stopping the machine, pulling the water inlet gate, pulling the tail water gate, the utilization of water diversion pipe and volute pressure elimination, to control the water jetting

IOP Conf. Series: Materials Science and Engineering **793** (2020) 012013 doi:10.1088/1757-899X/793/1/012013

Fig. 1 The picture of water jetting on the pressure steel pipe

After the pressure of the volute was eliminated, the volute manhole door was opened for inspection, and the specific location of the crack was determined (Fig. 2 and Fig. 3). The crack started at the right corner of the lower left corner of the square ear hole of the volute manhole, extended about 300mm upstream from the lower left corner of the square ear hole of the volute manhole door shaft pin in the reverse direction of water flow (Zone Ⅰ), and extended radially about 105 mm along the outer wall of the square ear hole in the door shaft (Zone II).

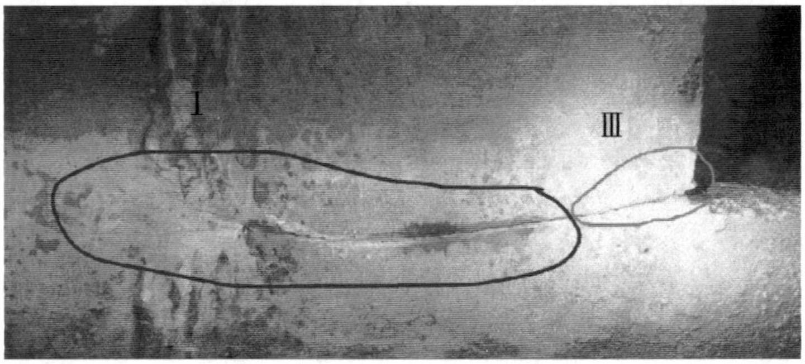

Fig. 2 The extension picture of the crack in inner wall of the pressure steel pipe

Fig. 3 The picture of the crack in outer wall of the pressure steel pipe

The defects of the door shaft baffle were cleaned up, and many welding defects were found (Fig. 4). Visual observation showed that the defects were linear, and some of the defects had corrosion products. According to the design drawings, it could be preliminarily determined that the above defects were in the gap between each welding plate. The defects were determined as follows:

IOP Conf. Series: Materials Science and Engineering **793** (2020) 012013 doi:10.1088/1757-899X/793/1/012013

1. The gap between the pressure steel pipe and the stiffener. According to the design drawing, the ring width of the reinforcement ring was 270mm. Between the reinforcement ring and the pressure steel pipe, only the edge of the stiffener was welded and fixed. The reinforcement ring and the pressure steel pipe were not completely welded (Fig. 5), so during the polishing process, the gap between the pressure steel pipe and the reinforcement ring was exposed.

2. The gap between the stiffener and the side plate of the door shaft. According to the design drawing, the side plate of the door shaft was welded to the stiffener, without groove in the welding process (Fig. 5). The attachment weld between the side plate of the door shaft and the stiffener was not fully welded, so the gap between the side plate of the door shaft and the stiffener was exposed during the polishing process.

3. The gap between the side plate of the door shaft and the support plate on the door shaft. According to the design drawing (Fig. 6), there was no groove in the welding of the two, and the attachment weld was not complete. The enlarged view of each defect is shown in Fig. 4. According to DL/T 869-2004 "Technical Specification for Welding in Thermal Power Plants", if the plate thickness is more than 20mm, the one-side bevel cutting process should be implemented at least. Therefore, the welding of each steel plate of the manhole door did not meet the standard requirements.

The position of the crack was inspected in detail, and it was found that the crack on the inner wall of the steel pipe started at the right angle of the square ear hole and was located above the support plate under the door shaft, extending upstream of the steel pipe; The crack position on the outer wall of the steel pipe was located on the support plate under the door shaft after inspection, extending upstream of the steel pipe. Since the crack was a through crack, it could be inferred that the crack started at the right angle to the square earhole.

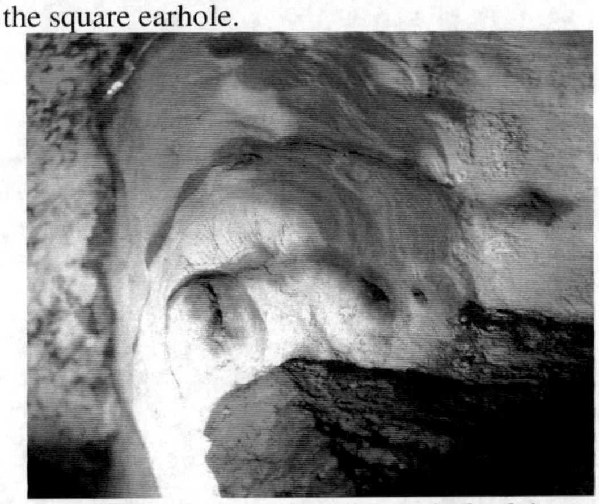

Fig. 4 The earhole defect on the door shaft of the manhole

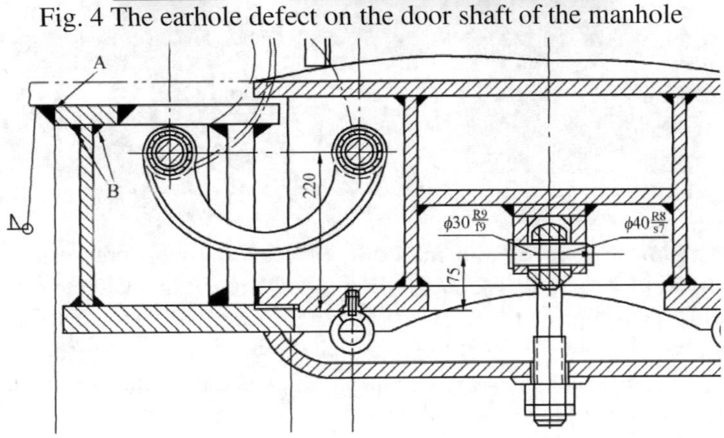

Fig. 5 Top view of the volute manhole

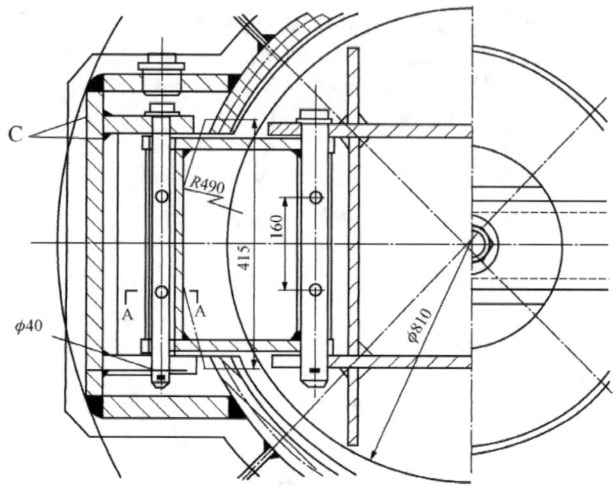

Fig. 6 Front view of the volute manhole

It could be judged from the extending direction of the crack on the inner wall of the pressure steel pipe that the crack started at the right angle of the square ear hole, which was the stress concentration position. The strength of the pressure steel pipe was decreased after the manhole door was opened, and it was necessary to adopt a stiffener to reinforce the structure of the manhole door. For the stiffener, it was not suitable to make holes in it, otherwise the reinforcement effect would be weakened. The square ear holes of the steel pipe were made on the stiffener and the pressure steel pipe at the same time, which reduced the local strength, resulting in the actual bearing stress on the steel plate around the square ear hole higher than that on the rest of the manhole door [5].

In addition, the quality of the square ear hole was poor, and the form of the square hole led to the obvious concentration of stress on the right angle of the square hole. From the perspective of local strength and local stress distribution, the square hole was an extremely unreasonable form. The manholes doors of other pressure steel pipes of the same type unit were trouble shot, and similar problems were found (Fig. 7 and Fig. 8).

Fig. 7 Cleaning area of welding

Fig. 8 Welding defects on the square earhole

There were two main reasons for the initiation of cracks in the pressure steel pipes: (1) The structure of the manhole door was unreasonable. Making square ear holes in the reinforcement ring not only reduced the reinforcement effect of the reinforcement ring, but also caused stress concentration. (2) The welding quality was poor. The weld between the door side plate of the shaft and the stiffener, and the weld between the side plate of the door shaft and the support plate on the door shaft, were not complete. The local stress was too high, causing the initiation and extension of cracks at the unwelded defect sites.

3. Conclusions

Combined with the above cases, the causes of the crack in the volute manhole were:

1. The volute manhole was made unreasonably, the reinforcement structure was damaged, and the reinforcement strength was insufficient.

2. The quality of the weld was poor, which caused the weld defect as a cracking source to expand into a crack.

The defect in the volute manhole of a hydropower plant could cause a serious accident such as unit damage and water flooding of the plant. Combined with the unit maintenance, the supervisors should conduct non-destructive testing on manholes and the welds around manholes, so as to find defects early. For the manholes with unreasonable structures and damaged stiffeners, repair and replacement work should be carried out in combination with maintenance as soon as possible, to ensure the safe operation of the unit.

References

[1] Qiling Zhang, Fu Dan, Hu Lei, Wu Hegao, Study on low-cycle fatigue of steel spiral cases embedded under pressurized conditions, J huazhong Univ. of Sci.& Tech. (Natural Science Edition) 2017 (45) 127-132.

[2] Three-dimensional simulation of unsteady flows in a pump-turbine during start-up transient up to speed no-load condition in generating mode, Zhongjie Li, Huili Bi, Zhengwei Wang, Proceedings of the Institution of Mechanical Engineers, 2016 (8).

[3] Manufacturing Quality Control of Gate Operating Mechanism of Baihetan 1 000 MW Hydro Unit, ZHANG Dongsheng, ZHANG Huaqing, Water Power, 2019 (45) 76-80.

[4] Analysis and treatment of an emergency shutdown of a hydroelectric unit, Haojie Chen, Huadian Technology, 2016(38) 53-55.

[5] Analysis and treatment of abnormal reversal of 75MW hydropower unit, Houyu Zhang, Wenfeng Lin, Industrial Technology, 2016 (31), 125.

IOP Conf. Series: Materials Science and Engineering **793** (2020) 012014 doi:10.1088/1757-899X/793/1/012014

Research on Ecological Driving Strategy of Urban Road Vehicles under Vehicle-road Collaborative Environment

SuDan, CongWanjuan, JiangKun

HeiheUniversity, Heihe, HeilongjiangProvince, 164300

Abstract: With the increase of motor vehicles, the energy resources available for vehicles have become scarce. In view of the harmfulness of traffic emissions, energy conservation and emission reduction have become strategic goals for the development of the transportation industry. Guiding students to realize ecological driving and reducing vehicle fuel energy consumption and emissions have increasingly attracted the attention of researchers. We are supposed to do a research on signalized intersections at important nodes of urban transportation, use ecological driving to simulate the operation of bicycles and vehicle queues, establish an ecological driving model, and control energy conservation and emission reduction goals from multiple perspectives. This paper combines the ecological driving technology and micro fuel consumption calculation model under the vehicle-road collaborative environment to study the fleets within the signalized intersection. At the same time this paper proposes street signs, models, and case applications that can effectively reduce the average fuel consumption and parking times of the vehicles passing the signalized intersection to ensure the smoothness of operation at signalized intersections has been reduced, and rapid acceleration and deceleration operations have been performed. After verification, the establishment of the fleet's eco-driving model has great effectiveness.

1. Introduction

With the economic development and the increasing number of motor vehicles in recent years, the annual energy consumption of the transportation industry has become a rapid growth trend. At the same time, our dependence on traditional fossil raw materials, such as petroleum, is increasing. From the perspective of improving the current situation of vehicle energy use, it is one of the strategic development goals of China's current transportation industry to control emissions and achieve energy conservation and emission reduction.

2. Ecological Driving Theory in the Context of Vehicle-Road Collaboration

In the current realistic environment, the fuel consumption of vehicles is changed by improving vehicle fuel, improving vehicle performance, and affecting driving behavior. The use of road-coordination technology to guide the driver to change driving behavior to prevent idling and rapid acceleration and deceleration. At the same time, in order to reduce vehicle fuel consumption and emissions, the vehicle-road collaboration technology is used to simulate the ecological driving of the vehicle to develop an ecological driving model, which has achieved good results in energy conservation and emission reduction.

The theory of vehicle-road coordination at the signalized intersection and the vehicle-road coordination theory at the signalized intersection are cutting-edge theories in the field of intelligent transportation. A large number of technical studies on vehicle-road coordination have been conducted abroad. The United States has proposed that vehicle road coordination can improve vehicle driving safety and reduce the impact of ground transportation on the environment. In Japan, it integrates various ITS functions, toll systems, and intelligent highway systems to establish an open public vehicle-road collaboration foundation platform [1]. China 's vehicle-road collaboration technology started relatively late. After continuous research and development, the vehicle-road collaboration technology was used to establish intersection speed guidance and path guidance simulation scenarios, which can effectively reduce the traffic on the road network. The average time of network traffic flow formation also has a good control effect. At present, vehicle-road coordination research is an important research direction of signalized intersections. Vehicle-road coordination technology is used to improve the safety factor and operating efficiency of vehicles passing through intersections. Combine vehicle speed induction to draw vehicle trajectory and signal timing, adjust the signal green light market as appropriate to reduce vehicle delays.

Research on eco-driving within signalized intersections is currently focused on improving vehicle energy use efficiency, forming environmentally-friendly speed driving, driving behaviors, etc., and removing additional structures to improve the accuracy of the preparation of traffic conditions in advance. This improves driving safety. Eco-driving technology can reduce the fuel consumption and emissions of vehicles at intersections. After research, the goals of acceleration, deceleration, magnitude, duration, and number of stops have been simplified. At the same time, the use of micro fuel consumption models, emission models, etc. can form a complete logical proof to minimize the amount of fuel consumed by vehicle traffic. Through the intersection of fuel consumption and carbon dioxide emissions, the optimization of the model significantly increased the formation time to varying degrees, effectively reducing emissions. For example, quantitative and qualitative analysis, vehicle driving behavior characteristics, based on the VSP emission model design is an ecological driving strategy, and simulation results are maintained for a long time to show that the strategy significantly reduces vehicle emissions. Based on VSP fuel consumption, calculation model, mathematical proof, ecological acceleration and deceleration, and constant-speed operating curve functions, the vehicle trajectory optimization method is established, and the vehicle ecological driving strategy is provided. The VISSIM simulation platform is used to verify the average fuel saving in the case of emissions and unlicensed pairs. rate. By adjusting the objective function, it can effectively guide and control the energy saving and emission reduction of the eco-driving technology at intersections, and provide various theoretical research results for the composition of the eco-driving model in the vehicle-road collaborative environment [2].

3. Induced Running Trajectory of Signalized Intersections under Vehicle-road Coordination

(1)The conditions of the vehicle pair study at the signalized intersection in the vehicle-road collaborative environment are explained, including the delay of information interaction between the vehicle and the infrastructure. In the vehicle-road collaborative environment, the vehicle speed, vehicle position, and signal between the vehicle and the infrastructure Time-sharing information sharing, vehicle trajectory and the situation of the intersection, the change of vehicles within the induction range and so on. For the vehicle-road collaborative environment, the identification of the fleet at signalized intersections must first identify the position, including the headway between vehicles. However, for speed identification and whether the vehicle speed is approximately the same, then we can deduct the head-to-head distance between the front and rear vehicles entering the induction range to identify the fleet, or use the speed and head distance to determine the fleet. This paper uses the two indicators of headway and instantaneous speed to identify the speed of vehicles in the vehicle-road cooperative fleet to identify vehicles at signalized intersections in the vehicle-road cooperative environment [3].

The first is the headway distance. At a certain time, the distance between the two head ends is the headway distance. The distance between the front car and the rear car is determined by the instant of the start line. The calculation formula is:

$$F_T = \sum_i \overline{F_i} * D_i * T$$

F is the total fuel consumption, D is the distribution frequency, and is the total length of time in the calculation range. Then calculate the headway of the vehicle through the calculation formula, use the headway to make the preliminary identification of the fleet, and determine the standard headway. After a large number of vehicles appear at the intersection, the calculation and statistics of the simulation operation data are performed to comprehensively consider the headway distribution of the vehicles near the induction start line. The instantaneous speed is the speed of the vehicle at a certain moment. The moment the directly-selected vehicle induces its realization, the moment of the instantaneous speed of the vehicle and subsequent vehicles is determined to complete the identification of the fleet, and the headway is satisfied. As the base speed value, the relative error between the base speed value and the instantaneous speed is calculated.

On this basis, the subsequent vehicles are judged in the same way, the relative error range is determined, a large number of vehicle operating data are established, and the statistical analysis reaches the instantaneous speed difference between adjacent vehicles near the starting line.

(2)The non-stop running time of the signalized intersection team in the vehicle collaborative environment. We set a scenario where the target team enters the induction range until the green light moment. Based on the state of the signal lights and the running conditions of the downstream vehicles, under various scenarios, such as the remaining time before the target team enters the induction range, the signal lights turn red, and the target team enters the induction range. Within the range, the time required for downstream vehicles to leave the intersection, and the target fleet entering the induction range, time, etc. at the moment of green light.

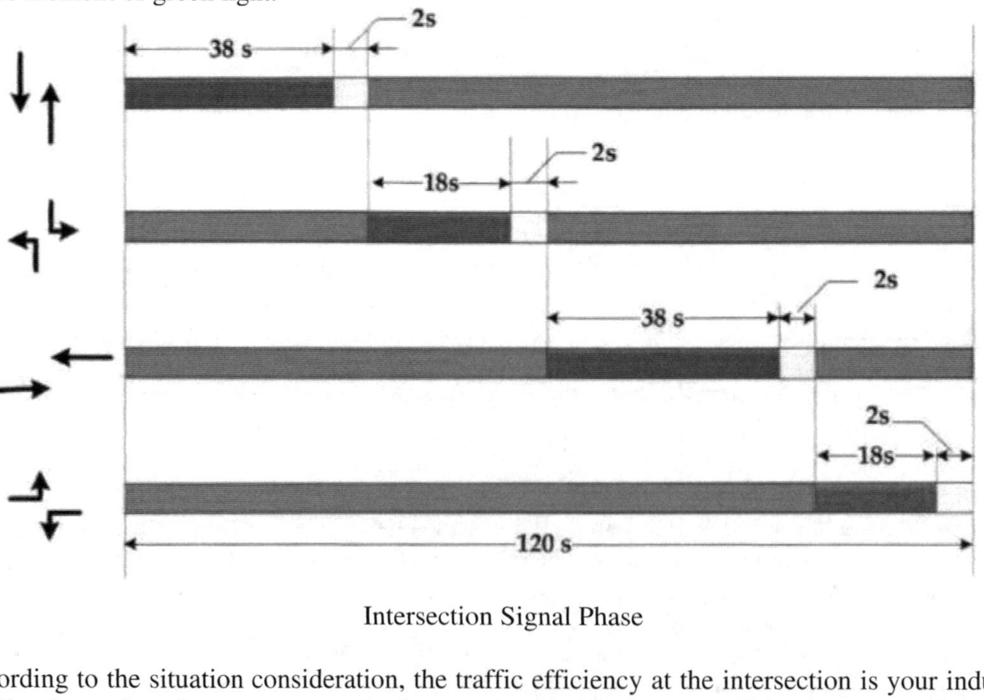

Intersection Signal Phase

According to the situation consideration, the traffic efficiency at the intersection is your induction target. The time when the team passes the intersection during the green light time, according to the setting of the induction range, the target team may pass the intersection at the green light time or when it reaches the parking line. In order to prevent conflicts with downstream vehicles, it may be necessary for the small fish to not stop at the green light time when they reach the parking line, the minimum time to pass the intersection, or after the target fleet enters the induction range, the downstream vehicles do not pass the front intersection , Has a certain impact on the operation of the target vehicle [4]. After the next green light starts, downstream vehicles will pass the intersection, and when they reach the parking line, the target fleet must avoid conflicts with downstream vehicles. At this time, the running time of the target vehicle may be longer than the stopping time, and the minimum running time of passing the intersection. In various scenarios, the target fleet enters the induction range. Based on the signal status and vehicle operating status, for the downstream vehicle operating status and different signal status, the running time range of the signalized intersection in the vehicle-road collaborative environment must be calculated. Discussion on the range of induced trajectory of vehicles at signalized intersections under the environment of vehicle-road coordination, to determine the feasibility of uniform induction of the target fleet. For the fleet of signalized intersections in the vehicle-road coordinated environment, maintain the current speed, pass through the signalized intersections at a constant speed, and take into account the ecological driving trajectory. Develop a uniform and minimal acceleration operation, so that the fuel consumption of the vehicle can be achieved by inducing the leading vehicle Speed-based. At the same time, the running speed of the vehicles in the target fleet is set. For the target fleet, the existing speed and uniform speed induction measures are adopted to introduce the retreat into the induction range, including the speed correction phase and the uniform speed operation phase. Discuss the range of the leading fleet of the target fleet, induce the running trajectory, determine the follow-up vehicles based on the leading fleet, complete the establishment of the fleet's leading operating trajectory, perform a scenario analysis on the leading fleet of the target fleet, and perform the reverse trajectory of the leading fleet in different scenarios calculation. For example, the leading car passes by at the lowest possible speed, the value of the signal light changes, and so on.

4. Construction of Eco-Driving Model for Vehicles at Signalized Intersections in Vehicle-Road Collaborative Environment

(1)The VSP micro fuel consumption measurement and calculation model is relatively simple and easy to use. The fuel consumption calculation model based on power can reasonably explain the fuel consumption. The types of models include the CM1M model and the VT-CPFM model, which are independent of the power in the car, comprehensively consider the driving function of the vehicle, and overcome air resistance. In the vehicle-road collaborative environment, the key part of the ecological driving trajectory selection is calculated to obtain the ecological driving trajectory that meets the expected value.

(2)The objective function of the vehicle ecological driving model is constructed. A large number of options are used to convert the energy saving and emission reduction goals. The vehicle-road cooperative induction strategy is developed. The fuel consumption objective function and the traffic efficiency function are used to determine the traffic efficiency at the intersection and the running time of the parking line as a response basis for traffic efficiency, a corresponding data model was designed [5].

(3)Construction of an eco-driving model for a signalized vehicle fleet at a signalized intersection. The specific implementation steps are to first identify the vehicle, generate a fleet number, then calculate the fleet induction running time range, determine whether the fleet accepts production driving guidance, determine the scenario of the induced vehicle, and follow the vehicle to calculate the leading vehicle. Invert the trajectory range, use genetic algorithm to calculate the objective function, and induce the fleet according to the selected ecological trajectory.

(4)The application of the eco-driving model at the signalized intersection vehicle uses the VISSIM simulation platform to collect vehicle operation data in the vehicle-road collaborative environment, set the fleet eco-driving model, and perform parameter calculations. The fuel consumption situation is based on the signalized intersection simulation platform to build a traffic simulation platform, which enables the simulation platform to describe the individual vehicle behavior in real time, accurately describe the geometric conditions of the signalized intersection, simulate the actual operation of the vehicle, and provide a visual interface for a output the driving condition data of the vehicle in real time.

5. Conclusion

So far, microscopic traffic simulation models have been applied in various fields of transportation research, and are commonly used simulation platforms. For the urban traffic road network and highway network simulation, you can simultaneously use the fields of traffic analysis and modeling to realize the simulation of vehicle operating conditions under various traffic conditions. Through the construction and calibration of the signalized intersection simulation platform, the signal distribution situation and vehicle traffic flow are determined, and finally the calibration results are followed. The universality of the simulation results is improved, and the characteristics of the traffic flow and driving behavior of module parameters in driving behavior are determined. The simulation and actual measurement results are compared before and after, and the total fuel consumption error is found to be within 10%, indicating that the parameter calibration of the simulation model basically meets the accuracy requirements.

Acknowledgment

Fund project: Basic scientific research-related expenses of provincial institutions of higher leaming of Heilongjiang Province Education Department — Scientific Research on Ecological Driving Strategy of Signalized Intersection in the Environment of Vehicle-Road Collaboration (Project No.: 2018-KYYWF-1249)

References:

[1] Chen Ze, Cai Mingmao, Gao Jie, et al. Zebra crossing safety warning system in the vehicle-road collaborative environment [J]. Journal of Qingdao Technological University, 2019, 40 (5): 106-112.

[2] Ma Lei, Chen Ke, Wang Minglu, et al. Research on driver behavior recognition method in vehicle-road collaborative environment [J] .Automotive Engineering, 2018,40 (11): 1330-1338.

[3] Yao Zhihong, Jiang Yangsheng, Wang Yi. Adaptive Signal Timing Optimization Model in Vehicle-Road Collaborative Environment [J]. Industrial Engineering, 2018, 21 (4): 8-14, 33.

[4] Zhang He, Guo Wenqian, Zhang Jiansong, et al. Research on DVRP of multi-distribution center in the vehicle-road collaborative environment [J]. Journal of Wuhan University of Technology (Transportation Science and Engineering Edition), 2017, 41 (6): 895-899.

[5] Jia Fengyuan, Li Zhongbing, Chen Qingdong, et al. Speed guidance strategies at signalized intersections in a vehicle-road collaborative environment [J]. Journal of Hebei University of Science and Technology, 2017, 38 (5): 432-437.

IOP Conf. Series: Materials Science and Engineering **793** (2020) 012015 doi:10.1088/1757-899X/793/1/012015

Thermal Analysis for Lithium-Ion Battery Pack based on Parameter Estimation based on Genetic Algorithm

Yong Wang[1], Yelin Deng[2*], Weiwei Liu[1], Kunkun Hao[1] and Hongchao Zhang[3]

[1] School of Mechanical Engineering, Dalian University of Technology, Dalian, Liaoning, 116024, China

[2] School of Rail Transit, Soochow University, Suzhou, Jiangshu,215006, China

[3] Industrial Engineering, Texas Tech University, Lubbock, TX 79409-3061, USA

[*]Corresponding author's e-mail: yelin.deng@hotmail.com

Abstract. Thermal analysis of Lithium-ion battery pack is the important portion of battery management for electric vehicles. The heat produced in charging and discharging will bring about impairment of the safety and service life of batteries. It is thus important to monitor battery temperature for prevention of the battery failure. This paper first sets up a simulation model based on the second-order RC circuit model of the heat generation and dissipation of the battery pack using SIMULINK. The temperature of the battery pack is tested under The New European Driving Cycle conditions. And by comparing with the experimental data of the battery temperature, the heat dissipation coefficient in the simulation model will be optimized by the genetic algorithm using MATLAB. The optimization result shows that the difference between the simulated temperature and the actual temperature is within one degree, so the model based on the optimization result can accurately reflect the actual temperature change.

1. Introduction

As the situation of environmental pollution and energy consumption being serious, it is urgent to find sustainable energy alternative development methods [1]. Lithium -ion battery is growing in popularity for the electric vehicles (EVs) due to its advantages of high energy density, little or no memory effect, and low self-discharge [2]. The temperature state of the battery directly affects energy and efficiency issues during use. Thus, for further growth of Li-ion battery business, thermal management should attract enough attention.

There are many types of battery models to estimate the battery performance parameters and its state at different operating conditions. Common battery models include mathematical models, electrochemical models, equivalent circuit models, and coupling models. The equivalent circuit model does not need to study the electrochemical reaction inside the battery, and can be written to specific mathematical equations for analysis and application. Typical equivalent circuit models include General Nonlinear（GNL）model, Thevenin model, and Partnership for a New Generation of Vehicles (PNGV) model etc [3]. Second-order Thevenin circuit model can balance both steady-state and transient characteristics in the charge and discharge process. High order Thevenin circuit model is too complicated, which will cause too much calculation [4]. This paper chooses the second-order Thevenin circuit model.

The heat dissipation coefficient in the simulation model needs to be optimized to get accurate simulation results [5]. The general iterative method is liable to fall into the local minimum trap and the phenomenon of "endless loop" appears, which will make iteration impossible. Genetic algorithm is a global optimization algorithm which can overcome this shortcoming. There are no restrictions on the derivation and the continuity of the function [6]. So the genetic algorithm is suitable for optimization analysis of equivalent circuit model.

2. The design of simulation model

2.1. second-order Thevenin battery model

The Thevenin model is simpler than other models and can well reflect the dynamic characteristics of the battery. The second-order Thevenin model adds an additional RC loop to the first-order Thevenin model, which can accurately simulate the charge and discharge behavior of the battery.

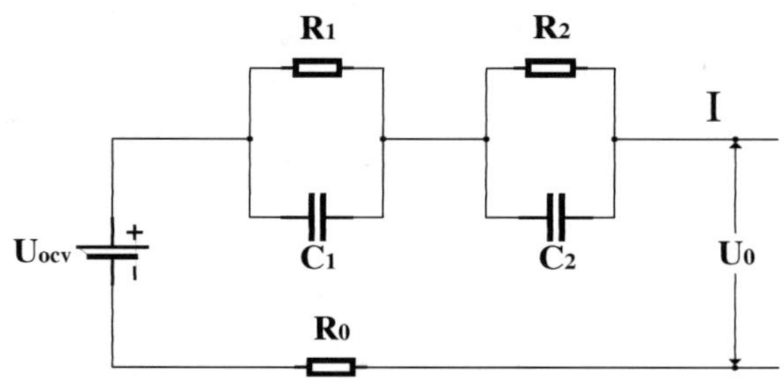

Figure 1. Second-order Thevenin Model.

The mathematical expression of the model is as follows:

$$U_{ocv} - U_1 - U_2 - IR_0 = U_0 \tag{1}$$

$$I = \frac{U_i}{R_i} + C_i \frac{dU_i}{dt} \tag{2}$$

$$\int I_{C_i} dt = U_i C_i = \frac{U_i \tau_i}{R_i} \tag{3}$$

$$I_{R_i} = \frac{U_i}{R_i} \tag{4}$$

$$I_{C_i} = I - I_{R_i} \tag{5}$$

$$\tau_i = R_i C_i \tag{6}$$

In these formulas, U_0 is the voltage across the battery, U_{ocv} is the battery internal voltage, R_0 is the ohmic resistance, R_i (*i*=1, 2) are the polarization resistances, *I* is the discharge current, C_i (*i*=1, 2) is the capacitance, τ_i (*i*=1, 2) is the time constant of the RC loop.

2.2. Analysis of heat generation and heat dissipation

Calculation of heat generation for battery pack is according to the equivalent circuit model. In the second-order thevenin battery model, the heat generation from internal resistance is looked as the main heat generation. The heat generation rate *Q* is calculated by the product of each internal resistance and the square of each electric current:

$$Q = I^2 R_0 + I_{R_1}{}^2 R_1 + I_{R_2}{}^2 R_2 \tag{7}$$

The heat exchanged between the battery pack and the environment mainly proceed through heat conduction and heat convection. Heat conduction occurs mainly through solids, while heat convection and heat conduction occur simultaneously in the fluid. In this study, the battery pack is located in the car. The heat dissipation coefficient between the battery pack and the environment can be represented by a parameter.

The heat balance formula is:

$$Q\Delta t - hS(T_{battery} - T_{env}) = cm(T_{battery} - T_0) \tag{8}$$

In this formula, h is the heat dissipation coefficient between battery pack and environment, S is the heat dissipation area of the battery pack, c is specific heat capacity *of the battery pack*, and m is the quality of the battery pack. $T_{battery}$ is the simulation temperature in model, T_{env} is the Ambient temperature, T_0 is the initial temperature of the battery pack.

And the values obtained in previous experiments are showed as follow: S is 1.5 m^2 for the pack, c is 1.026 J/g/K for one battery module, m is 2000g for one battery module. And there are 96 battery modules in the battery pack. The quality of the battery pack is 1.5 times as much as the 96 battery modules. And the ambient temperature during the experiment T_{env} is 25°C.

2.3. Simulation model and settings

The simulation model is built in SIMULINK by the analysis of heat generation and heat dissipation. The model inputs include current, SOC and battery pack temperature gotten from 1-D Look-up Tables, which is obtained through charge-discharge experiment by the The New European Driving Cycle (NEDC).

The experiment was performed by the NEDC cycle test. The initial SOC is 96% and the ending SOC is 7%. The initial battery pack temperature was 25 degrees Celsius, and the end temperature was 30 degrees Celsius.

The ohmic internal resistance(R_0), polarization internal resistance(R_i (i=1, 2)), and time constant(τ_i (i=1, 2))are effected by the SOC and temperature. And R_0, R_i (i=1, 2) will also be affected by charge and discharge. So these parameters should be gotten from 2-D Look-up Tables.

The simulation model diagram is shown in Figure 2:

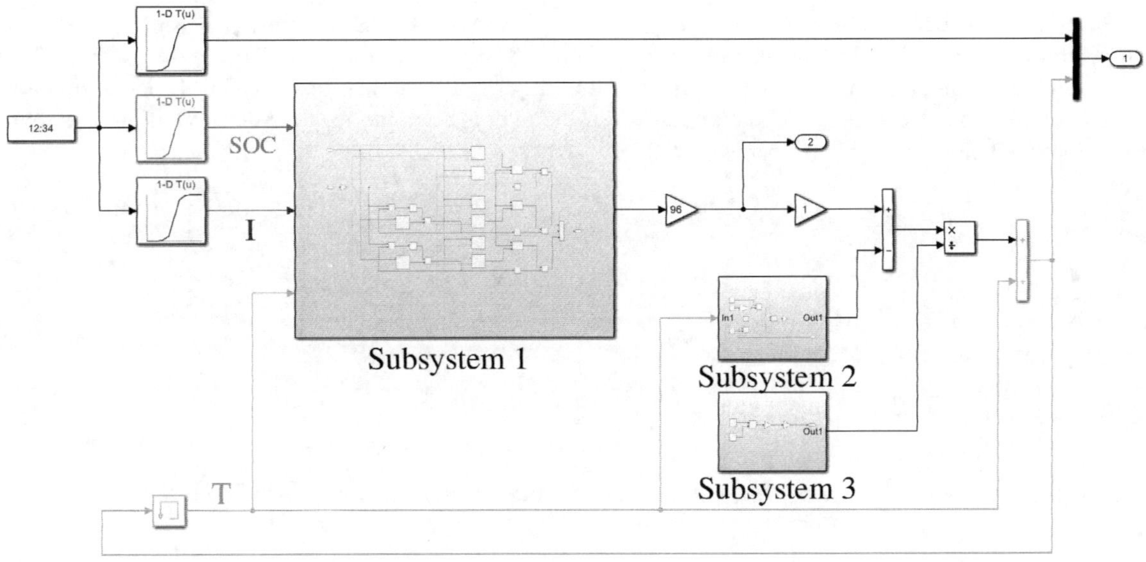

Figure 2. Model diagram of the heat generation and dissipation in SIMULINK.

In the subsystem 2 showed on the figure 3, the heat dissipation coefficient h is given a gain module G to be a variable, which will be optimized through genetic algorithm.

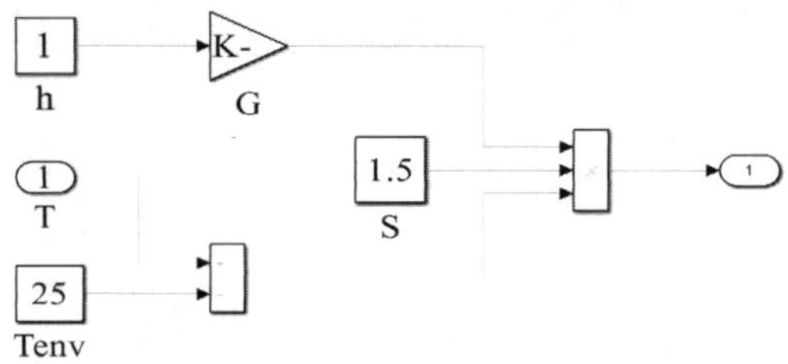

Figure 3. Subsystem 2 in the model.

3. Parameter Optimization by Genetic Algorithm

In the simulation model, there is an uncertain parameter which is the heat dissipation coefficient h, which needs to get the optimized value by comparing with the experimental data. Parameter optimization is achieved by the Genetic Algorithm. In this paper, the heat dissipation coefficient h is optimized by using the genetic algorithm toolbox that comes with MATLAB.

3.1. Design of fitness function

The selection of fitness function directly affects the convergence speed of the genetic algorithm and determines whether it can find the optimal solution. Generally speaking, the fitness function is transformed from the objective function. In this paper, the fitness function is designed as the 1/2 power of average value of the variance between the simulated value and the actual value. The formula is as follow:

$$F = \sqrt{\sum_{k=1}^{k=n} [(T_m(k) - T_e(k))^2] / n} \tag{9}$$

In this equation, $T_m(k)$ is the simulated temperature on every simulation moment, and $T_e(k)$ is the actual temperature.

3.2. Optimization of Genetic Algorithm

The connection between the Genetic Algorithm Toolbox and SIMULINK is implemented by programming in MATLAB. The variable in the Genetic Algorithm Toolbox is the gain module G in the simulation model, so the final result of the algorithm optimization is the heat dissipation coefficient h. The algorithm will be stopped when the average change in the fitness value is less than the Function Tolerance, which is set as 10^{-6}.

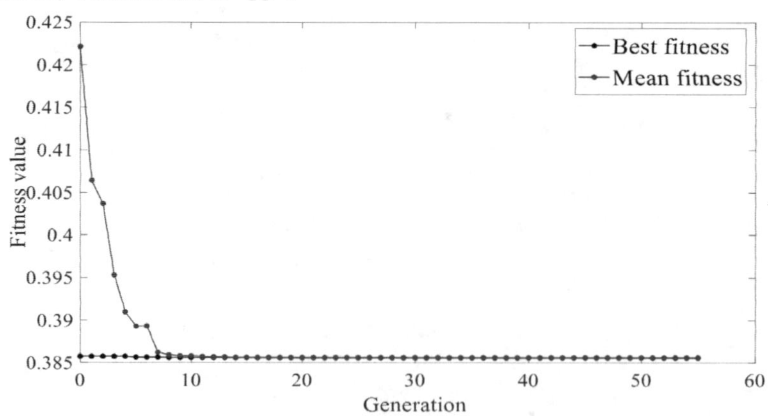

Figure 4. The result of the best fitness by the Genetic Algorithm Toolbox.

And the result of the best fitness from the entire optimization process is show as the figure 4. The fitness value shows that the basic genetic algorithm has evolved into the local optimal solution after about 10 generations. In the figure, the optimal solution is 0.385596 in the last generation, which is equal to the variable F in the formula (9). And it can be looked as the fitting degree between the simulation temperature and the actual temperature.

During the optimization process, individuals of each generation will be displayed in G of the Subsystem 2 on the figure 3. And final point of the best individual is 8.487, which is the final heat dissipation coefficient.

3.3. Result of the simulation

The heat dissipation coefficient h is 8.487 getting from the optimization result in the simulation model. The comparison of simulation result and experimental data is show on the figure 5. In the figure 5, the simulation curve can reasonably simulate the change of actual temperature.

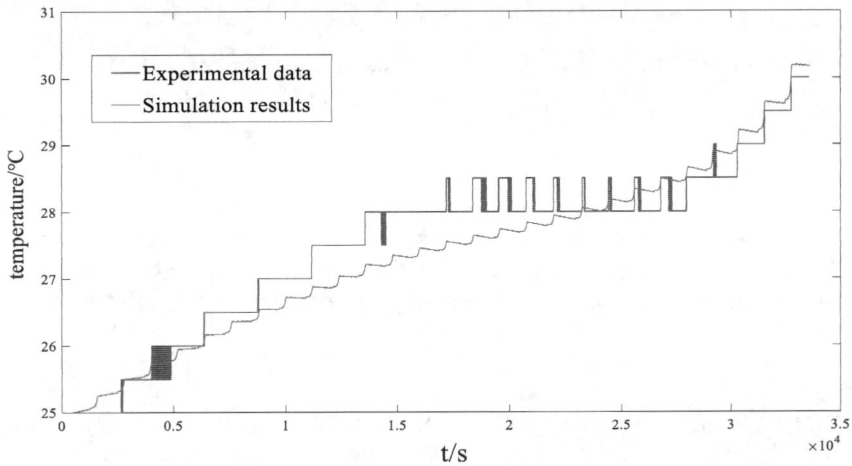

Figure 5. Comparison of simulation result and experimental data.

4. Conclusions

Based on the genetic algorithm, the uncertain heat dissipation coefficient in the thermal model is optimized in this paper. The optimization result in the model shows that the difference between the simulated temperature and the actual temperature is within one degree in the whole process, which makes the simulation model more accurately simulate the temperature change of the actual operating conditions of the battery pack. And this study has certain reference value for the calibration of thermal parameters in battery management systems.

References

[1] Lu L., Han X., Li J. (2013) A review on the key issues for lithium-ion battery management in electric vehicles. J. Power Sources., 226: 272-88.

[2] Abada S., Marlair G., Lecocq A., Petit M. (2016) Safety focused modeling of lithium-ion batteries: A review. J. Power Sources., 306: 178-192.

[3] Yang W., Ma C. (2015) SOC estimation algorithm based on PNGV improved model. J. Automotive engineering., 37(05): 582—586+598.

[4] Zhai H.Z. (2017) Modeling of lithium-ion battery for charging/discharging characteristics based on circuit model. J. Online Engineering., 13(06): 86-95.

[5] Bandhauer T.M., Garimella S. and Fuller T.F. (2011) A critical review of thermal issues in lithium-ion batteries. J. Electro-Chemical Soc., 158: 1-25.

[6] Li T., Zhang B., Liu Y. (2016) Design and implementation of performance testing system for lithium battery. J. Automation instrument., 05: 67—70.

IOP Conf. Series: Materials Science and Engineering **793** (2020) 012016 doi:10.1088/1757-899X/793/1/012016

Case analysis of mechanical maintenance supply scheme based on NSGA-II algorithm

Wei Xia, Quan Shi*, YaDong Wang

[1] Army Infantry Academy, Shijiazhuang, 050083, P.R. China;

[2]Army Engineering University, Shijiazhuang, 050003, P.R. China

Wei Xia: axwangel@126.com

ABSTRACT: Based on the maintenance strategy of mechanical equipment, the mechanical maintenance supply scheme is designed and studied. Combined with a specific case, the intelligent optimization algorithm is used to solve the model, and a series of Pareto optimal scheme sets are obtained. Compared with other classical intelligent algorithms, the effectiveness and robustness of the algorithm are verified.

1. Introduction

Under the condition of information battlefield, equipment users consume a lot of spare parts in a short time. In the whole process of spare parts supply, how to allocate the spare parts scientifically and reasonably, and how to dispatch and transport the comprehensive path information and transport capacity will directly affect whether the equipment users can obtain the guarantee of spare parts supply in time[1].

2. Condition setting and parameter setting

In a mission of Southwest campaign stage, according to the pre-prepared operational plan of the joint battle command department, a campaign Corps deployed its main participating forces in its operational area, with a total of four machine brigade to perform the mission. It is estimated that two key spare parts, A and B, will be consumed to complete the mission, of which the mass of unit quantity spare part A is fifteen kg, and the mass of unit quantity spare part B is twenty kg. In order to complete this task, four base level warehouses will support the equipment using unit with spare parts, and two base level warehouses in the rear will supply the base level warehouse with spare parts. See the table below for the specific spare parts supply information.

Table. 1 spare parts fill rate

	Equipment using unit 1	Equipment using unit 2	Equipment using unit 3	Equipment using unit 4
Spare parts satisfaction rate	0.8	0.8	0.8	0.8

3. Parameter setting

The chromosome is decoded and the candidate solution and its fitness value are generated according to the fitness function. Fitness value is a measure of the optimal degree of the solution of the original objective function and "infeasible quantity". The fitness function is formed by adding penalty to the original objective function. The original objective function consists of supply time, supply cost and satisfaction.

In terms of constraint processing, this paper uses penalty function method to transform constrained multi-objective optimization problem into unconstrained multi-objective optimization problem[2]. That is to say, the constraint is multiplied by a larger penalty coefficient to form a penalty term and added to the objective function to form the fitness function. For example, the fitness function of the first objective function is:

$$\text{fitness1}=\sum_{i=1}^{m}\sum_{j=1}^{n}T_{ij}+\sum_{j=1}^{n}\sum_{k=1}^{l}T_{jk}+\sum_{j=1}^{n}\sum_{j'=1}^{n}T_{jj'}+M(\sum(\max(0,\ \text{constraint})))\quad(1)$$

Among them, the penalty coefficient m is a large positive number, the greater the value of M, the greater the penalty. When the value of the decision variable is within the feasible range, the penalty term is 0, and the fitness function is the objective function value. When the decision variable is beyond the feasible range, the value of penalty term will be enlarged, and the value of fitness function will be greatly increased [3]. At this time, the individual's fitness function value is larger, which does not meet the conditions of the optimal individual and will not enter the next generation.

The basic parameter settings of this algorithm are as shown in Table 2

Table.2 Parameter setting of the NSGA-II

Population	Generations	Pool Size	Tour Size	ηc	ηm
200	10000	50	2	20	20

4. Supply plan and result analysis

4.1 Supply plan

Finally, 58 non dominated solutions are obtained as alternative solution sets. Each scheme in the scheme set satisfies the non-dominant relationship, and the decision variables of each scheme are within the feasible range. 72 decision variables and 3 target values of each scheme are shown in Table 3.

Table 3 Set of network supply schemes

	Decision variables										Objective function		
	x_{11}^1	x_{12}^1	x_{13}^1	x_{14}^1	x_{11}^2	x_{12}^2	x_{13}^2	x_{14}^2	…	s_{43}^2	T	C	θ
programme 1	6	10	5	40	40	9	10	8	…	1	12275	2017397	0.9
programme 2	6	10	5	41	39	8	1	7	…	8	12164	2041817	0.9
programme 3	7	11	6	46	23	5	9	8	…	1	12580	1721934	0.8
…	…	…	…	…	…	…	…	…	…	…	…	…	…
programme 58	25	1	1	1	34	8	3	1	…	1	10110	4360055	0.9

Because Pareto optimality is used to describe the optimal solution, it is not a single optimal solution but a set of solutions that satisfy Pareto optimality that can be obtained by solving the supply model. The scheme set provides 58 schemes for decision makers, all of which make the decision objectives reach the optimum as much as possible on the premise of satisfying all constraints. Decision makers can make further decisions according to the actual supply situation and their own preferences. For example, in peacetime, the scheme with the lowest cost and the highest satisfaction rate can be considered as the final scheme; while in wartime or emergency supply, the cost is not the most critical factor, but to make the spare parts demand of the combat unit meet to the maximum extent in the shortest time. Therefore, the scheme with the shortest supply time and the highest satisfaction rate can be selected as the final supply scheme.

4.2 Result analysis

This paper mainly measures the effectiveness and efficiency of the algorithm in solving the networked spare parts supply scheme model. The effectiveness of the algorithm is mainly reflected in convergence and distribution, that is, the distance between the obtained non dominated solution set and the real PF, and whether the obtained non dominated solution set is widely and evenly distributed in the real pf space. The efficiency of the algorithm is mainly reflected in the running time. Therefore, the following indicators are selected as the evaluation criteria[4].

(1) generation distance(GD). GD is usually used to measure the approximation degree of the non-dominated solution to the real optimal solution set of the multi-objective problem. Its mathematical definition is as follows:

$$GD = \sqrt{\sum_{i=1}^{n} d_i^2} / n \tag{2}$$

Where, n is the number of non-dominated solutions that have been obtained, and Di is the distance between the found non dominated solutions and the real non dominated solutions (in the target space). Obviously, GD = 0 means that all the non-dominated solutions found are in the real optimal solution set. The larger the GD, the worse the convergence of the algorithm is.

(2) Spacing(SP). SP index mainly measures the uniformity of the algorithm, that is, the distribution of the non-dominated solution in the target space should be uniform. Its mathematical definition is as follows:

$$SP = \sqrt{(\sum_{i=1}^{n} (\bar{d} - d_i)^2 / (n-1)} \tag{3}$$

Where, n is the number of non-dominated solutions that have been obtained. According to the value of N, the conclusion can be drawn as follows:

$d_i = \min_j(\left|f_1^i(\bar{x}) - f_1^j(\bar{x})\right| + \left|f_2^i(\bar{x}) - f_2^j(\bar{x})\right|), i, j = 1, 2, ..., n$ 。 Obviously, when SP = 0, it means that the non-dominated solution is equidistant in the target space. The larger SP value is, the worse the algorithm distribution is.

(3) RUNTIME. RUNTIME is the running time of the program under the same configuration computer. The smaller this index, the higher the efficiency of the algorithm is.

CSADMO, DMOPSO and DVEPSO are selected as the comparison algorithms on the basis of learning and reading a lot of evolutionary algorithm literature.

The parameter settings of each algorithm are set on the basis of the original literature and try to be consistent with the algorithm in this paper. See Table 4for the parameter settings.

Table.4 Parameter setting of the algorithm

Algorithm	Parameter setting
CSADMO	Population size: N=200
	Cloning proportion: n_c=3
DMOPSO	Population size: N=200
	Acceleration coefficients: c1=c2=0.494
	Inertia weight: ω=0.4
DVEPSO	Sub-swarm size: N=20
	Archive size: A=200
	Acceleration coefficients: c1=c2=1.49
	Inertia weight: ω=0.72

From the visual point of view of qualitative analysis, it can be seen from figure 1 that when solving the networked spare parts supply model, the individuals obtained by NSGA-II algorithm do not have obvious aggregation in the target space, which is more evenly distributed, and the individuals are all over the target space. Therefore, NSGA-II algorithm has the best distribution. The distribution of individuals in the target space obtained by CSADMO and DMOPSO algorithm does not cover the whole space, and there is a certain aggregation phenomenon, and its distribution is not as good as NSGA-II algorithm[5]. Although the individuals obtained by DVEPSO algorithm are all over the whole target space, it can be seen from the figure that they are more evenly distributed on the edge, but not evenly distributed in the interior, so their distribution is not as good as NSGA-II algorithm.

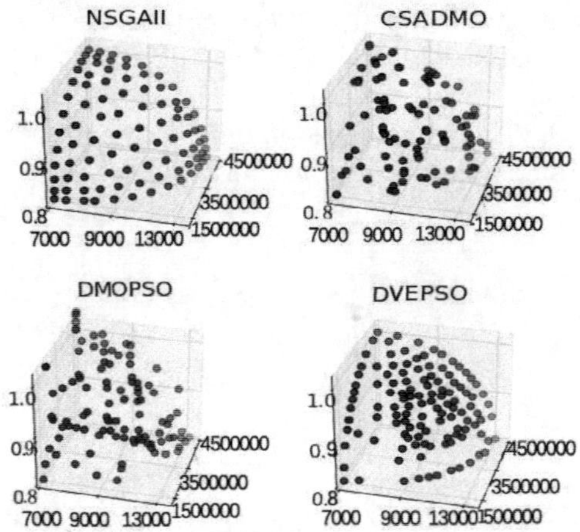

Fig.1 Distribution of non-dominant solutions in target space

In order to further quantify the evaluation index, the effect and efficiency of different algorithms in solving the model of network spare parts horizontal supply proposed by us are compared quantitatively. Each algorithm is used to perform 30 operations on the model, and the calculation results are counted. See table16 for the average value of each evaluation index. After each algorithm runs 30 times, the average value of each index is calculated.

Table 5 Evaluation index of each algorithm

	NSGA-II	CSADMO	DMOPSO	DVEPSO
$\overline{GD}$	8.2521e-3	3.2468e-2	1.0618e-2	2.7087e-2
$\overline{SP}$	2.1078e-1	3.0776e-1	5.5672e-1	4.0155e-1
$\overline{RUNTIME}$	5.6446e-1	1.3881e+1	5.3424e+0	2.4729e+0

From Table 5, we can see the mean values of GD, SP and runtime indexes. The results are consistent with those in Figure 7 and figure 8. Every index of NSGA-II algorithm is optimal, which shows better convergence and distribution. At the same time, NSGA-II algorithm has the shortest computing time and better computing efficiency on the same computer. To sum up, it can be proved that NSGA-II algorithm has a good effect and high efficiency in solving the network spare parts horizontal supply model.

5. Conclusion

The mathematical model of mixed nonlinear discrete integer programming is established by analyzing the scheme considering transverse supply. The model belongs to the multi-objective optimization problem, and the three objectives are conflicting, so it is impossible to achieve the shortest supply time, the lowest supply cost and the best satisfaction rate at the same time. In order to solve this problem, Pareto optimality is usually used. By using intelligent optimization algorithm to solve the model combined with the case, a series of Pareto optimal scheme sets are obtained. Compared with other classical intelligent algorithms, the effectiveness and robustness of the algorithm are verified. An economic, effective and safe supply scheme and its corresponding solution are put forward.

REFERENCES

[1] Amir M. Sharif and Zahir Irani. Supply chain leadership [J]. International Journal of Production Economics, 2012, 140(1): 57-68.

[2] Hu Qiwei, Boylan John E., Chen Huijin, Labib Ashraf. OR in spare parts management: a review[J]. European Journal of Operational Research, 2017

[3] Cao Wenbin, Song Wenyuan, Han Yucheng, Wu Yutao. Spare parts demand forecasting: a review [J]. Journal of Donghua University, 2016, 33(2): 340-344.

[4] Wang Wenbin, Syntetos Aris A. Spare parts demand: Linking forecasting to equipment maintenance[J]. Transportation Research Part E: Logistics and Transportation Review, 2011, 47(6): 1194-1209.

[5] Hong-Yu Gu., Research on the allocation decision of equipment maintenance resources in wartime [D]. Shijiazhuang Ordnance Engineering College, 2016.12.

IOP Conf. Series: Materials Science and Engineering **793** (2020) 012017 doi:10.1088/1757-899X/793/1/012017

Study on Test of Color Fastness to Rubbing of Textiles

Zhenyu Lei

(Yangjiang Supervision and Inspection Institute of Quality and Metrology ,
yangjiang529500,China)

Absrtact: The textile sample is rubbed with a dry friction cloth and a wet friction cloth respectively to evaluate the staining degree of the friction cloth. The tester for color fastness to rubbing provides two sets of test conditions through two friction heads of optional sizes: one for pile fabrics; The other is used for monochrome fabrics or large-area printed fabrics. Through the analysis of the original test method, the existing problems and defects of the original test method are obtained. A substitute material is used on the tester platform for testing samples, and the fixation method of the tester's samples is improved in a pawn type, so as to solve the problem of uneven staining of friction cloth caused by uneven surface of friction head or friction medium and unilateral wear during friction.

1.Introduction

Color fastness to rubbing of textiles is a common and necessary test method for staining and pigment firmness of textiles. Different fibers, different pigments, Its pigments have different staining and firmness, In the clothing of the daily public, there is often a large area of color fading of some fabrics. When the clothes are scrubbed and rubbed, the overall color tone is inconsistent, which greatly damages the overall beauty of the clothes. Therefore, it is very necessary to test the color staining degree and firmness of textile pigments caused by friction. At present, the method for testing the color fastness of textiles to rubbing is to test the friction resistance of textiles by wrapping friction cloth on the friction head and placing it on a testing instrument padded with 600 mesh water-resistant fine sandpaper or a metal test platform with a diameter of 1mm and a mesh width of about 20mm. The friction cloth tested by this method will often have a deep and a shallow stained surface, which not only cannot well reflect the color fastness to rubbing of textiles under fixed pressure, but also will cause misjudgment of the stained grade of the sample. In order to overcome the defects and existing problems of the existing detection methods and technologies, the author studied a substitute medium and a new design structure for testing the color fastness to rubbing of textiles, and explored and studied the detection method for the color fastness to rubbing of textiles.

2.Equipment and Materials for Testing Color Fastness to Rubbing of Textiles

2.1 Color fastness to rubbing tester (as shown in Fig. 1), with metal friction heads of two optional sizes for reciprocating linear friction motion.

2.1. 1 For pile fabrics: The size of the friction head on the rectangular friction surface shall be 19 mm × 25. 4mm. The friction head shall be applied to the downward direction of (9 ± 0.2) N and the linear reciprocating distance shall be (104 ± 3) mm.

IOP Conf. Series: Materials Science and Engineering **793** (2020) 012017 doi:10.1088/1757-899X/793/1/012017

2.1.2 For other textiles: The friction head consists of a cylindrical shape with a diameter of (16 ± 0.1) mm, applied with a downward pressure of (9 ± 0.2) N and a linear reciprocating distance of (104 ± 3) mm.

2.2 Cotton friction cloth, in accordance with the national standard GB/T7568.2, is cut into (50 mm ± 2 mm) × (50 mm ± 2 mm) squares for the friction head of 1.1. 2, and cut into (25 mm ± 2 mm) × (100 mm ± 2 mm) rectangles for the friction head of 1.1. 1.

2.3 Gray cards shall be used for evaluation of staining, in accordance with GB/T251.

2.4 Water-resistant sandpaper and metal mesh with a diameter of 1mm and a mesh width of about 20mm are replaced by 1cm thick rubber elastic blocks and 6 inverted conical minions with a diameter of 0.5 cm (as shown in Fig. 1).

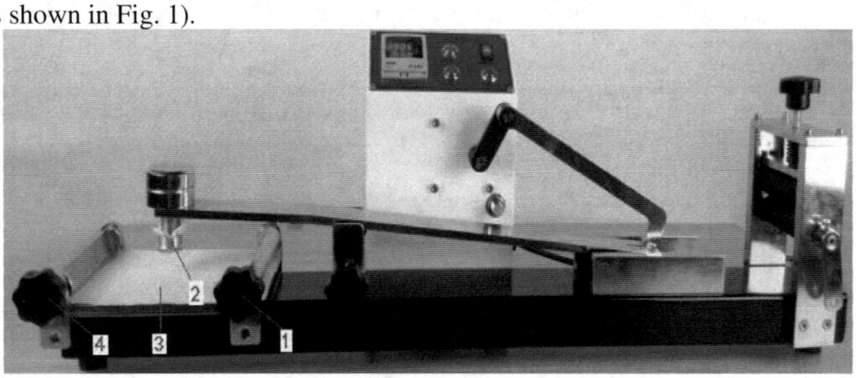

Fig. 1 Schematic diagram of tester for color fastness to rubbing

1-Inverted conical pawn right cam knob
2-Friction head
3-1cm thick rubber elastic block
4-Inverted Conical Pawn Left Cam Knead

3.Sample preparation

3.1 If the tested textile is fabric or carpet, two groups of samples with a size of 50 mm × 140 mm shall be prepared for dry friction test and wet friction test respectively. Each group has two samples, and the length direction of the samples is parallel to the longitude direction or the latitude direction respectively.

3.2 If the tested textile is yarn, weave it into a fabric with a sample size of not less than 50 mm × 140 mm. Or wrap the yarn parallel to the cardboard with the same sample size along the length direction of the cardboard, and make the yarn evenly spread into a layer on the cardboard.

3.3 Before the test, place the sample and friction cloth in the standard atmosphere specified in GB/T6529 for debugging for at least 4h. For fabrics such as cotton or wool, longer moisture conditioning time may be required.

4.Dry and wet friction test

4.1 Dry friction: Place the square of the humidity-adjusted friction cloth (as shown in 1.2) on the friction head so that the meridian direction of the friction cloth is consistent with the running direction of the friction head. The running speed is 1 reciprocating friction cycle per second, with a total of 10 cycles of friction. The motion of friction on the dried specimen was (104 ± 3) mm and the downward pressure applied was (9 ± 0. 2) N (as shown in 1.1. 1 and 1.1. 2). Remove the friction cloth, humidify it according to 2.3, and remove any excess fibers from the friction cloth that may affect the rating.

4.2 Wet friction: Weigh the friction cloth after humidity adjustment, immerse it completely in distilled water, and weigh the friction cloth again to ensure that the moisture content of the friction cloth reaches 95%-100%. Then follow 3.1.

5.Drying and evaluation

5.1 Dry the wet friction cloth for evaluation.

5.2 During evaluation, three layers of friction cloth shall be placed on the back of each evaluated friction cloth (as shown in 1.2); Under a suitable light source, use a gray sample card for evaluating staining (as shown in 1.3)

6. Disadvantages of water-resistant sandpaper and metal mesh with a diameter of 1mm and a mesh width of about 20mm

6.1 Colour fastness to rubbing Study on the Fixing Effect of Water-resistant Sandpaper and Metal Mesh with a Diameter of 1mm and a Mesh Width of about 20mm

6.1.1 Solidification Analysis of Water Resistant Sandpaper
The water-resistant sandpaper used for color fastness to rubbing is used to fix the sample and prevent it from moving during rubbing. Water-resistant sandpaper relies on increasing friction resistance to achieve the effect of fixing the sample. The water-resistant sandpaper is static friction relative to the friction head, and the friction head is sliding friction relative to the water-resistant sandpaper. Both of them are doing useful work on each other. The conditions for friction are: the first condition is that the two objects are in contact with each other and have pressure; The necessary condition is that the objects in contact with each other should have relative motion trend or have relative motion. Both the water-resistant sandpaper and the friction head meet the above conditions, That is to say, as the medium for fixing the sample, The water-resistant sandpaper is rubbed with the friction head and the specimen from the beginning to the end of the friction, However, water-resistant sandpaper relies on friction resistance to fix the sample. As the frequency of friction increases, the friction resistance of water-resistant sandpaper decreases due to friction loss. Therefore, water-resistant sandpaper is unstable and unideal as a fixing medium for textile color fastness to friction.

6.1.2 Fixation analysis of metal mesh with a diameter of 1mm and a mesh width of about 20mm
The metal mesh with a diameter of 1mm and a mesh width of about 20mm used for color fastness to rubbing is also used as a medium for fixing the sample to prevent movement during rubbing. Compared with water-resistant sandpaper, metal mesh is not suitable as a fixing medium for textile color fastness to rubbing. First of all, the surface of the metal mesh is smoother than that of the water-resistant sandpaper. Although there are many metal mesh holes, the adhesion to cloth and fibers is not as strong as that of the water-resistant sandpaper. Although the friction loss is slightly lower than that of the water-resistant sandpaper, the fixing effect of the metal mesh surface will be worse than that of the water-resistant sandpaper after long-term use and friction. Secondly, the metal mesh is used as the fixed medium. Due to the concave and convex mesh in the plane, the uniform reciprocating motion of the friction head is not very smooth, and the imagination of sticking occasionally appears. Therefore, it is not ideal to use metal mesh as the fixing medium for the color fastness to rubbing of textiles.

6.2 Colour fastness to rubbing Study on the staining effect of rubbing cloth with water-resistant sandpaper and metal mesh with a diameter of 1mm and a mesh width of about 20mm

6.2. 1 Analysis of color fastness to rubbing and staining effect using water-resistant sandpaper
Water-resistant sandpaper is a hard medium with abrasion resistance. If it is used to fix the sample or as a medium for friction movement with the friction head, it will wear the sample itself and the friction head. The friction head is of metallic nature, Water-resistant sandpaper is an abrasive hard medium with certain friction resistance, If the two hard media carry out sliding friction, the increasing number of friction times will inevitably lead to a rapid decline in the quality of the contact media of the friction, thus affecting the flatness of the friction plane, the downward pressure exerted by the friction, and the uniform stress of the sample on the friction head. According to the friction samples with different thicknesses, two friction cloth will be stained with different shades. The thinner the friction sample, the more obvious the shades of the same friction cloth will be stained. The specific situation is shown in Figure 3 below:

Table 1 Dark and Light Colors of the Same Friction Cloth

Resistance to dry and wet friction		Cotton corduroy (purplish red)	Wool knitted fabric (brown)		Cotton woven fabric (black)		Polyester woven fabric (black)	
Dry friction	Meridional direction	4	4	4	4	4-5	3	4
	Latitudinal direction	4	4	4-5	4	4-5	3	4-5
Wet friction	Meridional direction	2	3-4	4	2-3	3	2-3	4
	Latitudinal direction	2	3-4	4	2-3	4	2-3	4

To sum up, the thicker the specimen used in the friction test, Friction cloth after friction is not easy to produce the same friction cloth dark staining situation, The thinner the sample, the easier it is to produce different shades of staining on the same friction cloth. In fact, that is to say, the friction between two rigid friction media is more likely to lead to uneven and unbalanced stress on the friction surface, and it is easy to form friction points instead of friction surfaces. Therefore, the use of water-resistant sandpaper for the color fastness to friction of textiles is also defective.

6.2.2 Analysis of color fastness to rubbing and staining effect using metal mesh with diameter of 1mm and mesh width of about 20mm

As for that color fastness of textile to rubbing, it is not ideal to use metal mesh as the medium to fix the sample. In addition to the easy occurrence of jamming and jamming in the friction process, After rubbing, the rubbing cloth will often leave mesh marks, which is extremely easy to cause misjudgment of color fastness grade in the evaluation process. Therefore, the metal mesh with a diameter of 1mm and a mesh width of about 20mm is not ideal for the color staining effect of the rubbing head. Generally, it is not recommended to choose a diameter of 1mm and a mesh width of about 20mm Metal mesh for testing.

7.Improvement of Fixing the Sample of Textile Color Fastness to Rubbing and Improving the Full Contact between the Friction Head and the Sample and Uniform Stress

7.1 The pawn type fixed sample is adopted, i.e. 6 inverted conical pawns with a diameter of 0.5 cm are added to the fixed rotating rod of the sample (as shown in Fig. 2). The pawns can be pushed up and

IOP Conf. Series: Materials Science and Engineering **793** (2020) 012017 doi:10.1088/1757-899X/793/1/012017

down, and 6 inverted conical holes with a diameter of 0.5 cm are accurately distributed at the corresponding positions at the bottom, which coincide with the pawns in good condition. In this way, when the sample is clamped on the fixed rotating rod, the sample is locked through 6 inverted conical minions to enhance the fixing force of the sample on the test bench, so that the sample can be well locked in the friction reciprocating process and the phenomenon of sample displacement can be prevented.

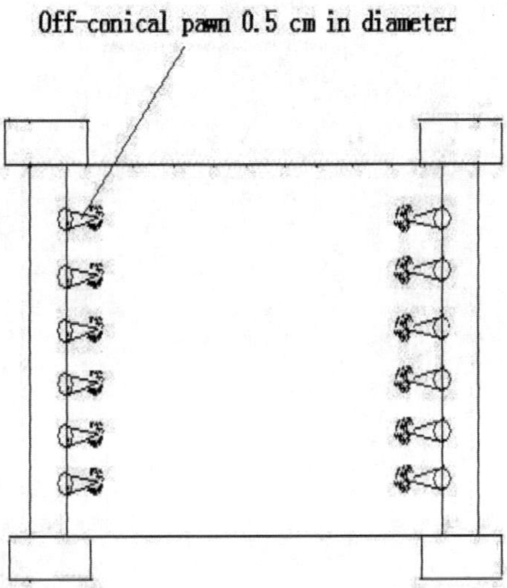

Figure 2 Schematic Diagram of Inverted Conical Pawn

7.2 Insert 1cm thick rubber elastic blocks (as shown in Fig. 3) into the surface layer of the original metal test bed, The adhesive force between the base of the test bed and the friction head is increased, Absorbing the uneven factors on the surface of the sample and the surface of the friction sleeve, Strengthen the uniform stress between the friction head and the sample, make the friction surface of the friction head contact with the sample more appropriately, make the friction cloth obtain more uniform sample staining, eliminate uneven point friction formed by factors such as the friction head, the sample, the test bench, etc., enhance the effect of surface friction, and solve the phenomenon of deep and light staining of the same friction cloth.

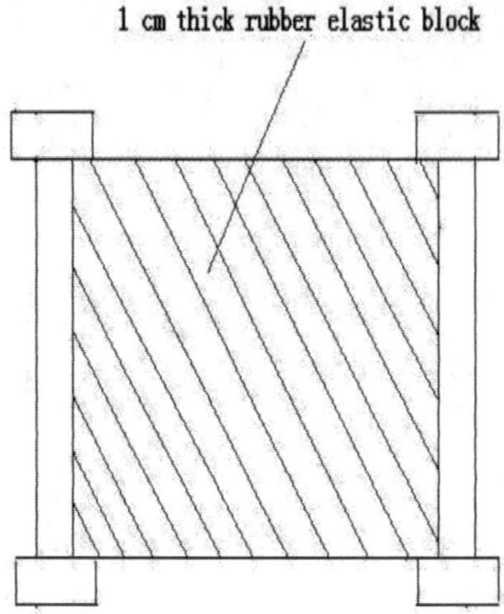

Figure 3 Schematic Diagram of Rubber Elastic Block

Table 2 Friction Test of Specimens with Different Materials

Resistance to dry and wet friction		Cotton corduroy (purplish red)	Wool knitted fabric (brown)	Cotton woven fabric (black)	Polyester woven fabric (black)
Dry friction	Meridional direction	4	4	4	3-4
	Latitudinal direction	4	4	4	3-4
Wet friction	Meridional direction	2	3-4	3	3
	Latitudinal direction	2	3-4	3	3

8.Conclusion

(1) The color fastness to rubbing of textiles uses water-resistant sandpaper as the medium for fixing samples. With the increasing frequency of use, the repeated rubbing between the friction head and the water-resistant sandpaper reduces the friction resistance of the water-resistant sandpaper, thus making the fixing effect show a decreasing effect; When a metal mesh with a diameter of 1mm and a mesh width of about 20mm is used as the medium for fixing the sample, jamming and settling will occur and the fixing effect is not ideal. Therefore, it is not ideal to use water-resistant sandpaper and metal

IOP Conf. Series: Materials Science and Engineering **793** (2020) 012017 doi:10.1088/1757-899X/793/1/012017

mesh with a diameter of 1mm and a mesh width of about 20mm as the medium for fixing the sample to reduce the movement in the friction process.

(2) Color fastness to rubbing of textiles using water-resistant sandpaper medium will cause different shades of color staining of the same rubbing cloth during rubbing, while using metal mesh with a diameter of 1mm and a mesh width of about 20mm will also cause mesh marks. Therefore, the above two conditions have certain influence on the evaluation of color fastness to rubbing of textiles.

(3) By using 1cm thick rubber elastic blocks and 6 inverted conical minions with a diameter of 0.5 cm (as shown in Fig. 2) instead of water-resistant sandpaper and metal mesh with a diameter of 1mm and a mesh width of about 20mm, it can play a very good role in fixing samples, and can solve the problems of deep and light staining and mesh marks of the same friction cloth.

Authors' background

Your Name	Title*	Research Field	Personal website
Zhenyu Lei	Master's Degree	Textile inspection	

References

[1] Si Ying, Chen Jianhua, Zhang Qiping, Zhao Chungui. "People's Republic of China National Standard Textile Color Fastness Test Color Fastness to Rubbing", China Standard Publishing House (2008)
[2] Qi Yamin, Xu Jieshou, Lu Wenbao, Tong Jinzhu, Li Ming. General Rules for Testing Color Fastness of Textiles, China Standard Publishing House (1997)
[3] Wu Zongze and Gao Zhi. Mechanical Design, Higher Education Press (2009)
[4] Yao Liang and Wang Guozhi. "Practical Technology for Formulation of General Rubber Materials", Chemical Industry Press (2019)
[5] Jili George Drobny. Handbook of Thermoplastic Elastomers, Chemical Industry Press (2018)

Design path of hybrid bean planting machine based on human-machine symbiosis aesthetics theory

ZHAO aili [1], GUO zhanbin[1], WANG liming[1*],QU qingfeng[1] ,QIN haibin[1]

[1]Heilongjiang Bayi Agricultural University DaQing 163000

[*]Corresponding author's e-mail:573053416@qq.com

Abstract: Based on the theoretical study on the creative design theory of hybrid bean planting machinery, and combined with the aesthetic characteristics of hybrid bean planting machinery modeling of perceptual engineering-promoted symbiosis mechanism, an innovative design path based on man-machine symbiosis aesthetics of hybrid bean planting machinery was constructed to realize the innovative design style of hybrid bean planting machinery under the background of intelligent age, and thus improving the design level of agricultural machinery finishing.

1. Introduction

The No.1 Central Document in February 2019 [1]actively advocated the development of related industries such as miscellaneous grains and hybrid beans. This policy signal indicates that with the development of the intelligent era, the miscellaneous grains and hybrid beans industry and planting machinery will achieve comprehensive development. At the same time, the "human-machine symbiosis" era of agricultural machinery is also approaching. Therefore, how to realize the harmonious symbiosis of human and machinery and optimize the experience is the trend and goal of the future development of hybrid bean planting machinery.

2. Research status of hybrid bean planting machinery

Considering the current status of domestic theoretical research, scholars have conducted related research on human-machine symbiosis and agricultural machine design, but the research on hybrid bean planting machinery is relatively inadequate. Through literature analysis and empirical research, Zhang Yue pointed out in the "*Study on Human-Computer Symbiosis Based on Intelligent Product Design*" that the rapid development of science and technology has accelerated the process of informationization, and more products are developing in the direction of intelligence. Humans and machines have also become inseparable, and from the background of intelligence, the key factors in product design and the physiological and psychological needs of modern people are studied to explore the symbiotic relationship between man and machine in products. Therefore, some key characteristics and general laws are summarized in the human-computer symbiosis relationship, which provides some ideas for future intelligent product design. [2] Aiming at the problems of small plots, difficult to operate large machinery, lack of small machinery, inconvenient purchase of fuel, and high labor intensity in the hilly and mountainous areas of Shanxi Province, Ma Li studied and designed the self-propelled machine in "*Design and Experiment of Self-propelled Small Grain Multigrain Seeding Fertilizer*" type electric miscellaneous grain sowing and fertilizing machine. The development of hybrid bean planting machinery lies in the co-ordination and trend of "function" and "aesthetics".[3] However, through the above analysis, it is found that the current research mainly focuses on how to develop functions, and how to combine engineering and aesthetics is blank. Moreover, the research on human-computer symbiosis is only at the theoretical level, which has not been applied to deeper content such as innovation practice and path exploration. Therefore, in-depth research is needed.

Considering the current status of foreign research, the research content is rich and diverse, but has similar characteristics to domestic research, mainly focusing on mechanical principles and other aspects. For example, *Research on 2BJYM-4 Type Precision Seeder for Interplanting Corn with Soybean* pointed out that intercropping of corn and soybean is an effective yield-increasing technology, but a seeder capable of simultaneously seeding corn and soybean has not yet been formed. [4] The innovative design of hybrid grain planting machinery started late in China, but has developed rapidly. Combined with the current rough design and poor experience of China's hybrid grain planting machinery, based on the concept of human-machine co-existence, exploring the aesthetic model of hybrid bean planting machinery and carrying out innovative design practice exploration is the only way for the development of agricultural machinery. Through the exploration of this innovative path, the bean planting machinery will enhance its emotional properties on the basis of realizing its intelligent function improvement, and activate the bean planting machinery through systematic and logical research in modeling design, user experience, aesthetics innovation and promote the development of hybrid bean planting industry.

3. Innovative design trends of hybrid grain planting machinery

3.1 Industry development adaptability trend

From the perspective of industry development, the innovative design of hybrid bean planting machinery based on the theory of human-machine symbiosis can systematically sort out its aesthetic rules and development direction, and innovate for independent hybrid bean planting machinery brands from aspects such as modeling, experience, and aesthetics. The design provides certain reference basis, so as to improve the overall level of the design of hybrid bean planting machinery, meet the development needs of the intelligent era, enable research and development products to be compatible with future development trends, and enhance the competitiveness and influence of hybrid bean planting machinery in China.

3.2 Market-oriented adaptation trend

From the perspective of marketization, the aesthetic model construction and human-machine analysis carried out by the innovative design of hybrid bean planting machinery based on the human-machine symbiosis theory can promote its rapid market-oriented response capabilities. This design can make the research and development of hybrid bean planting machinery not only have a certain function, but also reflect the corresponding aesthetic education, improve farmers' aesthetic level of the design of hybrid bean machinery, and thus promote the market of miscellaneous cereal machinery benign development.

4.Analysis of innovation path

4.1Path content

(1) Investigation and extraction of hybrid bean planting machinery

The research of bean planting machinery should be based on perceptual engineering, using field surveys, field research, etc. It is necessary to conduct in-depth research on the man-machine size and human-machine experience of hybrid bean planting machinery, classify and summarize the results, and extract data to provide a basis for subsequent transformation. In this way, theoretical analysis and practical exploration of the human-machine symbiosis aesthetic construction and innovation of hybrid bean planting machinery can be realized.

(2) Construction of a human-machine symbiosis aesthetic model for hybrid bean planting

By analyzing problems and designing "pain points" between various system environments related to the bean planting machinery, you can use design methods such as "scenario maps" and "scene descriptions" to conduct research, and explore its human-machine symbiosis logic to construct aesthetic models and determine the design direction. Based on the aesthetic principles, you can analyze the artistic characteristics of the research on the aesthetics of mechanical design of hybrid bean planting, explore the rules of mechanical design aesthetics, and interpret and analyze them.

(3) Innovation practice and analysis of hybrid bean planting machinery

Combining the previous human-machine measurement data and the aesthetic model architecture, you can carry out innovative design of hybrid bean planting machinery from the aspects of appearance design, human-machine experience, and aesthetic factors. Based on the psychological experimental platform, the sensible engineering test of the realization and transmission of its human-machine symbiosis aesthetics can fully realize the design case of hybrid bean planting machine in the new era, which can meet the needs of consumers and have the ability of rapid market response.

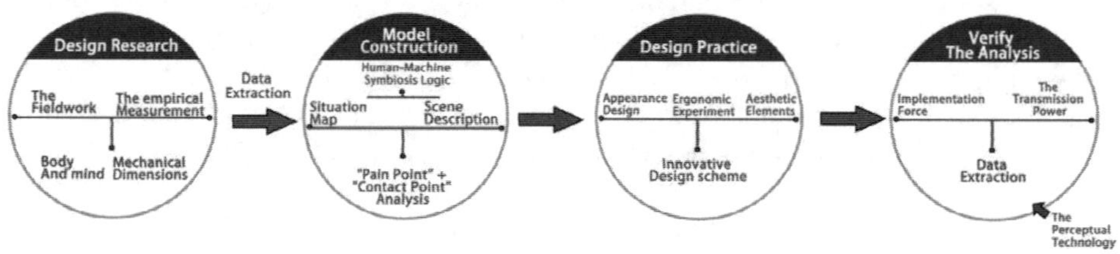

Figure 1. Design path analysis diagram

4.2 Innovation point analysis

By using the perceptual engineering theory to study the human-machine symbiosis aesthetic law of hybrid bean planting machinery, and use this as a basis for innovative practical exploration. And you can systematically study the basic factors such as the technology, art, and economy of the bean planting machine, introduce the creative thinking and research methods of the bean planting machine, and conduct scientific and rigorous research on the shape and experience of the grain planting machine.

First of all, this method can realize the fusion of the technical beauty and experience beauty of hybrid bean planting machinery in the process of human-machine symbiosis aesthetics construction and innovative design research of hybrid bean planting robots, and finally conduct perceptual engineering experiments to form a fusion of engineering and aesthetics. The research method of "service design" is introduced into the innovative design research of hybrid grain planting machinery to realize the innovative research method of human-machine symbiosis aesthetics.

Secondly, the psychological cognition of human-machine symbiosis aesthetics was proposed in the process of researching the innovative design style of hybrid bean planting machinery. Strengthen the "technical and artistic" style of hybrid bean planting machinery by realizing the design, human-machine experience and aesthetic path of hybrid bean planting machinery. This unique innovative practice path can adapt to the development of the intelligent era and promote the improvement of China's hybrid bean planting machinery.

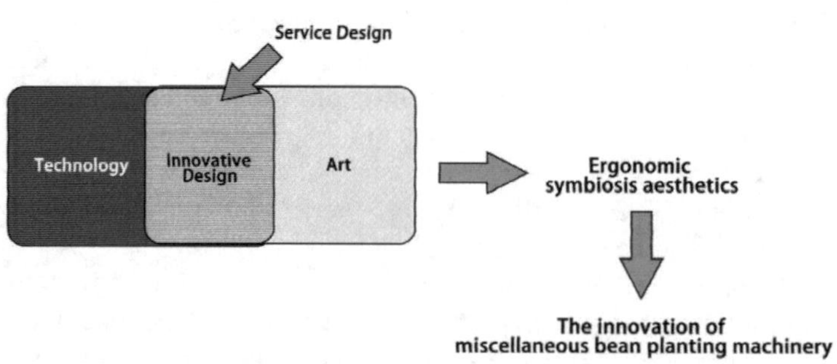

Figure 2. Analysis of mechanical innovation points of miscellaneous bean planting

5. Path achievement indicators

5.1 Construction index of man-machine aesthetic model of hybrid bean planting robot

When observing and measuring the bean planting machinery, collecting data, determining the quantitative relationship, designing the human-machine symbiosis aesthetic relationship, conducting analysis and inference, and clarifying the aesthetic creativity and transmission power indicators, you can make the interaction relationship reaches the optimal value. In terms of shape, function and technology, it achieves a balance, subverts and reproduces the rigid mechanical aesthetics, realizes flexible interaction, and has a higher service feedback mechanism.

5.2 Innovative design indicators of hybrid bean planting machinery

By constructing a complete innovative design system of human-machine symbiosis, and clarifying the realization goals of modeling design, user experience, aesthetics, etc., we can achieve the "innovative development" of product appearance modeling changes, the application of material surface texture and color " "Time trend", "emotional transmission" in human-machine experience between human and machine

6.Conclusion

The research strives to analyze the human-machine symbiosis aesthetic law and development direction of hybrid grain planting machinery through systematic combing, and provide a certain reference for the innovative design of China's independent hybrid grain machinery, thereby improving the overall level of the design of agricultural machinery in China.It aims to improve the aesthetic level of the innovative design of hybrid machinery by agricultural machinery users through theoretical and practical means of human-machine symbiosis, so as to promote the healthy development of the hybrid machinery market.

Acknowledgment

Fund Project: Characteristic discipline project of " coarse cereals production and processing"(Project Number:102/2042070008)

References:

[1] Ministry of Agriculture and Rural Affaris of the People's Republic of China. No.1 Central Document, [EB/OL], 2019-2. http://www.moa.gov.cn

[2] Zhang Yue. Study on Human-Computer Symbiosis Based on Intelligent Product Design[J]. Computer Products and Distribution.2018, (11) :124.

[3] Ma Li. Design and Experiment of Self-propelled Small Grain Multigrain Seeding Fertilizer [J]. Agricultural Engineering.2019, (06) :93-96.

[4] QU Zhe, YU Yong-chang. Research on 2BJYM-4 Type Precision Seeder for Interplanting Corn with Soybean [J]. Soybean Science.2014, (04) .

IOP Conf. Series: Materials Science and Engineering **793** (2020) 012019 doi:10.1088/1757-899X/793/1/012019

Brittle-Plastic Properties Evaluation of Rock based on Element Logging

Li Yi[1], Zhang Yan qi[1], Hu Zong Min [1], Geng Chang Xi[1], Yuan Bai Yan[1], Li Hao[2], Ligang Zhang[2]

[1]Daqing No.1 Geo-logging company, Daqing Drilling & Exploration Engineering corporation, Daqing, 163411, China

[2] Institute of Petroleum Engineering, Northeast Petroleum University, Daqing, 163318, China

*Corresponding author's e-mail: 18565369823@163.com

Abstract. Rock brittle-plasticity is an important factor which can influence the selection of rock breaking method and fracturing layers. It is very important to build up an appropriate brittle-plasticity characterization method and prediction method. In this paper, the point load method was used to test the deep rock load-displacement curve in Songliao basin, calculate the plasticity coefficient and the brittle-plasticity. Then the element content of the corresponding rock samples was tested. The relationship between plasticity coefficient and element content were estimated , and the influence of elements on brittle-plasticity of rocks was revealed. Finally, the calculation model of rock plastic coefficient based on element contents was built up. The research results can be used to evaluate the real-time brittle-plasticity of formation by using element logging data.

1. Introduction

Brittle-plasticity of rock refers to the deformation and failure characteristics of rock under different stress states [1]. There is no obvious plastic deformation when brittle rock is subjected to stress failure. Plastic rock has large plastic deformation before failure [2]. Brittle-plasticity of rock is the important criterion for optimization of crushing methods and selection of fracturing layers. It directly affects crushing efficiency, volume reconstruction efficiency and productivity of fractured wells [3]. At present, there is no unified characterization method and evaluation criteria for rock brittle-plasticity. There are two kinds of evaluation methods: rock mechanics characteristic method and rock mineral composition method [4]. The rock mechanics characteristic method evaluates the rock brittle-plasticity degree via various mechanical experiments based on the rock deformation failure mechanism. The mineral composition method is used to characterize the brittle-plasticity degree based on the contents of different minerals such as quartz, feldspar, calcite, gypsum, clay and so on [5]. Both of the evaluation methods have the problem of discontinuous sampling and high cost. It is very important to build up a real-time continuous method for evaluating rock brittle-plasticity. Therefore, this paper studies the relationship between element logging and plasticity coefficient, and forms a real-time continuous evaluation method of rock brittle-plasticity based on element logging data.

2. Rock plasticity coefficient

The evaluation of rock brittle-plasticity by mechanical characteristics method is mostly based on the characteristics of load-deformation curve. One of the typical method is hardness test, the plasticity

IOP Conf. Series: Materials Science and Engineering **793** (2020) 012019 doi:10.1088/1757-899X/793/1/012019

coefficient is determined according to the load-penetration curve. The test is done by pressing a flat cylinder of a certain diameter into the rock surface, with the increasing load, the penetration depth increases, until the cuttings occur the first volumetric fracture. The typical load-penetration curve is shown in figure 1.

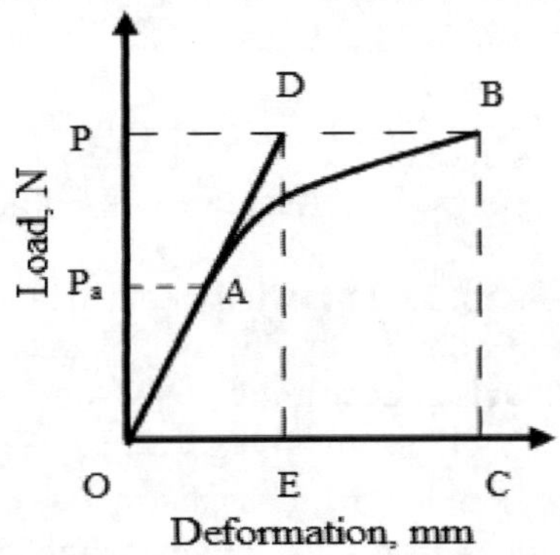

Fig.1.point loading test load-penetration depth curve

The rock plasticity coefficient can be derived by equation (1):

$$K = \frac{A_F}{A_E} = \frac{S_{OABC}}{S_{ODE}} \quad (1)$$

The plasticity coefficient is the ratio of the total work consumed by crushing to the work of elastic deformation during the first volume fracturing of rock. The rock brittle-plasticity is graded according to the plasticity coefficient, shown in table 1.

Tab. 1. Classification criteria of rock plasticity coefficient

Category	brittle	Brittle-plasticity low plasticity→high plasticity				plastic
Grade	1	2	3	4	5	6
Plasticity coefficient	<1	1~2	2~3	3~4	4~6	>6~∞

3. Rock-forming mineral elements

Minerals are composed of elements and compounds that are relatively stable in the crust. Different petrogenic minerals have their own chemical element types, common elements in minerals are shown in table 2.

Tab.2. General Petroleum Formation Minerals contain chemical elements

Mineral	Molecular formula	Major elements
Quartz	SiO_2	O,Si
Calcite	$Ca[CO_3]$	O,Ca,C
Dolomite	$CaMg[CO_3]_2$	O,Ca,Mg,C
Orthoclase	$(K,Na)[AlSi_3O_8]$	O,Si,Al,K,Na
Muscovite	$KAl_2[AlSi_3O_{10}](OH)_2$	O,Si,Al,H,K
Illite	$K_{0.75}(Al_{1.75},Mg,Fe)[Si_{3.5}Al_{0.5}O_{10}](OH)_2$	O,Si,Al,K,Mg,Fe

Kaolinite	$Al_4[Si_4O_{10}](OH)_8$	O,Si,Al,H
Montmorillonite	$Na_{0.33}(H_2O)_4(Al_{1.67}Mg_{0.33})[Si_4O_{10}](OH)_2$	O,Si,Al,H,Na,Mg
Chlorite	$(Mg,Fe,Al)_3[OH]_6\{(Mg,Fe,Al)_3[(Al,Si)_4O_{10}][OH]_2\}$	O,Si,Al,H,Mg,Fe
Gypsum	$Ca[SO_4]$	O,Ca,S

Different minerals have different elements, the contents of different minerals can be estimated by the contents of elements. Therefore, the element contents can reflect the rock brittle-plasticity by the mineral contents.

4. Relationship between element content and plasticity coefficient

This paper focuses on the deep rocks of Songliao basin, rock plasticity coefficient is measured by point load method, the content of rock elements is determined by X-ray fluorescence analysis, then analysis the relationship in between. Si, K, Na element content vs. Plasticity coefficient is shown in figure 2. Silicon has the highest content in rock sample, which is distributed at 43%-72%, it mainly reflects the content of quartz. As the content of silicon increases, the plasticity coefficient decreases obviously, and the brittleness increases. Potassium mainly exists in orthoclase and muscovite, sodium exists in pyroxene, amphibole and feldspar minerals. With the increase of sodium and potassium, the plasticity coefficient decreases and brittleness increases.

Al, Fe, Mg, Ti element content vs. Plasticity coefficient is shown in figure 3. Aluminum is widely exists in minerals, which is distributed at 10%-17%, as the content of aluminum increases, the coefficient increases obviously, is a typical plastic element. Iron is often found in dark minerals, the higher iron content, the greater plasticity of the rock. Magnesium and titanium content are very low, with the increase of the content of magnesium and titanium, the plasticity coefficient tends to increase.

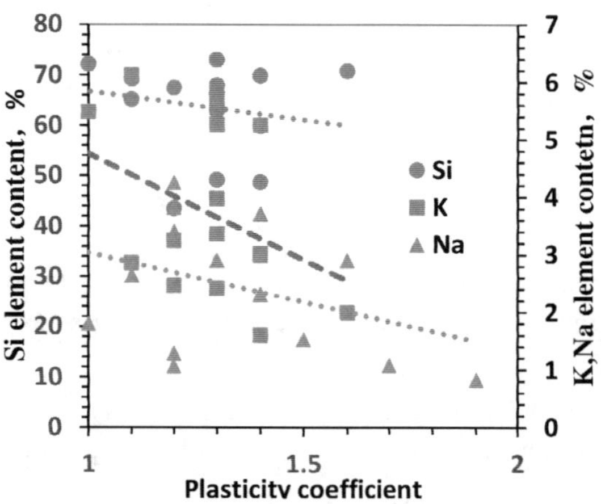

Fig. 2. Si, K, Na element content vs. Plasticity coefficient

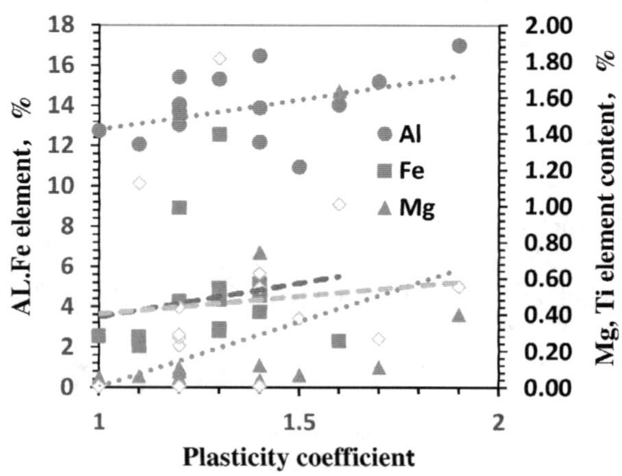

Fig.3. Al, Te, Mg, Ti element content vs. Plasticity coefficient

5. Conclusion

The point load method and X-ray fluorescence analysis method are used to measure the plasticity coefficient and element content in deep layer rocks of Songliao basin which has a correlation between each other. With the increase of silicon, sodium and potassium, the plasticity coefficient decreased and brittleness enhanced. With the increase of the content of aluminum, iron, magnesium and titanium, the plasticity coefficient tended to increase, and plastic enhanced.

Acknowledgments
This work was supported by Youth Science Fund of Northeast Petroleum University (Grant No. 15011030105).

Reference

[1] Qinghui L , Mian C , Yan J , et al. Rock Mechanical Properties and Brittleness Evaluation of Shale Gas Reservoir[J]. Petroleum Drilling Techniques, 2012.
[2] Lawn B R , Marshall D B . Hardness, Toughness, and Brittleness: An Indentation Analysis[J]. Journal of the American Ceramic Society, 1979, 62(7-8):347-350.
[3] Cong G , Chong Z , Lin-Qi Z , et al. Method and Application of Shale Brittleness Evaluation Based on Elemental Capture Spectroscopy Logging[J]. Bulletin of Mineralogy Petrology & Geochemistry, 2017，6（36）：1040-1047.
[4] Liu X G, Sun J M, Guo Y F. Application of Elemental Capture Spectroscopy to Reservoir Evaluation[J]. Well Logging Technology, 2005, 29(3):236-239.
[5] Feng Z , Li X T , Wu H L , et al. Multimineral optimization processing method based on elemental capture spectroscopy logging[J]. Applied Geophysics, 2014, 11(1):41-49.

IOP Conf. Series: Materials Science and Engineering **793** (2020) 012020 doi:10.1088/1757-899X/793/1/012020

Structural analysis of performance optimization of high g accelerometer chip

Jiajun Chen[1], **Haojie Li**

[1]No. 200 Xiaolingwei Street, Xuanwu District, Nanjing, China

Abstract. Accelerometers based on micro-electromechanical systems (MEMS) technology are widely used due to their high cost performance and easy integration. However, there are still weaknesses in the mutual limitation of sensitivity and natural frequency in ordinary MEMS acceleration-sensitive structures. In this paper, a new type of piezoresistive acceleration-sensitive structure is optimized. The specific parameters are optimized by finite element simulation. Finally, the natural frequency of the sensitive unit reaches 2.039MHz and the sensitivity reaches 0.546μV / g.

1. Introduction

High-g acceleration sensors are mainly used in various types of collision and blast tests, especially in the field of weapon fuze . First, it is required to have a higher natural frequency, thereby improving its anti-interference ability; second, it is required to have a higher range while ensuring its sensitivity. But high natural frequency and high sensitivity are contradictory. With the development of micro-electromechanical systems (MEMS) technology, in order to alleviate the possibility brought about by this contradiction, this paper will optimize a three-axis high-g acceleration-sensitive chip to have a higher natural frequency while ensuring its sensitivity[1].

2. Sensor sensitive chip and its working principle

Acceleration sensors can be divided into piezoresistive, piezoelectric and capacitive based on their working principles[2]. A piezoresistive high-g acceleration sensor is designed in this paper. The specific working principle will be introduced below.

2.1. Sensitive cell structure

The structure of the sensitive unit is shown in Figure 1. The sensitive units all adopt a double-end fixed beam structure with microbeams. The prominent features of this structure are small volume, high sensitivity, and simple process. Compared with the traditional double-end clamped model, this structure adds a microbeam to measure its acceleration signal, which not only retains its advantages of high natural frequency, but also improves its sensitivity.

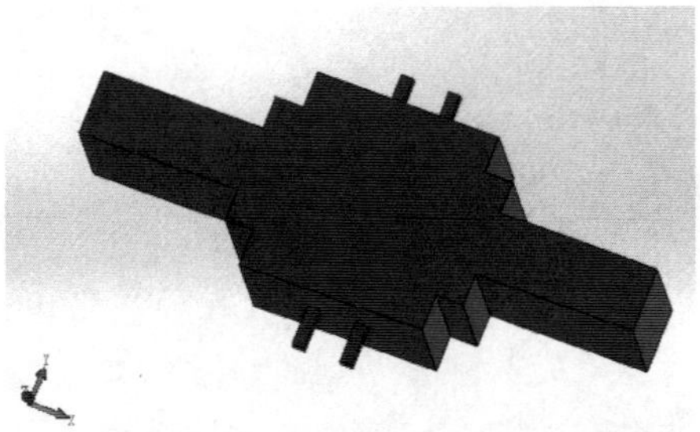

Figure 1. Sensor sensitive unit

2.2. Sensor output circuit

When the sensitive unit is accelerated in the sensitive direction, the two microbeams on one side of the mass are compressed and the two microbeams on the other side are pulled. The corresponding two varistor resistances become larger, and Each of them becomes smaller, and the power supply of the sensor can convert the change of the varistor of the sensitive unit into the output voltage of the sensor. According to the principle of the output circuit, the sensor sensitivity can be derived[3]. The output circuit is shown in Figure 2.

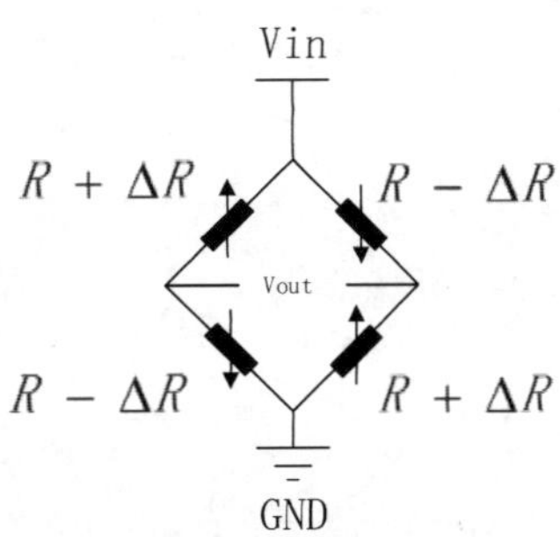

Figure 2. Wheatstone full bridge circuit schematic

When the sensitive unit is stressed, its output voltage is[4]:

$$V_{out} = \frac{(R+\Delta R)(R+\Delta R)-(R-\Delta R)(R-\Delta R)}{(2R)(2R)} Vin \tag{2.1}$$

The sensitivity calculation formula is:

$$S = \frac{V_{out}}{a} = \frac{\pi_{44}|\sigma_l - \sigma_t|}{2a} Vin \tag{2.2}$$

Where S is the sensitivity, Vin is the bridge voltage. $\sigma_l - \sigma_t$ is the equivalent stress of the microbeam.

3. Optimal design of sensitive units

The main dimensions of the sensitive unit are three parts: mass, microbeam and support beam. In order to ensure the stability of the structure, the mass thickness of the mass and the supporting beam is equal. The last eight parameters to be optimized are mass length, mass width, support beam width, main structure thickness, microbeam length, microbeam width, Beam thickness. Its main performance indicators are first-order natural frequency and microbeam stress. In order to measure its comprehensive performance, we use the P value (the product of the two values) to measure its simulation performance[5].

3.1. Impact of scaling on performance

Change the scaling ratio of the sensitive unit. The ratio is basically uniformly distributed from 0.5 to 1.5. A total of 7 sets of simulations have been done. The relationship between the expiration P value and the scaling ratio is shown in Figure 3.

IOP Conf. Series: Materials Science and Engineering **793** (2020) 012020 doi:10.1088/1757-899X/793/1/012020

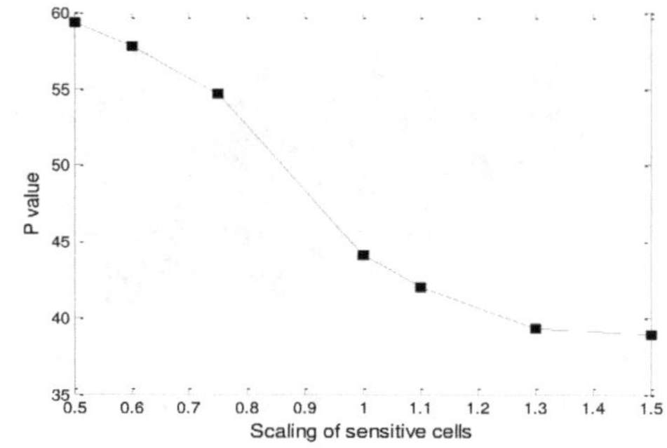

Figure 3. Relationship between P value and scaling

It can be seen from the figure that the change amplitude is the largest near the zoom ratio of 0.85, and the change amplitude at both ends tends to be gentle. In order to meet the requirements of better sensor performance and consider the processing difficulty of silicon substrates, the data after the scaling factor of 0.75 is selected as the sensitive cell size.

3.2. Effect of mass size on performance

Use the same method as in section 3.1 to determine the specific size of the mass, and study the effect of the length and width changes of the mass on the P value. The curve of P value as a function of length and width is shown in Figures 4 and 5.

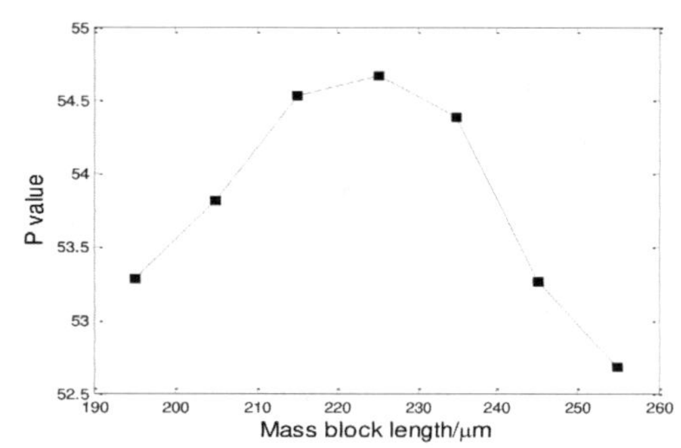

Figure 4. Relationship between P value and mass length

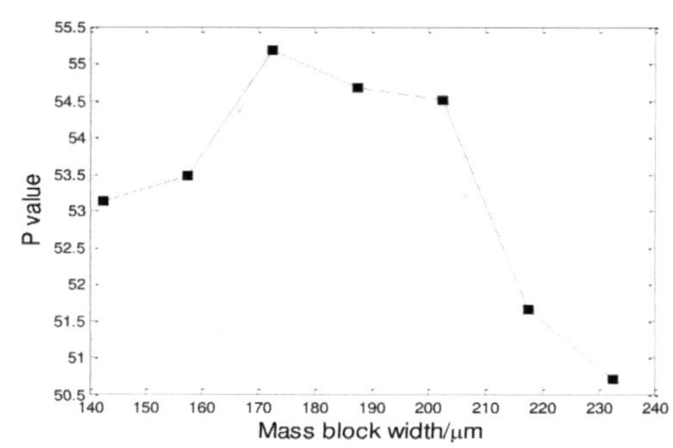

Figure 5. Relationship between P value and mass width

As can be seen from Figure 3, the P value first increases and then decreases with the increase of the length of the mass, and its change range is small on both sides, so in this case the ideal length of the mass is between 210μm and 235μm. . Figure 4 shows the relationship between the P value and the width of the mass. From the figure, it can be seen that the change law is similar to that in Figure 4. The P value is small on both sides and large in the middle. Considering the size of the P value in Fig. 4and Fig. 5 the width of the selected mass is 187.5μm and the length is 225μm.

3.3. Influence of thickness of main structure and size of support beam on performance

The overall dimensions of the sensitive unit and the length and width parameters of the mass have been determined. This section will determine the influence of the thickness of the subject structure and the width of the supporting beam on the performance of the sensitive unit. During the simulation, it is found that the thickness of the main structure and the width of the supporting beam have a greater impact on the natural frequency of the sensor. The two parameters will change under different proportions Its mode shape changes, so this section will study the effect of different support beam width on the performance of cotton stalk unit under different thickness of the main structure. Six sets of data were uniformly taken as the thickness of the main structure from 75 μm to 225 μm, and four sets of data were uniformly taken as the thickness of the main structure from 45 μm to 67.5 μm. The obtained P value results are shown in Table 1.

Table 1. Relationship between P value and support beam width under different thicknesses of main structures

Body structure thickness /μm	Support beam width /μm			
	45	52.5	60	67.5
75	28.78	27.35	26.38	25.35
105	44.84	41.77	39.75	37.48
135	49.78	49.83	50.31	49.03
165	52.61	53.86	54.14	53.84
195	57.11	56.49	56.50	55.48
225	58.49	58.94	58.28	56.93

In the size range given above that meets the natural frequency of 2 MHz, a set of sizes with a higher P value is selected. From the P value data in Table 1, it can be seen that the largest P value in this range is 56.50, the corresponding main structure thickness is 195 μm, and the support beam width is 60 μm. At this time, its natural frequency is 2.039 MHz, the equivalent stress of the microbeam is 29.08 MPa, and the sensitivity is 0.546 μV / g from formula (2.2).

Finally, a set of dimensions of the sensitive unit is given according to the simulation process. The specific dimensions and performance parameters are shown in Table 2.

Table 2. Final optimization results of structural parameters of sensitive units

parameter name	unit	value
Mass length	μm	225
Mass block width	μm	187.5
Body structure thickness	μm	195
Support beam width	μm	60
Support beam length	μm	150
Microbeam section side length	μm	12
Microbeam length	μm	30

4.Conclusion

The structure of the high g impact-resistant piezoresistive acceleration-sensitive chip with microbeam was simulated and optimized by the finite element method. The results show that the microbeam structure can effectively improve the natural frequency and overload capacity of the chip.

References

[1] Zhang Zhenhai, Li Kejie, Ren Xianren, et al. Design simulation and optimization of high overload 3D MEMS acceleration sensor sensitive chip [J]. Acta Armamentarii, 2008, 29 (6): 690 ~ 696.
[2] Li Yuan, Dong Peitao, Wu Xuezhong, et al. Test method and experimental research of triaxial high-g accelerometer [J]. Journal of Transduction Technology, 2008, 21 (11): 1845 ~ 1847.
[3] Deng Tao. Research on Air Cannon Acceleration Overload Test Technology [D]. Nanjing: Nanjing University of Science and Technology, 2011.
[4] Tang Zhaoqian, Huang Wenhu. Handbook of Vibration and Shock [M]. Beijing: National Defense Industry Press, 1990: 84 ~ 89.

IOP Conf. Series: Materials Science and Engineering **793** (2020) 012020 doi:10.1088/1757-899X/793/1/012020

[5] The 13th Research Institute of China Electronics Technology Group. A silicon MEMS piezoresistive acceleration sensor [P]. Chinese invention patent, CN101118250A. 2008 ~ 09 ~ 13.

IOP Conf. Series: Materials Science and Engineering **793** (2020) 012021 doi:10.1088/1757-899X/793/1/012021

Design and terrainability analysis of a novel mobile robot with variable-diameter wheels

Feng Gao [1], Wen Zeng [1,*], Hui Jiang [1], and Guoyan Xu [1]

[1]School of Transportation Science and Engineering, Beihang University, Beijing 100191, China

* Correspondence: zengw@buaa.edu.cn

Abstract. Mobile robots are usually demanded to enter into unstructured terrain. Thus, terrainability of a robot is specially important in cluttered environments. A novel mobile robot with variable-diameter wheels is designed here. This robot can adjust its attitude by variable wheel diameters to improve its terrainability. For terrainability, stability and obstacle-surmounting capability of the robot are studied. Furthermore, variable wheel diameters' effects on these performances were researched. For lateral and longitudinal stabilities, parametric relationships between the maximum slope that the robot can climb and stability requirements were obtained. To avoid stability failures, appropriate failure criterions were proposed. Stability of the robot can be enhanced by changing wheel diameters. The obstacle-surmounting performances of front and rear wheels were analyzed by building two quasi-static models. The results showed that increased wheel diameters enhanced the obstacle-surmounting height. The analytical calculations and the simulations were compared. The simulations verified the reliability of the theoretical calculations. Based on tip-over failure criterions, appropriate strategies can be used to enhance the stability. These analysis presented also can provide theoretical basis for performance optimization design of the robot.

1. Introduction

Mobile robots have been applied to various applications such as planetary explorations, agriculture, mining industries, and reconnaissance in dangerous situations [1-3]. Such tasks may result in terrainability failures, which mainly includes loss of obstacle-surmounting capability, loss of stability and even tipover. We define terrainability as the ability to negotiate uneven terrain without losing stability and to navigate through obstacles in complicated environments. Loss of terrainability may bring about robot damage and total mission failure [4]. Therefore, terrainability in rugged terrain is a critical criterion for mobile robot design. Excellent terrainability requires the design of innovative locomotion systems. Because reconfigurable mobile robots can adjust their attitudes to improve terrainability [5-7], there has been a growing interest among researchers in these robots.

By controlling the articulated suspension of a rover, its stability can be adjusted [4]. A wheeled-legged hybrid mobile robot was developed to optimize stability [8]. An actively articulated rover with rocker-bogie suspension can increase rollover stability and adhesion by a posture control method [9]. By controlling the attitude of a planetary rover with active suspensions, lateral slips of the wheels are prevented [10]. A reconfigurable rover is proposed to traverse slopes with less slippage by properly changing its configuration [11]. Above researches indicate that reconfigurable mobile robots with active suspension system be capable of enhancing their terrainability by adjusting their configurations.

However, trafficability of these robots may be insufficient. By increasing wheel diameter, a mobile robot can enhance both the tractive performance on soft terrain and terrainability [12]. Hence, an innovative mobile robot with four variable-diameter wheels is introduced here. The robot can actively adjust its posture by four variable-diameter wheels. Configuration of the robot and concept of variable-diameter wheels are presented. Then, lateral stability, longitudinal stability, and obstacle-surmounting capability of this robot are analysed. Analytical stability failure criterions are introduced here. Corresponding simulations are used to verify these criterions. Furthermore, the static equilibrium equations describing obstacle-surmounting capability of the front and rear wheels are established, respectively.The quasi-static obstacle-surmounting simulation of the robot was performed.

2. Design of mobile robot with variable-diameter wheels

2.1. Configuration of mobile robot

As shown in figure 1, a new mobile robot is designed. This robot consists of four variable-diameter wheels, a rocker-bogie suspension and a body frame. Moreover, the robot is equipped with four-wheel drive/four-wheel steering (4WD/4WS) actuators. The total mass of the robot is 200 kg.

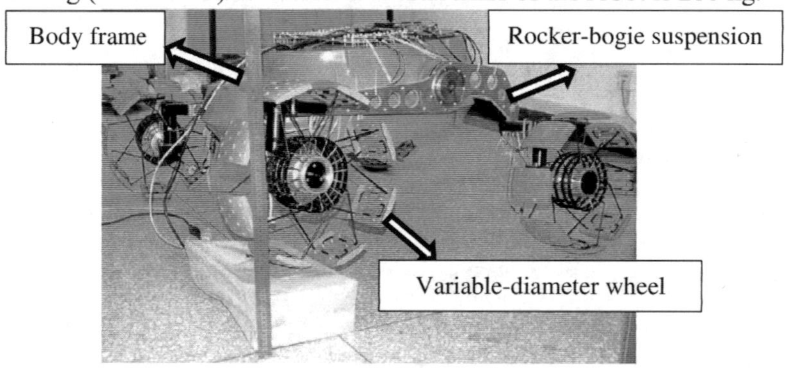

Figure 1. Configuration of mobile robot prototype.

2.2. Concept of variable-diameter wheels

For a variable-diameter wheel, a compliant variable-diameter mechanism allows the wheel to transform its structure using expansion-retraction motion and to possess excellent terrainability. This compliant mechanism consists of helical torsion springs [13]. According to reference [13], when the two wheel hubs rotate at different speeds, which causes the angular deflections of the springs, the changes in the wheel diameter occur, as shown in figure 2. The change range of each wheel diameter is 240-400mm.

(a) Normal configuration (b) Adjusted configuration

Figure 2. Two configurations of mobile robot with variable-diameter wheels.

3. Trafficability analysis

3.1. Stability

3.1.1. Lateral stability. As shown in figure 3, a mobile robot laterally traverses a slope of angle α. Assume that the traversing speed is low, and its motion remains in a steady state. The forces acting on the front and rear wheels are assumed to be equivalent. When the angle α beyond the limit value, the tip-over instability of the rover occurs. Generally, the contact point between the downhill wheels and the slope is regarded as the rotation point of tip-over. Hence, a mechanical model of the robot on a side slope is proposed here.

When there are no forces from ground acting on the uphill wheels, the moment equilibrium equation of the robot with normal wheels is

$$G\cos\alpha\left(B/2\right)=Gh\sin\alpha \tag{1}$$

where α is the slope angle; B (1090mm) is the distance between the downhill and uphill wheels along the slope surface; h (360mm) denotes the distance between the center of gravity (COG) of the robot and the slope surface; r_i is the ith wheel radius; G is the gravity of the robot.

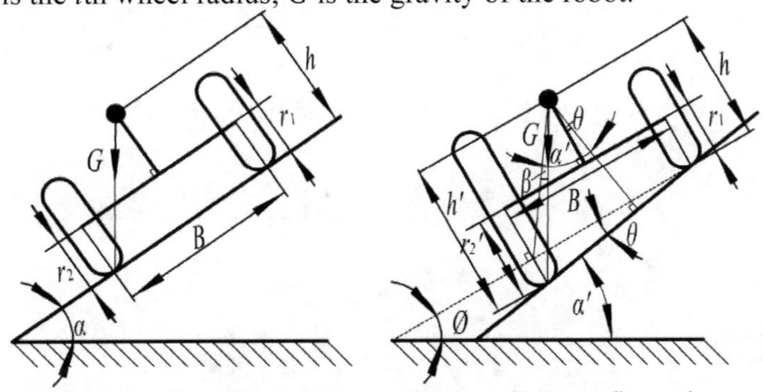

(a) Normal configuration (b) Levelled configuration

Figure 3. Mechanical models of lateral stability.

Thus, the condition that a lateral tip-over does not occur is

$$\alpha\le\tan^{-1}\left(B/2h\right) \tag{2}$$

The mobile robot with variable-diameter wheels can change the downhill wheel diameter to rotate its body, as shown in figure 3 (b). The moment equilibrium equation is established by

$$G\cos\alpha'\left[\left(B/2\right)\cos\theta+h'\sin\theta\right]=G\sin\alpha'\left[h'\cos\theta-\left(B/2\right)\sin\theta\right] \tag{3}$$

where α' is the slope angle; ϕ is the imaginary slope angle; h' denotes the distance between the COG of the robot and the bottom of the downhill wheels; r_2' is the adjusted wheel radius.

With the change in downhill wheel diameter, the maximum slope angle becomes

$$\alpha'=\tan^{-1}\left[\frac{\left(B/2\right)+h'\tan\theta}{h'-\left(B/2\right)\tan\theta}\right] \tag{4}$$

Note that

$$\theta=\tan^{-1}\left[\left(r_2'-r_1\right)/B\right]=\tan^{-1}\left[\left(h'-h\right)/B\right] \tag{5}$$

Thus,

$$\alpha' = \tan^{-1}\left[\frac{B^2 + 2h'(h'-h)}{B(h'+h)}\right] \tag{6}$$

When $r_2' = nr_1$, there is

$$h' - h = (n-1)r_1 \tag{7}$$

Equation (6) becomes

$$\alpha' = \tan^{-1}\left\{\frac{B^2 + 2(n-1)^2 r_1^2 + 2h(n-1)r_1}{B\left[(n-1)r_1 + 2h\right]}\right\} \tag{8}$$

According to figure 3 (b), there is

$$\alpha' - \theta = \tan^{-1}(B/2h) - \beta \tag{9}$$

That is,

$$\alpha' - \tan^{-1}(B/2h) = \theta - \beta \tag{10}$$

If $\theta > \beta$, it is indicated that when the downhill wheel diameter increases, the maximum slope angle that the robot can traverse will enlarge.

$$\beta = \tan^{-1}(B/2h') - \tan^{-1}(B/2h) = \tan^{-1}\left(\frac{(B/2h') - (B/2h)}{1 + B^2/(4h'h)}\right) \tag{11}$$

Obviously,

$$(h'-h)/B - \frac{(B/2h') - (B/2h)}{1 + B^2/(4h'h)} = \frac{(h'-h)\left[4h'h + 3B^2\right]}{B\left(4h'h + B^2\right)} > 0 \tag{12}$$

Hence,

$$\alpha' > \tan^{-1}(B/2h) \tag{13}$$

(a) Lateral stability (b) Longitudinal stability

Figure 4. Simulation analysis of stability.

Therefore, the increase in downhill wheel diameter can improve the lateral stability of the robot. Moreover, the simulation result also illustrates the improvement in figure 4 (a). Based on the simulation, the robot can traverse a slope at a greater limit angle, compared with a robot with normal wheels.

3.1.2. Longitudinal stability. As presented in figure 5, a mobile robot goes up or down a slope of angle β. When the angle β beyond a limit value, the tip-over instability of the robot occurs. The forces acting on the left and right wheels are assumed to be equivalent. The contact point between the downhill wheels and the slope is regarded as the rotation point of tip-over. Hence, a mechanical model of a mobile robot on a slope is proposed in Figure 5.

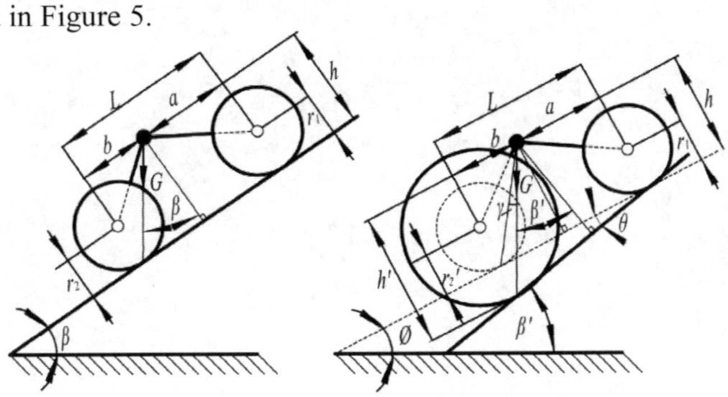

(a) Normal configuration (b) Levelled configuration
Figure 5. Mechanical models of longitudinal stability.

The moment equilibrium equation of the robot with normal wheels is established by

$$N_1 L = Gb\cos\beta - Gh\sin\beta \tag{14}$$

where β is the slope angle; L (1055mm) is the distance between the downhill and uphill wheels along the longitudinal slope surface; the COG of the mobile robot is located at distances b (650mm) and a (340mm) from the downhill and uphill wheels along the slope surface; h (360mm) denotes the distance between the COG of the robot and the slope surface; N_1 is the supporting force acting on the underside of the uphill wheels; r_i is the ith wheel radius.

When N_1=0, the condition that a longitudinal tip-over does not occur is

$$\beta \le \tan^{-1}(b/h) \tag{15}$$

According to equation (15), when the distance b increases, the maximum slope angle will enlarge. Thus, this change can improve the longitudinal stability of the robot.

With the change in wheel downhill diameter, when N_1=0, the moment balance equation of the robot is established by

$$-G\cos\beta'(b\cos\theta + h'\sin\theta) + G\sin\beta'(h'\cos\theta - b\sin\theta) = 0 \tag{16}$$

where β' is the slope angle; ϕ is the imaginary slope angle; h' denotes the distance between the COG of the robot and the bottom of the downhill wheels; r_2' is the adjusted wheel radius.

Therefore,

$$\beta' = \tan^{-1}\left(\frac{b + h'\tan\theta}{h' - b\tan\theta}\right) \tag{17}$$

Note that

$$\theta = \tan^{-1}\left[(r_2' - r_1)/L\right] = \tan^{-1}\left[(h' - h)/L\right] \tag{18}$$

Thus,

$$\beta' = \tan^{-1}\left[\frac{bL + h'(h' - h)}{h'L - b(h' - h)}\right] \tag{19}$$

When $r_2' = nr_1$, there is

$$h' - h = (n-1)r_1 \tag{20}$$

Equation (19) becomes

$$\beta' = \tan^{-1}\left[\frac{bL + (n-1)^2 r_1^2 + h(n-1)r_1}{hL + (n-1)r_1(L-b)}\right] \tag{21}$$

According to Figure 4 (b), there is

$$\beta' - \theta = \tan^{-1}(b/h) - \gamma \tag{22}$$

That is,

$$\beta' - \tan^{-1}(b/h) = \theta - \gamma \tag{23}$$

If $\theta > \gamma$, it is indicated that when the downhill wheel diameter increases, the maximum slope angle that the robot can climb will enlarge.

$$\gamma = \tan^{-1}(b/h') - \tan^{-1}(b/h) = \tan^{-1}\left(\frac{(b/h') - (b/h)}{1 + b^2/(h'h)}\right) \tag{24}$$

Obviously,

$$(h'-h)/L - \frac{(b/h') - (b/h)}{1 + b^2/(h'h)} = \frac{(h'-h)(h'h + b^2 + bL)}{L(h'h + b^2)} > 0 \tag{25}$$

Hence,

$$\beta' > \tan^{-1}(b/h) \tag{26}$$

The adjustment in downhill wheel diameter can enhance the longitudinal stability of the robot. Moreover, the simulation result also illustrates the improvement in figure 4 (b). Based on the simulation, the robot can climb a slope at a greater limit angle without tip-over, compared with a robot with normal wheels.

3.2. Obstacle-surmounting capability

The obstacle-surmounting capability of a mobile robot is elevated by the vertical height of an obstacle or step. Assume that the obstacle-surmounting speed of the robot is low, and its movement remains in a steady state. The forces acting on the left and right wheels are assumed to be equivalent. The quasi-static models that a mobile robot climbs a step are presented here. The first case is that the front wheels of the robot climb a step. Static equilibrium equations in the case are as follows.

$$\varphi F_1 \cos\alpha + F_1 \sin\alpha + F_2 - G = 0 \tag{27}$$

$$-\varphi F_1 \sin\alpha + F_1 \cos\alpha - \varphi F_2 = 0 \tag{28}$$

$$-\varphi F_1 R_1 - F_2\left(\varphi R_1 - L\cos\beta\right) - Ga\cos\beta - Gh_0 \sin\beta = 0 \tag{29}$$

where F_1 is the radial force (from the step) acting on the front wheels; F_2 is the radial force (from the ground) acting on the rear wheels; the static friction coefficient, φ, is 0.7; R_i is the ith wheel radius; G is the total weight of the robot.

By eliminating F_1, G, and F_2, a dimensionless equation is obtained by

$$\left(\frac{\varphi R_1}{L\cos\beta} - 1\right)\sin\alpha - \left(\frac{R_1}{L\cos\beta} - \frac{1}{\varphi} + \frac{a + h_0\tan\beta}{L\varphi} + \frac{\varphi\left(a + h_0\tan\beta\right)}{L}\right)\cos\alpha = \frac{\varphi R_1}{L\cos\beta} \tag{30}$$

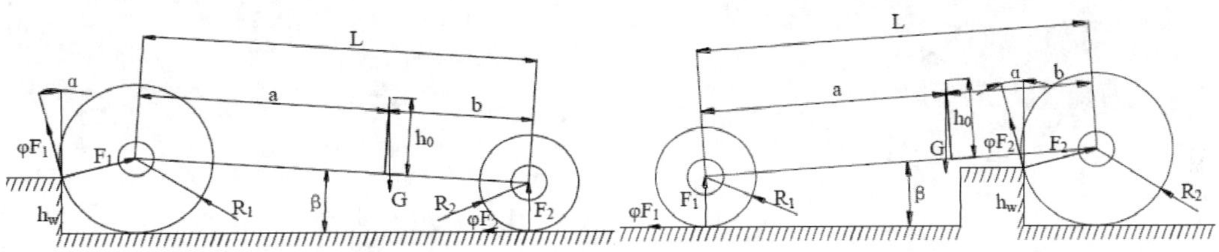

(a) Obstacle-surmounting front wheel (b) Obstacle-surmounting rear wheel

Figure 6. Obstacle-surmounting mechanical models.

The second case is that the rear wheels of the robot pass over a step. Static equilibrium equations in the case are as follows.

$$\varphi F_2 \cos\alpha + F_2 \sin\alpha + F_1 - G = 0 \tag{31}$$

$$-\varphi F_2 \sin\alpha + F_2 \cos\alpha - \varphi F_1 = 0 \tag{32}$$

$$-\varphi F_2 R_2 - F_1\left(\varphi R_2 + L\cos\beta\right) + Gb\cos\beta + Gh_0 \sin\beta = 0 \tag{33}$$

By eliminating F_1, G, and F_2, a dimensionless equation is obtained by

$$\left(\frac{\varphi R_2}{L\cos\beta} + 1\right)\sin\alpha - \left(\frac{R_2}{L\cos\beta} + \frac{1}{\varphi} - \frac{b + h_0\tan\beta}{L\varphi} - \frac{\varphi\left(b + h_0\tan\beta\right)}{L}\right)\cos\alpha = \frac{\varphi R_2}{L\cos\beta} \tag{34}$$

When $R_1 = R_2$, β is zero. The two cases correspond to an obstacle-surmounting mobile robot with normal wheels.

Equations (30) and (34) become

$$\left(\frac{\varphi R_1}{L}-1\right)\sin\alpha-\left(\frac{R_1}{L}-\frac{1}{\varphi}+\frac{a}{L\varphi}+\frac{\varphi a}{L}\right)\cos\alpha=\frac{\varphi R_1}{L} \tag{35}$$

$$\left(\frac{\varphi R_2}{L}+1\right)\sin\alpha-\left(\frac{R_2}{L}+\frac{1}{\varphi}-\frac{b}{L\varphi}-\frac{\varphi b}{L}\right)\cos\alpha=\frac{\varphi R_2}{L} \tag{36}$$

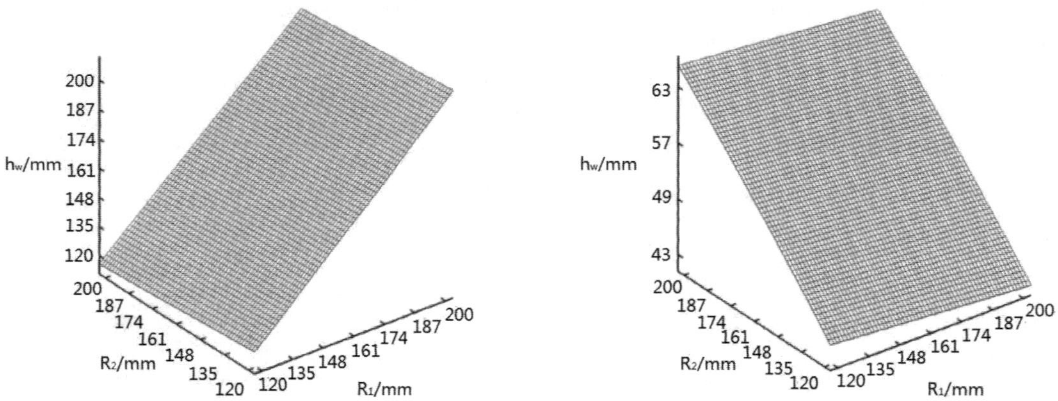

(a) Obstacle-surmounting front wheels (b) Obstacle-surmounting rear wheels

Figure 7. Relationships between step height and obstacle-surmounting wheel diameter

With the increases in wheel radiuses, obstacle height increases, as represented in figure 7. The obstacle-surmounting front wheels can pass over much higher obstacles than that of the obstacle-surmounting rear wheels. Based on dimension parameters in Section 3.1, quasi-static obstacle-surmounting simulation of the robot was performed. The simulation results showed that when the front wheel radius R_1 was 120mm, the step height that the robot can climb was 102 mm; when the front wheel radius R_1 was 200mm, the step height was 186 mm; when the rear wheel radius R_2 was 120mm, the corresponding step height was 43 mm; when the rear wheel radius R_2 was 200mm, the step height was 67 mm. In general, very good agreement is found for the simulation results and the analytical results.

4. Conclusions

Unlike conventional mobile robots, most of the weight of the novel mobile robot is concentrated in the variable-diameter wheels, which reduces the height of COG. This robot with variable-diameter wheels can adjust its body attitude to adapt to rough terrains so that it can cross over obstacles or climb a slope. For the folded limit status, the wheel becomes a circular wheel. This provides good ride performance for the robot on flat terrain. Furthermore, variable-diameter wheels can enhance the trafficability of the robot on soft terrain by increasing the wheel diameters.

Both the analytic models and the simulation results represent that with the increase of the downhill wheel diameters, lateral and longitudinal stabilities of the robot are improved. Failure criterions of stability were introduced. The robot can traverse or climb a slope at a greater limit angle, compared with a robot with normal wheels. Considering variable wheel diameters, the static equilibrium equations describing obstacle-surmounting performances of the robot are proposed here. Diameters of the obstacle-surmounting wheels are linear with the limit obstacle height that the robot can cross over. The obstacle-surmounting front wheels can climb much higher steps than that of the obstacle-surmounting rear wheels. The simulation results show that the predictions made by the analytical model are in good agreement.

Acknowledgments

This work was supported in part by the National Natural Science Foundation of China under Grant No. 51675027.

References

[1] Bruzzone L and Quaglia G 2012 Locomotion systems for ground mobile robots in unstructured environments *Mech. Sci* **3** 49-62
[2] Tarokh M, Ho D H and Bouloubasis A Jan. 2013 A systematic kinematics analysis and balance control of high mobility rovers over rough terrain *Robot. Auton. Syst* **61** 13-24
[3] Luo Z, Shang J, Wei G and Ren L Oct. 2018 A reconfigurable hybrid wheel-track mobile robot based on Watt II six-bar linkage *Mech. Mach. Theory* **128** 16-32
[4] Iagnemma K, Rzepniewski A, Dubowsky S and Schenker P Jan. 2003 Control of robotic vehicles with actively articulated suspensions in rough terrain *Auton. Robot* **14** 5-16
[5] Aoki T, Murayama Y and Hirose S Feb. 2014 Development of a transformable three-wheeled lunar rover: Tri - Star IV *J. Field. Robot* **31** 206-223
[6] Edwin E L, Denhart D J, Gemmer R T, Ferguson M S and Mazzoleni P May. 2014 A performance analysis and technical feasibility assessment of a transforming roving-rolling explorer rover for mars exploration *J. Mech. Design* **136** 1-11
[7] Zheng J, Gao H, Yuan B, Liu Z, Yu H, Ding L and Deng Z Oct. 2018 Design and terramechanics analysis of a Mars rover utilising active suspension *Mech. Mach. Theory* **128** 125-149
[8] Grand C, Benamar F, Plumet F and Bidaud P Oct. 2004 Stability and traction optimization of a reconfigurable wheel-legged robot *Int. J. Robot. Res* **23** 1041-58
[9] Kubota T and Naiki T 2011 Novel mobility system with active suspension for planetary surface exploration in *2011 Aerospace Conf.* (Big Sky Montana American: IEEE) p 1-9
[10] Chen F and Genta G Dec. 2012 Dynamic modeling of wheeled planetary rovers: a model based on the pseudo-coordiates approach *Acta. Astronaut* **81** 288-305
[11] Inotsume H, Sutoh M, Nagaoka K, Nagatani K and Yoshida K Sep. 2013 Modeling, analysis, and control of an actively reconfigurable planetary rover for traversing slopes covered with loose soil *J. Field. Robot* **30** 875-896
[12] Siddiqi A and de Weck O L Nov. 2009 Reconfigurability in planetary surface vehicles *Acta Astronaut.* **64** 589-601
[13] Zeng W, Gao F, Jiang H, Huang C, Liu J and Li H 2018 Design and analysis of a compliant variable-diameter mechanism used in variable-diameter wheels for lunar rover *Mech. Mach. Theory* **125** 240–258

IOP Conf. Series: Materials Science and Engineering **793** (2020) 012022 doi:10.1088/1757-899X/793/1/012022

Analysis of the influence of the vane width on the fluid transmission performance in the vane-type tank

Baotang Zhuang[1,2], Yong Li [1,2 *], Hailin Pan[1] and Wei Rui[3]

[1] Beijing Engineering Research Center of Efficient and Green Aerospace Propulsion Technology, Beijing 100190, China;

[2] Institute of Mechanics, Chinese Academy of Sciences, Beijing 100190, China;

[3] Corresponding author's e-mail: bt_zhuang@126.com

Abstract. Fuel tanks are a core component in satellites that manage the propellant. This study numerically analyzed the fluid transport with parallel guide vanes in a vane type surface tension tank. In this paper, the numerical simulation analysis is carried out for the fluid transmission performance of the parallel vane of the new generation plate surface tension tank, and the influence of vane width on the transmission performance of vane is studied. The results show that, in the whole capillary flow stage, the liquid flow rate increases with the increase of the width a of the guide plate.

1. Introduction

The literature has many studies of propellant transport in microgravity. For example, Li et al. studied liquid sloshing in a vane type tank in microgravity [1]. Lapilli et al. studied the microgravity fluid dynamics in a sphere slosh experiment. Li et al. numerically analyzed the fluid mechanics of capillary flow in microgravity in a fan-shaped asymmetric interior corner [2]. They analyzed the effect of the contact angle on the rise height in the corner and developed a Concus-Finn condition for the capillary flow in the corner. Han and Chen studied the effect of the geometry on merged droplet formation in a double T-junction microchannel using three-dimensional numerical simulations using the level-set method. The fluid motion in a spherical tank in microgravity conditions was investigated by Dalmon et al. through comparisons of data from the FLUIDICS experiment in the ISS with Direct Numerical Simulations of the two-phase flow [3]. Gaulke et al. numerically studied capillary transport of a liquid between parallel perforated plates in microgravity with various geometries to show how the capillary transport capability correlated with the perforation diameter and plate porosity [4]. Bolleddula et al. reviewed capillary rise flows and drop tower investigations and developed an analytical solution for flows along planar interior edges[5]. Kang et al. described a microgravity experiment investigating thermocapillary convection in an open cylindrical annular pool to be done on the SJ-10 satellite [6]. Ground experiment results showed the temperature oscillations, surface oscillations, and flow patterns. Li et al. presented the governing equation for capillary driven flow in a cylindrical interior corner with an approximate analytical solution and showed that capillary driven flow in cylindrical interior corners satisfied the Concus-Finn condition in microgravity [7]. The also presented a relationship for the variation of the liquid front position with time which was then compared with the results of drop tower experiments and numerical simulations. Zhang et al. experimentally investigated the migration and interaction of two axisymmetric drops in a vertical temperature gradient on the ground with a discussion of the influence of the dimensionless initial distance between the drop centers on the drop migration [8].

2. Computational Model

Fig. 1 shows the vane dimensions. The flow channel is based on the dimensionless flow channel ratio $m=\varepsilon/\alpha$. The vane width is α and the clearance between the midpoint of the vane and the tank wall is ε. Anhydrous ethanol was used as the medium with the constant properties listed in Table 1. The fluid was assumed to be an incompressible Newtonian fluid. The fluid was assumed to be laminar with consideration of the effect of viscosity.

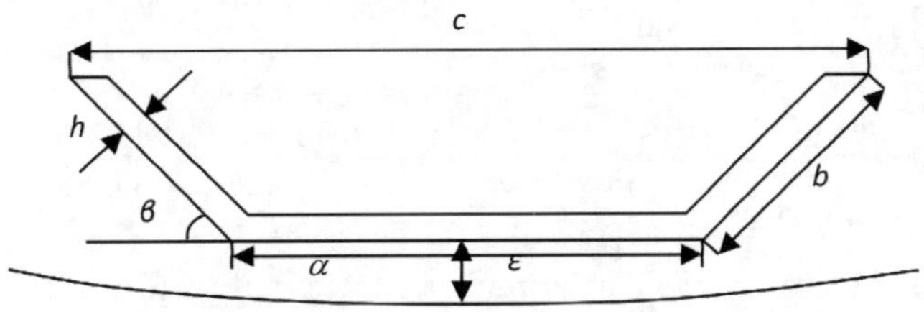

Figure 1. Vane middle cross section.

Table 1. Physical parameters for an anhydrous ethanol tank at 20°C.

Name	Molecular formula	ρ (kg/m³)	σ (N/m)	μ (Pa·s)	Contact angle (PMMA)	Grade
Anhydrous ethanol	C_2H_5OH	789	0.02246	0.01096	0	analytical purity

3. Influence of different vane width on transmission performance

Increasing the number of baffles will undoubtedly increase the liquid transfer capacity of the baffles, but it will also increase the overall weight of the tank. Another way to increase liquid transport is to increase the width of the vane. By simulating the capillary flow in the tank under different vane width, the influence of different vanes width on the transmission efficiency of a single vane is studied. The number of vanes is 4. When discussing the influence of the width a of the vanes on the transmission performance, the parameters m and h are kept constant. Table 2 shows the parameters used in the simulation.

Fig. 2 shows the time-varying relationship between the leading edge position L^2 of the vane and the time interval of 0.1s when the width of the vane is a = 4.8mm, 6.0mm, 7.5mm and 9.0mm. It can be seen from the figure that the leading edge climbing speed increases with the width of the vane *a*. When the width *a* increases from 4.8mm to 6.0mm, the climbing speed increases obviously. When the width *a* increases from 6.0mm to 9.0mm, the speed increase decreases.

Table 2. Simulation parameters with different width *a*

Working condition	1	2	3	4
Vane width *a*/mm	4.8	6.0	7.5	9.0
Clearance ε/mm	1	1	1	1
Angle *β*/°	45	45	45	45
Analog medium	Absolute ethanol			

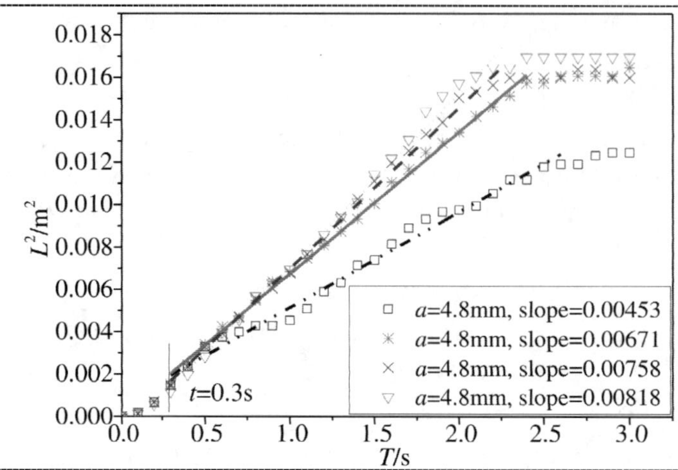

Figure 2. Square of lip location as a function of t with width a.

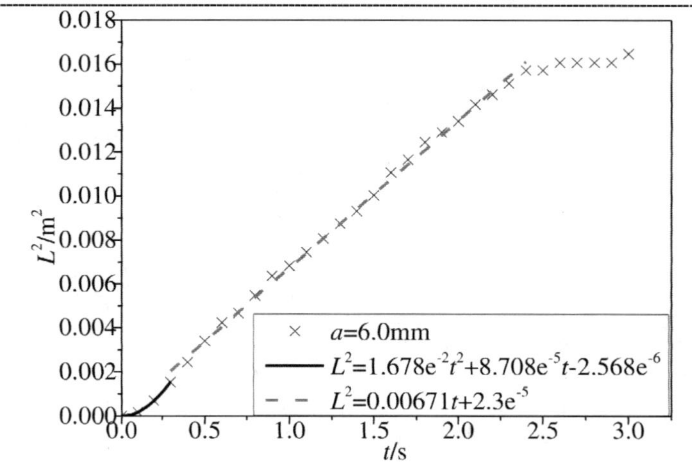

Figure 3. Square of lip location as a function of t, a =6mm

In order to further investigate the movement law of the leading edge L of the vanes in the tank, the relationship between the leading edge L^2 and the time change was made when a = 6mm. Combined with Fig. 2 and Fig. 3, we draw the following conclusions:

There are two main stages in the liquid climbing process of the vanes in the tank. That is the liquid level relocation stage and the stable transfer stage of the vanes after the sudden disappearance of gravity.

When $0 < t < 0.3$s is the first stage. In this stage, the leading edge positions of different vanes width a basically coincide. This is because during the repositioning process, there will also be liquid climbing due to the sudden disappearance of gravity on the tank wall. By fitting the equation, it can be found that when t = 0.1s (minimum time), the primary term and constant term of t can be ignored compared with the secondary term of t. The fitting equation shows that the liquid front L^2 is in direct proportion to time t^2, that is, the front L is in direct proportion to t.

When t > 0.3s, it is the second stage, in which the leading edge capillary is mainly driven by the capillary force between the vanes and the wall, and the position of the leading edge varies with the vane width a. By fitting the curve, it is found that there is a linear relationship between L^2 and t, that is, L and $t^{1/2}$ are in direct proportion.

In the simulation, the last two points of the climbing law of the front edge of the liquid on the vane and the capillary flow law in the cylindrical vessel proposed by stange have been observed all the time.

Because the time interval is not small enough, no flow stage with l-approximation proportional to t^2 has been observed.

4. Conclusion

There are two main stages in the liquid climbing process of the vanes in the tank. That is the liquid level relocation stage and the stable transfer stage of the vanes after the sudden disappearance of gravity.

In the whole capillary flow stage, the liquid flow rate increases with the increase of the width a of the guide plate.

Acknowledgements

This research was financially supported by the National Natural Science Foundation of China (Grant No. 51776017).

References

[1] Liu J.T, Chen X.Q, Huang Y.Y.(2012) The Review of the Interior Corner Flow Research in Microgravity. Procedia Engineering, 31,331 – 336

[2] Li Y.Q, Chao W.H, Liu L.(2017) Numerical Simulation of Capillary Flow in Fan-Shaped Asymmetric Interior Corner Under Microgravity. Microgravity Science and Technology, 29, 65–79

[3] Alexis D, Mathieu L, Sébastien T. (2019) Comparison Between the FLUIDICS Experiment and Direct Numerical Simulations of Fluid Sloshing in Spherical Tanks Under Microgravity Conditions. Microgravity Science and Technology, 31, pp 123–138

[4] Daniel A. B, Chen Y.K, Ben S. (2010) Compound Capillary Flows in Complex Containers: Drop Tower Test Results. Microgravity Science and Technology. 22, pp 475–485

[5] Diana G, Michael E.D. (2015) CFD Simulation of Capillary Transport of Liquid Between Parallel Perforated Plates using Flow3D. Microgravity Science and Technology, 27, 261–271

[6] Kang Q., Duan L., Zhang L. (2016) Thermocapillary Convection Experiment Facility of an open Cylindrical Annuli for SJ-10 Satellite. Microgravity Science and Technology. 28, 123–132

[7] Li Y.Q, Hu M.Z, Liu L.. (2015) Study of Capillary Driven Flow in an Interior Corner of Rounded Wall Under Microgravity. Microgravity Science and Technology. 27, 193–205

[8] Zhang S.T, Duan L., Kang Q. (2017) Experimental Research on Thermocapillary-Buoyancy Migration Interaction of Axisymmetric Two Drops by Using Digital Holographic Interferometry. Microgravity Science and Technology. 30, 183–193

IOP Conf. Series: Materials Science and Engineering **793** (2020) 012023 doi:10.1088/1757-899X/793/1/012023

Numerical analysis of the effect of turbulence on particle agglomeration

Weikuo Zhang[1*], Guanqing Liu, Chongen Zou

[1]China Huadian Research Institute of Science and Technology. Beijing, 100170, China

[*]Corresponding author's e-mail: weikuoz@163.com

Abstract. The turbulent agglomeration characteristics of a mechanical elements were investigated by numerical simulation using the SST k-ω model. The vorticity field, turbulent dissipation rate and pressure drop in the flow field were obtained. The results show the bluff body flow phenomenon occurs when the incoming flow passes through the spoiler element. On the back of the element, a periodic vortex similar to Karman vortex street is formed. After the airflow passes through the turbulence element, the turbulent dissipation rate in the main flow area is higher, and the particle agglomeration effect is significant. In the vortex street period, due to the viscous dissipation, the dissipation rate decreases exponentially with the increase of the propagation distance, and the agglomeration effect decreases. Due to the resistance of the turbulence element, the pressure drop between the inlet and outlet is up to 143pa, resulting in a large pressure loss.

1. Introduction

Particles in the atmosphere have become one of the main hazards affecting the world's environment and human health[1], and the removal of particles has always been the focus and difficulty of pollution control in coal-fired power plants, especially the fine particles have strong penetration and are difficult to capture. In the past, the main particle pollution control technologies include the following methods[2-3]: mechanical, electric, wet, fiber bag and agglomeration dust removal, etc. Among them, agglomeration dedusting refers to promote the agglomeration of fine particles into larger particles, which are then removed by the dust collector. Turbulent agglomeration technology is an effective method for particle removal due to its simple structure and low cost. Turbulent agglomeration technology has attracted more and more attention, and related researches have been carried out.

Turbulent agglomeration exists widely in natural environments and industrial processes, such as the formation of soot in internal combustion engines[4], the growth of nano-grains, and the condensation and growth of raindrops in high clouds. The turbulent agglomeration device has been studied and partly put into practical application. Zhang et al.[5] carried out experimental and numerical simulation research on a structure of Indigo's commercial agglomerator based on Fluent using the RANS + DPM model, and explored the effect of component arrangement and particle concentration on agglomeration efficiency. Liu et al.[6] studied the influence of particle trajectory and particle charge in a agglomeration device similar to the Indigo structure using k-ε combined with DPM model. Kalt[7] pointed out that in shear turbulence with appropriate scale and intensity, significant agglomeration between coarse and fine particles could be achieved, but the agglomeration effect was not obvious when only small particles existed. Liu[8] studied the effect of cylindrical turbulence device on agglomeration effect based on group equilibrium method, and found that although the pressure loss was low, the turbulent dissipation rate was small, resulting in limited agglomeration effect. However, there are still a lot of problems to be

IOP Conf. Series: Materials Science and Engineering **793** (2020) 012023 doi:10.1088/1757-899X/793/1/012023

solved in the research on placing of turbulence generated devices in flow field to improve the effect of turbulent agglomeration and reduce the energy loss.

2. Numerical model

The study adopts the numerical simulation method. The velocity of incoming flue gas is 15m/s and the temperature is 403K. The fluid can be treated as incompressible. The theoretical model is as follows:
Continuous equation

$$\frac{\partial \rho}{\partial t}+\nabla.(\rho\vec{V})=0$$

Momentum equation

$$\frac{D\vec{V}}{Dt}=\vec{f}-\frac{1}{\rho}\nabla P+\nu\nabla^2\vec{V}+\frac{1}{3}\nu\nabla(\nabla.\vec{V})$$

The SST k-ω model combines the advantages of the k-ω model and the k-ε model, including the modified turbulent viscosity formula and effects of turbulent shear stress; SST k-ω model generally can more accurately simulate the separation point and separation zone caused by the adverse pressure gradient. The specific details of this model can be found in the literature [9].

The structure of the spoiler element is shown in Figure 1. The height of the element used in calculation is 0.045m and the headwind angle is 90 degrees. In order to save calculation time, the device is simplified into a two-dimensional structure, and the flow field of flue is approximately treated as two-dimensional. Only one element is selected for numerical simulation. The calculation area is 1.9m*0.19m, in which the width of 0.19m is the height of four elements. In order to prevent the influence of backflow on the flue turbulent flow field, the computational domain length L=1.9m. The periodic boundary condition is used at the upper and lower boundaries, the inlet uses average velocity, and the flow velocity is equal to the wind speed in the actual power plant flue. The equivalent pressure gradient boundary condition is used at outlet. and the wall function is used for the spoiler element wall.

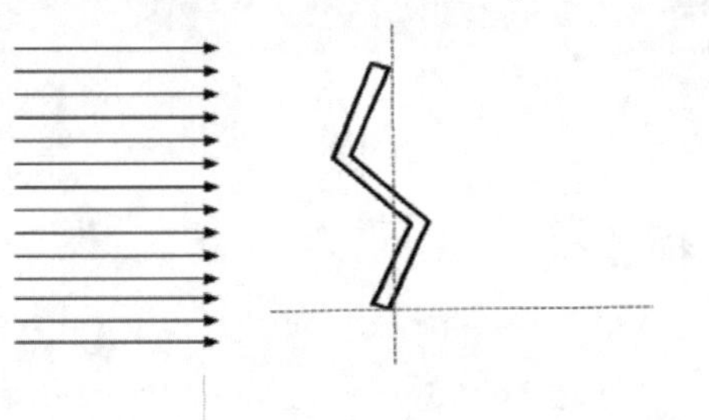

Figure 1. Schematic of the mechanical element

Because of the special shape of the element, the unstructured grid is used in the simulation. In order to accurately capture the turbulence vortices generated by the element shadowing effect, the mesh is partially encrypted near the element, and the sparse grid is used in the area far from the element. The total number of grids is about 330,000 and the time step is $5*10^{-5}$s. The Fluent software is used to calculate the flow field.

3. Results and discussion

We first study the transient process of turbulent flow formed by the element shadowing effect. Figure 2 shows the transient vorticity field of a single spoiler element with 90° headwind angle. It can be seen that the bluff body flow phenomenon occurs when the incoming flow passes through the spoiler element. The incoming flow is affected by the resistance of the bluff body, and the energy is lost during the flow. A part of the kinetic energy must be taken out to increase the pressure, which causes the speed to gradually decrease to zero, that is the separation point. The airflow accumulates here. The pressure drop continues to rise after this point, forcing the stagnant fluid to reverse the flow and squeeze the mainstream away from the object wall, causing the boundary layer flow to separate, and the vorticity also reaches the maximum. A series of vortices are formed at the downstream of the bluff body, and the vortex intensity decreases with the flow propagation. In addition, by comparing the vorticity field in [5] with the same initial conditions, it is found that the overall tendency of the flow field distributions is in good agreement with our results, so the numerical simulation results in this paper are varied.

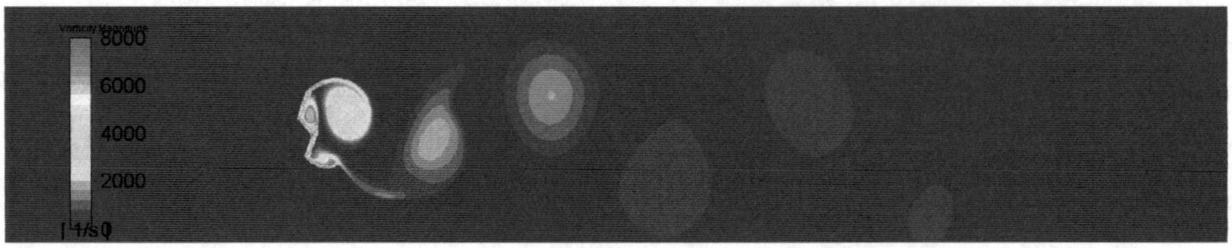

Figure 2. Instantaneous vorticity field

In order to analyze the effect of turbulence on particle agglomeration efficiency, Figure 3 shows the transient turbulent dissipation rate in the flow field. The parameter related to the flow field of the classical collision kernel function[10] for turbulent agglomeration is the turbulent dissipation rate, which plays a major role in controlling the particle collision rate. The turbulent dissipation rate is usually measured by the turbulence energy lost per unit mass of fluid in one second. The turbulent velocity fluctuates randomly in space, thus forming a significant velocity gradient. Influenced by molecular viscous force, turbulence kinetic energy is continuously transformed into kinetic energy of molecular motion through internal friction.

It can be seen from Figure 3 that after the airflow passes through the spoiler element, the region with the highest turbulent dissipation rate is at the back of the element, and then gradually diffuses to the mainstream area. The turbulent dissipation rate of the mainstream area is also relatively higher, and the distribution range of the high turbulent dissipation area is wide, which can effectively increase the particle agglomeration effect.

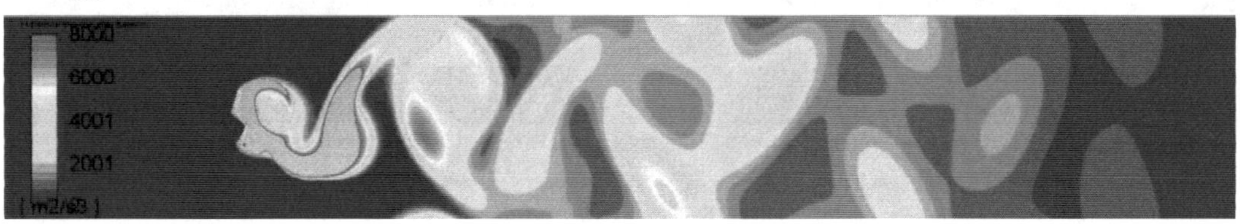

Figure 3. Instantaneous turbulent dissipation rate

It can be seen from Figure 1 that the vortex distribution is similar to Karmen vortex street, which is periodic. By monitoring the lift of the flow field, it is found that the cycle of the vortex street is t = 0.0145s. we studied the mass weighted average turbulent dissipation rate of the flow field. At the beginning, due to the influence of the unsteady inlet flow, the monitoring time is greatly different from the vortex period, so the average dissipation rate fluctuates with the flow time; when the monitoring time is equal to the vortex street period, the average turbulent dissipation rate of the flow field approaches constant.

The average turbulent dissipation rate in the vortex period is relatively rough, so it is necessary to reasonably divide the entire area into a series of small regions to analyze the accurate distribution of the average turbulent dissipation rate. It can be seen from Figure 4, when the distance to the spoiler is very short, due to the strong vortex produced by boundary layer separation, the turbulent velocity is fast, lead to large dissipation rates. With the increase of propagation, because of viscous dissipation, the turbulent vortex velocity is greatly reduced, resulting in an exponential decrease of the dissipation rate.

IOP Conf. Series: Materials Science and Engineering **793** (2020) 012023 doi:10.1088/1757-899X/793/1/012023

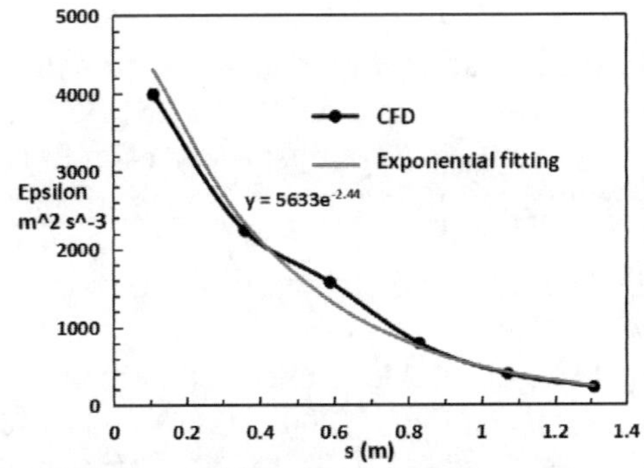

Figure 4. The distribution of the mass weighted average turbulent dissipation rate

The optimal design guideline for the structure and position of the spoiler element is to enhance the agglomeration effect of the main flow area and reduce the pressure drop. So we need to consider not only the effect of turbulent agglomeration, but also the effect of spoiler elements on pressure drop.

Figure 5 shows the spatial distribution of the pressure. It can be seen that due to the obstruction of the turbulent element, the airflow accumulates in front of the element, causing the highest pressure here, and the pressure in the rear region of the element is negative, thus forming a large pressure differential. As a result, the airflow is reversed at the back of the element, forming a turbulent vortex, and spreading downstream. Therefore, obvious vortex shedding can be observed in the following process. In addition, the pressure difference between the inlet and outlet is 143Pa, so the pressure drop caused by the spoiler is relatively high and energy loss is large.

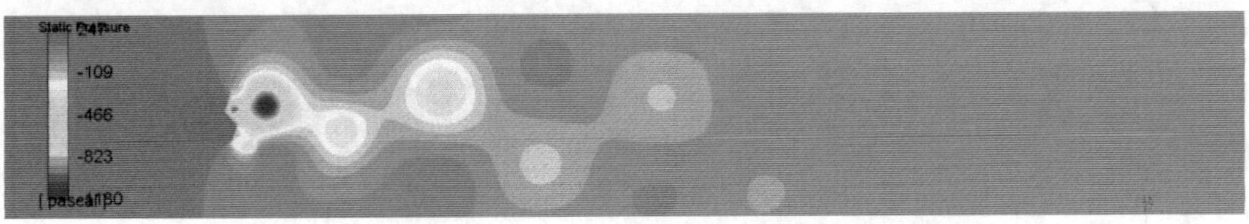

Figure 5. The contour of the static pressure

4. Conclusions

In this paper, the SST k-ω model is used to solve the incompressible and transient turbulent flow field. The influence of the spoiler elements on the turbulent agglomeration is numerically studied. The conclusions are as follows:

(1) The airflow passes through the spoiler with a velocity of 15m/s and the phenomenon of bluff body recirculation flow occurs. A series of vortices are formed at the back of the element. It is found that these vortices are similar to the periodic Karman vortex street, and the periodic time $t = 0.0145$s.

(2) After the airflow passes through the element, the turbulent dissipation rate in the mainstream region is higher. The distribution range of high turbulent dissipation area is relatively wider, which can effectively increase the particle agglomeration. In the vortex street period, the mass weighted average turbulent dissipation rate of the whole flue flow field approaches a constant. Because of viscous dissipation, the dissipation rate decreases exponentially with the increase of the propagation, and the agglomeration effect decreases.

(3) The pressure drop between inlet and outlet is up to 143Pa, and the pressure loss is relatively high.

Acknowledgments
This work has been supported by the Chinese National Key Research and Development Project under Grant No.2017YFB0603203.

IOP Conf. Series: Materials Science and Engineering **793** (2020) 012023 doi:10.1088/1757-899X/793/1/012023

References

[1] Wu, G.P., Hu, W., Teng, E.J., et al. Analysis of the effect of air pollution on the adult's respiratory health[J]. Environmental Monitoring in China, 2001, 17(S): 33-38.

[2] Yi, H.H., Hao, J.M., Duan, L., et al. Influence of dust catchers on PM10 emission characteristics of power plants[J]. Chinese Journal of Environmental Science, 2006, 27(10): 1921-1927.

[3] Xiong, G.L., Li, S.Q., Chen, S., et al. Development of advanced electrostatic precipitation technologies for reducing PM2.5 emissions from coal-fired power plants[J]. Proceedings of the CSEE, 2015, 35(9): 2217-2223.

[4] Wen, Z., Yun, S., Thomson, M.J., et al. Modeling soot formation in turbulent kerosene/air jet diffusion flames[J]. Combustion and Flame, 2003, 135(3): 323-340.

[5] Zhang, P.F., Mi, J.C., Pan, Z.M., et al. Influences of elemental arrangement and particle concentration on fine particle amalgamation[J]. Proceedings of the CSEE, 2016, 36(6): 1625-1632.

[6] Liu, Z., Liu, H.X., Feng, X.X., et al. Flow field and particle trajectory numerical simulation of turbulent coalescer [J]. Journal of Chinese Electrical Engineering Science, 2012, 32(14): 71-75.

[7] Kalt, P., Nathan, G., Kelso, R., et al. Assessing the Significance of the Indigo Aerodynamic Agglomeration Technology using Mie/Lif Laser Diagnostics[C]. Proceedings of the Tenth International Conference on Electrostatic Precipitation, Australia, 2006, Paper No. 9B1.

[8] Liu, G.Q., Li, S.Q., Zhang, W.K., et al. Group equilibrium simulation of turbulent agglomeration of fine particles by partition method [J]. in press.

[9] Wu, J., Gu, Z.Q., Zhong, Z.H., The application of turbulence model in vehicle flow simulation [J] . Automotive engineering, 2003, 25(4): 326- 327.

[10] Saffman, P.G., Turner, J.S., On the collision of drops in turbulent clouds[J]. Journal of Fluid Mechanics, 1956, 1(1): 16-30.

IOP Conf. Series: Materials Science and Engineering **793** (2020) 012024 doi:10.1088/1757-899X/793/1/012024

Studies on Overheating Fault of Disconnecting Switch

LAN Zhijun*, WANG Yaping, WU Jianling , FAN Yongqiang

Inner Mongolia EHV Power Supply Bureau, Hohhot, Inner Mongolia, 010080, China

*Corresponding author: HUO Feng (1972-), male, senior engineer, master candidate. Main direction: safety operation and maintenance management of power transmission and transformation equipment. E-mail: huofeng@impc.com.cn.

Abstract: This paper mainly presents typical solutions to disconnecting switch overheating and mechanical problems of power grid in Extra High Voltage (EHV). In recent years, many faults have happened in the EHV power grid due to overheating and mechanical problems of disconnecting switches in Inner Mongolia. Dust storm weather and the special structure are main reasons which induced disconnecting switch lower operation reliability. For the overheating problems, the solutions include using conducting metal belt, replacing material of knee-arm wheel and modifying the structure of rolling-contact assembly; for the mechanical problems, the solutions include using dust shield, renovating operating mechanism and decreasing mechanically-driven part. Meanwhile suggestions are proposed as well for the design and application of isolator in the power networks of Western Inner Mongolia and the Northwester China which have similarly peculiar natural environments.

1. Introduction

In recent years, the lower operation reliability for disconnecting switch in EHV power grid in Inner Mongolia is due to dust storm. For the impact of different climate and environment on the normal operation of electrical equipment, there has been a large number of studies and coherent results, such as climates of alpine, high altitude, snow and so on[1-5]. However, the relatively rare concerning is the effect of dust storm environment on equipment reliability. In this paper, based on the basic structure and working principle of electric rolling-contact-type disconnecting switch, we studied the effect of dust storm environment on the reliability of disconnecting switches. Also, some practical solutions were proposed to improve the operation reliability of the disconnecting switch, which is of great significance for power grid construction, operation and maintenance.

2. Overheating fault

There had been totally 25 times of overheating fault disconnecting switches in the past three years at the 500kV-substation of West Inner Mongolia Power Grid. 10 times of those were happened in Wuhai-Substation, where the dust storm is one of the most serious areas. More than 70% of the overheating disconnecting switch is electric rolling-contact-type. The knee-arm position is the mainly overheated parts, such as rolling contact, gear box, knee-arm wheel and composite sleeve (junction position of rod and even plate). Overheating picture is shown as ‘figure 1’ and ‘figure 2’.

IOP Conf. Series: Materials Science and Engineering **793** (2020) 012024 doi:10.1088/1757-899X/793/1/012024

Figure 1. Overheating of knee-arm wheel.

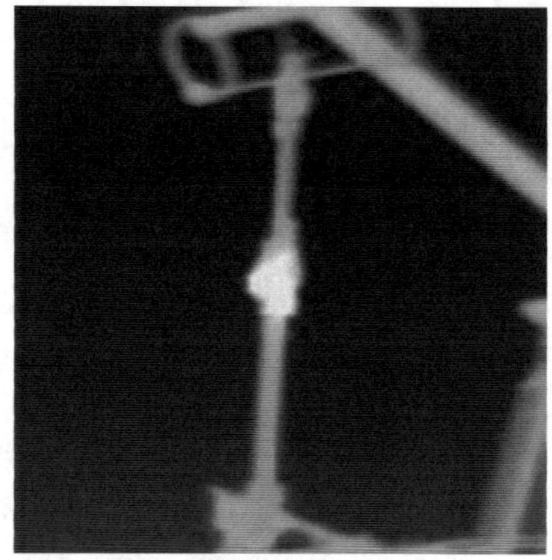

Figure 2. Overheating of rolling contact assembly.

Because of the sand environment, the mechanical jam of high-voltage disconnecting switch was happened. For example, un-flexible operation, increased or decreased operation torque, close and open improperly, or out of place seriously, insulator breakdown by additional stress, etc. Those accidents are seriously threat the safe operation of power grids, and even cause the possibility of large-scale grid blackouts.

3. Fault cause analysis

3.1. Analysis of overheating failure

Overheating faults are the main factors that caused the lower reliability of disconnecting switches in dust storm climate. The overheating parts of the rolling-contact disconnecting switch is mainly on the knee-arm position, such as rolling contact, gear box, the knee-arm wheel and composite sleeve (rod and even the plate junction). For the structure and working principle of disconnecting switch, it is indicated in 'figure 3' and 'figure 4'. Overheating spots of the disconnecting switch are not only located on the normally conductive channel, such as gear box and rolling contact, but also located on the non-conductive channel, such as disconnecting switch knee-arm wheel and composite sleeve, overheating of knee-arm wheel was shown as 'figure 5(a)'. For the overheating rolling contact of the disconnecting switch, we found that some rollers are burned severely and the dust deposition is great, as shown in 'figure 5(b)'. Thus, the severe dust storm climate is the main reason for heat disconnecting switch.

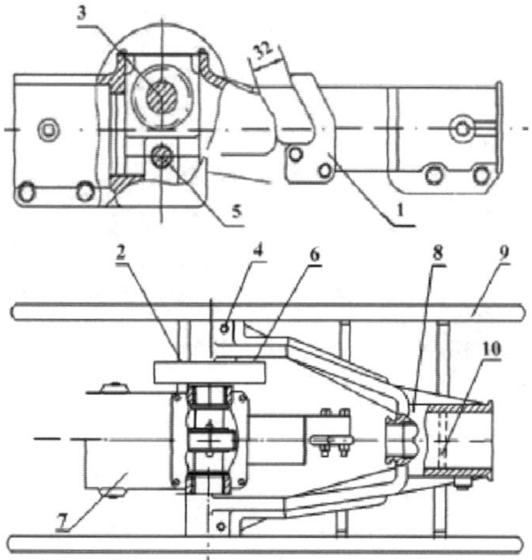

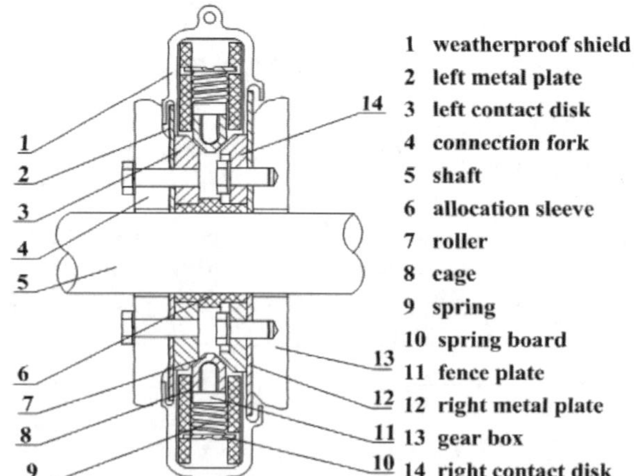

1 ice break hook; 2 rolling contact; 3 rotated shaft;
4 elastane cylindrical; 5 fulcrum; 6 gear; 7 gear box
8 connection fork; 9 screw rod 10 screw

Figure 3. Knuckle assembly.

Figure 4. Electric rolling-contact-type assembly.

(a) Roller overheating and burning (b) Overheating and burning of crank arm roller

Figure 5. Burning position of overheating disconnecting switch.

3.2. Analysis of mechanical fault

The mechanical jam is another important factor that caused the lower reliability of disconnecting switch in dust storm climate. A majority of fault disconnecting switches were CJ7 type operating mechanism. Its mechanical principle is as follows: the motor start-up firstly, and output shaft is obtained by operating torque through the deceleration of gears, screw and screw nut. Next, the double four-bar mounted on the top of institutions boxes drive the rotating porcelain of middle Phase, then both sides of the adjustment rod are driven by a pair of bevel gears installed on wiring bed, and the lower conductive rod starts to move. The upper conducting tube rotates 180° with respect to the lower conducting tube through the joint between the gear and rack. When near closing and starting gate time, the small roller at the connection fork starts to move on the incline plate of the gear box, and the reset spring and clamp spring and other components of the upper conductive rod start to moving. The dynamic touch-piece is closed or separate so that the main switch is controlled to close or separate.

The operation structure above contains several mechanical steps. However, sandstorms and poor sealed condition of operating mechanism are lead to entering of tiny grains of sand. When the sand caulk is in mechanical transmission parts such as gear, rack, etc., mechanical output efficiency will be affected, which lead to mechanical jam, division improperly and the other failures.

4. Treatment methods

4.1. Overheating fault treatment

Faults caused by overheating which decrease the reliability, the reason of which is due to that the structural design of disconnecting switch cannot adapt to special sand climate and environment. At the basis of those reasons and the structure principle of disconnecting switch, some transform programs are put forward as follows:

Aimed at the overheating causes of disconnecting switch, the transform method included： replacing the roller material of rolling contact assembly, improving roller coating process, increasing the through-flow capacity of roller. Meanwhile, adding enough dust cover to prevent dust entering the rolling contact of the disconnecting switch. The characteristics of those transform methods not only is able to reserve the merits that the upper and lower disconnecting switch conductive tube rotate convenient, but also has the simple and nice-looking structure and convenient maintenance procedure. Moreover, the overheating problem of disconnecting switch is controlled to the certain extent. However, the method involves replacing the roller material of rolling contact assembly and improving roller coating process, which required that manufacturers of disconnecting switch have to upgrade the current products and completely replace rolling contact assembly upgraded. The structure transformation cycle will be extended and more difficult for those having been operated. Thus, the method is suitable to transform those disconnecting switch still not in practice. The other method is to install the soft conductive connection between the upper and lower conductive tube junction. The current can be conducted mainly by soft conductive connection in lieu of the knee-arm.

The rolling contact of disconnecting switch improved has function of rotation and supplementary diversion. As shown in 'figure 6', the soft conductive connection with one end fixed on the reliable cross connected to the upper conductive tube and the other end fixed on the gear box sleeve connected reliable to the lower conductive tube. The application of soft conductive connection is very good solution to the overheating incidents of rolling contact assembly. Also, the soft conductive connection is so easy to produce and install that it is able to accelerate the process of transformation. However, the frequent close and open of disconnecting switch may lead to the soft conductive connection cutting off. Regular observations are necessary. Therefore, it is not easy to maintain. For the disconnecting switch having been putting into operation, it is suitable for applying the above transformation programs.

Figure 6. Installed position of conducting metal belt.

In order to completely solve the structure switch overheating fault occur, the disconnecting switch joints of roller can be replaced at the same time. Insulating resin wheel is instead of the original stainless steel wheel that insulates the part, and avoids the breakdown of composite bushing, and form the new diversion channel, and cause the occurrence of overheating fault.

4.2. Mechanical Fault Treatment

The dust storm environment cause a large number of mechanical faults was mainly due to inadequate consideration on structural design of disconnecting switch that under the special operating environment. There are more mechanical parts that can easily lead to overheating. On the basis of structural principle of the disconnecting switch, we can replace CJ7-type operating mechanism to CJ7A-type. the double four-bar drive of CJ7 have many mechanical parts and many possible point of failure. Thus, the efficiency and output is affected by dust storm climate. On the contrast, The CJ7A give up the double four-bar, turbines are fit at the output shaft directly. The output shaft is connected with vertical rod device through the transfer angle. When it is close or open, robs drive the turbine, the vertical output shaft rotate, the sub-vertical move. In this structure, the middle linked parts decreases, and output torque increases, the volume is also relatively small. Therefore, the operation is easy and flexible. Moreover, the disc spring is installed at both ends of retard box is improving, which controlled the impact to the minimum. Meanwhile, sealing performance required fully consideration against sandy environment in order to increase the anti-wind ability of mechanical body.

5. Conclusion

Using the aforementioned programs, all disconnecting switches of Mengxi 500kV power grid are gradually transformed. After transformation, the equipment temperature has maintained the normal level, and machinery jam is controlled fundamentally. These have greatly improved the operation reliability of disconnecting switch. The Electrical equipment manufacturers should take care of the geographical characteristics in order to make the design more applicable to the specific local environmental characteristics. As a result, it will not only significantly reduce the cost of manufacturers' inputs after-sales service, but also is paramount to the safe and reliable operation of the power grid.

Reference

[1] Semedo, S.M.V., Oliveira, J.E.G., Cardoso, F.J.A. (2014) Research on BLDCM actuator intelligent control strategy of the HV disconnecting switch. C. Industrial Electronics (ISIE), 2014 IEEE 23rd International Symposium.
[2] Yuanyi, D. (2014) Design of Temperature Measurement System for High Voltage Disconnecting Switch. C. Intelligent Human-Machine Systems and Cybernetics (IHMSC).
[3] Graf, A., Weihermüller, L., Huisman, J.A. (2008) Measurement depth effects on the apparent temperature sensitivity of soil respiration in field studies. J. Biogeosciences, 5(4):1175.
[4] Verscharen, D., Marsch, E., (2011) Apparent temperature anisotropies due to wave activity in the solar wind. J. Annales Geophysicae, 29(5):909
[5] Ryan, M., Craig, C., Olga, D. (2019) Why the disconnect Problems with a clamp style catheter connector. J. Regional Anesthesia & Pain Medicine, 44(2):281-282.

Study on Prediction of Binder Distribution in the Drying Process of the Coated Web of Positive Electrode for Lithium ion Battery

Yongjian Su[1], Kai Zhou[1], Yingchun Yuan[3], Weiwei Liu[1] and Yelin Deng[2*]

[1]School of Mechanical Engineering, Dalian University of Technology, Dalian, Liaoning Province, 116024, China

[2]School of Rail Transit, Soochow University, Suzhou, Jiangsu Province, 215006, China

[3]Case Western Reserve University 10900 Euclid Ave Cleveland, OH 44106 USA

[*]Corresponding author's e-mail: yelin.deng@suda.edu.cn

Abstract. The slow and high energy consumption of drying process of the coated web of positive electrode for automotive lithium ion battery have become the bottleneck in the manufacturing process of cathode coating. However, high drying rate of solvent (commonly N-Methyl-2-pyrrolidone (NMP)) result in higher concentration of binder (commonly polyvinylidene fluoride (PVDF)) at the evaporating surface than at the bottom. The uneven distribution of binder between coating layers will lead to the decrease of adhesion between active layer and substrate, and even an increase in internal resistance and the associated deteriorated rate capability, thus seriously damaging the mechanical and electrochemical properties of the electrode. In order to improve the drying efficiency of the coated web, and regulate the concentration distribution of binder, this study mainly calculate the evaporation rate under different drying parameters. and determine the binder distribution at the different evaporation rates.

1. Introduction

In the latest research, Deng has analyzed in detail the manufacturing process of the mainstream Lithium Nickel Manganese Cobalt (NCM) cathode and graphite anode of battery pack, and the results show that cathode coating manufacturing is the most important energy and time consuming link in the manufacturing process of battery cells [1, 2].

The high time consumption of cathode coating comes from its slow drying process [1, 2]. However, simply increasing the drying temperature/wind speed to accelerate the drying speed will cause the concentration on the evaporation surface to be higher than that on the bottom [3]. On the one hand, the uneven distribution of binder between coating layers will lead to the decrease of adhesion between the active layer and the substrate, affecting its mechanical properties and then damage battery life. Experiments have shown that less distribution of adhesives at the bottom causes decreased adhesion strength of coating foil [4, 5], and the corresponding battery life will be affected. On the other hand, it will cause the increase of internal resistance and capacity attenuation. During fast charging, uneven PVDF interlayer concentration distribution can cause electrode capacity loss up to 40% [6].

How to improve the coating production speed and ensure the uniform binder distribution between layers is one of the most important issues in the positive coating manufacturing [7].

IOP Conf. Series: Materials Science and Engineering **793** (2020) 012025 doi:10.1088/1757-899X/793/1/012025

2. NMP evaporation rate

NMP evaporation in porous media can be divided into two stages: the first stage is a constant-rate evaporation stage, and the second one is a reduced-rate evaporation stage [8, 9]. Since the first stage accounts for most of the evaporation process, this study focuses on the first stage of evaporation.

First, we can figure out the coating temperature (T_s) and the surface mainstream flow rate (v_∞) by:

$$q = h(T_{in} - T_s) \tag{1}$$

$$v_\infty = v_{in} \cdot k \tag{2}$$

where q is the heat flux, h is the convective heat transfer coefficient, T_{in} is the inlet temperature. v_{in} is the wind speed at the inlet of the drying oven, k is the ventilation coefficient.

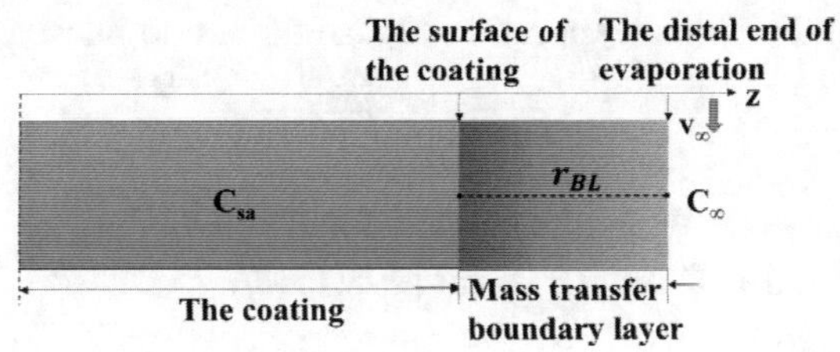

Figure 1. Mass transfer resistance during NMP evaporation.

As shown in figure 1, the evaporation of NMP can be considered as driven by the difference between the saturated vapor concentration (c_{sa}) at T_s and the distal vapor concentration (c_∞).

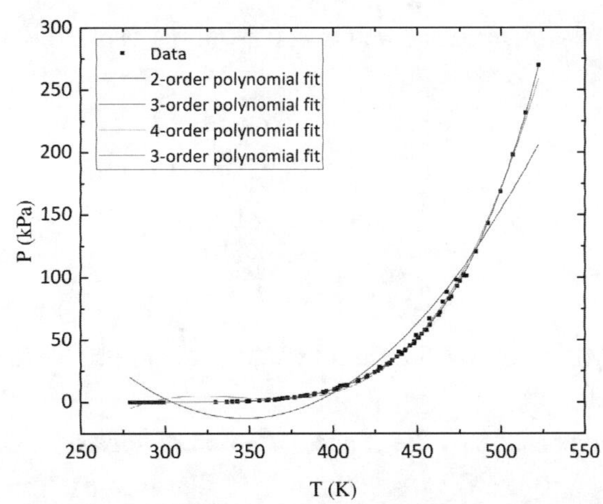

Figure 2. Equilibrium vapor pressure of NMP at different temperatures.

Evaporation rates of NMP during evaporation (e) can be calculated according to:

$$e = \frac{c_{sa}(T_s) - c_\infty}{\Upsilon_{BL}} \tag{3}$$

$$c_{sa}(T_S) = \frac{P(T_S)}{RT_S} \tag{4}$$

$$P(T_S) = 2.67159\times10^{-7}T_S^{\ 4} - 3.7468\times10^{-4}T_S^{\ 3} + 0.1989T_S^{\ 2} - 47.29151T_S + 4243.28404 \tag{5}$$

$$\Upsilon_{BL} = \frac{\delta_m}{D_{NMP}} \tag{6}$$

$$\delta_m = a \cdot v_\infty^{-1/2} \tag{7}$$

$$D_{NMP} = \frac{D_{298}}{(\frac{298}{T_S})^n \cdot (\frac{P}{101.325})} \tag{8}$$

for c_∞, must be less than the explosive limit concentration of NMP, which is very low, about 1.1%. So $c_\infty \ll c_{sa}(T_s)$, c_∞ was considered as 0. The equilibrium vapor pressure of NMP (P) at different temperatures is fitted by MATLAB [10, 11], as shown in figure 2. δ_m is the thickness of the mass transfer boundary layer, and D_{NMP} is the diffusion coefficient of NMP in air. a =2.26 corresponds to turbulent flow [12]. For equation (8), we take n as 2, P=101.325 kPa, D_{298}=1.00×10^{-5} m^2/s [13].

The relation between e and γ (the evaporation flux of the liquid phase during evaporation) is:

$$\gamma = \frac{e \cdot M_{NMP}}{\rho_{NMP}} \tag{9}$$

where M_{NMP}=99.133 g/mol (PVDF molecular weight), ρ_{NMP}=0.9259×10^3 kg/m^3 (half of the sum of 0.824×10^3 kg/m^3(398K) [13] and 1.0279×10^3 kg/m^3 (298.15K)).

From equations (3), (4), (5), (6), (7), (8), (9), the relationship between γ and T_S , v_∞ can be obtained:

$$\gamma = 6.4162\times10^{-10} v_\infty^{1/2}(2.67159\times10^{-7}T_S^5 - 3.7468\times10^{-4}T_S^4 + 0.1989T_S^3 - 47.29151T_S^2 + 4243.28404T_S) \tag{10}$$

3. Distribution of PVDF binder in coating during drying

3.1. Binder transport model

As shown in figure 3, the film can be considered to consist only of solid-liquid two phases.

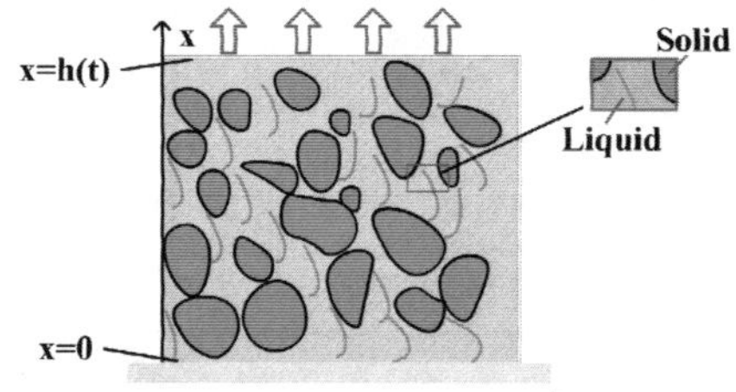

Figure 3. Migration mechanism of PVDF binder during evaporation.

The mass transfer equation describing the concentration of dissolved binder in the solvent (c) is [3]

$$\frac{\partial(\varepsilon_l c)}{\partial t} + \frac{\partial(u_l c)}{\partial x} = D_{eff}\frac{\partial^2 c}{\partial x^2} \tag{11}$$

where ε_l is the liquid volume fraction, u_l is the average flux velocity of liquid phase, the effective diffusion coefficient $D_{eff} = D \cdot \varepsilon_l^{3/2}$, D is the diffusivity of PVDF in the solvent.

Suitable boundary conditions on (11) is

$$\begin{aligned} x = 0, \quad & u_l c - D_{eff}\frac{\partial c}{\partial x} = 0 \\ x = h(t), \quad & u_l c - D_{eff}\frac{\partial c}{\partial x} = \varepsilon_l c\frac{dh(t)}{dt} \end{aligned} \tag{12}$$

where $h(t)$ is the initial thickness of the electrode film, as shown in figure 3.

And a suitable initial condition to close (11) is

$$c(x,0)=c_0 \tag{13}$$

In addition, some parameters need to be set as Table 1, where ε_s^0 is the initial solid volume fraction, h_0 is the initial thickness of the film.

Table 1. Parameters of binder mass transfer process.

Parameters	Units	Values
ε_s^0		0.2792
D	$m^2 \cdot s^{-1}$	1.14×10^{-10}
$\mathbf{h_0}$	μm	114
γ	$m \cdot s^{-1}$	1.16×10^{-6}

3.2. Results and discussion

The software FlexPDE7.07 was used to solve the numerical equation of binder in section 2, and the total dimensionless time $\gamma t_{total}/h_0$=0.3, γ = 1.16×10^{-6} m/s. Finally, the concentration distribution of PVDF between $x/h(t)$ =0 to 1 (that is, x is between 0 and $h(t)$) at $\gamma t_{total}/h_0$=0.3 is obtained, as shown in figure 4. At the same time, we also track the PVDF concentration changes in different sections ($x/h(t)$ =0.1, 0.2, 0.3, 0.4, 0.5, 0.6, 0.7, 0.8, 0.9, 1) from $\gamma t/h_0$=0 to 0.3, as shown in figure 5.

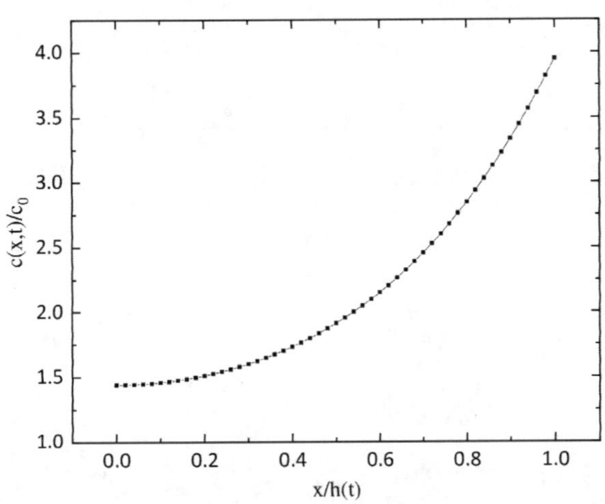

Figure 4. Distribution of PVDF at the end of the drying process ($\gamma t_{total}/h_0$=0.3). (γ=1.16×10^{-6} m/s)

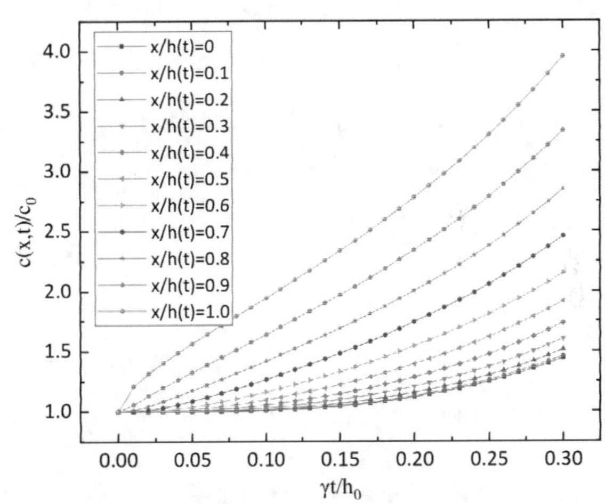

Figure 5. Evolution of concentration of PVDF at different height sections. (γ=1.16×10^{-6} m/s)

Affected by evaporation, diffusion, sedimentation and other processes, binder distribution is usually dense at the top and sparse at the bottom [14]. Therefore, the ratio of binder concentration at the top ($x/h(t)=1$) and the bottom ($x/h(t)=0$) of the coating, $\eta=\dfrac{c(x=h(t))}{c(x=0)}$, can be used to roughly characterize the uniformity of the binder distribution. As shown in figure 6, the binder distribution at the end of the drying process ($\gamma t_{total}/h_0$=0.3) under different γ was studied. As γ increases gradually, η also increases gradually, that is, the binder distribution becomes more and more uneven. It can be found that when γ is between 1.0×10^{-7} m/s and 1.0×10^{-6} m/s, η is relatively close to 1, indicating that

PVDF can obtain a more uniform distribution. However, it increases sharply with the increase of γ when $\gamma > 1.0\times10^{-6}$ m/s. Figure 7 and figure 8 respectively show the evolution of concentration of PVDF at different height sections as drying progressed when $\gamma = 1.0\times10^{-7}$ m/s and $\gamma = 5.0\times10^{-6}$ m/s.

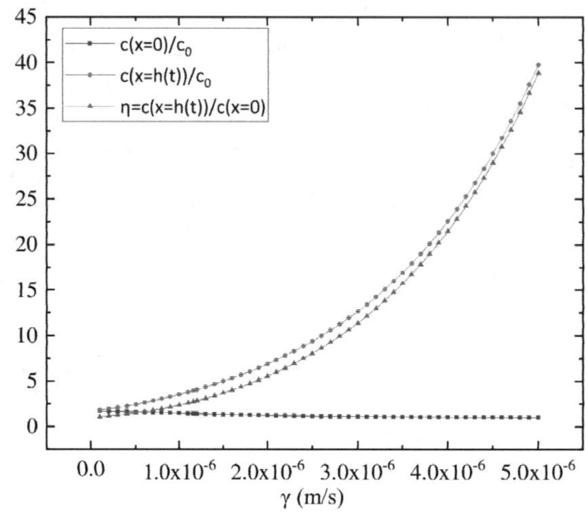

Figure 6. Distribution of PVDF at the end of the drying process ($\gamma t_{total}/h_0$=0.3) under different γ.

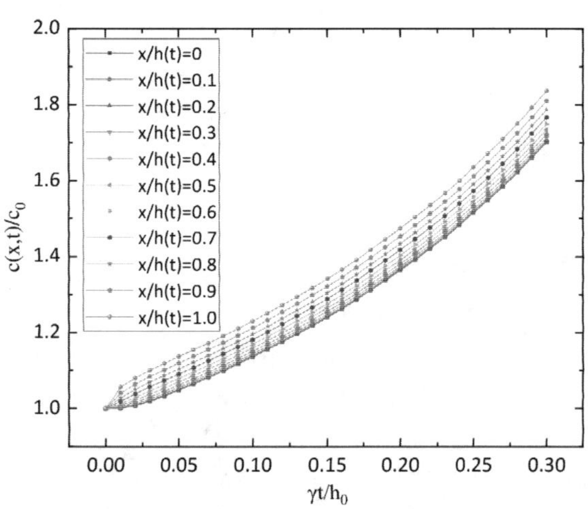

Figure 7. Evolution of concentration of PVDF at minimum evaporation rate. ($\gamma=1.0\times10^{-7}$ m/s)

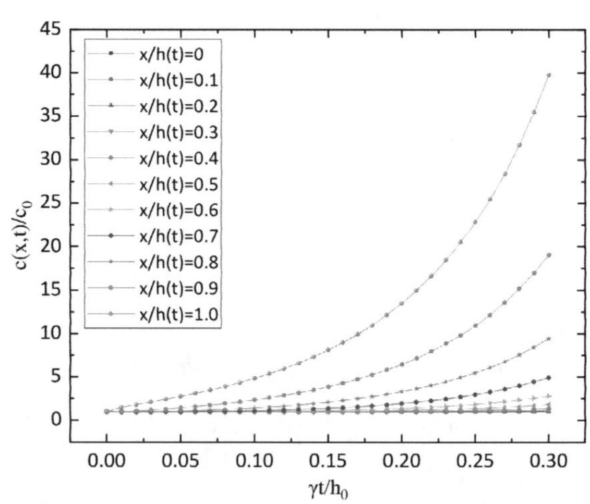

Figure 8. Evolution of concentration of PVDF at maximum evaporation rate. ($\gamma=5.0\times10^{-6}$ m/s)

4. Conclusions

First, in this study, the drying rate solving model during evaporation was given. Then, we proposed a model to calculate the interlayer concentration of PVDF. FlexPDE was used to simulate the PVDF interlayer concentration distribution at different drying rates and the concentration distribution at different heights at low ($\gamma=1.0\times10^{-7}$ m/s) and high drying rate ($\gamma=5.0\times10^{-6}$ m/s) respectively, as well as the binder concentration at the top ($x = h(t)$) and bottom ($x = 0$) of the coating and the ratio of them (η) at different drying rates. As can be seen from the figure, as the drying process proceeds, the binder concentration at the top and bottom of the coating will increase, and the concentration at the top of the coating increases much faster than at the bottom, which will lead to a significant concentration gradient of PVDF between the top and bottom of the coating, and this concentration gradient will increase sharply with the increase of drying rate.

Acknowledgments
We would like to acknowledge the supports from the National Science Foundation of China (Grant No.51905361), Fundamental Research Funds for the Central Universities (No. DUT18RC(3)008), Talent Project of Revitalizing Liaoning (No. XLYC1802106) and Dalian Innovation and Entrepreneurship Project for High-level Talents (No. 2018RD04) on this project.

References

[1] Deng, Y., Li, J., Li, T., Gao, X, Yuan, C. (2017) Life cycle assessment of lithium sulfur battery for electric vehicles. JOURNAL OF POWER SOURCES, 343: 284-295.
[2] Yuan, C., Deng, Y., Li, T, Yang, F. (2017) Manufacturing energy analysis of lithium ion battery pack for electric vehicles. CIRP ANNALS-MANUFACTURING TECHNOLOGY, 66: 53-56.
[3] Font, F., Protas, B., Richardson, G, Foster, J M. (2018) Binder migration during drying of lithium-ion battery electrodes: Modelling and comparison to experiment. JOURNAL OF POWER SOURCES, 393: 177-185.
[4] Baunach, M., Jaiser, S., Schmelzle, S., Nirschl, H., Scharfer, P, Schabel, W. (2016) Delamination behavior of lithium-ion battery anodes: Influence of drying temperature during electrode processing. DRYING TECHNOLOGY, 34: 462-473.
[5] Kim, K., Byun, S., Choi, J., Hong, S., Ryou, M, Lee, Y M. (2018) Elucidating the Polymeric Binder Distribution within Lithium-Ion Battery Electrodes Using SAICAS. CHEMPHYSCHEM, 19: 1627-1634.
[6] Morasch, R., Landesfeind, J., Suthar, B, Gasteiger, H A. (2018) Detection of Binder Gradients Using Impedance Spectroscopy and Their Influence on the Tortuosity of Li-Ion Battery Graphite Electrodes. JOURNAL OF THE ELECTROCHEMICAL SOCIETY, 165: A3459-A3467.
[7] Jaiser, S., Mueller, M., Baunach, M., Bauer, W., Scharfer, P, Schabel, W. (2016) Investigation of film solidification and binder migration during drying of Li-Ion battery anodes. JOURNAL OF POWER SOURCES, 318: 210-219.
[8] Jaiser, S., Funk, L., Baunach, M., Scharfer, P, Schabel, W. (2017) Experimental investigation into battery electrode surfaces: The distribution of liquid at the surface and the emptying of pores during drying. JOURNAL OF COLLOID AND INTERFACE SCIENCE, 494: 22-31.
[9] Jaiser, S., Kumberg, J., Klaver, J., Urai, J L., Schabel, W., Schmatz, J, Scharfer, P. (2017) Microstructure formation of lithium-ion battery electrodes during drying An - ex-situ study using cryogenic broad ion beam slope cutting and scanning electron microscopy (Cryo-BIB-SEM). JOURNAL OF POWER SOURCES, 345: 97-107.
[10] Linek, J., Wichterle, I, Marsh, K N. (1996) Vapor-liquid equilibria for N-methyl-2-pyrrolidone plus benzene, plus toluene, plus heptane, and plus methylcyclohexane. JOURNAL OF CHEMICAL AND ENGINEERING DATA, 41: 1212-1218.
[11] Kneisl, P, Zondlo, J W. (1987) VAPOR-PRESSURE, LIQUID DENSITY, AND THE LATENT-HEAT OF VAPORIZATION AS FUNCTIONS OF TEMPERATURE FOR 4 DIPOLAR APROTIC-SOLVENTS. JOURNAL OF CHEMICAL AND ENGINEERING DATA, 32: 11-13.
[12] Ben Neriah, A., Assouline, S., Shavit, U, Weisbrod, N. (2014) Impact of ambient conditions on evaporation from porous media. WATER RESOURCES RESEARCH, 50: 6696-6712.
[13] Susarla, N., Ahmed, S, Dees, D W. (2018) Modeling and analysis of solvent removal during Li-ion

battery electrode drying. JOURNAL OF POWER SOURCES, 378: 660-670.
[14] Buss, F., Roberts, C C., Crawford, K S., Peters, K, Francis, L F. (2011) Effect of soluble polymer binder on particle distribution in a drying particulate coating. JOURNAL OF COLLOID AND INTERFACE SCIENCE, 359: 112-120.

IOP Conf. Series: Materials Science and Engineering **793** (2020) 012026 doi:10.1088/1757-899X/793/1/012026

Application of High-Speed Photography in Amplitude Measurement of a Satellite Antenna

Zhibin Liu[1,2]*, Gang Sun[1,2], Yu Wang[1,2], Ruizhao Du[1,2], Zhe Liu[1,2]

[1] Beijing Institute of Spacecraft Environment Engineering, Beijing 100094, China

[2] Beijing Engineering Research Center of the Intelligent Assembly Technology and Equipment for Aerospace Product, Beijing 100094, china

Email of all the authors: raindown@126.com

* Corresponding Author: Zhibin Liu; email: raindown@126.com; phone:(+86)010-68746483; fax:010-68746751.

Abstract: Spacecraft antenna is an antenna loaded on the spacecraft and an indispensable part of the spacecraft system. It is particularly important to measure the assembly accuracy of the antenna. A certain type of satellite needs to test the amplitude of antenna A and B in sinusoidal vibration test. Because the distance between antenna A and antenna B is very close to the fairing, it is necessary to measure the amplitude of antenna A to verify the spatial compatibility of antenna A with carrier fairing and antenna B in the test process. By comparing four kinds of antenna measurement technology, this paper chooses high-speed photography technology to measure the amplitude of the antenna, uses recognition processing computing technology to binarize the image, changes the image into calculable coordinate value, obtains the amplitude data of X-direction and Y-direction, and verifies the space composite compatibility requirements between antenna A, carrier fairing and antenna B under vibration conditions. There will be interference and collision.

1. Introduction
Spacecraft include many kinds, such as satellites, space probes, spacecraft, space shuttles, space stations and so on. Spacecraft antenna is the antenna loaded on the spacecraft, that is, spaceborne antenna. In space activities, spacecraft should communicate between space and ground, between spacecraft (space and space, constellation) and even between deep space or interstellar space, as well as wireless transmission of various information.Spacecraft antenna is to construct corresponding wireless transmission channels for the spatial wireless transmission of these information. Spacecraft antenna is the equipment that undertakes the electromagnetic energy conversion between space wave and guided wave field in the process of information wireless transmission.Spacecraft antenna is an indispensable part of spacecraft system. According to the different tasks of spacecraft, spacecraft antenna has many shapes and types. It is the link between ground system and satellite. Therefore, the measurement of antenna assembly accuracy is particularly important.

2. Current status of antenna measurement technology
Because of the different configurations and sizes of spacecraft carriers, the functions of various spacecrafts are different, and the antennas of spacecrafts are of various types and forms. Therefore, the measurement systems used are different. At present, there are four main types:

1) Electronic theodolite measurement system: The electronic theodolite measurement system takes more than two high-precision electronic theodolites as the core measuring equipment, combines with other accessories and measuring software, realizes non-contact measurement of the measured object (antenna) based on the principle of angle intersection. The measuring speed is fast, the measuring range is from several meters to tens of meters, and the measuring accuracy is (+0.05~0.2) mm.Because of its outstanding advantages, the system has been widely used in the manufacture and assembly of spacecraft antenna.[1]

2) Electronic Total Station Measurement System: The measurement system of electronic total station takes a high precision electronic total station as the core measuring equipment, equipped with the equidistant measuring mark of reflector, and measures the measured object (antenna) according to the principle of polar coordinates. The measurement range is from several meters to two hundred meters, and the measurement accuracy is (+0.2-0.5) mm.Compared with the electronic theodolite measurement system, the single electronic total station measurement system has the characteristics of low cost, easy operation, fast establishment of measurement coordinate system and long maintenance time.

3) Photogrammetric System: Photogrammetric system takes single or multiple high precision measuring cameras as the core measuring equipment, and uses intersection measuring principle to carry out fast non-contact measurement. It has large measuring range and high speed, and is especially suitable for dynamic measurement occasions. With the upgrading of Arecibo telescope in the United States, the accuracy of total station can no longer meet the requirements. For this reason, photogrammetry is used to measure the total station, and its surface measurement accuracy is better than (+2mm). Wang Baofeng and others in China have used photogrammetry system to measure and adjust the surface of large antenna with high accuracy. The point coordinate measurement accuracy is better than 0.11mm, and the surface calculation accuracy reaches 0.4mm.[2][3]

4) Laser Tracker Measurement System: Laser tracker measurement system takes a single laser tracker as the core measurement equipment, and uses the measuring principle of spherical coordinates for fast tracking measurement. In the range of 10m, the precision of coordinate repetition measurement can reach 15 micron. As long as the reflector (cat eye reflector, corner reflector or tool ball reflector) is moved on the surface of the measured object, the rapid measurement of the surface can be realized, especially for the measurement and accuracy evaluation of a single antenna panel.[4]

3. Antenna Amplitude Measurement Scheme

3.1. Test target

A certain type of satellite has three antennas, antenna A, antenna B and antenna C. The configuration of transmitting and closing states is shown in Figure 1.According to the analysis of satellite configuration layout, under static conditions, the minimum clearance between antenna A and carrier fairing is 79 mm and that between antenna B is 99mm. The space compatibility of antenna A with carrier fairing and antenna B can be verified by measuring the amplitude of antenna A in sinusoidal vibration test. At the same time, because the boundary conditions of mechanical test of antenna A and B are different from those of on-board satellite, the amplitude test can also provide a basis for the formulation of mechanical environment conditions of components.

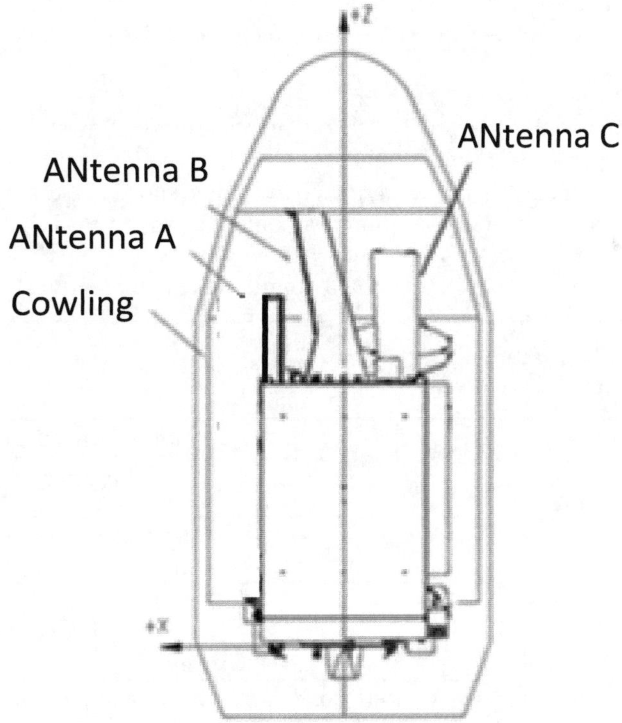

Figure 1. Configuration Diagram of Closing State of a Satellite.

The main objectives of antenna A and B amplitude measurement in sinusoidal vibration test are as follows:

(1) Verify the spatial compatibility of antenna A with carrier fairing and antenna B under vibration conditions;

(2) Provide basis and reference for the formulation of mechanical environment conditions of antenna A and B.

3.2. Test method

Based on the above two test objectives, this paper chooses high-speed photography technology in photogrammetric system for measurement.

Based on the above two test objectives, this paper chooses high-speed photography technology in photogrammetric system for measurement.

The digital high-speed camera used in high-speed photography technology belongs to industrial camera. It is a professional fast and continuous acquisition of images. Its essential function is to convert optical signals into electrical signals. Since 1851, Fox Talbot of England made the world's first high-speed camera, the high-speed camera has experienced 160 years of development. Unlike ordinary cameras, high-speed cameras have higher image stability, which can take pictures quickly, transmit and save pictures quickly, and grasp the moment that the naked eye can not see. When high-speed cameras store results, they can choose many different formats, such as image format and video format, to facilitate the preservation of research results.

The main parameters of high-speed camera include resolution, pixel depth, maximum frame rate, exposure mode, pixel size and spectral response characteristics. The image sensitive chips of high-speed cameras can be divided into two kinds: CCD chips and CMOS chips. High-speed cameras on CMOS chips mostly use electronic curtain shutters. Compared with CMOS chips, the cost of CCD chips is slightly higher. The global shutter is usually used in high-speed CCD cameras. Sampling and holding function is added at each pixel to keep the stored data unchanged.[5,6]

3.3. Test configuration

The test is carried out on a shaking table. The test contents are as follows: while the acceptance level and appraisal level of the satellite sinusoidal vibration test in X and Y directions are tested, the swing of the feed tower at the top of antenna A is photographed by a high-speed camera. Specifically divided into four working conditions, each working condition test content as shown in Table 1.

Table 1. Antenna a amplitude test content.

Whole star state	Whole Star Vibration Direction	working condition	Test content
Vertical state	X direction	Working condition 1	Amplitude measurement in vibration test process of acceptance stage
		Working condition 2	Amplitude measurement in vibration test process of appraisal level
	Y direction	Working condition 3	Amplitude measurement in vibration test process of acceptance stage
		Working condition 4	Amplitude measurement in vibration test process of appraisal level

A total of two high-speed cameras are equipped to shoot the swing of the top of antenna A from -X direction and -Y direction. Target 1, target 2, target 3 and target 4 are respectively pasted on the top of antenna A +X direction, +Y direction and -X direction. The photographic direction and target pasting schematic are shown in Figure 2. The two cameras are basically at the same altitude with the target. It ensures that high-speed camera A can measure target 1, target 2 and target 4 simultaneously from photographic direction A. High-speed camera B can measure target 1, target 3 and target 4 simultaneously from photographic direction B.

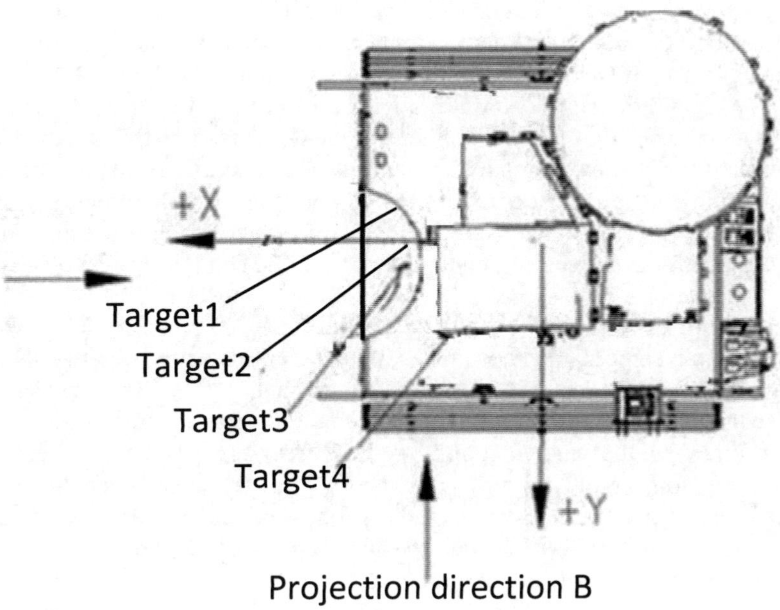

Figure 2. Diagram of antenna an amplitude test direction.

The composition of the equipment used in this test is shown in Table 2, and the main performance parameters of the high-speed camera are shown in Table 3.

Table 2. Hardware composition of high-speed camera.

Serial number	Name	Model	Num-ber
1	High Speed Camera Host	Phantom V641	2
2	Tripod		2
3	Computer	DELL	2
4	High light		3

Table 3. Main performance parameters of high-speed camera.

Resolving power	2560 x 16004 mega-pixels
Maximum resolution shooting speed	1450Frame / sec
Maximum shooting speed	219,000Frame / sec
pixel size	10micron
Minimum exposure time	1Microsecond
Image depth	12position

3.4. Test configuration

The X-direction vibration test uses a high-speed camera to shoot from the direction B of the camera at a frequency of 750 frames per second. The shooting time is 49.856 seconds for each working condition and 37393 frames are taken. The coordinate origin is located in the upper left corner of the graph. During the vibration, the position of each target relative to the coordinate origin is measured.

Two high-speed cameras are used in the Y-direction vibration test. The left camera is located in the direction A in Figure 2 and the right camera is located in the direction B in Figure 2. The shooting frequency is 600 frames per second. The front image of the left camera is tracked by the rightmost point of the three points, and the Y-direction displacement of the antenna A is taken. The side image of the right camera is tracked by the left-most point of the two points, and the X-direction displacement of the antenna A is taken. The left camera takes 35S and 21000 frames at each working condition. The right camera takes 25 seconds for each working condition and 15,000 frames.

4. Scanning Recognition Processing Computing Technology

To change an image into a computable coordinate value, it is necessary to binarize the image. The gray value of the image is set to a total of 256 brightness levels ranging from 0 to 255.The most important thing is to extract edges from images into recognizable markers. The accuracy of the extraction directly determines the accuracy of the final recognition. Because edge has been widely used in motion reconstruction, stereo vision, target tracking, target recognition and other fields, there are many kinds of detection algorithms based on edge industry, including Sobel operand corner detection algorithm, Harris corner detection algorithm and SUSAN corner detection algorithm. At present, the common step is to detect the edge contour of the target first, and then to binarize it. In general, the Sobel operands often generate pseudo-boundaries, and often form unclosed regions. Unclosed regions have a great impact on edge extraction. Therefore, this paper adopts a more accurate algorithm than Sobel, Carmy operator method, and makes some improvements accordingly.The basic steps of image edge detection are shown in Figure 3.

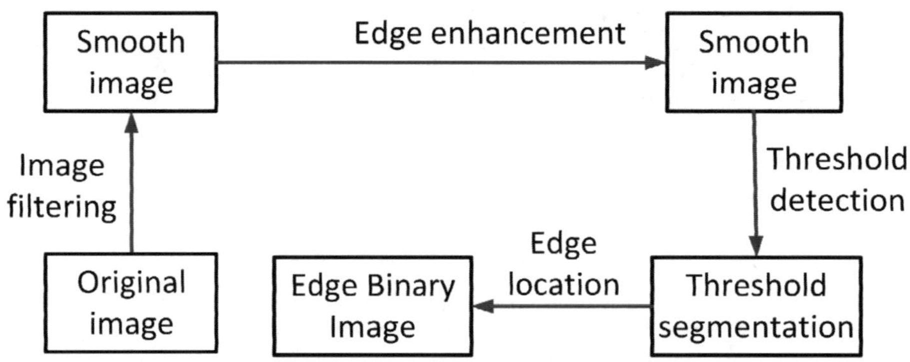

Figure 3. Basic steps of image edge detection.

5. Concluding remarks

Through the amplitude test, it can be found that the minimum distance between satellite antenna A and carrier net envelope is about 55 mm, and between antenna A and antenna B is about 87 mm. Under the vibration condition, the space composite compatibility between antenna A and carrier fairing and antenna B is required, and there will be no interference and collision.And the successful application of high-speed photography technology in the field of satellite antenna amplitude measurement is realized.

References

[1] Li G.Y, Li Z.C. (2000) Theodolite industrial measurement system for installation and adjustment of large multi-beam antennas. Surveying and mapping bulletin. 10:41-44.

[2] Goldsmith P. (2001) Resetting the Arecibo Primary Reflector Surface. The Arecibo Observatory Newsletter. 32: 1-4.

[3] Wang B.F, Li G.Y, Li Z.C. (2006) Study on the application of photogrammetry to detect the

working status of large antennas. Journal of China Academy of Electronic Sciences. 1(5): 435-454.
[4] Zhang X, Chen W. (2006) Research on Uncertainty of Surface Measurement by Laser Tracker. Journal of Metrology. 27 (2): 107-112.
[5] Tan X.X. (1990) High-speed photography. Beijing Atomic Energy Publishing House, Beijing.
[6] Tan X.X. (1992) Optical high-speed photographic testing technology. Beijing: Science Press, Beijing.

Laser Trimming Method for Reducing Frequency Split of Cylindrical Vibrating Gyroscope

Youwang Hu[1], Hongmin Zhong[1], Yunpeng Lu[1], Ji'an Duan[1], Xiaoyan Sun[1*]

[1] State Key Laboratory of High Performance Complex Manufacturing, College of mechanic engineering, Central South University, Changsha 410083, China

*Corresponding author's e-mail: sunxy@csu.edu.cn

Abstract. Frequency split is the frequency difference between the driving mode and sensing mode of a vibratory gyroscope, which significantly influences the performance of cylindrical vibratory gyroscopes. In order to reducing frequency split, a method of trimming by drilling micro-holes through laser has been introduced in this paper. Firstly, the processing method of the trimming hole is determined by the finite element simulation, and then the feasibility of the method is verified by the experiment. In the experiment, the frequency split of gyroscope is reduced from 2.090Hz to 0.119Hz.

1. Introduction

A gyroscope is a device that senses the rotation of a body. Gyroscopes are widely used in many fields, such as industrial robot, consumer electronics and aerospace navigation [1-4]. The cylindrical vibratory gyroscope (CVG), is a novel, solid state gyroscope, whose work is based on the inertia effect of elastic waves[5]. Its cylindrical shell simplifies the fabrication of the resonator [6]. Without a rotary structure, this type of gyroscope proves to be high accuracy, better impact resistance and stability [7-9]. The gyroscope studied in this paper has a constant elastic alloy as its shell material. This material is more easily manufactured than fused silica, which some are made from, yet still has acceptable accuracy [6,10].

Frequency split is an important measure to evaluate the quality of a gyroscope [11]. It affects both the sensing accuracy and stability of the gyroscope, causing potential hazards in areas with high accuracy requirements [12,13]. Ideally, the frequency of the driving mode and sensing mode of the cylindrical vibratory gyroscope are equal [12]. However, frequency split is unavoidable due to manufacturing errors and disproportionate variation in density and stress [14,15]. Therefore, it is imperative to develop an effective method to minimize the frequency split.

Several studies on frequency split trimming have been reported. For example: trimming by processing grooves on the rim of the resonant ring, and holes in the base plate [16]; trimming by processing grooves on the suspension and base plate [9] etc.

In these methods, several grooves are made on the cylindrical wall of the resonator, which reduce the stiffness of the resonator. Additionally, there are many processing steps in these methods, increasing the chance of manufacturing errors. Also less attention is paid to the correlation between frequency split trimming and such factors of geometrical dimensions, removed mass and so on. The purpose of this paper is to find a method that can reduce the frequency split while keeping the gyroscope structure as intact as possible.

IOP Conf. Series: Materials Science and Engineering **793** (2020) 012027 doi:10.1088/1757-899X/793/1/012027

This paper explores a way to trim the frequency split while simplifying the processing procedures and minimizing the mass removed from the gyroscope. It discusses the main factors that influence the trimming of frequency split. The values of frequency split in trimming processes are presented based on FEM simulation. In addition, relevant experiments have also been carried out, which validated the analysis.

2. Structure and Working Principle

The structure of a typical CVG is shown in Figure 1. The resonator is the core of gyroscope which is used to produce gyroscopic effect [8]. It is made of elastic alloy and is fixed on the base by the supporting bar. Piezoelectric electrodes can be divided into two groups of driving electrodes and sensing electrodes, which are used to excite driving mode and detect sensitive mode respectively [6].

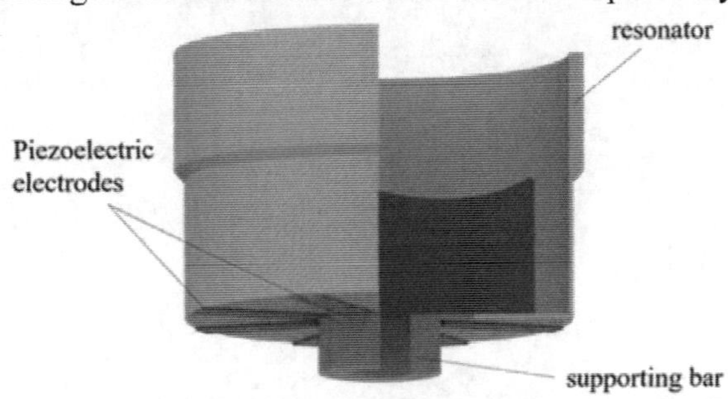

Figure 1. The structure of cylindrical vibratory gyroscope.

The driving mode and the sensing mode are standing wave vibration with wave number of 2, and they form 45 degree angles to each other, as show in Figure 2. The Eigen frequency difference between these two modes is defined as frequency split.

The working principle of a CVG can be briefly described as follows [17,18]: Utilize the inverse piezoelectric effect, AC voltage is input to the driving electrodes to excite the vibration of the driving mode. When the gyroscope has no angular velocity input, the vibration of the sensing mode direction should be 0. When gyroscope has angular velocity input, Coriolis effect results in vibration in sensing mode direction. Because of the piezoelectric effect, the vibration in the direction of the sensing mode is converted to the voltage output of the sensing electrodes, and the input angular velocity can be calculated by this voltage.

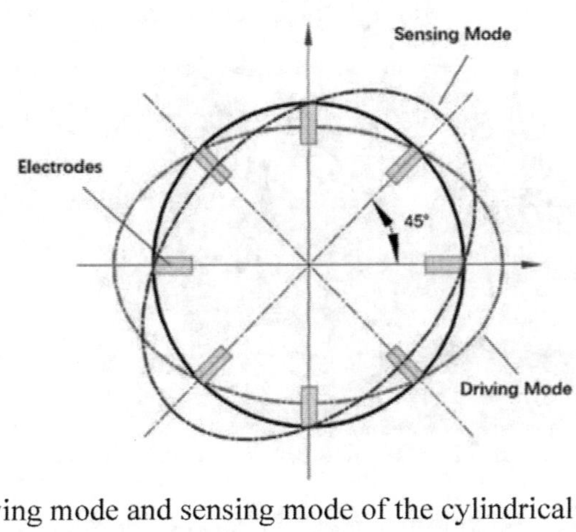

Figure 2. The driving mode and sensing mode of the cylindrical vibratory gyroscope.

3. Simulations and Discussion

3.1. Structural and material properties

In the experiment, in order to increase the frequency split reduction, we can increase the depth or diameter of the hole. In order to compare the difference between them and find out the appropriate trimming method, the finite element software is used for simulation analysis. A resonator model has been built as shown in Figure 3. The geometry parameters are listed in Table 1.

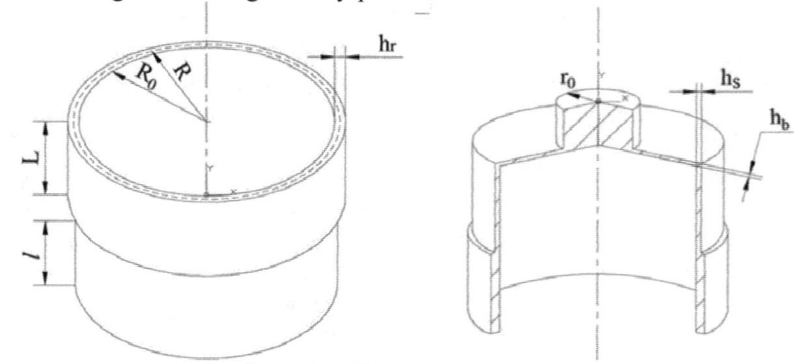

Figure 3. The resonator model.

Table 1. Geometric parameters of the resonator model.

Parameters	Value [1]
Height and thickness of resonant ring L and hr	8.4 and 1
Height and thickness of suspension ring l and hs	9.6 and 0.3
Internal radius of resonant R_0	12
External radius of supporting bar r_0	4
Thickness of bottom h_b	0.3

To validate the effectiveness of trimming measures, a finite element model of gyroscope with defects is needed to simulate the actual situation. Frequency split can be purposefully fabricated by creating several defect blocks Symmetrically distributed at the top of the resonant ring with different density from others [13]. The height and angular width of the defect block are 1400 um and 4°, as show in figure 4.

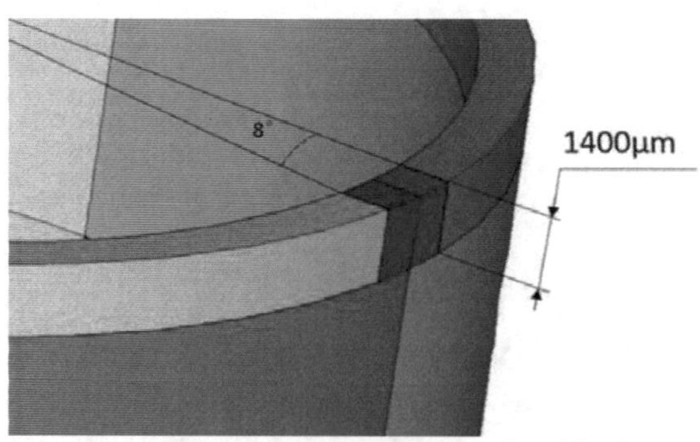

Figure 4. The structure of defect block.

The resonator material possesses isotropic and homogeneous properties which are listed in Table 2.

IOP Conf. Series: Materials Science and Engineering **793** (2020) 012027 doi:10.1088/1757-899X/793/1/012027

Table 2. Material characteristics of gyroscope.

Parameters	Value
Original Material Density	8030 kg/m^3
Defect Block Density	7065 kg/m^3
Young's Modulus	210 GPa
Poisson Ratio	0.3

3.2. *Trimming Method*

Frequency splitting can be reduced by changing the stiffness or mass in the direction of working mode and its principle can be briefly described by the following equation [13]:

$$\omega=\sqrt{\frac{k^*+\Delta k}{m^*+\Delta m}} \tag{1}$$

Where ω is Eigen frequency of working mode after trimming. k^* and m^* are equivalent stiffness and mass of the working mode direction, respectively. Δk and Δm are the stiffness and mass changes of k^* and m^* in trimming, respectively.

As show in figure 5, in order to reduce the frequency split, four holes are processed in the mode A or mode B direction by laser. And the trimming holes are located on the middle of the resonant ring.

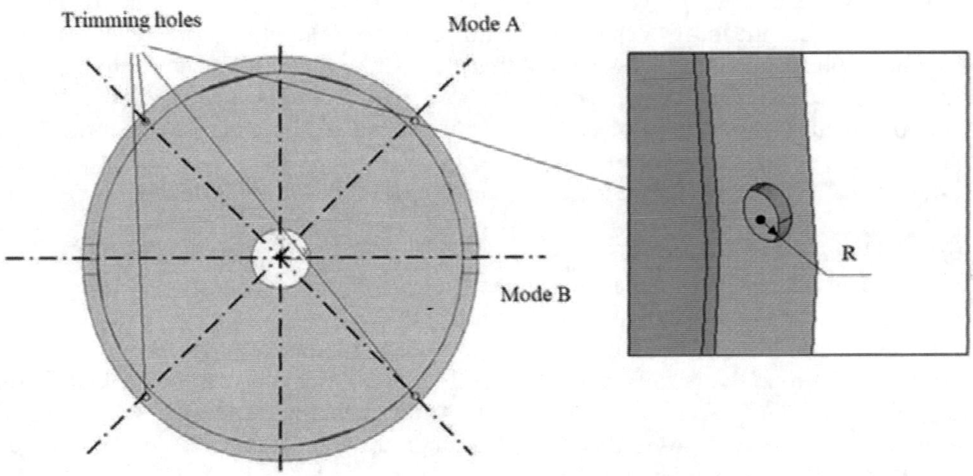

Figure 5. The trimming holes.

3.3. *Effect of Hole Depth on Frequency Split*

The original frequency split is set to 3.50995Hz by modifying the density of defect blocks. In order to reduce frequency splitting, the trimming holes is located in the direction of low frequency mode axis (mode B).

In order to study the effect of hole depth on frequency splitting, a group of simulation experiments (group A) are carried out as follows: the radius of the hole is always 250mm, and the hole depth increases gradually. Simulation results are plotted in Figure 6.

IOP Conf. Series: Materials Science and Engineering **793** (2020) 012027 doi:10.1088/1757-899X/793/1/012027

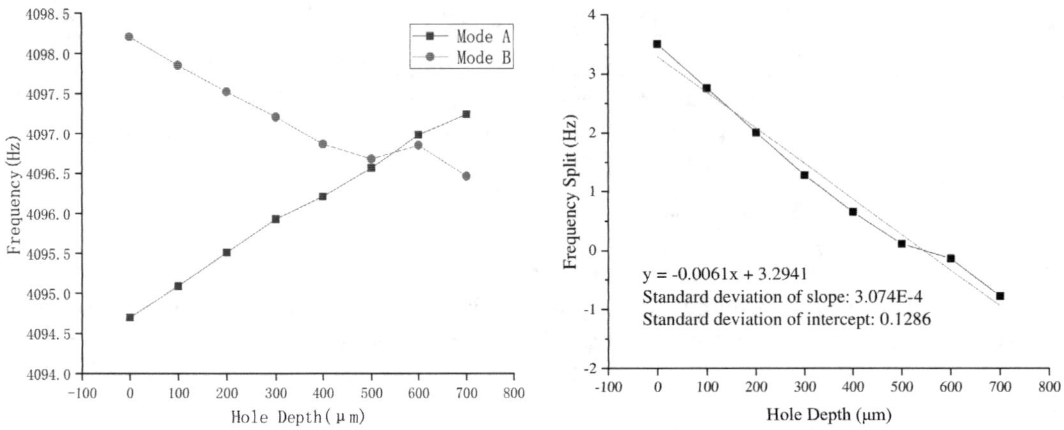

(a) (b)

Figure 6. The variation of frequency with the increase of hole depth. (a) Mode frequency. (b) Frequency split.

As shown in Figure 6a, with the increase of hole depth, the frequency of mode B (trimming mode) decreases gradually, while the frequency of mode A increases gradually. This causes the frequency split decrease gradually and reach the ideal value when the hole depth is 500μm to 600μm, as shown in Figure 6b. The simulation results show the effectiveness of this method. Negative frequency split indicates that the conversion occurs between high frequency axis and low frequency axis due to over trimming.

There is a good linear relationship between frequency splitting and hole depth. As shown in Figure 6b, the results of linear fitting show that the slope is -0.0061 (Hz/μm), the slope standard deviation is 3.074×10-4, the intercept is 3.2941 (Hz) and the intercept standard deviation is 0.1286.

3.4. *Effect of Hole Radius on Frequency Split*

The method of reducing frequency split by increasing the hole depth is described above. In fact, a similar effect can be achieved by increasing the radius of trimming hole.

In order to study the influence of hole radius on frequency splitting, the following group of simulation experiments (group B) are carried out: the depth of each hole is set to a constant value: 400μm, and their radiuses gradually increase. In order to acquire better comparative data between group A and group B, the volume of the holes in group B equal to that in group A at depth of 400μm, 500μm, 600μm and 700μm one by one. Results are plotted in Figure 7.

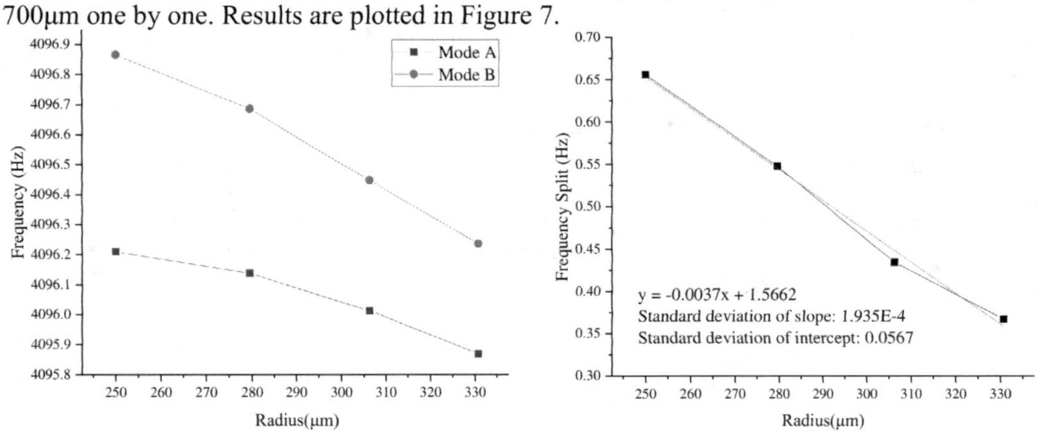

(a) (b)

Figure 7. The variation of frequency with the increase of hole radius. (a) Mode frequency. (b) Frequency split.

As shown in Figure 7a, with the increase of the hole radius, the frequencies of mode B (trimming mode) and mode A decrease at the same time, which is different from the simulation results of group A. The frequency reduction of mode B is greater than that of mode a, resulting in the decrease of frequency split, as shown in Figure 7b.

When the depth of the hole is constant, there is also a good linear relationship between the hole radius and the frequency splitting. As shown in Figure 7b, the results of linear fitting show that the slope is -0.0037 (Hz/μm), the slope standard deviation is 1.935×10-4, the intercept is 1.5662 (Hz) and the intercept standard deviation is 0.0567.

In order to compare the different effects of hole radius and depth on frequency split, the removal mass is used to measure the hole size to plotting the simulation results of group A and group B on one chart, as shown in Figure 8.

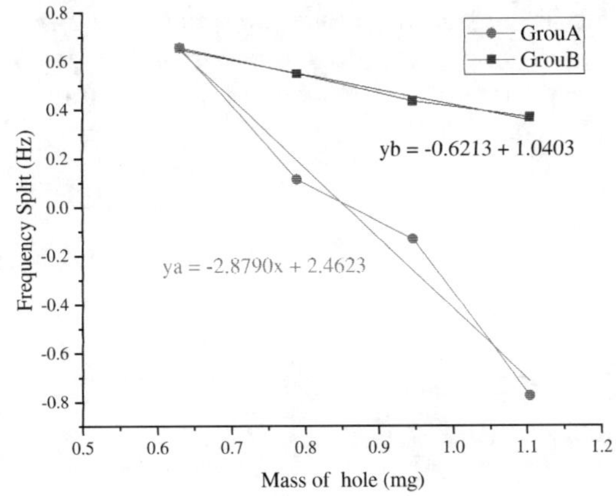

Figure 8. Comparison of frequency split in Group A and Group B.

As show in figure 8, by comparing the trimming methods of increasing the hole depth and radius gradually, we can see that the former method can trim the frequency split to zero faster. Through the linear fitting of the two sets of data, it can be seen that the trimming efficiency is about 2.8790Hz/mg by increasing the hole depth, and 0.6213Hz/mg by increasing the hole diameter. This shows that the effect of the depth of the hole on the frequency split is greater than that of the radius of the hole.

3.5. Effect of Trimming Holes with Different Depth-Diameter Ratio on Frequency Split

In order to further compare the different effects of hole depth and hole radius on the trimming process, group C simulation experiments are carried out: the mass of the hole remained unchanged, and the depth-diameter ratio increased gradually. The simulation results are shown in Table 5 and Figure 9.

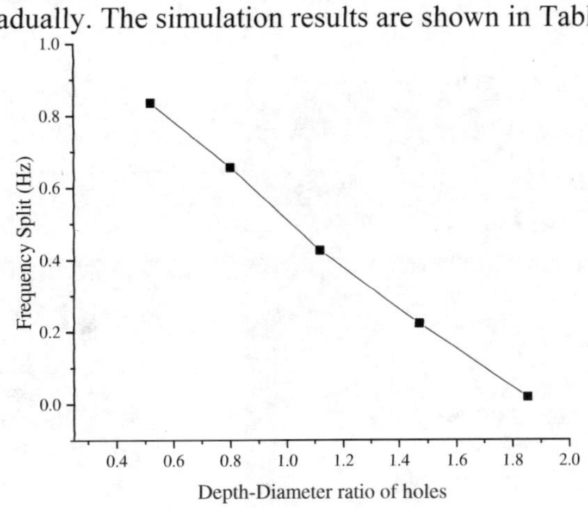

Figure 9. Effect of trimming holes with different depth-diameter ratio on frequency split.

It can be seen that the frequency split decreases with the increase of hole depth-diameter ratio. This shows that trimming holes with larger depth-diameter ratio have higher trimming efficiency. So the effect of hole depth on frequency splitting is greater than that of hole diameter.

Therefore, in the actual trimming process, in order to quickly reduce the frequency split, the trimming method of increasing the hole depth can be preferred.

4. Experiments and Discussion

4.1. Experimental Set Up and Method

The size of resonator in experiments is identical with that in simulation. Before the trimming process, the frequency split of the gyroscope is measured. The original working mode frequency is 3585.1213Hz (mode A) and 3585.0359Hz (mode B), and the frequency split is 0.0854Hz.

A laser is employed. By modifying the pulses number and energy, it can remove micro-mass accurately [19]. Such characteristic is advantageous in precise trimming. The experimental set up is plotted in Figure 10. The focal length of the lens is 50mm.

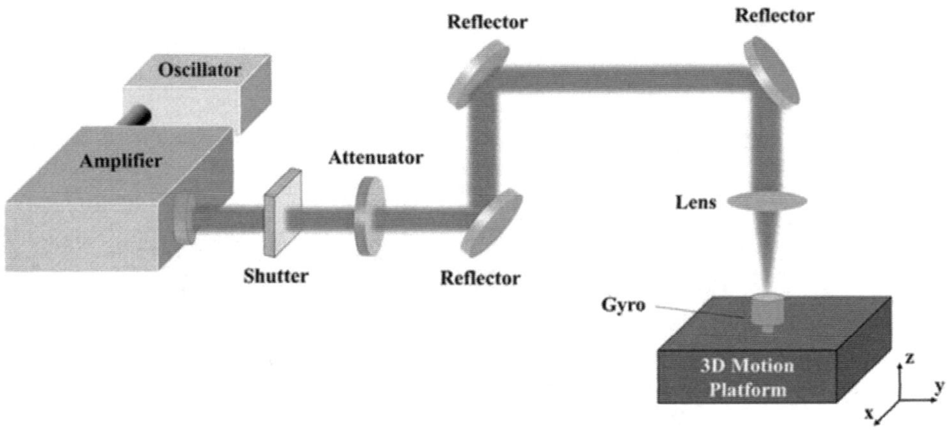

Figure 10. Schematic of experimental set up.

4.2. Trimming Experiments

To reduce the frequency split, the trimming hole is located at the direction of low frequency axis (mode B). As show in Figure 11, four holes are ablated at the top of the resonator. As increasing the hole depth is beneficial to reduce the frequency split quickly, Z-direction feed is added to increase the hole depth as much as possible during drilling.

Figure 11. Ablation holes in experiment.

The experiment was carried out in 8 steps, and each step ablating was in the same position to increase the hole depth. And in each step, laser energy, Z-direction feed and ablation time was increase in order to increase the hole depth.

In order to measure the diameter and depth of the hole to compare with the simulation results, the following methods are used to measure the diameter and depth of the hole: first, the same processing parameters are used on the samples of the same material, and then the samples are cut open to measure the cross section of the hole, as shown in Figure 12.

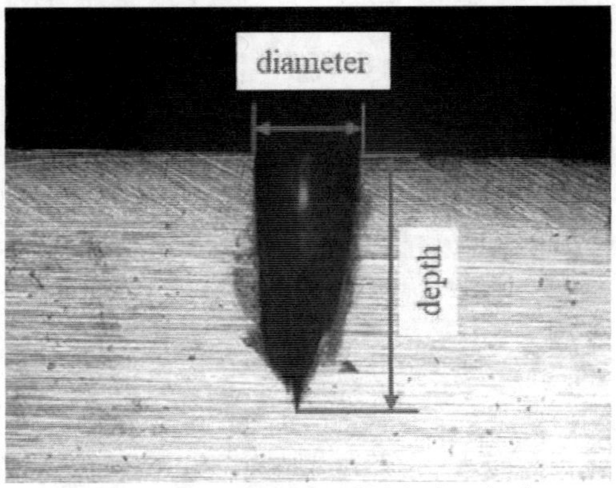

Figure 12. Section of ablative hole.

The experimental results are shown in Table 3 and figure 13. Figure 13a shows the relationship between mode frequency and hole depth in trimming process, figure 13b shows the relationship between frequency split and hole depth, and figure 13c shows the relationship between frequency split and hole mass. The mass of trimming hole is calculated as follows:

$$m = k \cdot \pi r^2 h \tag{2}$$

Where, m is the mass of trimming hole, r is the radius of trimming hole, h is the depth of trimming hole, and the coefficient k depends on the shape of hole. As the hole is similar to the cone (As show in Figure 12), k = 1 / 3 is taken.

Table 3. The experimental result.

Steps	Hole diameter (μm)	Hole depth (μm)	Mass (mg)	Frequency of Mode A (Hz)	Frequency of Mode B (Hz)	Frequency split (Hz)
0	0	0	0	3872.602	3870.512	2.090
1st	67	327	0.003	3872.539	3870.626	1.913
2nd	88	442	0.007	3872.531	3870.756	1.775
3th	152	494	0.024	3872.526	3870.811	1.715
4th	205	544	0.048	3872.495	3870.825	1.670
5th	252	632	0.084	3872.456	3870.912	1.544
6th	381	1283	0.392	3872.013	3871.012	1.001
7th	459	1707	0.756	3871.523	3871.012	0.511
8th	500	1987	1.044	3871.093	3870.974	0.119

As shown in Figure 13a, with the increase of the hole depth, the frequency of the trimming mode increases gradually, while that of the untrimmed mode decreases gradually. The frequency splitting decreases with the increase of the hole depth, and has a good linear relationship with the hole depth, as shown in Fig. 13b. These results are similar to those in Fig. 6, which shows that the hole depth is an important factor affecting frequency split. Compared with the hole depth, the linear relationship between

IOP Conf. Series: Materials Science and Engineering **793** (2020) 012027 doi:10.1088/1757-899X/793/1/012027

the hole mass and frequency split is worse, especially the first several points, as shown in Fig. 13c. This shows that the quality of the trimming hole is not the decisive factor to determine the frequency split.

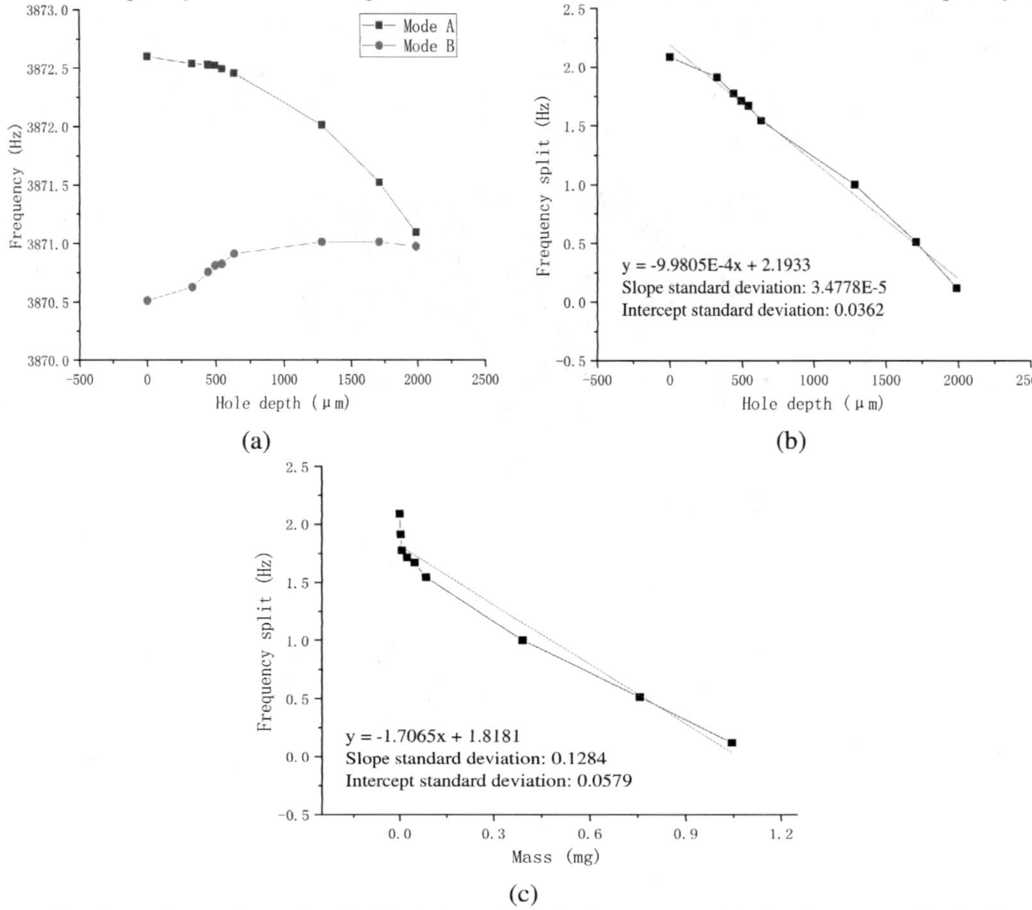

Figure 13. Experimental results. (a) Variation of mode frequency with the increase of hole diameter. (b) Variation of frequency split with the increase of hole diameter. (c) Variation of frequency split with the increase of hole mass.

By linear fitting the data in Fig. 13c, it can be seen that the average trimming efficiency is 1.7065Hz/mg, which is smaller than the trimming method of increasing hole depth in the simulation (2.8790Hz/mg), as shown in Fig. 8. This is because in the experiment, when the hole depth increases, the hole diameter also increases, while in the simulation, only the hole depth is increased. In addition, the difference of hole shape also leads to this result.

The experiment reduces the frequency split from 2.090Hz to 0.119Hz, which verifies the feasibility of this method.

5. Conclusion

In this paper, a method of reducing frequency split by drilling micro-holes using laser was introduced. The influence of the hole diameter and depth on the trimming process was analyzed by finite element software. The simulation results show that the effect of hole depth on frequency split is greater than that of hole diameter. In the trimming process, in order to reduce the frequency split quickly, the trimming method of increasing the hole depth can be preferred.

The experimental results show that the hole depth is an important factor affecting the frequency split. At the same time, the experiment reduces the frequency split from 2.090Hz to 0.119Hz, which verifies the feasibility of the method.

Acknowledgments
This work is supported by the projects (Grant Nos. 51875585, 51875584, 51475481, 51475482) of the National Natural Science Foundation of China.

References

[1] Lammel G, Ieee (2015) THE FUTURE OF MEMS SENSORS IN OUR CONNECTED WORLD. In: 2015 28th Ieee International Conference on Micro Electro Mechanical Systems. Proceedings IEEE Micro Electro Mechanical Systems. pp 61-64
[2] Shen Q, Li H, Hao Y, Yuan W, Chang H (2016) Bias Contribution Modeling for a Symmetrical Micromachined Coriolis Vibratory Gyroscope. Ieee Sensors Journal 16:723-733.
[3] Xia D, Huang L, Zhao L (2019) A New Design of an MOEMS Gyroscope Based on a WGM Microdisk Resonator. Sensors (Basel) 19.
[4] Yang B, Wu L, Lu C, Wang G (2019) A Digital Mode-Matching Control System Based on Feedback Calibration for a MEMS Gyroscope. Journal of Sensors.
[5] Shkel AM (2006) Type I and Type II Micromachined Vibratory Gyroscopes.
[6] Chikovani VV, Okon IM, Barabashov AS, Tewksbury P A set of high accuracy low cost metallic resonator CVG. In: Position, Location & Navigation Symposium, IEEE/ION, 2008.
[7] Langmaid C (1996) Vibrating structure gyroscopes. Sensor Review 16:14-17.
[8] Luo Y, Qu T, Zhang B, Pan Y, Xiao P (2017) Simulations and Experiments on the Vibrational Characteristics of Cylindrical Shell Resonator Actuated by Piezoelectric Electrodes with Different Thicknesses. Shock and Vibration 2017:1-6.
[9] Xi X, Wu Y, Wu X, Tao Y, Wu X (2013) Modeling and analysis of the stress effects for trimmed cupped resonator under varying temperature. Sensors and Actuators A: Physical 189:429-440.
[10] Chikovani VV, Yatsenko YA, Barabashov AS, Marusyk PI, Umakhanov EO, Taturin VN (2009) Improved accuracy metallic resonator CVG. Aerospace & Electronic Systems Magazine IEEE 24:40-43.
[11] Xi X, Wu Y, Wu X, Tao Y, Wu X (2012) Investigation on standing wave vibration of the imperfect resonant shell for cylindrical gyro. Sensors and Actuators A: Physical 179:70-77.
[12] Bisegna P, Caruso G (2007) Frequency split and vibration localization in imperfect rings. Journal of Sound & Vibration 306:691-711.
[13] Lu K, Xi X, Li W, Shi Y, Hou Z, Zhuo M, Wu X, Wu Y, Xiao D (2019) Research on precise mechanical trimming of a micro shell resonator with T-shape masses using femtosecond laser ablation. Sensors and Actuators A: Physical.
[14] Birman V (1987) On stability of axisymmetric forced vibration of imperfect cylindrical shells. Zeitschrift Für Angewandte Mathematik Und Physik Zamp 38:129-136.
[15] Mcwilliam S, Ong J, Fox CHJ (2005) On the statistics of natural frequency splitting for rings with random mass imperfections. Journal of Sound & Vibration 279:453-470.
[16] Yi T (2012) Precision balance method for cupped wave gyro based on cup-bottom trimming. Chinese Journal of Mechanical Engineering 25:63-70.
[17] Loveday PW, Rogers CA (1998) Modification of piezoelectric vibratory gyroscope resonator parameters by feedback control. IEEE Trans Ultrason Ferroelectr Freq Control 45:1211-1215.
[18] Yao P, Qu T, Wang D, Wu S, Liu J, Tan Z, Yang K, Hui L (2017) Observation and analysis of the quality factor variation behavior in a monolithic fused silica cylindrical resonator. Sensors & Actuators A Physical 260:81-89.
[19] Xia B (2013) Femtosecond Laser Drilling of Micro-Holes. Chinese Journal of Lasers 40:1-12.

Development of a New Wax Scraper

XieLu[a], Liu Yangyang[b]

No.3 Oil Production Plant, Saltu District, Daqing, Heilongjiang Province
[a]lxie@petrochina.com.cn
[b]395111677@qq.com

Abstract: Wax removal and prevention in oil wells is an important work in field management. This paper introduces the process of developing a new type of loose-leaf wax scraper by using TRIZ method. At the same time, 48 schemes, 33 feasible schemes and 5 patent plans have been obtained by using this method, and certain economic benefits have been obtained.

1.Preface

Wax deposition in oil wells is the most common problem encountered in Daqing oilfield since its development. During the upward migration of crude oil along the tubing, the composition, temperature and pressure of crude oil change. Part of the wax is precipitated from the oil. Then they gather and adhere to the tubing wall. Waxing on the tubing will reduce the production of the oil well and even block the tubing to stop production. Therefore, oil well wax removal and prevention work is an important work for field management.[1]

At present, the wax removal method for electric pump wells in oilfields is mainly mechanical wax removal: the wax scraper connecting weighting rod moves downwards to scrape wax under the action of gravity, and the scraped wax is taken out of the wellbore by liquid flow.After the wax scraper goes down to the well section without wax deposition, start the winch to pull the wax scraper up to the wellhead to complete wax removal

With the development of oilfields, the implementation of new oil extraction technologies such as polymer flooding and ASP flooding, wax deposition in oil wells is becoming more and more serious. The wax removal period of oil wells is becoming shorter and shorter. And the wax removal time of single wells is also gradually increasing. How to prolong wax removal period and shorten wax removal time has become an urgent problem to be solved.

2.System Analysis

2.1 System analysis-Nine Screen diagram[2]

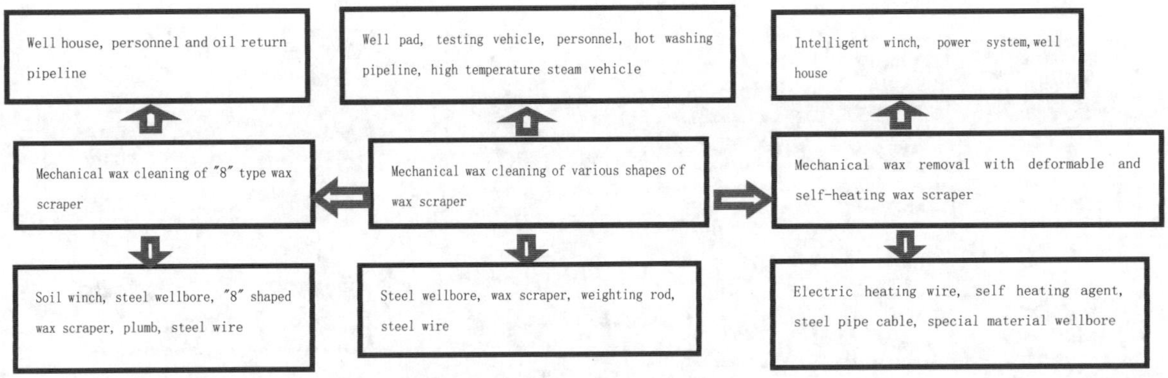

Figure1.Nine Screen diagram

According to the method of nine screen diagram, we put forward the technical scheme of installing intelligent paraffin removal system to realize unmanned paraffin removal.

2.2 System analysis-Causal chain analysis[3][2]

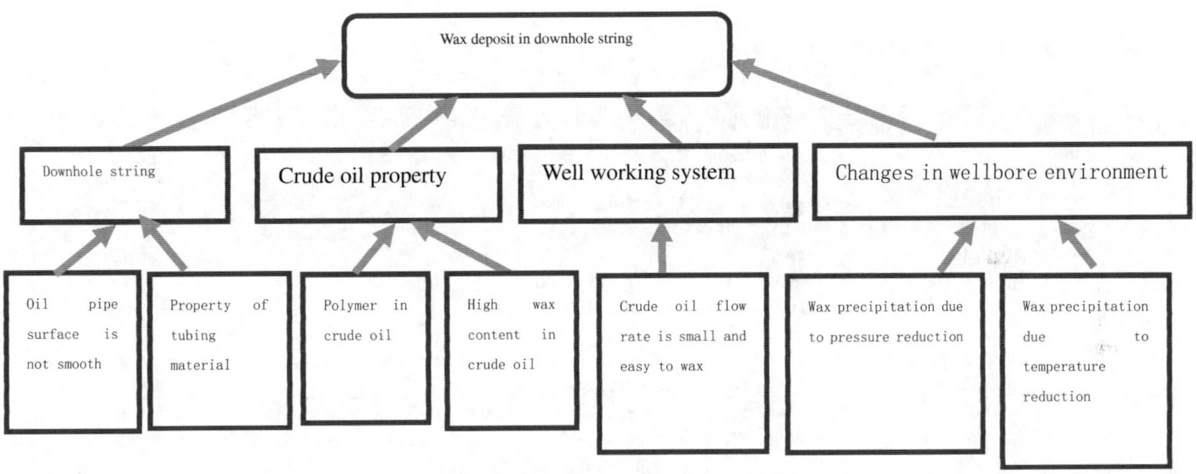

Figure2.Causal chain analysis

According to the cause and effect chain analysis, we put forward five technical schemes: using smooth or coated tubing, using tubing that is not easy to wax, reducing polymer injection, speeding up the flow speed of crude oil in the pipeline, and improving the temperature of produced fluid.

2.3 System analysis-Function analysis and cutting[2]

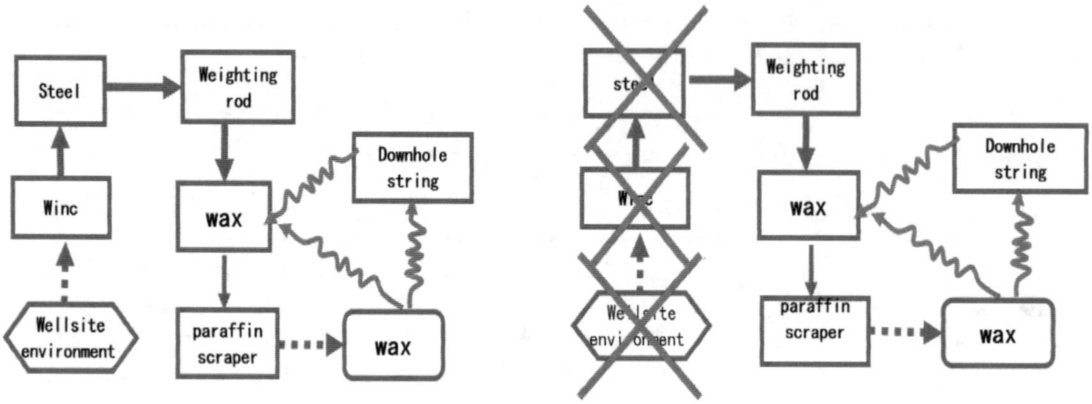

Figure3.Function and cutting model of mechanical paraffin removal system

According to the function analysis and cutting, we propose to change the diameter of the wax scraper to prevent it from being stuck by wax or downhole tools; increase the weight of the weighted rod to speed up the speed of wax scraping; change the shape of the wax scraping piece to speed up the speed of wax scraping.

2.4 System analysis-Resource analysis[2]

Table 1.Resource analysis

Species	Material resources	Energy resources	Information resources	Space resources	Time resources	Functional resources
System resources	wax scraper, high temperature cleaning car, lubricator,	mechanical energy, potential energy, thermal energy	patent, professional technology,	well pad, wellbore space	wax scraping time	wax scraper, thermal medium hardening
Subsystem resources	driving car, steel wire, cable, weighting rod, wax scraper, wax scraping slice	mechanical energy, potential energy	driving technology, steel wire bearing, weighting rod, shape of wax scraper	lubricator,a nnular space of oil sleeve,	wax scraper time	wax scraper, lifting and lowering of wax scraper
Super system resources	wellbore environment, personnel	mechanical energy, thermal energy, electric energy	production, water content, physical properties of crude oil	well pad space, well bore space	well production time	crude oil lifting

According to the resource analysis, we put forward three technical schemes: winch capacity to speed up the speed of paraffin scraper; using optical cable well test car to electrify and heat the paraffin scraper to speed up the speed of paraffin scraper; using oil well power resource to change to electric winch.

3. Using TRIZ method to solve problems

3.1TRIZ Tool-Ultimate ideal solution[2]

Table 2. Ultimate ideal solution

IFR Implementation deduction 6 Steps	Analysis results
1. What is the ultimate goal of design?	Downhole string does not need wax removal
2.What is the final understanding of the problem?	The downhole string does not wax itself in the production process

3. What are the obstacles to achieving the ideal solution?	1. Wax in crude oil 2. Wax Precipitation Caused by Environmental Changes 3. The string material is easy to adsorb wax.
4. What is the result/reason of this obstacle?	1. Property of crude oil 2. The temperature and pressure drop during the process of crude oil lifting from underground to surface. 3. The pipe string is made of steel and its surface is not smooth.
5. What are the conditions for the absence of such obstacles?	During the process of lifting crude oil to the ground, maintaining a high temperature, and using a pipe string made of non-wax material
6. What resources are available to create these conditions?	Downhole string, crude oil, mining equipment, thermal insulation equipment

According to the final ideal solution, we put forward five plans: changing the nature of crude oil, keeping wax not separated from crude oil in the process of exploitation; maintaining the pressure of exploitation system, preventing wax from precipitation in crude oil; ensuring the temperature of crude oil in the process of exploitation, preventing wax from precipitation; changing the material of downhole pipe string, making wax unable to be adsorbed on the pipe wall; designing the inner wall of pipeline as smooth surface, preventing wax from adhering.

3.2TRIZ Tools-Scientific effects base[2]

Table 3. Scientific effects base

1.What is the function of the problem?	Control the movement of the wax scraper and increase the temperature of the wellbore and the wax scraper
2. How to find the function code?	F6 control object displacement, F3 raise temperature
3.What are function and phenomenon?	E15、E2、E64、E91、E93、E44、E92、E98、E49、E75、E76、E82、E24、E26、E57、E42、E25、E84、E39、E73、E67、E72、E80、E71、E69、E101
4.Screening effect phenomena	E49 Inertial Force, E82 Displacement, E57 Joule-Lenz's law, E73 Thermal Radiation, E71 Thermoelectricity phenomenon, E98 Vibration, E92 Hydrodynamics
5.The use of screening effect and phenomenon	E49 inertial force, E73 thermal radiation, E71 Thermoelectricity phenomenon, E98 vibration, E92 hydrodynamic

According to the scientific effect library, we put forward to increase the inertia force in the downward process of the paraffin scraper, speed up the paraffin scraper; power up the oil pipe, increase the temperature to melt the wax; use resonance to make the wax vibrate, shake down the wax; use thermal medium power to wash away the wax; put self heating agent into the paraffin scraper before it goes down the well, heating the paraffin scraper can speed up the paraffin scraper; make the paraffin scraper with heat accumulating materials After heating on the ground, it goes down to the underground to continue heating and scraping wax.

3.3 TRIZ Tool-Information Interaction Theory[2]

According to the theory of information interaction, we put forward five ways to remove wax: using tungsten weighting rod, mechanical paraffin scraper with chemical agent; using tungsten weighting rod, mechanical paraffin scraper with heat wash pump; using tungsten weighting rod, mechanical paraffin scraper with steam car; using tungsten weighting rod, mechanical paraffin scraper with glass inner liner string; using tungsten weighting rod with movable paraffin scraper Technical proposal.

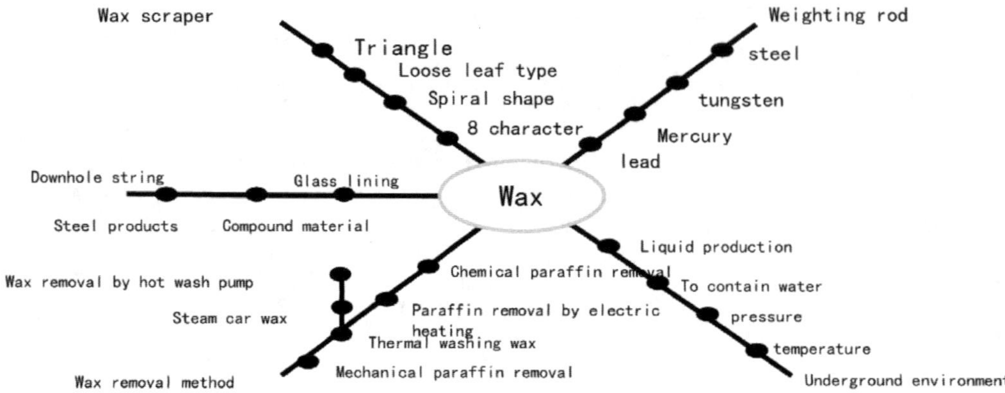

Figure4 Wax

3.4 Using TRIZ method to solve problems-Technical contradictions[2]

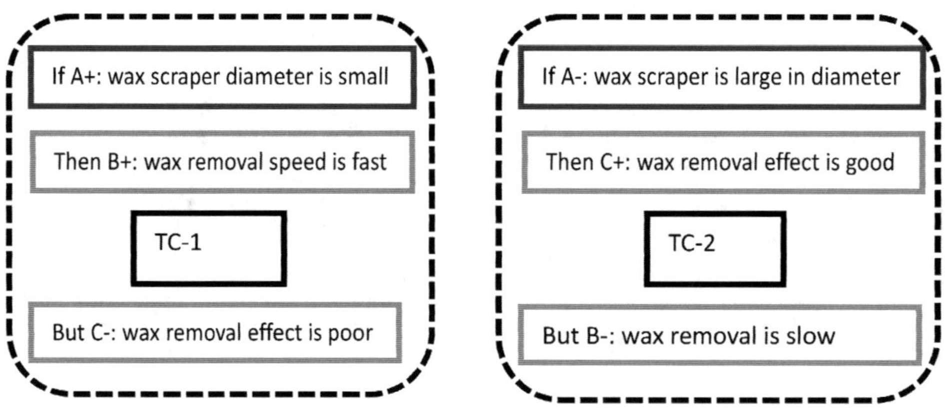

Figure5.Technical contradictions

According to the technical contradiction diagram, we determined that the technical contradiction to be solved is TC-1, which occurs between wax removal speed and wax application effect, and under the condition of changing the diameter of the wax scraper.

In the model of this problem, the parameter we improve is the quality of wax scraping; the parameter we deteriorate is the speed of wax scraping. The solution can adopt the performance conversion method, dynamic method, self generated and self abandoned method and mechanical vibration method in TRIZ scientific effect library. We put forward the technical scheme of lowering the finer paraffin scraper to the bottom of the well, increasing the diameter of the paraffin scraper, and using the mechanical power of the winch to lift the paraffin scraper to scrape the wax.

3.5 Using TRIZ method to solve problems-Physical contradictions[2]

According to the physical contradiction diagram, we determined that the diameter of the wax scraper should be large to meet the requirements of wax removal effect, and the diameter of the wax scraper should be small to meet the requirements of wax removal speed. According to the principle of space separation, time separation, condition separation, whole and part separation, we put forward to put the tubing out of the well bore and put it into the bottom of the well after paraffin removal, paint special coating inside the well bore to delay paraffin deposition, install automatic paraffin scraper in advance in the well bore, push the paraffin scraper with air or hydraulic pressure, inject fluid to push the paraffin scraper upward, and increase the speed of paraffin removal. , after separating the paraffin scraper from

IOP Conf. Series: Materials Science and Engineering **793** (2020) 012028 doi:10.1088/1757-899X/793/1/012028

the steel wire rope, put it into the shaft to scrape the wax by itself, and put the steel wire rope fishing and other technical schemes after the paraffin scraper reaches the bottom of the well.

3.6 TRIZ tool-Goldfish method[2]

According to the goldfish method, we put forward five technical schemes: changing the string to be anti wax material; powering up the downhole string, using the electric thermal effect to remove wax; increasing the weight of the wax scraper, using the gravity of the wax scraper to speed up the wax removal speed; releasing the wax melting agent in the process of wax scraper scraping; adding the automatic heating device in the wax scraper, and using the heat energy to cooperate with the wax scraping in the process of wax scraping.

Table 4. Goldfish method

1. Divide the problem into two parts: reality and fantasy	Reality: the wax deposition of downhole string affects oil well production Fantasy: no wax on down hole string
2. Why the fantasy part can't be realized?	The nature of crude oil leads to wellbore wax deposition
3. Under what conditions can the fantasy part become reality?	Crude oil wax cannot be separated out or does not adhere to the pipe wall after precipitation
4. List the available resources in subsystem, system and super system	Downhole string, high temperature steam truck, sucker rod, mechanical energy, electrical energy, thermal energy and gravity
5. From the perspective of available resources, put forward a possible solution to the problem	Use the electric heat flow effect to remove wax, and use the scraper and weighting rod to cooperate with gravity to remove wax

3.7 TRIZ tool—STC algorithm[2]

Table 5.STC algorithm

1. Imagine the solution of infinite size (S ∞) of the research object	The wellbore diameter is infinite and wax removal is carried out by large mechanical equipment
2. Imagine the solution of infinitesimal size (S 0) of the research object	Propose to remove wax from the surface outside the well
3. imagine the solution of infinitesimal (T ∞) of working process time or object movement speed	The downhole string shall be made of wax-free material
4, imagine the working process of time or object movement speed infinity (T 0) solution thinking	Downhole string has been electrified heating to prevent wax precipitation and adhesion
5. imagine that the cost of expenditure (allowable expenditure) is infinite (C ∞)	Improve the temperature of wellbore produced fluid
6. imagine that the cost of expenditure (allowable expenditure) is infinitesimal (C 0) the solution	Put a low-cost wax scraper into the wellbore and then flow out automatically with the produced fluid

According to STC operator, we put forward five technical schemes: removing the downhole string and removing the wax; using the non wax material to avoid the wax adhesion; improving the temperature of the produced fluid in the wellbore to prevent the wax precipitation; the downhole string has been electrified and heated; putting the paraffin scraper into the wellbore and scraping the wax by gravity, and flowing out with the produced fluid after scraping the wax.

3.8 Using TRIZ method to solve problem-Matter-Field Model[2]

According to the matter field model, we put forward two technical schemes, which are electrifying the paraffin scraper, electric paraffin cleaning and mechanical paraffin scraping at the same time, adding self heating material into the paraffin scraper, putting it into the wellbore to scrape wax, and speeding up the paraffin cleaning speed.

IOP Conf. Series: Materials Science and Engineering **793** (2020) 012028 doi:10.1088/1757-899X/793/1/012028

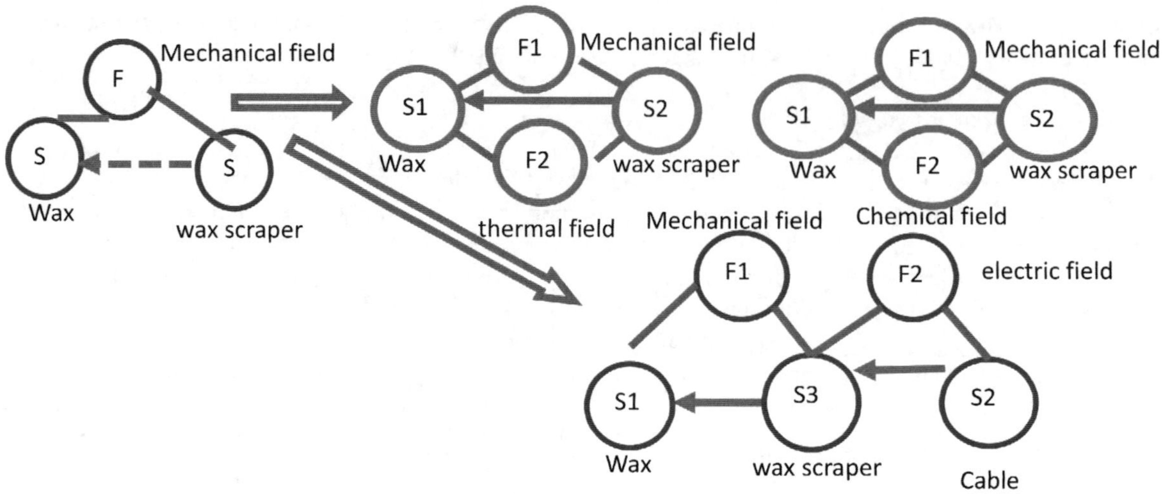

Figure 6. Matter-Field Model

3.9 Using TRIZ method to solve problems-Little People method[2]

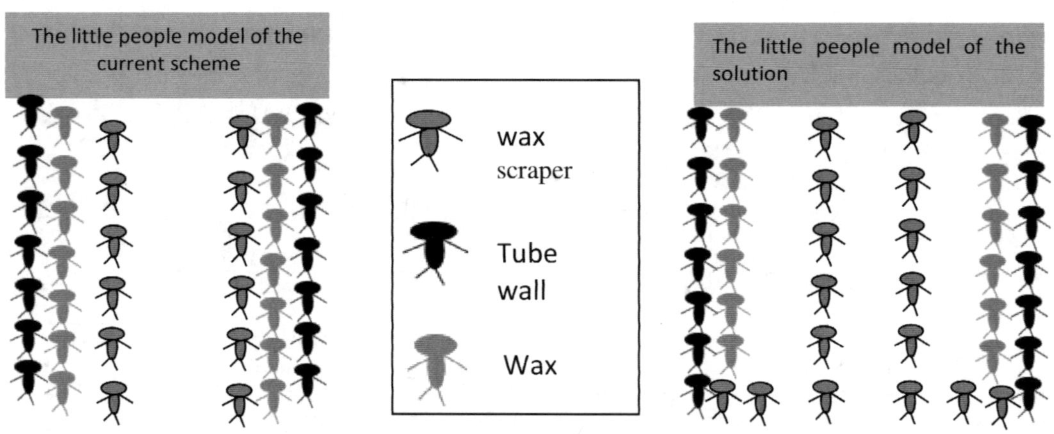

Figure 7. Little People method

According to the Lilliputian method, we put forward the technical scheme of reducing the diameter of the paraffin scraper, enlarging the diameter of the paraffin scraper after running into the bottom of the well, and using the pulling force of the winch to remove the wax.

4. Technical Route and Summary of Scheme

4.1 Technical Route[5]

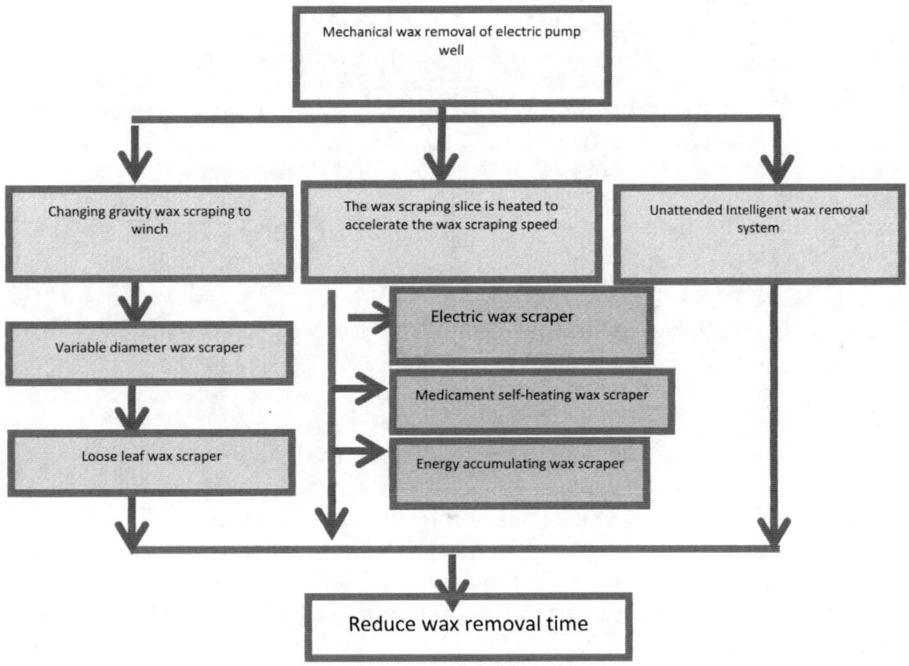

Figure 8.Technical Route

4.2 Summary of Scheme

Table 6 Summary of Scheme

Serial number	Innovation tool	Scheme (number)	Feasibility scheme (number)	Patent Intention(number)
1	Nine Screen diagram	1	1	1
2	Causal Chain analysis	5	2	
3	Function analysis and cutting	3	3	
4	Resource analysis	5	1	1
5	Ultimate ideal solution	5	1	
6	Scientific effects base	1	1	
7	Information interaction theory	6	3	1
8	Technical contradictions	5	5	1
9	Physical contradictions	5	3	
10	Goldfish method	5	5	
11	STC algorithm	5	3	
12	Matter-Field model	2	2	
13	Little People method	1	1	1
total		48	33	5

5.Application of achievements

We have adopted the movable leaf structure of the wax scraper to reduce the resistance in case of wax retraction when going down the well. After going down to the specified depth, the wax scraper can be removed under the action of the lifting force of the spring and winch. The overall design of the wax scraper and the weight bar is that the front section is a cone-shaped guide to minimize the downward

IOP Conf. Series: Materials Science and Engineering **793** (2020) 012028 doi:10.1088/1757-899X/793/1/012028

resistance. The three wax scraper pieces are set at 120 degree intervals, and each wax scraper piece is provided with a reverse flow slope to make the wax scraper open In the process of lifting, the device rotates to enhance the effect of wax scraping.[4]

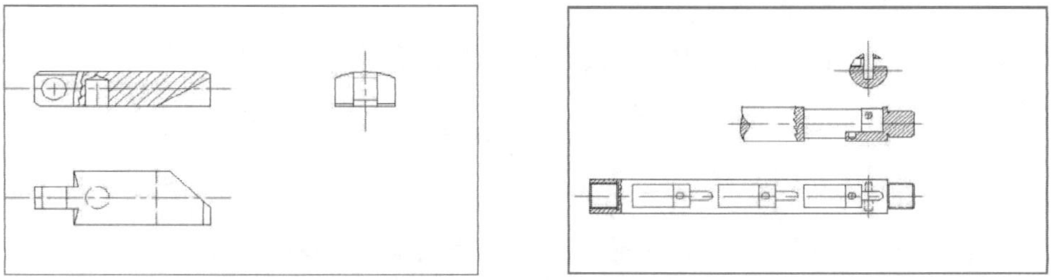

Figure 9. Design drawing of movable leaf wax scraper

6.Conclusion

After using the paraffin scraper, the working efficiency is improved. First, the working time of single well paraffin removal is reduced, and the average working time of single well paraffin removal team is shortened by 14 minutes. The second is to extend the paraffin removal period, and the average paraffin removal period of the whole mine electric pump well is 0.77 times longer. It also reduces costs. Due to the shortening of paraffin removal time and the prolongation of paraffin removal cycle in single well, the cost is effectively saved. The whole mine can save 160000 yuan of labor cost, 5000 yuan of fuel cost and more than 30000 yuan of other costs.

This achievement won the second prize of technical innovation of the third oil production plant and the second prize of major technical innovation achievement of Daqing Oilfield Co., Ltd.

Author brief introduction

XieLu(1975-),male,Heilongjiang people of the people's Republic of China,Oil production engineer of PetroChina Daqing Oilfield Co.,Ltd, Bachelor of Engineering,Tel:13936965892,
LiuYangyang(1987-),female,Heilongjiang people of the people's Republic of China,Assistant oil production engineer of PetroChina Daqing Oilfield Co.,Ltd,Bachelor of Engineering, Tel:15846938059.

References

[1] Wan Renpu Oil Production Engineering Manual [M]. Petroleum Industry Press, 2003.3
[2] Research Committee on Innovative Methods, China Agenda 21 Management Center Compiles Course on Innovative Methods (Primary) Higher Education Press, May 2012
[3] Luo Yingjun Production technology of hard to develop reserves. Petroleum Industry Press, 2005.3
[4] Yang kezhen Fundanmentals of mechanical design Petroleum Industry Press, 1994.5
[5] Sun Yongwei TRIZ:The golden key to innovation,Science Press,2015.11

IOP Conf. Series: Materials Science and Engineering **793** (2020) 012029 doi:10.1088/1757-899X/793/1/012029

Supply network Optimization Based on Supper Efficiency Non-dominated Sorting Genetic Algorithm

Yadong Wang, Quan Shi, Zhifeng You, Lei Li, Fang Zhang

[1]Department of Equipment Command and Management, Army Engineering University of PLA , Shijiazhuang 050003

[2]Science and Technology Innovation Research Center, No.32178 troops of PLA, Beijing 100012

Quan Shi, E-mail: xwzj0003@163.com

Abstract: In order to optimize the supply network, a multi-objective constraint optimization model is established with the constraints of demand satisfaction, lead time and inventory capacity, aiming at minimizing total supply time and the supply costs. By using the proposed upper efficiency non-dominated sorting genetic algorithm, the set of non-dominant solutions are obtained. During optimization process, the upper efficiency data envelopment analysis (SE-DEA) is used to calculate the supper efficiency of the non-dominant solutions. In this way, on the one hand, the algorithm is guided to converge to the optimal efficiency individuals. On the other hand, the non-dominant solutions are sorted to select the optimal one. The example shows that 16 non-dominate solutions are obtained by using proposed algorithm, and the solution with supper efficiency of 0.9538 is determined as the optimal scheme.

1. Introduction

The supply network optimization should take into account the factors of time and cost comprehensively. Thus, the problem is a multi-objective optimization problem and is difficult to model and solve. At present, many scholars have carried out the relevant research on the supply network optimization. Fazli et.al. minimized the total supply cost and transit time between the facilities while maximizing the supply reliability to establish an emergency supply three-objective optimization model [1]. Zhang et.al. developed a multi-objective three-stage stochastic programming model to minimize the transportation time, the transportation cost and the non-satisfaction degree, and the multi-objective model is processed by using the substitute single objective model based on the membership fuzzy variable [2]. Mohammadi et.al. established a multi-objective random planning model to maximize the total expected demand coverage, the total expected cost, and the satisfaction rate among the nodes. A multi-objective particle swarm optimization algorithm was used to solve the model [3]. Su et.al constructed a double-objective multiplication-constrained integer linear programming model based on total response time and emergency resource cost, and used the differential evolution algorithm to search for optimal solution of the model [4].

It can be seen that most of the current research uses multi-objective optimization model. However, the multi-objective optimization solution is difficult to be solved, and cannot obtain the unique optimal solution. In order to solve the problem of supply network optimization, a multi-objective optimization model of supply network is constructed to minimize the supply time and cost. The super efficiency non-dominant sorting genetic algorithm is used to solve the algorithm.

IOP Conf. Series: Materials Science and Engineering **793** (2020) 012029 doi:10.1088/1757-899X/793/1/012029

2. Supply network optimization model

2.1 Problem description

A three-level supply network consists of supply centers, distribution centers and customers. The supply network optimization is based on the demands and the information of the nodes to determine the best product allocation scheme, thereby ensuring that the demands are met with the shortest time and the minimum cost.

The model is built on the following assumptions: (1) the transportation cost and transportation time between nodes are known and fixed, and the transportation capacity between nodes is sufficient; (2) the open cost, inventory cost and maximum capacity of distribution centers are known and fixed; (3) the demands for customers and the shortage costs are known.

2.2 Parameter declaration

I : the index of supply centers, $i = 1, 2, \cdots, I$; J : the index of distribution center, $j = 1, 2, \cdots, J$; K : the index of customers, $k = 1, 2, \cdots, K$; U_j : the maximum capacity of distribution center j ; C_j^{open} : open cost of distribution centers j ; C_j^{inven} : inventory cost of distribution center j ; d_k : demand of customer k ; C_k^{short} : shortage cost of customer k ; T_k^{lead} : the maximum lead time of customer k ; T_{ij}^{trans} : transport time form supply centers i to distribution centers j ; T_{jk}^{trans} : transport time form distribution centers j to customers k ; C_{ij}^{trans} : transport cost form supply centers i to distribution centers j ; C_{jk}^{trans} : transport time cost distribution centers j to customers k .

Decision variables:

x_{ij} :quantity of production supplied from supply centers i to distribution centers j ; x_{jk} :quantity of production supplied from distribution centers j to customers k ; y_j :binary variables, when distribution centers j is operated, $y_j = 1$, otherwise, $y_j = 0$.

2.3 Modeling

Objective function 1: maximizing the total supply time :

$$\min \sum_{i\in I}\sum_{j\in J} T_{ij}^{trans} \cdot x_{ij} + \sum_{j\in J}\sum_{k\in K} T_{jk}^{trans} \cdot x_{jk} \tag{1}$$

Objective Function 2: minimizing the total supply cost of product.

$$\min \; C = C^{open} + C^{trans} + C^{invent} + C^{short} \tag{2}$$

where, C^{open} indicates the open cost of the distribution center, C^{trans} indicates the transport cost, C^{invent} indicates inventory cost , and C^{short} indicates shortage cost, which are formulate as follows:

$$C^{open} = \sum_{j\in J} C_j^{open} \cdot y_j \tag{3}$$

$$C^{trans} = \sum_{i\in I}\sum_{j\in J} C_{ij}^{trans} \cdot x_{ij} + \sum_{j\in J}\sum_{k\in K} C_{jk}^{trans} \cdot x_{jk} \tag{4}$$

$$C^{invent} = \sum_{j\in J} C_j^{invent} \cdot (\sum_{i\in I} x_{ij} - \sum_{k\in K} x_{jk}) \tag{5}$$

$$C^{short} = \sum_{k\in K} C_k^{short} \cdot \left| d_k - \sum_{j\in J} x_{jk} \right| \tag{6}$$

The following constraints are met:

s.t.

$$\sum_{i\in I} x_{ij} \le y_j \cdot U_j \qquad j=1,2,\cdots,J \tag{7}$$

$$\sum_{k\in K} x_{jk} \le y_j \cdot U_j \qquad j=1,2,\cdots,J \tag{8}$$

$$\frac{d_k}{\sum_{j\in J} x_{jk}} \le 1 \qquad k=1,2,\cdots,K \tag{9}$$

$$\sum_{k\in K} x_{jk} \le \sum_{i\in I} x_{ij} \qquad j=1,2,\cdots,J \tag{10}$$

$$\begin{aligned} &\max\left\{T_{ij}^{trans}\cdot \mathrm{sgn}(x_{ij}) \middle| i\in I, j\in J\right\}+ \\ &\max\left\{T_{jk}^{trans}\cdot \mathrm{sgn}(x_{jk}) \middle| j\in J\right\} \le T_k^{lead} \qquad k=1,2,\cdots,K \end{aligned} \tag{11}$$

$$x_{ij}\in N^+ \; x_{jk}\in N^+ \; y_j=\{0,1\} \tag{12}$$

where, the constraint (7) and (8) indicate that the closed distribution centers does not participate in the product supply, and also specifies that the number of products supplied should not exceed the maximum capacity of the distribution centers. Constraint (9) stipulates that the product demand of customers must be met; The constraint (10) specifies that the amount of output of the distribution centers shall not exceed the input; Constraint (11) specifies that the lead time for each customers shall not exceed the maximum allowable deadline.

3. Supper efficiency non-dominated sorting genetic algorithm

In order to solve the optimal solution of multi-objective optimization model of supply network, an supper efficiency non-dominated sorting genetic algorithm (SENSGA) is proposed in this paper. The algorithm adopts the basic framework of genetic algorithm. In each iteration, the non-dominant solutions of the population are stored into archive, and the elite individual is selected from the archive to guide the evolution of the population. Different from traditional genetic algorithm, SENSGA algorithm uses sorting mechanism to select elite individuals. The non-dominant individuals are sorted based on their efficiency values and the individual with the maximum efficiency is selected as the elite individual. The efficiency can reflect the comprehensive performance of the individual in the optimization model, which has practical significance. The efficiency is calculated by using an improved Data Envelopment Analysis (DEA). Each non-dominant individual is a Decision Making Units (DMU).

3.1 Evolution strategy

The evolution strategy is the process of the mutation, crossover, selection. By evolution, the population is constantly improved to obtain a better solution.

Step 1 Mutation

$\boldsymbol{x}_i = x_{i1}, x_{i2}, \cdots, x_{iD}$ is the i th parent individual, and $\boldsymbol{v}_i = v_{i1}, v_{i2}, \cdots, v_{iD}$ is the corresponding mutation individual:

$$\boldsymbol{v}_i = \boldsymbol{x}_i + F\times(\boldsymbol{x}_{elite} - \boldsymbol{x}_i) + F\times(\boldsymbol{x}_{rand} - \boldsymbol{x}_i) \tag{13}$$

where, $\boldsymbol{x}_{elite}$ is the elite individual selected from archive, $\boldsymbol{x}_{rand}$ is the random individual selected form the archive randomly, $\boldsymbol{x}_i \ne \boldsymbol{x}_{elite} \ne \boldsymbol{x}_{rand}$, F is mutation rate.

Step 2 Crossover

The crossover individual $\boldsymbol{u}_i$ is produced by parent individual $\boldsymbol{x}_i$ and the mutation individual $\boldsymbol{v}_i$:

$$u_{ij}=\begin{cases} v_{ij}, & if\ (rand_{ij}\le CR) \\ x_{ij}, & otherwise \end{cases} \tag{14}$$

where, $rand_{ij}$ is uniformly distributed random number between $[0,1]$, $CR\in[0,1]$ is crossover rate.

Step 3 Selection.

Compared the the fitness function of the parent and the crossover individual. The non-dominant one is selected as offspring individual.

Step 4 Update the archive

Mix the non-dominant solution of the offspring population to the archive, the non-dominant solution of the mixed population is selected as a new archive.

3.2 Fitness function

The multi-objective optimization model constructed in this paper is a constrained optimization problem. In this paper, the dynamic penalty function method is used to deal with the model constraints [5]:

$$f(\vec{x})=O(\vec{x})+(M_0-\frac{iter}{\max iter}\cdot M_K)\cdot P(\vec{x}) \tag{15}$$

where, $f(\vec{x})$ is fitness function, $O(\vec{x})$ is objective function, M_0 is the initial penalty factor, $iter$ is the current iterations, $\max iter$ is the maximum iterations, M_K is penalty coefficient, and $P(\vec{x})$ is the constraint violation.

3.3 Metrics

According to the nature of the DEA model, the evaluation metrics is divided into the input metrics and the output metrics, the smaller the input metrics, the higher the efficiency of the corresponding DMU is, and the higher the value of the output metrics, the higher the efficiency of the DMU is.

The input metrics in this paper include:

(1) Supply cost: the calculation formula is the objective function (2);

(2) Supply time: the calculation formula is the objective function (1).

The output metrics includes:

(2) Timeliness metrics: The inverse of the lead time is taken as timeliness metrics, the calculation formula of the delay time is the constraint (11);

(3) Fill rate: the calculation formula is the constraint (9);

(4) Constraint violation: because the model also involves constraints such as capacity and flow balance, the reciprocal of the overall constraint violation of the model is used as the metrics.

3.4 Super efficiency sorting

The DEA model is used to calculate the individual efficiency [6]:

$$\max\ E_{dd}=\frac{\sum_{r=1}^{s}\mu_{rd}\cdot y_{rd}}{\sum_{i=1}^{m}\upsilon_{id}\cdot x_{id}} \tag{16}$$

$$s.t.\quad E_j=\frac{\sum_{r=1}^{s}\mu_{rj}\cdot y_{rj}}{\sum_{i=1}^{m}\upsilon_{ij}\cdot x_{ij}}\le 1 \qquad j=1,2,\cdots,n \tag{17}$$

$$\mu_{rj}\ge\varepsilon\quad \upsilon_{ij}\ge\varepsilon\qquad r=1,2,\cdots,s\quad i=1,2,\cdots,m \tag{18}$$

where, y_{rj} indicate the r th output metrics of the j th DMU, μ_{rj} is the weight of y_{rj}; x_{ij} indicate the

i r th input metrics of the j th DMU, υ_{ij} is the weight of x_{ij}. E_{dd} is the self efficiency of DMU_d. Constraint (17) specifies that the efficiency value of all DMU should be between 0 and 1. Constraint (18) stipulates that the weight of ownership shall be greater than 1, ε is a non-Archimedes number smaller than any positive number.

According to the self-evaluation efficiency, the traditional DEA model can only distinguish the effective units (the efficiency value is equal to 1) and the ineffective unit (the efficiency value is less than 1) , and the effective units cannot be sorted [7]. SENSGA algorithm can evaluate and sort all DMUs by calculating the super efficiency of each DMU.

The calculation steps for super efficiency are as follows:

First of all, the dual form of the traditional DEA model is obtained. Then, on the basis of the dual model, the super efficiency DEA model will exclude the evaluated DMU, and its input and output will be replaced by the linear combination of the input and output of all other DMEs. In this way, the efficiency value is no longer limited to between 0 and 1 but is allowed to exceed 1. Therefore, the model of the super-efficiency DEA is as follows:

$$\begin{cases} \max\left[\theta - \varepsilon(\hat{e}^T s^{-0} + \hat{e}^T s^{+0})\right] \\ s.t. \sum_{\substack{j=1 \\ j \neq j_0}}^{n} \lambda_j x_j + s^- = \theta x_{j_o} \\ \sum_{\substack{j=1 \\ j \neq j_0}}^{n} \lambda_j y_j - s^+ = y_{j_o} \\ \lambda_j \geq 0, j = 1,2,\cdots,n \\ s^- \geq 0 \\ s^+ \geq 0 \end{cases} \tag{19}$$

4. Numerical examples

4.1 Example description

A supply network consists of 2 supply centers, potential distribution centers, and 6customers. The product is shipped to distribution centers from supply centers, and distributed to customers based on their demands. The relevant data for supply centers, distribution centers and customers are shown in tables 1 to 3.

Table 1 Unit transportation time/cost of spare parts between nodes

	distribution center1	distribution center2	distribution center3	distribution center4
supply centers1	56/325	60/520	32/230	50/300
supply centers2	28/530	53/280	45/170	36/260
customers1	5/80	5/82	5/76	4/60
customers2	3/65	3/65	5/65	5/65
customers3	5/58	5/58	2/57	5/80
customers4	2/76	5/65	6/65	6/65
customers5	6/75	6/58	5/58	5/65
customers6	5/80	6/55	3/62	3/75

Table 2 Capacity, unit inventory cost, and opening cost of distribution centers

	distribution centers1	distribution centers2	distribution centers3	distribution centers4
Capacity	55	35	30	20
Inventory cost	30	35	25	15
Opening cost	9000	5500	7500	6000

Table 3 Spare parts demands, shortage cost, and the maximum lead time of customers

	Demands	Shortage cost	Maximum lead time
customers1	18	550	35
customers2	15	600	50
customers3	16	500	65
customers4	15	600	35
customers5	10	550	45
customers6	15	700	50

4.2 Optimization results

The total of 16 non-dominant solutions are obtained by the algorithm, and each solution represents a supply scheme. Table 4 shows the values of the input metrics, output metrics, super efficiency and self efficiency of all the DMUs. It can be seen that, there are 9 effective DMUs. Of which self efficiency are all equal to 1. The optimal solution cannot be further selected from these effective DMUs only according to their self efficiency. By calculating and comparing the super efficiency of each DMU, the one with supper efficiency equal to 0.9538 is selected as the optimal solution.

Table 4 Results of SENSGA algorithm

DMU	Input metrics		Output metrics								Supper efficiency	Self efficiency
	Cost	Time	Timeless	Fill rate1	Fill rate2	Fill rate3	Fill rate4	Fill rate5	Fill rate6	Constraint violation		
1	10755	377	0.1667	1.33	0.70	0.89	1.00	0.63	0.87	0.0058	0.9538	1
2	18401	383	0.1667	1.42	0.70	0.83	1.00	0.63	1.00	0.0069	0.9421	1
3	18905	378	0.1667	1.33	0.70	0.83	1.00	0.63	1.00	0.0069	0.9230	1
4	23898	388	0.1667	1.50	0.70	0.83	1.00	0.63	1.00	0.0072	0.9325	1
5	25354	388	0.1667	1.33	0.75	0.89	1.00	0.63	1.00	0.0078	0.8865	1
6	31461	388	0.1667	1.42	0.70	0.89	1.00	0.63	1.00	0.0081	0.8889	1
7	24811	397	0.1667	1.58	0.75	0.89	1.00	0.63	0.87	0.0072	0.9258	1
8	10846	459	0.1667	1.75	0.70	0.94	1.00	0.63	1.00	0.0052	0.9286	1
9	11425	457	0.1667	1.75	0.70	0.94	0.80	0.63	1.00	0.0053	0.9215	1
10	19276	462	0.1667	1.75	0.75	0.94	1.00	0.63	1.00	0.0062	0.8819	0.9739
11	20386	452	0.1667	1.75	0.65	0.94	1.00	0.63	1.00	0.0063	0.8895	0.9924
12	16846	459	0.1667	1.75	0.70	0.94	1.00	0.63	1.00	0.0057	0.8904	0.9852
13	25230	452	0.1667	1.75	0.75	0.83	1.00	0.63	1.00	0.0069	0.8772	0.9827
14	25276	462	0.1667	1.75	0.75	0.94	1.00	0.63	1.00	0.0068	0.8628	0.9624
15	32776	462	0.1667	1.75	0.75	0.94	1.00	0.63	1.00	0.0070	0.8371	0.9558
16	39299	683	0.0179	1.92	0.85	0.83	1.20	0.63	1.20	0.0023	0.5640	0.7212

5. Conclusion

In this paper, the multi-objective optimization model for supply network optimization is established, and the optimization decision-making problem of the supply network is solved. The main contributions of this paper are as follows: firstly, the proposed multi-objective optimization algorithm and constraint processing method can solve the multi-objective constrained optimization problem of supply network; secondly, the proposed efficiency sorting strategy can further sort the non-dominant solutions, and guide the convergence of the algorithm to the most efficient individuals. Thirdly, by calculating the super efficiency, the problem that the effective DMUs cannot be sorted according to their self efficiency is overcome. In summary, the model and algorithm in this paper provide a framework for supply network optimization of and decision making.

References

[1] Fazli Khalaf M., Khalilpourazari S., Mohammadi M. Mixed robust possibilistic flexible chance constraint optimization model for emergency blood supply chain network design [J]. Annals of Operations Research. 2017, 12, 1-31.

[2] Zhang J., Liu H., Yu G., et.al. A three-stage and multi-objective stochastic programming model to improve the sustainable rescue ability by considering secondary disasters in emergency logistics [J]. Computers & Industrial Engineering. 2019. 135, 1145-1154.

[3] Mohammadi R., Ghomi S. M. T. F., Jolai F. Prepositioning emergency earthquake response supplies: A new multi-objective particle swarm optimization algorithm [J]. Applied Mathematical Modelling.

2016, 40(9-10): 5183–5199.

[4] Su Z., Zhang G., Liu Y., et.al. Multiple emergency resource allocation for concurrent incidents in natural disasters [J]. International Journal of Disaster Risk Reduction.2016, 17: 199–212.

[5] Panda A., Pani S. A Symbiotic Organisms Search algorithm with adaptive penalty function to solve multi-objective constrained optimization problems [J]. Applied Soft Computing, 2016, 46, 344–360.

[6] Coello C A C . Constraint-Handling Techniques used with Evolutionary Algorithms.[C]// Conference Companion on Genetic & Evolutionary Computation Conference: Late Breaking Papers. ACM, Philadelphia, 2012: 2445-2466

[7] Ramanathan R. An Introduction to Data Envelopment Analysis [J]. Journal of the Operational Research Society, 2013, 56(6):751-752.

IOP Conf. Series: Materials Science and Engineering **793** (2020) 012030 doi:10.1088/1757-899X/793/1/012030

Control-oriented representation method of assembly process for intelligent assembly system

Yan Liu[1], Zhijing Zhang[1*], Pengyue Guo[1] and Lei Wang[1]

[1] Department of Mechanical Engineering, Beijing Institute of Technology, Beijing, Beijing, 100081, China

[*]Email: zhzhj@bit.edu.cn

Abstract. This work put forward a new idea of reconfigurable control of assembly process with a modular assembly control schedule (ACS) as system input and a series of automatic assembly actions of terminal execution units as system output. Based on this, an assembly process representation mechanism for reconfigurable control of intelligent assembly system is proposed, which provides an important basis and support for the further optimization and control of assembly actions. Firstly, the concepts of process meta-action (PMA) and control meta-instruction (CMI) of assembly were proposed, and the assembly process was decomposed into multiple PMAs by PSA (Procedure-Step-Action) decomposition method. Then, oriented to the control of assembly actions, PMAs were symbolized. Finally, the assembly of a micro gear was taken as a case study, and this work has been used to schedule the assembly actions of an actual micro-assembly system.

1. Introduction

The object of traditional automatic assembly system is one or several changeless products or components [1], and the control of assembly process is usually realized by running program codes, which are embedded in the software. If the assembly object is changed, in addition to adjusting the hardware of the system, the control program of the assembly process must be rewritten or modified, due to the change of assembly process flow. It must be completed by the programmer, and it is difficult for the on-site process engineer to complete. As for the intelligent assembly system, it must be able to meet the assembly requirements of products or components with multiple species and variable batches. Therefore, the intelligent assembly system should be reconfigurable not only at the mechanical level but also at the control level [2]. In other words, it should be able to adjust the control of the assembly process conveniently and simply, which can be completed by the assembly process engineer.

In view of this demand, this paper put forward an idea of reconfigurable control of assembly process with a modular assembly control schedule (ACS) as system input and a series of automatic assembly actions of terminal execution units as system output, as shown in figure 1. ACS refers to a sequence that carries assembly action control information corresponding to the assembly process and can be directly used for the control of automatic assembly. It is the basis for the automatic assembly system to perform the assembly task and the bridge between the assembly task layer and the assembly control layer. When the assembly requirements change, the corresponding ACS can be adjusted quickly, and the system can meet the new assembly task requirements only by changing the input. This is of great significance for the realization of intelligent assembly of multi-variety, small-batch and variable-batch precision micro-products or micro-components.

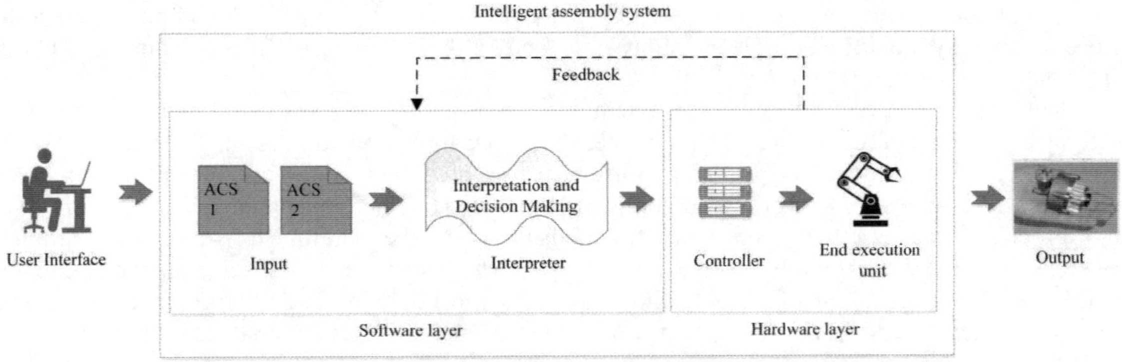

Figure 1. Reconfigurable control of intelligent assembly system.

To obtain the corresponding ACS through the assembly process flow is the basic premise to realize reconfigurable control of intelligent assembly system. Automatic assembly process is the key to guide the assembly tasks of automatic assembly system and the direct basis for compiling the corresponding ACS. Therefore, to build an effective sequence representation of the assembly process is the main task of this paper.

In the literature, several representation schemes have been proposed to represent the assembly sequences. It can be roughly divided into two categories, one is process-oriented virtual assembly planning, and the other is object-oriented product information construction. The purpose is to optimize and evaluate the assembly sequence. Liu J H and Liu Z Y respectively proposed a representation model of hierarchical assembly task Link (HATL) [3] and an approach of multi-level representation of product information [4]. But these schemes are not combined with the assembly action control, only for virtual assembly planning. Johnson P Thomas proposed a type of Petri nets to represent the different levels in the assembly hierarchy, namely the assembly plan level and the control plan level [5]. Su Q proposed and optimizing a hierarchical ASP approach, focusing on assembly precedence relations (APRs) reasoning, assembly sequence planning (ASP) [6]. Niels L and Li H Q respectively built an ontology for the definition and validation of assembly processes for evolvable assembly systems [7] and an ontology-based modelling and reasoning framework for assembly sequence planning [8]. These representation schemes were for sequence optimization, not for assembly process control. So the assembly processes were described from the perspective of the feature information of the assembly object, not the automatic assembly action of the system. Based on these efforts, an assembly process representation mechanism for reconfigurable control of intelligent assembly system is proposed in this paper.

2. Process representation of intelligent assembly

The assembly process is determined by the manner and the sequence in which the product parts are put together into a complete product. It should arrange the sequence of the assembly activities of the assembly system, and also solve the problem of task assignment of the system to the mechanical equipment. It is very important for the early design, debugging and actual operation of automatic assembly equipment, and even for the assembly-oriented product design [9]. So, the design of automatic assembly process is really important. It is usually designed by process engineers and can be well understood by users, but can't be read and understood directly by computers. Therefore, in order to make it readable, it is necessary to symbolize the assembly process. Decomposition of automatic assembly process is the premise of its symbolization.

IOP Conf. Series: Materials Science and Engineering **793** (2020) 012030 doi:10.1088/1757-899X/793/1/012030

2.1 Decomposition of process flow

For the reason above, this paper proposed a decomposition method named Process-Step-Action (PSA) for the assembly process of intelligent assembly system. As shown in figure 2, an automatic assembly process generally includes multiple assembly tasks and sub-tasks, each of which is completed by one or several terminal execution units of the system.

The intelligent assembly system can assemble several parts into a whole efficiently and quickly with a required assembly accuracy. This process is called the automatic assembly process, which is composed of several assembly procedures. In the whole process of automatic assembly, the system continuously completes a series of assembly operations on same parts or other parts added for assist, and produces one assembly connection relationship with the assembled parts or one contact relationship with the assembly platform for fixing, which can be called an automatic assembly procedure. The completion of an automatic assembly procedure requires the cooperation of each necessary execution module of the system, which consists of a series of automatic assembly steps. The system is divided into executive modules by the assembly function, such as the robot module for loading and unloading parts, the computer module for control and display, the assembly manipulator module for assembling, the precision adjustment module for position and posture slight adjustment, the image processing module for image acquisition, processing and matching, and etc. A series of assembly operations continuously be completed by the same execution module of the system on same parts is called an automatic assembly step. The system execution module also includes one or several terminal execution units. It is the smallest function unit, such as a motion axis, that can perform an action in the automatic assembly system. According to the different control modes of the system on the terminal execution units, the automatic assembly step is divided into several smallest action units which can't be further divided, and it is called the automatic assembly process meta-action (PMA).

The PMA is defined from the perspective of control and system input, not from the perspective of assembly action and system output. Therefore, some of the response effects of the PMAs are dynamic, such as axial motion, and some are static, such as image processing, and some include one action, and some include more than one action, such as the adjustment of precision adjustment module. The difference of control mode here refers to the difference in communication mode, communication port and communication content.

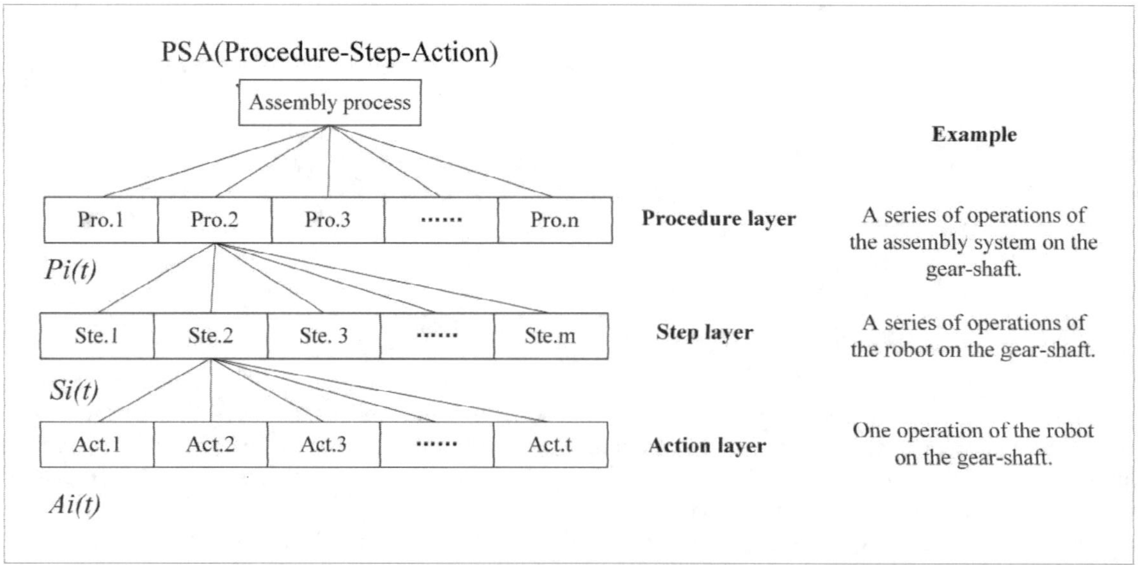

Figure 2. Decomposition mothed of PSA of assembly process.

2.2 Symbolization of PMA.

The symbolization of PMA means standardization, modularization and mathematical description. Each PMA carries specific assembly information, which is called PMA information. As follows:

$$A(i) = \{Mec, Exe, Pos.Par, Ope, Tim, Fun, Mod, Para\}$$

Where:

Mec — The mechanical execution module which the PMA execution unit belongs to.
Exe — The terminal execution unit which executed the PMA.
Pos — Location of static PMA or the target location of dynamic PMA.
Par — The part being operated in the PMA.
Ope — The operation of the part in PMA.
Tim — Time needed in the PMA.
Fun — The function of the PMA.
Mod — The control mode of the PMA.
$Para$ — Parameters of the PMA which control needed.

For example:

$$A_{2.2}(t) = \{Robot; Robot; Home; Part1; Held; 7.2; Robot; RobotMove; Coordinate = Cartesian;$$
$$Route = MoveJ; Speed = 100; X = 0; Y = 0; Z = 0; R = 0\}$$

The formula above means that the robot carries part 1 and moves to position $(0,0,0,0)$ in Cartesian coordinates at a speed of 100mm/s, with the goal of returning to the original position, which takes 7.2 seconds.

The PMA information contained in $Ai(t)$ can be divided into two parts, namely external attribute information and internal control information. The former is used to describe the time-space characteristics of the occurrence of the PMA, which is used in the process optimization stage. The first six elements belong to it. The latter is used for the control of the PMA, which is the key information for the successful communication between the industrial computer and the terminal execution unit, and also the basic content of the control meta-instruction (CMI). The last three elements belong to it. Each CMI can be composed of several different level instructions and several parameters. A CMI with full information can be used to control a PMA, that is, to control the terminal execution unit to execute an action unit, as shown in Figure 3.

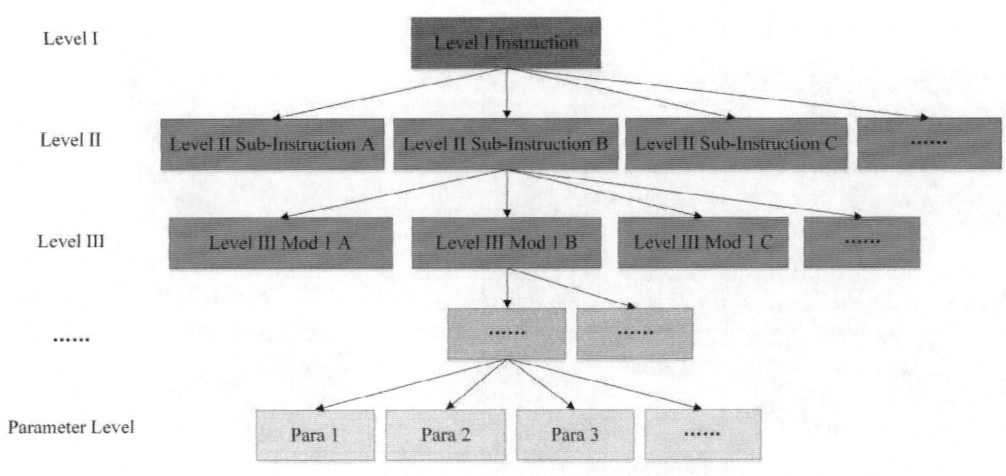

Figure 3. Composition of Control Meta-Instruction (CMI).

IOP Conf. Series: Materials Science and Engineering **793** (2020) 012030 doi:10.1088/1757-899X/793/1/012030

According to the internal control information of PMA information, the corresponding CMI can be easily obtained, which can be used for the control of the PMA, as shown in figure 4. It is the basic module that constitutes the ACS.

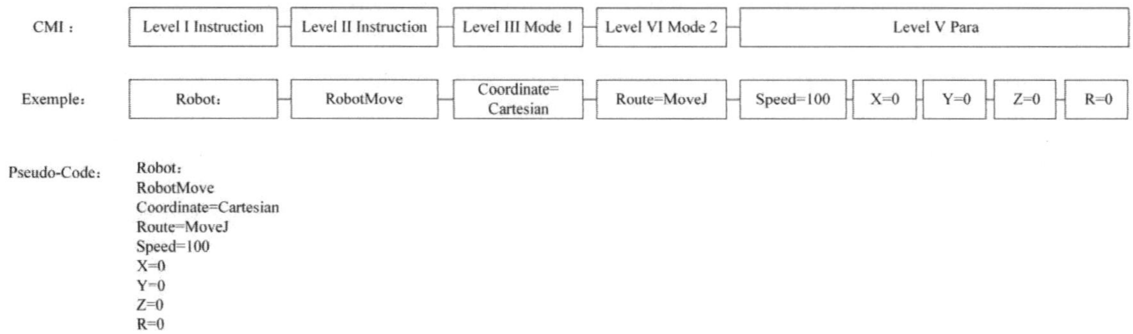

Figure 4. CMI instance and its pseudo-code.

3. Case study.

3.1 Assembly system composition.

In this paper, a specific flexible micro-assembly system is taken as a case study with reconfigurability in mechanical level [10]. The device has five basic modules: SCARA robot module, coaxial alignment module, assembly manipulator module, precision adjustment platform module, and industrial computer module as shown in Figure 5. Among them, the coaxial alignment module [11] includes a CCD camera and two light sources, with two degrees of freedom in the X direction. The SCARA robot module has three degrees of freedom of rotation and two degrees of freedom of movement, with the end equipped with six positions of fixtures, holding parts by vacuum adsorption and rigid clamping. The assembly manipulator module has one degree of freedom of rotation and another one in the Z direction, and also is equipped with six positions of fixtures. The precision adjustment platform module has three degrees of freedom: X and Y directions and rotation around Z axis. Industrial computer module plays the role of manager, such as display of assembly progress on the screen and coordination of assembly pace.

The interference assembly of micro gear-piece and micro gear-shaft is taken as an example. The structure and dimensions of parts are shown in Figure 6.

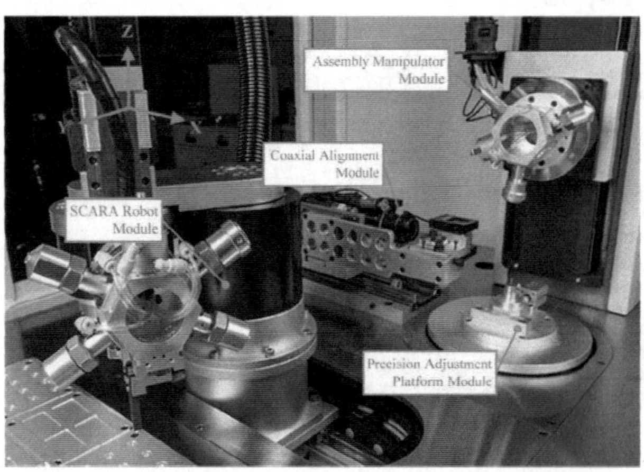

Figure 5. System configuration.

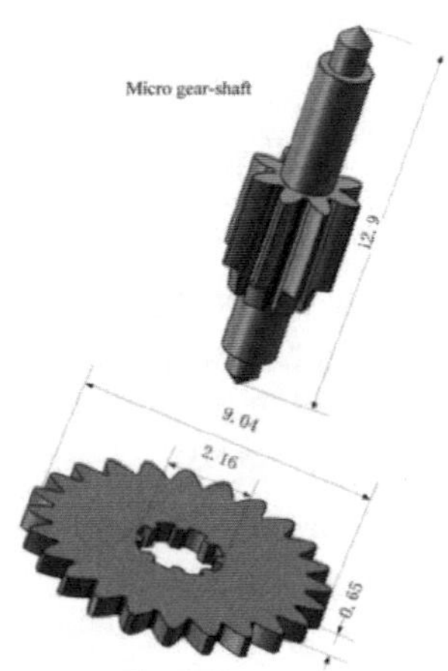

Figure 6. Micro gear parts.

IOP Conf. Series: Materials Science and Engineering **793** (2020) 012030 doi:10.1088/1757-899X/793/1/012030

3.2 Representation of assembly process flow of the micro gear.

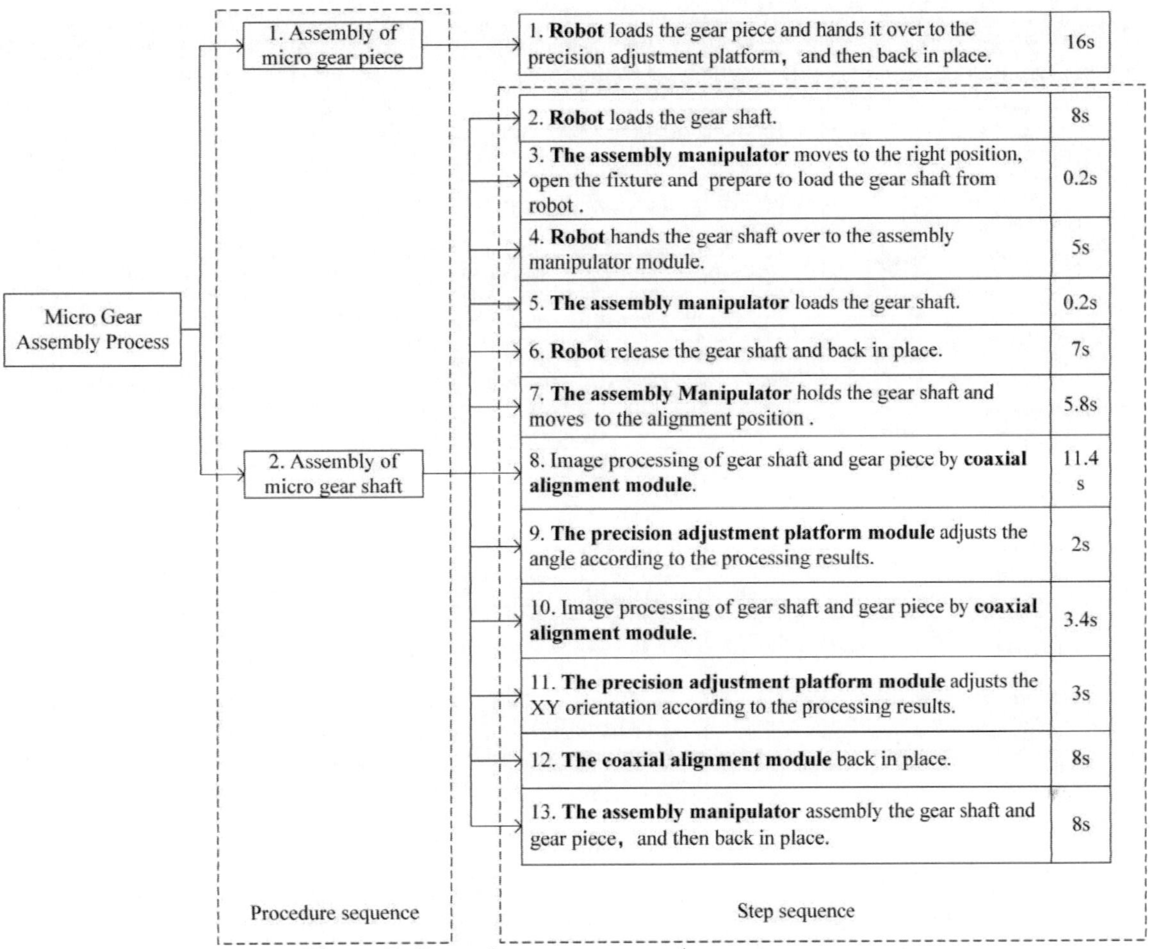

Figure 7. Procedure sequence and step sequence.

The micro gear consists of two parts: gear-piece and gear-shaft. Due to producing a contact relationship between the gear-piece and the assembly platform and a connection relation between gear-shaft and gear-piece, the automatic assembly process can be decomposed into two automatic assembly procedures. Due to the assembly tasks contained in those assembly procedures need to be assigned to different execution modules to complete, each assembly procedure can be decomposed into multiple assembly steps according to the division of task, as shown in figure 7.

According to the different control modes of the system for each terminal execution unit, each assembly step can be decomposed into multiple PMAs. Then the initial PMA sequence can be obtained, as shown in left of figure 8, which can be symbolized and represented according to the definition of PMA information. Finally the symbolized PMA sequence was obtained, as shown in the right of figure 8.

Specific ACS for this micro gear assembly can be obtained according to the internal control information in symbolized PMA sequence, which is the key for the future work of modular and reconfigurable control of assembly process.

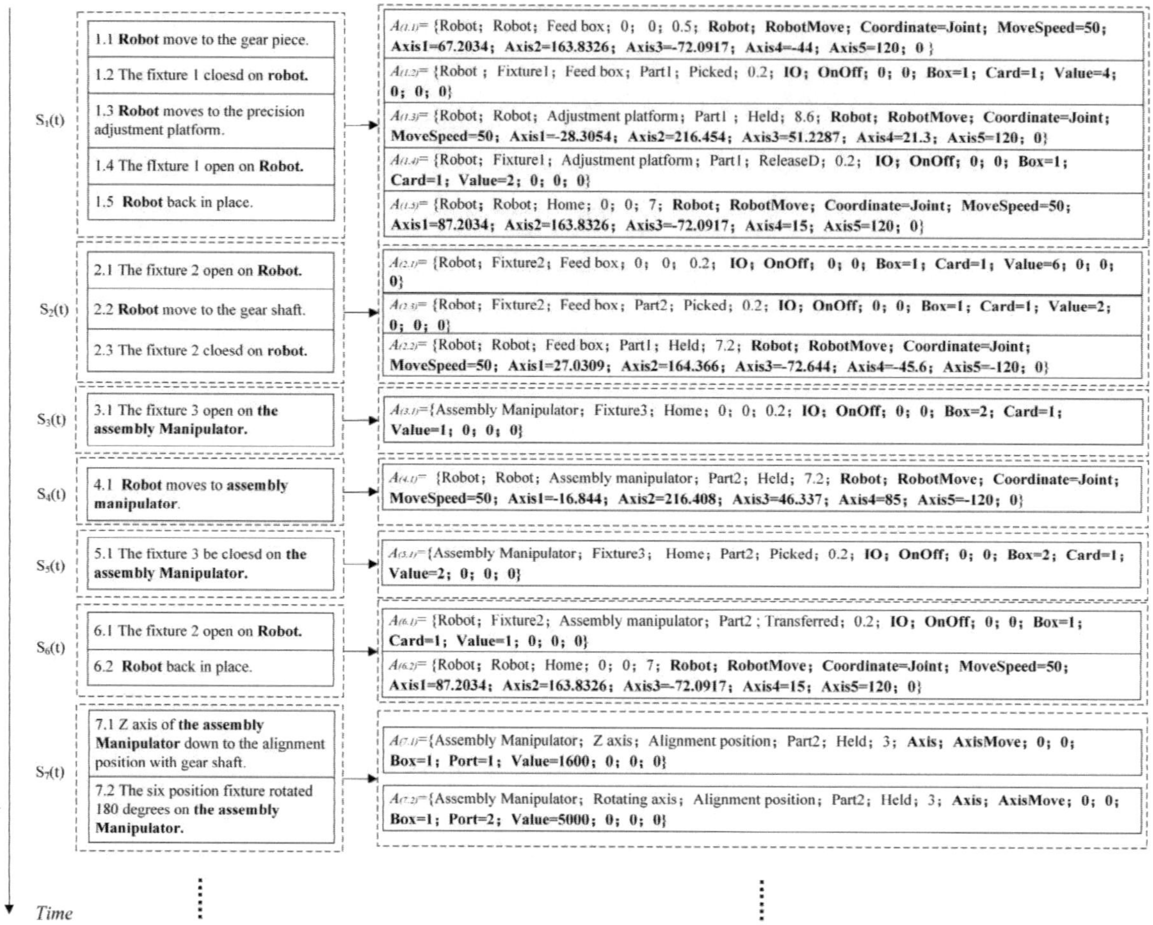

Initial PMA sequence Symbolized PMA sequence

Figure 8. Initial PMA sequence and Symbolized PMA sequence.

4. Conclusion

In order to meet the need of intelligent assembly of multi-variety, small-batch and variable-batch precision micro-products, this paper mainly studies the problems of poor scalability, insufficient reconfigurability and insufficient response to the changes of assembly object in the current micro-assembly systems. Firstly, a reconfigurable assembly process control idea is proposed, which takes the ACS as the system input and the automatic assembly actions of the terminal execution units as the system output. Based on this, a representation mechanism of assembly process for reconfigurable control of intelligent assembly system is proposed, which provides an important basis and support for further optimization and reconfigurable control of assembly process. For automatic assembly process, PSA decomposition method is used to decompose it into multiple PMAs. Then, oriented to assembly process control, these PMAs are symbolized. Finally, the assembly of micro gear is taken as a case study, and the above methods are used to represent the assembly process flow, which paves the way for the next step of assembly sequence optimization and the transformation to ACS. There can be no doubt that ACS is more complex than assembly process flow, which involves serial and parallel actions. The future work will focus on optimization of assembly process and the realization of modular and reconfigurable control for intelligent assembly system.

Acknowledgments
The work presented herein is funded by the National Defense Basic Scientific Research Project of Science and Industry Bureau of China (JCKY2016208B004).

References

[1] Liu JH., Sun QC., Cheng H., et al. (2018) Connotation and Developing Trends of the Products Assembly Technology. Journal of Mechanical Engineering. 54:1-28.
[2] Zhakypov, Z., Uzunovic, T., Nergiz, A. O., et al. (2017) Modular and reconfigurable desktop microfactory for high precision manufacturing. The International Journal of Advanced Manufacturing Technology. 90: 3749-3759.
[3] Liu JH., Ning RX., Yao J., et al. (2005) Research on Assembly Task Based Virtual Assembly Process Model. JOURNAL OF SYSTEM SIMULATION. 17(9):2163-2166.
[4] Liu Z.Y., et al. (2001). The multi-level representation of product information for virtual assembly. Journal of Computer Aided Design & Computer Graphics.13(3):223-228
[5] Thomas, J. , Nissanke, N. , & Baker, K. . (1996). A hierarchical petri net framework for the representation and analysis of assembly. IEEE Transactions on Robotics and Automation, 12(2), 268-279.
[6] Su, Q. (2009). A hierarchical approach on assembly sequence planning and optimal sequences analyzing. ROBOTICS AND COMPUTER-INTEGRATED MANUFACTURING. 25(1) pp. 224-234
[7] Lohse, N., Hirani, H., Ratchev, S., & Turitto, M. (2005). An ontology for the definition and validation of assembly processes for evolvable assembly systems. The 6th IEEE International Symposium on Assembly and Task Planning: From Nano to Macro Assembly and Manufacturing. Montreal, Que.
[8] Qiao, L., Qie, Y., Zhu, Z., Zhu, Y., et al. (2018). An ontology-based modelling and reasoning framework for assembly sequence planning. The International Journal of Advanced Manufacturing Technology, 94(9-12), 4187-4197.
[9] Tietje, C. , & Ratchev, S. . (2007). Design for Microassembly - A Methodology for Product Design and Process Selection. In: 2007 IEEE International Symposium on Assembly and Manufacturing. Ann Arbor. pp. 178-183
[10] Shao S., Zhang ZJ., Ye X., et al. (2018) Modular Design and optimization for Intelligent Assembly System, The 7th CIRP Conference on Assembly Technologies and Systems. 76(1), pp.67-72
[11] Ye, X. , Gao, J. , & Zhang, Z. J. . (2014). A microassembly system with coaxial alignment function. Applied Mechanics and Materials, 487, 678-681.

IOP Conf. Series: Materials Science and Engineering **793** (2020) 012031 doi:10.1088/1757-899X/793/1/012031

Thermal control technology for the space station adjoint modular satellite based on new thermal control materials

Guojie Hu*, Yujun Chen, Min Wang, Bailin Liu and Senhao Yan

Institute of Telecommunication Satellite, China Academy of Space Technology, Beijing 100094, China

hgj919@163.com

Abstract. A modular satellite could perform various functions with various modules and the function of the satellite could be expanded easily. Therefore, the modularization design is an important development direction of space station adjoint satellites and others. However, every module of a modular satellite is structurally independent and the heat dissipation is unevenly distributed, while the interface between modules should support repeatable connection-separating. The traditional thermal control design could not satisfy the thermal control demand. In this paper, the thermal control technology for the space station adjoint modular satellite based on the thermal interface of carbon nanotube array on copper substrate, graphene coating and smart thermal control coating is proposed. By using the new technology, the thermal connection of the assembly and reconstruction system is built and the synergistic heat dissipation of the whole satellite is achieved. As to validate the proposed technology, the finite element model of the space station adjoint modular satellite is established and the whole flight process is simulated. The result indicates that the thermal control technology proposed in this paper can satisfy the thermal control demand of the modular satellite.

1. Introduction

The space station adjoint satellite is a kind of multi-purpose spacecraft that flies around and services the space station. It has the unique advantage of monitoring, defending and technically supporting the station. The space station adjoint satellite has been studied and developed since the 1980s. The typical projects include Nanosat-D[1], Inspector, and the Shenzhou-7 spaceship adjoint satellite[2]. However, the existing adjoint satellites are micro-nano satellites whose volumes are small and functions are rather simplex. Considering the increase demand of the space exploration, the space station adjoint satellite needs to undertake more complicated and flexibly configurable mission. A modular satellite could perform various functions with various modules and the function of the satellite could be expanded easily. Therefore, the modularization design is an important development direction of space station adjoint satellites and others.

The concept of modular spacecraft[3] has been proposed by NASA in 1970s. A modular satellite consists of several structurally independent modules which are connected by uniform mechanical, electrical and thermal interface. The modular satellite could be reassembled to satisfy the need of various missions. The Orbital Express[4] and Phoenix projects[5], which are the in-orbit verification projects, have been launched in 2007 and started up in 2011 separately. But the modular satellite with complex functions and large heat dissipation hasn't been reported.

Different from traditional satellites, a modular satellite consists of several modules whose structure is independent and heat dissipation is unevenly distributed, while the interface between modules should

IOP Conf. Series: Materials Science and Engineering **793** (2020) 012031 doi:10.1088/1757-899X/793/1/012031

support repeatable connection-separating. Considering the particular conditions, the cross-module heat transfer based on the traditional thermal control technologies is inefficient. Meanwhile, the surface area of the module that is integrated and miniaturized is limited so that the heat flow of the large power payload equipment cannot be dissipated by the module itself. Therefore, the thermal control technology based on which the effective synergistic cross-module heat dissipation can be realized is the key technology for modular satellites.

In this paper, the thermal control technology for the space station adjoint modular satellite based on new thermal control materials is proposed. The cross-module thermal interface is designed by using the carbon nanotube array on copper substrate[6-7]. The inner faces of modules are coated by graphene film[8] while the outer faces are painted by smart thermal control coating[9-10]. So that the thermal connection of the assembly and reconstruction system can be built and the satellite can be adapted to the variety of the orbital heat fluxes.

2. Space station adjoint modular satellite

Configuration

Considering the basic demand of the space station adjoint satellite, the modular satellite is configured with four standard modules, including guidance and control module (GNCM), chemical propulsion module (CPM), electric power module (EPM) and satellite management module (SMM), as shown in Figure 1. The standard module adopts miniaturized design with side length of 0.5m and is equipped with standard docking interfaces integrating thermal, mechanical, electrical functions, as shown in Figure 2. The modules connected by standard docking interfaces could be assembled by space robot and support repeated installation and removal on orbit.

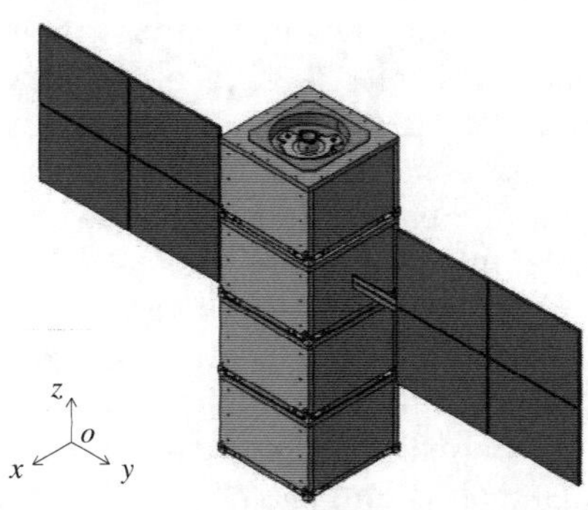

Figure 1. Configuration of the space station adjoint modular satellite.

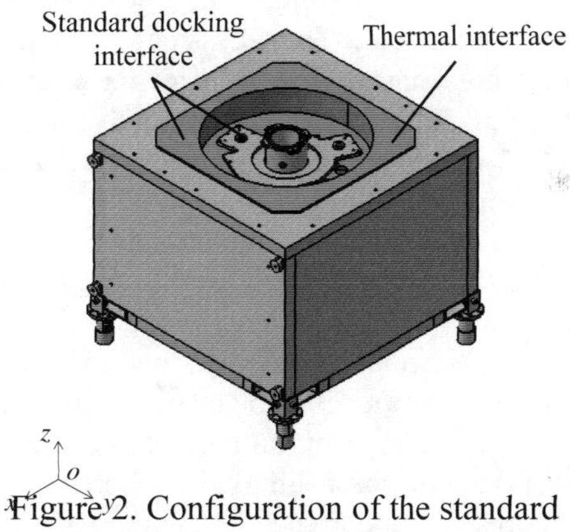

Figure 2. Configuration of the standard module

The heat dissipation of modular satellite is unevenly distributed in four modules. Specifically, the heat dissipation of GNCM, CPM, EPM and SMM is 80, 50, 120 and 60, respectively. The range of temperature control requirements is from -15°C to +55°C.

Orbital heat fluxes

The orbital heat fluxes, which are the main heat source accepted by the satellite in space, are one of the main factors affecting the temperature of the satellite equipment and the basis for the thermal design and thermal analysis of the satellite. The orbital heat fluxes of the space station adjoint modular satellite are determined by the space station orbit parameters that takes the international space station[11] ones, the adjoint orbit parameters which is an elliptical orbit relative to the space station with a perigee height of 0.5km and a apogee height of 1km, the position of the sun that takes the winter solstice as an example and the attitude of the satellite. In the whole flight process, the missions of the space station adjoint modular satellite are mainly divided into three stages including release, orbit and capture. The attitude of satellite is various in different stages.

Based on the given settings, the dependence of the incident heat flux on time is simulated, as shown in Figure 3. It can be seen that all six surfaces are exposed to sunlight and there is no surface with stable heat flow that can be set as the heat rejection surface with traditional thermal coating.

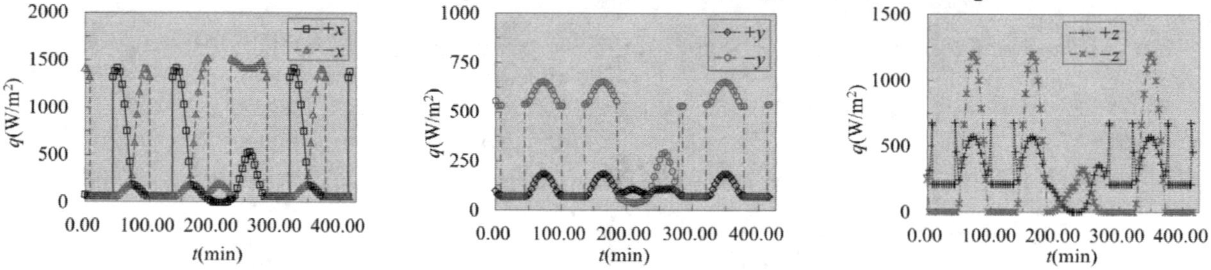

Figure 3. Dependence of the incident heat flux of every surface on time.

3. Thermal control technology and simulation verification

Considering the configuration characteristics, mission requirements and orbital heat fluxes variations, the thermal control design needs to meet the uneven heat dissipation demand, adapt to the unstable orbital heat fluxes, and support the on-orbit assembly and replacement of modules. The traditional thermal control technology, which supports the independent heat dissipation of a single satellite system, cannot meet the thermal control requirements of the modular satellite. Therefore, it is necessary to develop new thermal control technology.

The module of the satellite is equipped with standard docking interfaces that support repeatable connection-separating. The cross-module heat transfer is realized by compressing the thermal interfaces. The separable contact heat transfer between the two interfaces becomes the key of the whole heat transfer path. The traditional methods of solid-solid contact heat transfer enhancement have the problems of non separation, destructive separation or low thermal conductivity, which are not suitable for the modular satellite. The separable thermal disk based on the carbon nanotube array on copper substrate[6-7] is proposed in recent years. Under the contact pressure, the carbon nanotubes which are higher than the copper matrix are inserted into the copper matrix on the other side during the contact process, so that the contact heat transfer is enhanced and the separable and efficient heat transfer between modules is realized. Thus, the thermal disk can be used as the thermal interface of the modular satellite. The influence of the contact heat transfer coefficient at the thermal interfaces on the temperature response is calculated by numerical analysis, which indicates that as long as the coefficient reaches 1500W/(m^2·K) by controlling the properties of carbon nanotubes, optimizing the disk configuration and pressing pressure, it can meet the engineering requirements.

Besides, the module adopts miniaturized integrated design with limited internal space. The traditional method of internal heat transfer enhancement is to arrange the heat pipe network, which needs to occupy large space and then is not suitable for the modular satellite. Graphene is a new type of thermal conducting material with the theoretical thermal conductivity of 5000W/(m·K)[8]. Coating graphene film on the internal surface of the module can significantly increase the thermal conductivity of the deck, and realize the efficient internal thermal conduction. The thermal conductivity of graphene film coating prepared via modified Hummers method with the thickness of 25μm is about 800W/(m·K), which is 5.6 times higher than that of the conventional aluminum skin of the deck.

The ±y surfaces of all the modules are set as heat rejection surfaces, since their orbital heat flux variations are relatively small and the ±y surfaces of a single module cannot meet its own heat dissipation demand, especially for EPM. In addition, as to adapt to the unstable orbital heat fluxes, the smart thermal control coating is used at all the heat rejection surfaces. In this paper, $Ag/VO_2/SiO_2/VO_2$ multi-layer smart thermal control coating is used. The emissivity varies from 0.38 to 0.74 in the temperature range of 40~60°C, and the solar absorptance is only 0.32[10]. By using the smart thermal control coating, the heat rejection surface has the ability to adapt to the orbital heat flux variations.

In order to verify the thermal control technology, the temperature responses of the satellite in the whole flight process by using both the new and traditional technology are modeled, as shown in Figure 4. The results show that by using the new technology the temperature of equipment in every module is within the temperature control range, which proves that the thermal control technology for modular satellites based on new thermal control materials proposed in this paper is available.

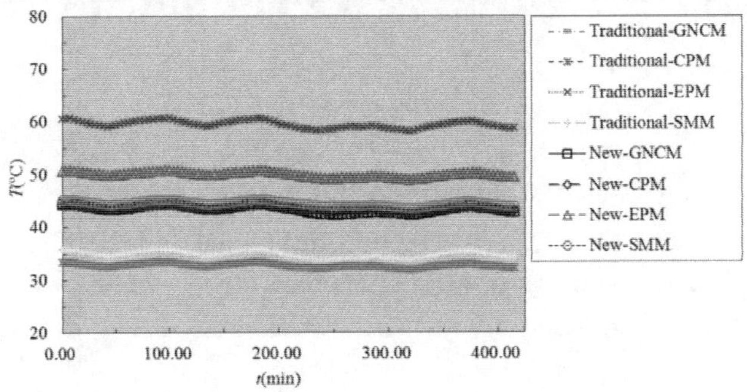

Figure 4. Dependence of the equipment temperature on time

4. Conclusion

The thermal control technology for modular satellites has been studied by taking the space station adjoint modular satellite as an object in this paper. Considering the characteristic of structural independence of every module and requirement of cross-module synergistic heat dissipation for the modular satellite, the thermal control design of repeatable connection-separating interface, internal heat transfer enhancement and adaptive heat rejection surface based on the carbon nanotube array on copper substrate, graphene film and smart thermal control coating is proposed. With this design, the thermal connection of the assembly and reconstruction system is built and the heat flow distributed dispersedly in every module can be synergistically dissipated through the whole satellite. As to verify the thermal control technology, the temperature response of the space station adjoint modular satellite in the whole flight process is modeled. The result shows that the temperature of equipment in every module is within the temperature control range and the thermal control technology for modular satellites based on new thermal control materials proposed in this paper is available.

References

[1] Alhorn D C and Scheierl J M 2010 *Feathersail – Design, Development and Future Impact* (Huntsville: Marshall Space Flight Center)
[2] Chen H, Zhu Z, Wang J, Yu Y and Wu H 2011 *Manned Spaceflight* **5** 19
[3] Li X, Zhang Y and Jiang N 2014 *J. Eqpt. Acad.* **25** 70
[4] Coppinger R 2007 *Flight Int.* **171** 23
[5] Chen L J, Hao J H, Yuan C Z and Fu D Y 2013 *Spacecraft Eng.* **22** 119
[6] Wang H, Feng J, Hu X and Ng K M 2007 *J. Phys. Chem. C* **111** 12617
[7] Sugime H, Esconjauregui S, D'Arsié L, Yang J, Makaryan T and Robertson J 2014 *ACS Appl. Mater. Interfaces* **6** 15440
[8] Xiang J and Drzal L T 2011 *Carbon* **49** 773
[9] Rajeswaran B, Pradhan J K, Ramakrishnas S A and Umarji A M 2017 *J. Appl. Phys.* **122** 163107
[10] Benkahoul M, Haddad E, Kruzelecky R, Wong B, Jamroz W, Soltani M, Chaker M and Poinas P 2010 *Proc. Int. Conf. on Environmental Systems* vol 5 (Barcelona: Spain)
[11] Kitmacher G H 2010 *Reference guide to the international space station* (Burlington: CreateSpace Independent Publishing Platform)

Electrical control design of CNC equipment fault based on part quality

HE Yi, CHEN Xuezhen*, ZHAO Guobo, Li Ying

AVIC Chengdu Aircraft Industrial (Group) CO., LTD., 610092, Chengdu, Sichuan, China（*Corresponding author: Dr. Chen Xuezhen, Senior Engineer, email: 532015901@qq.com.）

Abstract. Aeronautical structure component has high value and complex shape, normally manufacturing by advanced five-axis CNC machine tools, but sometimes random failures also occur. Therefore, this research focuses on how to predict the failure, imply redundancy design technique and minimize part quality risk as much as possible. This article first analyses quality risk caused by device failure in manufacturing and introduces several axis movements during alarm occurring, then studies the influence of accuracy in movements and optimizes by the electrical and external power grid redundancy design. At last this article carries out engineering verification though practice application and reduces the quality risk from device failure in order to prepare quality insurance system in intelligent manufacturing.

1. Introduction

With the rapid development of domestic manufacturing industry, computerized numerical control (CNC) machine tools continuously develop to flexible production line or intelligent manufacturing[1,2]. It does not only need higher reliability, efficiency and performance, but also much higher quality requirement. If high value parts are scrapped due to equipment failure, besides economic loss, the customer will also receive reputation damage.

At present, most of the original equipment manufacturers (OEMs) do not consider the influence of machine reaction to part quality when the equipment fails. The fundamental method to solve the problem is to design preventive measures in the design stage. However, the customer often realises the need of equipment failure control in the time of suddenly facing the part quality problem caused by the alarm. Its design has been finalized after the machine installed, so some improvement methods are urgently needed to avoid quality risk.

In this paper, according to the quality risk caused by the equipment failure in the parts machining process, the action and state of the machine tool when the alarm occurs and is cleared are analysed based on the view of machine alarm design. Then the corresponding design principles are carried out. The examples are given for electrical and external auxiliary design.

2. The effect of alarm on parts machining

At present, there are very few researches on alarm control of CNC machine, mainly focusing on the diagnosis and processing of fault alarm[3]. Some scholars study programmable logic controller (PLC) control system [4] and other methods to protect device safety and prevent operation errors, but there is almost no research on alarm design for part quality. In order to study alarm control, the alarm classification method is analysed first.

IOP Conf. Series: Materials Science and Engineering **793** (2020) 012032 doi:10.1088/1757-899X/793/1/012032

2.1 Alarm classification

Alarm information can be divided into system alarm and OEM alarm. CNC system manufacturers such as Siemens, Fanuc and other companies classify the fault information according to the structure or function of the system. The OEMs do the same thing. According above, it is obviously that the design of the alarm information has nothing to do with the safety or quality assurance. The alarm information is only used for reminding, and will not bring direct damage to the quality.

2.2 Alarm state

During the alarm, the machine tool may have five action states: “alarm prompt”, “pause”, “emergency stop”, “fault stop” and “abnormal stop”. At present, there is no design standard for them.

“Alarm prompt” means that the operation panel displays alarm information, but the machine status is not affected and still in normal. It is often used for safety door opening and closing, abnormal liquid level warning or overtemperature warning, etc. This kind of information will not cause damage in a short time or need to be dealt with immediately. Some process informations, such as normal tool change, generating opening and closing information do not even need to be considered.

“Pause” means the pause of the machine tool axis movement, including the machine coordinate axis, the spindle and some auxiliary axes. Common axis pause methods include activating the spindle or coordinate axis pause key in the panel, or switching the feedrate to zero and so on. The common reasons of pauses are such as no flow of coolant, the spindle temperature reaching the warning value, etc. Some “Pause” alarms require later operation or maintenance by authorized person.

“Emergency stop[5]” is an alarm generated when the machine tool is in an emergency situation. All the axes immediately stop and all the power to drive devices are cut off. It can be activated by manually pressing the emergency stop button or by some faults from the machine. Emergency stop normally means there is serious situation such as people in danger, spindle bearing damage or servo drive error, etc. At this time, the axis movement is forced to stop, so the coordinate axis with brake and spindle should immediately stop, but the coordinate axis without brake may still move because of inertia.

“Fault stop” is the stop caused by equipment failure, which may be controlled by CNC system, relay or drive module. Because it does not force the machine to stop or turn off power, some subsystems may still work. The response may be slower, so there is a certain part quality risk in “fault stop” state.

“Abnormal stop” refers to the sudden machine action stop due to the failure of machine tool or external reasons. At this time, the CNC system is out of control and unable to control mechanical system. For example, if the coordinate axis with damaged brake suddenly stops, failing to brake at this time will inevitably lead to out of control, resulting in serious quality and safety risks。

To sum up, when the machine tool is in the four states of “alarm prompt”, “pause”, “emergency stop” and “fault stop”, the control of the equipment is still effective and their actions are under the control of the system. When the equipment stops abnormally, the system will lose control, so the quality and safety risk of abnormal stop is the highest.

2.3 The influence of alarm state on the part quality

The machining quality of parts includes machining accuracy and surface quality. The machine tool failure mainly affects the accuracy. If the alarm affects the parts, it must first be in the condition that alarm occurs in the machining process. At this time, the tool is cutting the parts or the distance between them is small. Once the axis or spindle deviates abnormally due to fault may damage the parts.

In the moment of alarm during part machining, if only alarm prompt information appears, the machine tool status will hold and still execute the numerical control (NC) program, which will not affect part machining, so there is no quality risk. But in order to clear alarm prompt information the accompanying alarm action and reset action may cause the machine tool axis movement, leading to the part accuracy risk.

IOP Conf. Series: Materials Science and Engineering **793** (2020) 012032 doi:10.1088/1757-899X/793/1/012032

The keys on the operation panel such as the axis feedrate control are controlled by the PLC system. If the "pause" state is activated, the sudden change in the cutting process may form a trace which is normaly less than 0.02mm, which only has a slight impact to the surface quality. However, if the relay or servo drive is used to activate the "pause", some large marks may be largely formed due to the inertia or the axis brake, so there is a quality risk.

After troubleshooting, the alarm information must be cleared. Some information must be cleared through machine restart or system reset. Then there may be some influences such as the impact on the coordinate axis during the hydraulic start-up, the axis adjustment caused by the resynchronization, the offset of the coordinate axis caused by the brake releasing, etc. All the influences may cause slight axis movement. The effect is between the "pause" and the "emergency stop".

If emergency stop appears, the coordinate axis with brake will stop immediately. The coordinate axes without brake may move which would have a impact on the parts accuracy due to inertia. If fault stop occurs, the machine response is controlled by the system and hardware components. It is slower than emergency stop and has a greater impact on the part quality. When abnormal stop appears, the coordinate axis may not be able to brake, so the abnormal axis movement could reach several millimetres due to inertia. The gravity axis without balance pressure may even fall to the lowest due to gravity, resulting in parts scrapping and machine damage.

To sum up, the alarm prompt has no direct impact on the part quality. In most cases, pause and return do not affect part accuracy, but in high-precision machining such as boring, shape accuracy may be affected and surface pattern or steps may appear. Emergency stop has a great influence on the machining that the coordinate axes abnormally move due to inertia or brake delay, resulting in accuracy problem. Abnormal stop is most likely to creat scrapped parts, which shall be avoided with all efforts. In short, the ability to protect the part quality from high to low is followed by alarm prompt, pause, return, reset, emergency stop, fault stop and abnormal stop. There are some cases to guarantee the part quality shown in the examples of electrical alarm design and external auxiliary alarm design.

3. Electrical alarm design case

The electrical design can work if the control system of the machine tool is effective. In the state of "alarm prompt", "pause", "return", "emergency stop", "reset" and "fault stop", the CNC system is still under control, so they can be optimized through the control of electrical components or PLC.

A five-axis horizontal machining centre includes a linkage mechanism between machine and guard door. The Y axis balance pressure was at the low limit position then. When the operator opened the door , the Y axis dropped about 0.2mm at the moment, leading to part damage. The guard door switch controls the enable signal in the servo drive. When the guard door is opened, the enable signal is cut off and its message is generated. The drives do not output the driving power and the enable display on the panel is cancelled. Then the physical delay from enable relay and brake relay may produce axis deviation due to preload. When the driving power is turned off in Y axis, there may be a slight drop or rise due to the fluctuation of the balance pressure. This time the original control of the guard door opening and closing is applied by the design method of fault stop.

The machine tool is under the control during alarming, so the guard door control program can be modified in the PLC system. First disconnecting the door opening switch wirings from the drive hardware, then by modifying the PLC program the door opening switch signal links axis and spindle feedrate enable keys. When the door is closed, the system clears the "spindle and axis stop" message and other messages. Both the before and after optimized control flow charts are as shown in Figure 1:

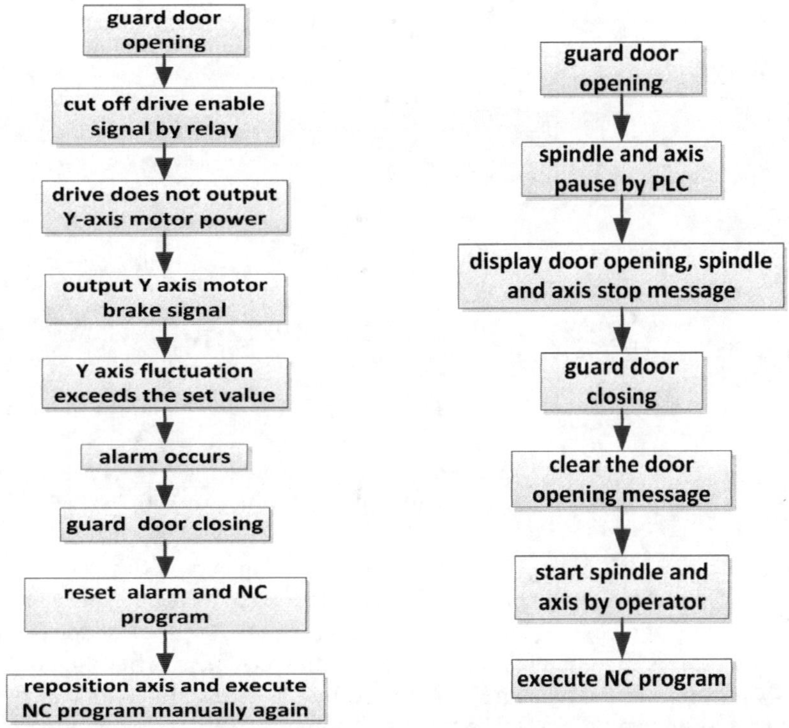

Figure 1. The door control flow chart before and after improvement

It can be seen from the above flow chart that after the safety door is opened, the PLC system activates the spindle and axis pause and display pause message. when the door is closed, the pause message is automatically cleared and the operator can manually start the spindle and axis again. In this case, stop control by the hardware drive with high risk is successfully changed into stop control by the PLC program with low risk, which reduces the risk of part quality.

4. External auxiliary design case

Some external factors such as compressed air supply or abnormal power grid may still cause equipment failure and abnormal stop alarm. In this condition it is difficult to guarantee the machining accuracy only by applied electrical or mechanical hydraulic design, so it is necessary to consider adding external auxiliary devices. The below takes power problem because it is hard to deal with.

The main causes of abnormal power grid are harmonic, voltage flash and power failure, which may be caused by lightning stroke, short circuit, reclosing or misoperation, etc. The harmonic problem is small due to good quality inductors and filters. The main problems are voltage flash and power failure.

Voltage flash means that the supply voltage deviates from the rated voltage large in a short time due to system failure, which belongs to the transient voltage quality problem[6]. Voltage flash includes voltage sag, voltage short-time interruption and voltage transient rise. It has been found that the main cause of CNC equipment failure is the short-term interruption. If there is a serious voltage flash, dynamic voltage restorer (DVR) is needed to ensure the quality.

DVR[7] is a voltage source power electronic compensation device which is connected in series between the power supply and the equipment. It has exceqllent dynamic performance. When the power grid is down, it can restore the power supply at the load side. When the voltage flash change to power cutting, a signal is sent to the machine tool. The NC equipment start emergency stop or return action to maintain machining accuracy of parts. The electrical diagram is as shown in Figure 2:

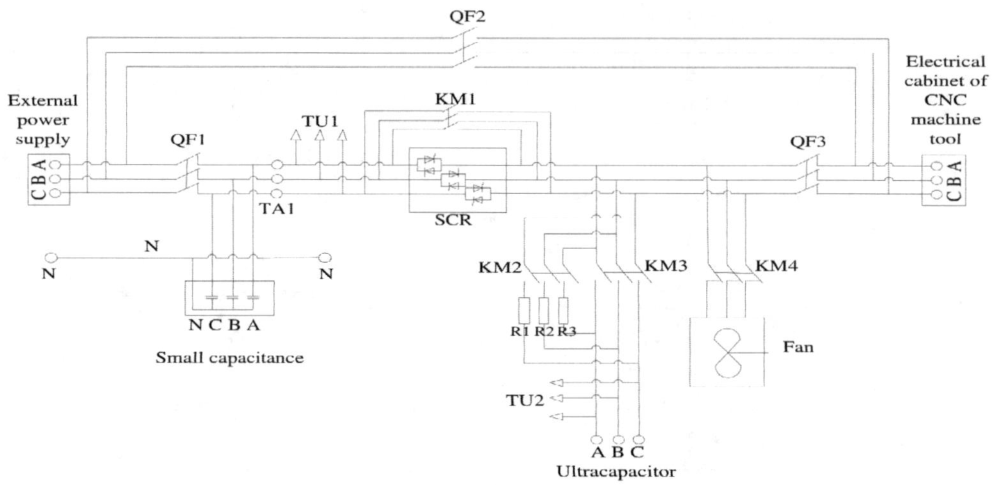

Figure 2. Dynamic voltage restorer (DVR)[8]

When the power grid is normal, the external power supply provides power to the CNC machine tool through QF1, SCR rectifier and QF3. DVR is equivalent to a current source at that time. When the power grid is abnormal and beyond the DVR setting value, DVR will close the relay KM3 and activate the ultracapacitor. At this time, the DVR equals to a voltage source. After the power grid voltage flash disappears, DVR stops and the machine is powered by the external network again. When the voltage transient becomes to power cutting, DVR will send a signal to the machine tool after the setting time. Then the machine tool will respond based on the signal. The control flow chart is as shown in Figure 3:

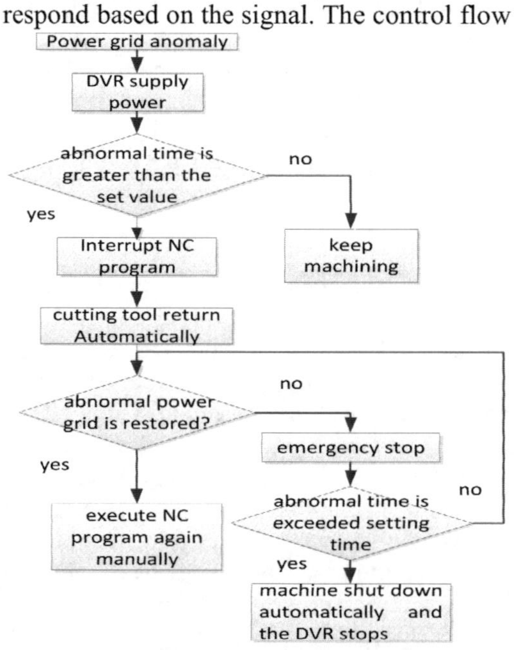

Figure 3. Control flow chart of DVR

Because the design is carried out and the above DVR application fixs the power grid problems such as power cutting and voltage flash, the parts quality problems are basically solved in a large horizontal five-axis machining centre. meanwhile, the high-value equipment itself also has been protected.

5. Summary

CNC equipment is developing rapidly towards intelligent manufacturing and flexible production line, and the quality insurance ability of equipment directly determines the products quality. In CNC equipment there must always be a certain amount of failures, which would directly determine the machining quality, so the design of equipment failure control needs higher requirements.

This paper mainly analyses the machining accuracy of parts in the cutting process when alarm occurs. Through the electrical design and the external auxiliary design of the machine tool, the NC equipment is fully ensured and the part quality problems caused by the equipment failure are minimized.

The fault control design of machine tool based on the machining accuracy of parts is to eliminate the quality risk by preventive methods. It is a important basic technology for the future flexible production line to ensure product quality. However, the manufacturing facilities do not only include machine tools, but also logistics equipment, robots and other device. These equipments may also have many contacts with parts, so there may be quality risks if an alarm occurs. Fault control design requirements must be needed then. This paper does not give the example for mechanical hydraulic design which is needed in future research. The fault design and analysis in this paper can be used as a reference to other equipment fault design.

Funding information: This work is supported by the Major Project of National Science and Technology (Grant No.2017ZX04002001).

References

[1] Ray Y.Zhong;Xun Xu;Intelligent Manufacturing in the Context of Industry 4.0: A Review, 2017,003(5) pp 616-630
[2] Zhou Ji, Li Peigen, Toward New-Generation Intelligent Manufacturing, 2018,004(1) pp 11-20
[3] Felix C. Gomez de Leon HijesMaintenance strategy based on a multicriterion classification of equipments, Reliability Engineering & System Safety, 2006,91(4) pp 444-451
[4] SU Qian, ZHANG Qunsheng, Security Control Design For Computer Numerical Control Machine Tool Based On PLC, Machine Tool & Hydraulics, 2012, 40(11) pp 169-171.
[5] Dragan Kuzmanovićđorđe Aćimović, Emergency Stop, Starting And Stopping Drives Of Cnc Controlled Machine Tools, Tehnika , 2005,(1) pp 11-19
[6] Zheng, Xiang; Feng, Shiwei, A voltage-transient method for characterizing traps in GaN HEMTs, Microelectronics & Reliability, 2019,93(2) pp 57-60
[7] LEI He, Study on Control Technique of Dynamic Voltage Restorer, Huazhong University of Science & Technology, 2014
[8] LUO Xinghua, WANG Qiang, A protection method for parts processing in prohibiting CNC machining power supply defect, Manufacturing Technology & Machine Tool, 2017, (5) pp 159-161

IOP Conf. Series: Materials Science and Engineering **793** (2020) 012033 doi:10.1088/1757-899X/793/1/012033

Backstepping DTC control of PMSM with robust adaptive law

Xiaoming Zheng*, **Lei Xue, Peng Wang, Jia Li and Zeyuan Shen**

Planning Review Center, Economic and Electrical Research Institute of Shanxi Electrical Power Company of SGCC, Taiyuan, Shanxi, 030002, China

*Corresponding author's e-mail: ysuzhxm@126.com

Abstract. The traditional direct torque control (DTC) of permanent magnet synchronous motor (PMSM) is characterized by the feature of inconstant switching frequency and large ripples of flux and torque which is due to the hysteresis controllers adopted in the system. Besides, the modification of motor parameters, especially the rise of stator temperature, and load torque disturbance further leads to the speed steady-state error and fluctuation. A novel DTC method of interior PMSM based on the adaptive backstepping control with load torque and stator resistance identification is proposed in this paper. In order to reduce the ripples of the traditional DTC, a speed backstepping controller is designed to replace PI speed controller, and torque and flux backstepping controllers are designed to replace torque and flux hysteresis controllers in this algorithm. Owing to the real-time and accurate estimation of the stator resistance and load torque, the control variables can be instantly and simultaneously modify to ensure the accurate and fast response of the speed. The effectiveness and correctness of the presented method are verified through multiple simulation experiments, and the results indicate that the dynamic and static performance of the PMSM can be significantly improved by using the method.

1. Introduction

Permanent magnet synchronous motor (PMSM) is utilized widely in various industrial applications. The traditional direct torque control (DTC) method of PMSM is proposed by Professor Zhong L, Rahman MF from Australian and Professor Hu Yuwen from Nanjing University of Aeronautics and Astronautics of China[1, 2]. Compared with field oriented control (FOC) which is another important method in the field of motor control, DTC owns the advantages such as simpler implantation, less machine parameter dependence and quicker dynamic torque response[3]. However, the traditional DTC uses flux and torque hysteresis regulator to control the system, which has the disadvantages of large fluctuation of control variables and unstable inverter frequency [4, 5].

Backstepping control is a new nonlinear system control strategy, which can realize the complete decoupling and simplification of controller design, and ensure the system stability by designing Lyapunov function. In [6], the adaptive control of resistance based on backstepping control was realized. Reference [7] improved the system speed tracking performance by the fuzzy controller adjusting backstepping parameters of the system. In [8] an adaptive backstepping control system was proposed to control the rotor position of PMSM drive using recurrent fuzzy neural network. But the methods proposed above were mainly focused on FOC based on backstepping control, the application of DTC based on backstepping control for PMSM[9, 10] was few and the adaptive backstepping control based on DTC with the consideration of deviation of stator resistance and the disturbance of load torque was fewer.

IOP Conf. Series: Materials Science and Engineering **793** (2020) 012033 doi:10.1088/1757-899X/793/1/012033

The influence of these factors of the system is relatively large, such as the winding temperature increase of 25 degrees, the resistance will rise 6% when the motor is running, and the performance of system will become degradation owing to the parameters can not adapt to the change. In the high precision control system, the influence of these parameters should be considered. This paper introduce the backstepping control to DTC of PMSM combined with adaptive control algorithm to design the stable function for subsystem on the purpose of the adaptive control of stator resistance and load torque. The algorithm proposed in this paper has 3 advantages: Firstly, The backstepping controller is used instead of the hysteresis controller and the PI controller, which is fast and greatly reduces the ripples of the system. Secondly, the influence of stator resistance change and load mutation on servo precision is solved by adaptive control algorithm. Finally, the SVM modulation method is used to keep the switching frequency of the inverter constant. The method can not only enhance the performance of traditional DTC but also suppress the influence of deviation of parameters to improve the robustness of the system.

2. System Description

2.1. Current Model of Interior PMSM

Define:

$$\varphi_s = \varphi_\alpha^2 + \varphi_\beta^2, a = \frac{1}{J}, b = \frac{B_m}{J}, \alpha = 1/L_d, \beta = (L_d - L_q)/L_d \tag{1}$$

A new current model of interior PMSM can be established, as shown in formula (2) and (3):

$$\begin{pmatrix} \frac{di_\alpha}{dt} \\ \frac{di_\beta}{dt} \end{pmatrix} = \begin{pmatrix} -\alpha R & -\beta w \\ -\beta w & -\alpha R \end{pmatrix} \begin{pmatrix} i_\alpha \\ i_\beta \end{pmatrix} + \begin{pmatrix} -\alpha & 0 \\ 0 & -\alpha \end{pmatrix} \begin{pmatrix} E_\alpha \\ E_\beta \end{pmatrix} + \alpha \begin{pmatrix} u_\alpha \\ u_\beta \end{pmatrix} \tag{2}$$

$$T_e = \frac{3}{2} n_p (\varphi_\alpha i_\beta - \varphi_\beta i_\alpha), \ \frac{\mathrm{dw}}{\mathrm{dt}} = a n_p (T_e - T_L) - bw \tag{3}$$

Where φ_s is the square of the stator flux linkage, φ_α , φ_β are the α axis and β axis flux linkage of the stator windings, J is the moment of intertia, B_m is the viscous friction coefficient, L_d and L_q are the equivalent inductances of stator windings in d axis and q axis, i_α and i_β are the components of stator current in α axis and β axis, R is the phase resistance of the stator windings, w is mechanical angular velocity of the rotating rotor, E is the EMF of PMSM, u_α and u_β are the components of stator voltage in α axis and β axis, n_p is the number of rotor pole pairs, T_L is the load torque, T_e is the electromagnetic torque.

2.2. Backstepping Control Design

For the PMSM system, w^*, T_e^*, $\varphi_s{}^*$ are the reference signal, it is assumed that the tracking error e_w, e_T, e_φ are:

$$e_w = w - w^*, e_T = T_e - T_e^*, e_\varphi = \varphi_s - \varphi_s{}^* \tag{4}$$

Lyapunov function V_1 is selected as follows:

$$V_1 = \frac{1}{2} e_w^2 + \frac{1}{2} e_T{}^2 + \frac{1}{2} e_\varphi{}^2 \tag{5}$$

The estimated value are $\hat{T}_L$, $\hat{R}$, then the estimations error ΔT_L, ΔR can be expressed as follows:

$$T_L = \hat{T}_L - \Delta T_L, R = \hat{R} - \Delta R \tag{6}$$

The α, β axis voltage of PMSM can be chosen as:

$$u_\alpha^* = \frac{2\varphi_\beta}{3n_p[\varphi_\beta(i_\beta - \varphi_\beta \alpha) + \varphi_\alpha(i_\alpha - \varphi_\beta \alpha)]} \cdot \Big[1.5 n_p (\varphi_\alpha \alpha i_\beta + \varphi_\beta \alpha i_\alpha) \hat{R} - 1.5 n_p (w\beta i_\alpha - \alpha E_\beta) \varphi_\alpha - a n_p e_w - 1.5 n_p (\varphi_\beta \alpha - i_\alpha) \left(\frac{\hat{R} \varphi_\alpha i_\alpha}{\varphi_\beta} + \hat{R} i_\beta - \frac{k_\varphi e_\varphi}{2\varphi_\beta} \right) - 1.5 n_p (w\beta i_\beta + \alpha E_\alpha) + (b - k_w) T_e + (b - k_w) \hat{T}_L + \frac{bw(b - k_w)}{a n_p} + \frac{\ddot{w}^*}{a n_p} + \frac{k_w \dot{w}^*}{a n_p} + \dot{\hat{T}}_L - k_T e_T \Big] \tag{7}$$

$$u_\beta^* = \frac{2\varphi_\alpha}{3n_p[\varphi_\beta(\varphi_\beta\alpha - i_\beta)+\varphi_\alpha(\varphi_\beta\alpha - i_\alpha)]}\cdot\Big[1.5n_p(\varphi_\alpha\alpha i_\beta + \varphi_\beta\alpha i_\alpha)\hat{R} - 1.5n_p(w\beta i_\alpha - \alpha E_\beta)\varphi_\alpha - 1.5n_p(w\beta i_\beta + \alpha E_\alpha)\varphi_\beta - a n_p e_w - 1.5n_p(i_\beta - \varphi_\beta\alpha)\left(\frac{\hat{R}\varphi_\beta i_\beta}{\varphi_\alpha} + \hat{R} i_\alpha - \frac{k_\varphi e_\varphi}{2\varphi_\alpha}\right) + (b - k_w)T_e + (b - k_w)\hat{T}_L + \frac{bw(b-k_w)}{a n_p} + \frac{\ddot{w}^*}{a n_p} + \frac{k_w \dot{w}^*}{a n_p} + \dot{\hat{T}}_L - k_T e_T\Big] \tag{8}$$

By substitution of (6), (7) and (8) into the derivative of (5), we can obtain

$$\dot{V}_2 = -k_w e_w^2 - k_T {e_T}^2 - k_\varphi {e_\varphi}^2 + a n_p e_w \Delta T_L + (b - k_w)\Delta T_L + 1.5n_p(\varphi_\alpha\alpha i_\beta + \varphi_\beta\alpha i_\alpha)\Delta R + e_\varphi(2\Delta R\varphi_\alpha i_\alpha + 2\Delta R\varphi_\beta i_\beta) \tag{9}$$

Where k_w, k_T, k_φ represents the positive control gain.

2.3. Adaptive Controller Design

Lyapunov function V_2 is chosen as follows:

$$V_2 = V_1 + \frac{\Delta R^2}{2r_1} + \frac{\Delta T_L^2}{2r_2} \tag{10}$$

Where r_1, r_2 represents positive adaptive gains. Adaptive law is obtained:

$$\begin{cases} \Delta\dot{T}_L = r_2(k_w - b - a n_p e_w) \\ \Delta\dot{R} = -1.5n_p(\varphi_\alpha\alpha i_\beta + \varphi_\beta\alpha i_\alpha) - 2e_\varphi r_1(\varphi_\alpha i_\alpha + \varphi_\beta i_\beta) \end{cases} \tag{11}$$

By substitution of (11) into the derivative of (10), we can obtain

$$\frac{dV_2}{dt} = -k_w e_w^2 - k_T {e_T}^2 - k_\varphi {e_\varphi}^2 \le 0 \tag{12}$$

Since $V_2(\infty)$ and $V_2(0)$ is bounded, in accordance with Barbalat's Lemma, the following equation can be obtained:

$$\lim_{t\to\infty} e_\omega = 0, \lim_{t\to\infty} e_T = 0, \lim_{t\to\infty} e_\psi = 0 \tag{13}$$

For PMSM system model with internal and external disturbances, design of adaptive control laws (11), selection of suitable controller gains k_w, k_T, k_φ and adaptive law gains r_1, r_2, the proposed adaptive robust backstepping controller (7) and (8) based DTC can ensure the tracking error signals (4) of PMSM system are asymptotically stable. That is to say, adaptive robust backstepping control based DTC for PMSM can quickly track the given reference signals and restrain the influence of outside interference at the same time.

3. Simulation Results

First of all, a comparison of steady-state performance with reference speed ω = 20 rad/s and load torque T_L=28N·m is carried out among the method in this paper. The backstepping controller parameters of the method in this paper are k_ω = 0.01, k_T = 230, k_ψ = 2000, the adaptive identification parameters t_1 = 4.52, t_2 = 4.78.

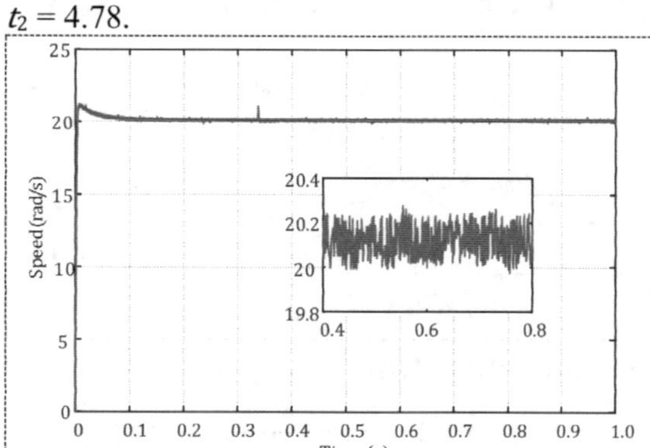

Figure 1 Speed wave of DTC.

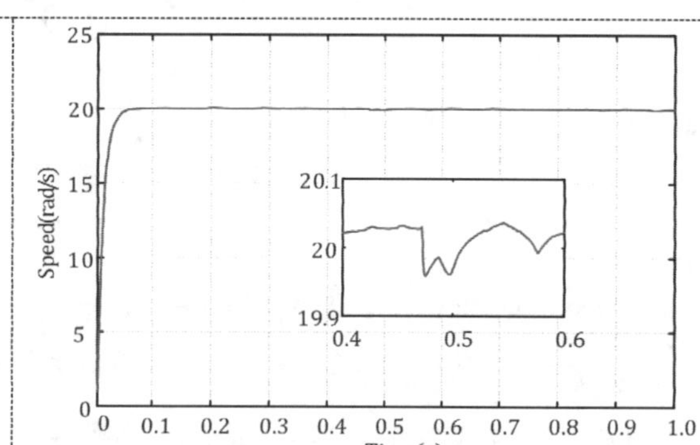

Figure 2 Speed wave of backstepping DTC.

IOP Conf. Series: Materials Science and Engineering **793** (2020) 012033 doi:10.1088/1757-899X/793/1/012033

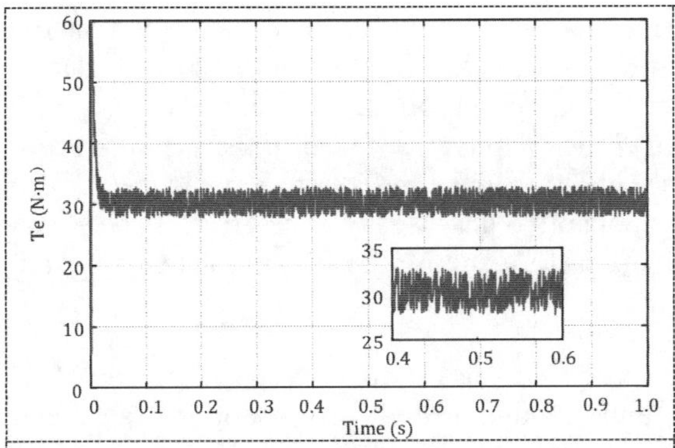

Figure 3 Torque wave of DTC.

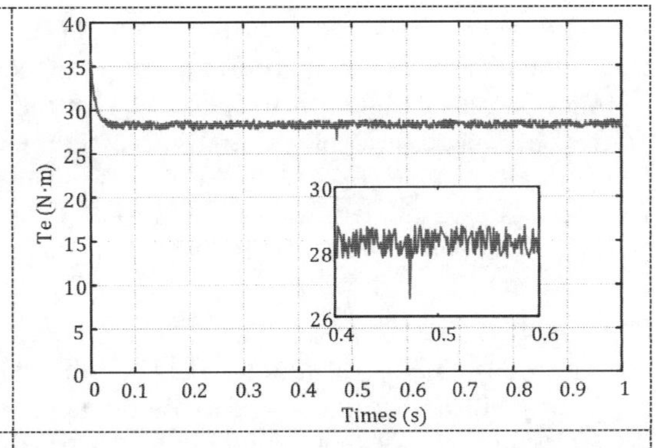

Figure 4 Torque wave of backstepping DTC.

The simulation results given in Figure 1 and Figure 3 show the speed, torque and flux ripples are apparent due to hysteresis controllers in traditional DTC. Figure 2 and Figure 4 indicated the speed, torque and flux ripples are dramatically improved, meanwhile, the adjustment time is faster and the overshoot is smaller in proposed method compared to the traditional DTC.

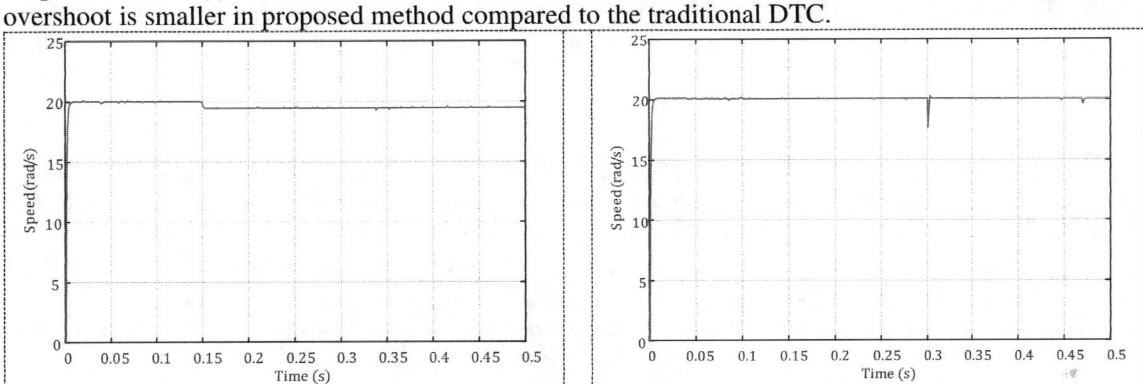

(a) (b)

Figure 5 Speed wave of backstepping DTC without adaptive law

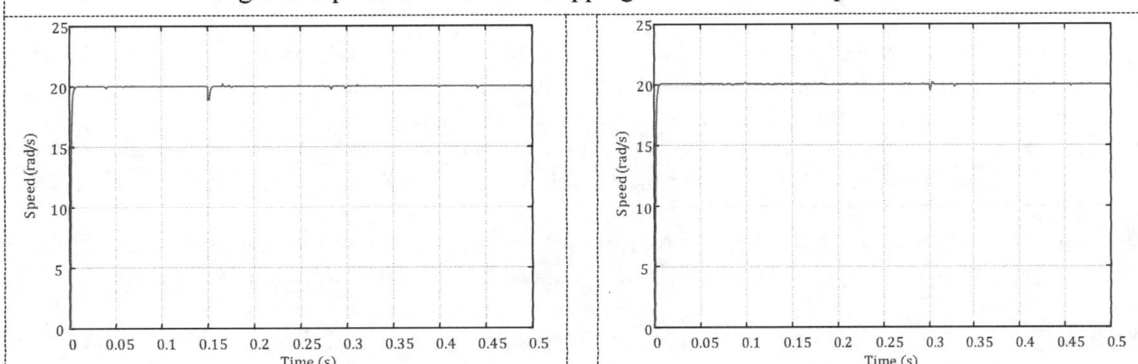

(a) (b)

Figure 6 Speed wave of backstepping DTC with adaptive law

Figure 5 shows the speed in backstepping DTC without adaptive law for PMSM while stator resistance increases to 0.4Ω at 0.15s and load torque T_L increases to 42N·m at 0.3s. Figure 5 (a) shows that the speed has dropped down to 19 rad/s from 20 rad/s while the stator resistance R_s of PMSM increases to 0.4Ω at 0.15s. Figure 5 (b) shows that the speed has a big spike which reaches 17rad/s while the load torque T_L increases to 42N·m at 0.3s.

Figure 6 shows the speed in backstepping DTC with adaptive law for PMSM while the parameters has the same change as in Figure 5. Figure 6 (a) indicates that the speed can be hold to 20 rad/s due to the estimation of R_S. Figure 6 (b) indicates that the speed of the proposed algorithm can be hold to 20 rad/s and only has a tiny spike due to the estimation of T_L by adaptive law used in the method. From Figure 5 and 6, the variation of R_S and T_L could cause steady-state error and a transient and great spike without parameters identification of adaptive law. Using adaptive law, steady-state error of speed can be eliminated and the great spike of can be reduced, then the performance of the system can be improved.

4. Conclusion

Aiming at the problems of traditional DTC, this paper proposes a novel adaptive backstepping DTC strategy for PMSM. The improved method can estimate the change of the motor stator resistance and the load torque of the system in real time, and the tracking error of the system is globally consistent and convergent. The method has a good effect on suppressing the change of parameters, and ensures that the robustness and fast tracking performance of the system are guaranteed. Simulation results show that:

(1) This method can significantly reduce the torque and flux ripples, and make the inverter work at a constant frequency.

(2) This method has excellent dynamic and static performance.

(3) The influence of parameter perturbation on the system is obviously reduced and the method is more robust.

References

[1] Zhong, L, Rahman M F, Hu W Y, Lim KW. (1993) Analysis of direct torque control in permanent magnet synchronous motor drives. IEEE Transactions on Power Electronics, 12(3): 528-536.

[2] Rahman M F, Zhong L, Haque M E, Rahman MA. (2009) A direct torque-controlled interior permanent-magnet synchronous motor drive without a speed sensor. Information and the modern corporation/ MIT Press, 16(2): 137-147.

[3] Dwivedi, S, and B. Singh. (2010) Vector Control vs Direct Torque Control comparative evaluation for PMSM drive. Joint International Conference on Power Electronics, Drives and Energy Systems IEEE, 18(8): 1-8.

[4] Rosic M M, Bebic M Z. (2015) Analysis of Torque Ripple Reduction in Induction Motor DTC Drive with Multiple Voltage Vectors. Advances in Electrical & Computer Engineering, 11(1): 105-114.

[5] Tang L, Zhong L, Rahman M F, Hu W Y, (2004) A novel direct torque controlled interior permanent magnet synchronous machine drive with low ripple in flux and torque and fixed switching frequency. IEEE Transactions on Power Electronics, 19(2): 364-356.

[6] Guo Y, Xi Z, Cheng D. (2007) Speed regulation of permanent magnet synchronous motor via feedback dissipative Hamiltonian realisation. Iet Control Theory & Applications, 1(1): 281-290.

[7] Liu D, Cui Y, Zhao X. (2014) Fuzzy Control of Speed of Permanent Magnet Synchronous Motor Based on Backstepping Control. Transactions of China Electrotechnical Society, 29(11): 38-44.

[8] Lin C H, Wu R C, Chong C . (2011) Adaptive backstepping control for a PMSM drive using RFNN uncertainty observer. Industrial Electronics and Applications. 18(13): 62-67.

[9] Xu Y, Lei Y, Ma L. (2015) A Novel Direct Torque Control of Permanent Magnet Synchronous Motors based on Backstepping Control. Transactions of China Electrotechnical Society, 30(10): 83-89.

[10] Xu Y, Lei Y Z, Ling-Zhi M A. (2015) Backstepping Direct Torque Control of PMSM Based on Fuzzy Parameters Optimization, Electric Drive: 18(16): 13-19.

Cosine Gear Planetary Transmission with Small Teeth Number Difference

Chuanrui Si*, Zhitong Chen, Hongyao Zhang, Yun Zhang and Chuanhui Cui

School of Mechanical Engineering and Automation, Beihang University, Beijing, 100191, China

*Corresponding author's e-mail: sichuanrui@buaa.edu.cn

Abstract. This paper presents a model of cosine gear planetary transmission with small number teeth difference (CG_PTSTD). Firstly, the mathematical model of the CG_PTSTD is established based on the theory of engagement, including the equations of the conjugate internal gear and the meshing line. Then an example of the CG_PTSTD is developed. In this example, the 2D and 3D models of the CG_PTSTD are established, and the bending and contact stresses are analyzed by finite element software ABAQUS. The results show that the cosine gear can be successfully applied to the PTSTD without tooth tip interference and there are multi teeth simultaneously participating in the mesh.

1. Introduction

PTSTD is widely used in robot joints, aerospace, weapons and other fields due to its large transmission ratio, compact structure and high transmission accuracy [1]. The harmonic reducer [2] and RV reducer [3] are the most typical PTSTD which are widely used in robot joints. Compared with harmonic reducer, RV reducer has better stiffness and larger bearing capacity due to the absence of flexible components. Hence it is widely used in joints with high bearing capacity requirements such as waist and big arm of robot [4]. F. Litvin et al. [5] established the cycloid tooth profile equation and deduced the conditions for avoiding singularity of the cycloid curve. Shuting Li [6] developed the design software of cycloid gear reducer, and conducted contact analysis of this reducer based on the presented mechanical model and finite element software. Xuan Li et al [7] comprehensively considered the influences of clearances, tooth profile modification and elastic deformation when establishing the load distribution model of cycloid gear reducer. Ta-shi Lai [8] proposed the geometric design flow of roller drives from the perspective of reducing processing cost and difficulty. Chiu-fan Hsieh [9] proposed to replace the pin tooth with an internal cycloid gear to overcome the output speed fluctuation and the uneven stress distribution. Bingkui Chen et al. [4] introduced a new cycloidal engaged meshing pair which is characterized that there has two contact points simultaneously in a specific meshing area. The above studies belong to cycloid type PTSTD. However, the cycloid tooth profile has a large pressure angle [10], leading to a large stress on the support bearing, and the cycloid pin tooth meshing pair is very sensitive to machining and assembly errors.

The involute PTSTD is also a commonly used PTSTD type [1]. Shuting Li [11] studied the contact problem of involute PTSTD. R. Maiti et al. [12] modified the involute tooth profile from the perspective of avoiding tooth tip interference. Wen-yi Lin et al. [13] optimize the involute tooth profile comprehensively considered the two important factors of pressure angle and contact ratio. However, as mentioned in the above literature, since the whole involute gear consists of several curves, complex interferences are prone to occur in involute internal meshing, especially in the PDSTD. Only through complex and accurate calculations can the valid design parameters be obtained.

From the above literature, it can be concluded that gear tooth profile, as one of the vital factors affecting gear transmission performance, has not been studied enough in the PTSTD. Shanming Luo et al. [14] proposed a novel cosine gear and introduced a corresponding processing method [15]. In external gearing drive, the cosine gear drive has smaller sliding coefficient and larger bearing capacity but lower contact ratio compared with the involute gear and mainly used in external gear pumps. Since the cosine tooth profile is generated by the cosine curve and the shape is similar to "S", it is theoretically suitable for the internal gear transmission. However, to the best of our knowledge, no research has been done on the cosine internal gear drive, especially in the PTSTD. This paper establishes the CG_PTSTD mathematic model according to the meshing theory. An example of CG_PTSTD solid model is presented. The results indicate that the cosine gear can be successfully applied to the PTSTD without tooth interference and with multi-tooth elastic meshing.

2. Mathematical model of cosine internal gear pairs

Figure 1 shows the established coordinate systems. g_1 and g_2 respectively denote the cosine gear and its conjugate internal gear. The moving coordinate systems C_1 (X_1, Y_1, Z_1) and C_2 (X_2, Y_2, Z_2) (abbreviated as C_1 and C_2) are respectively established at the centres of these two gears O_1 and O_2. The centre of the fixed coordinate system C_f (X_f, Y_f, Z_f) (abbreviated as C_f) is established on the pitch point P. At the initial position, X_1, X_2 and X_f are horizontal. The pitch circle radii of the g_1 and g_2 are r_1 and r_2, the number of teeth is z_1 and z_2, respectively. The centre distance between the two gear centres is d. Assuming the pivoted arm O_1O_2 is fixed. If the cosine gear rotates θ_1 anticlockwise around the Z_1 axis, the conjugate internal gear will rotate θ_2 anticlockwise around the Z_2 axis and $\theta_1/\theta_2=z_2/z_1$.

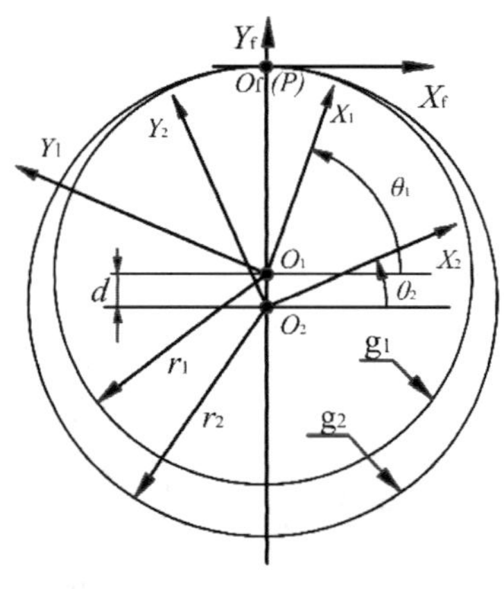

Figure 1. Coordinate systems

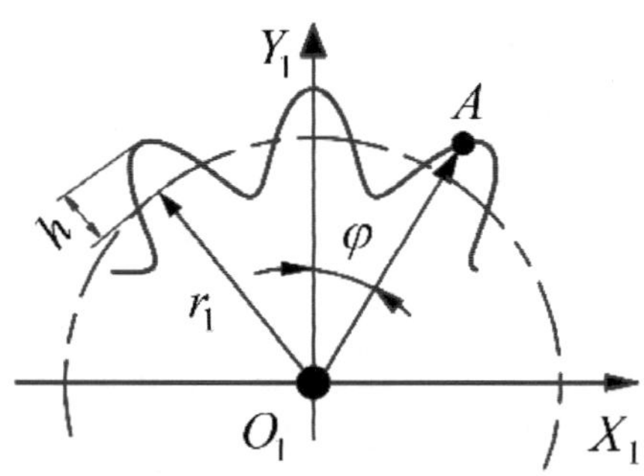

Figure 2. Cosine tooth profile

According to [19], as shown in Figure 2, the cosine gear in C_1 can be expressed as: $\mathbf{G}^{(1)}=[x_1 \quad y_1 \quad 1]^{\mathrm{T}}$, where x_1 and y_1 represent the abscissa and ordinate of the cosine gear, respectively, which are equal to:

$$\begin{cases} x_1=[r_1+h\cos(z_1\varphi)]\sin\varphi \\ y_1=[r_1+h\cos(z_1\varphi)]\cos\varphi \end{cases} \tag{1}$$

where φ and h represent the inherent parameter and the tooth dedendum height of the cosine gear, respectively, and $h=m\,h^*$ where m and h^* denote the gear modulus and tooth height coefficient, respectively.

According to the meshing principle [16], the locus of the contact point in C_2 is the conjugate tooth profile, which is determined by the following equation:

$$\begin{cases} \mathbf{G}^{(2)} = \mathbf{M}_{21}\mathbf{G}^{(1)} \\ f(\varphi,\theta_1) = \mathbf{n}_1 \cdot \mathbf{v}_1^{(12)} = 0 \end{cases} \tag{2}$$

where $\mathbf{M}_{21}$ represents the transformation matrix of the contact point from the C_1 to the C_2, which can be expressed as:

$$\mathbf{M}_{21} = \begin{bmatrix} \cos(\theta_1-\theta_2) & -\sin(\theta_1-\theta_2) & d\sin\theta_2 \\ \sin(\theta_1-\theta_2) & \cos(\theta_1-\theta_2) & d\cos\theta_2 \\ 0 & 0 & 1 \end{bmatrix} \tag{3}$$

The function $f(\varphi,\theta_1)=0$ represents the meshing equation of the cosine gear drive. It represents the angular displacement of a point on the cosine gear turning from its initial position to the position where it becomes a contact point. From [14], it can be obtained as:

$$\theta_1 = \arcsin\left\{\frac{1}{r_1}[r_1 + h\cos(z_1\varphi)]\sin(\varphi+\psi)\right\} - \psi \tag{4}$$

where ψ denotes the angle from the positive direction of the X_1 axis to the tangent to the point on the cosine gear. From the equation (1), it can be obtained:

$$\psi=\arctan\left(\frac{dy_1}{d\varphi}\right)\left(\frac{dx_1}{d\varphi}\right)^{-1}=-\frac{hz_1\sin(z_1\varphi)+\left[r_1+h\cos(z_1\varphi)\right]\tan\varphi}{\left[r_1+h\cos(z_1\varphi)\right]-hz_1\sin(z_1\varphi)\tan\varphi} \tag{5}$$

Substituting equation (5) into equation (4), the relationship between θ_1 and the parameter φ can be obtained. Substitute equations (1), (3) and (4) into equation (2) to obtain the expression of conjugate internal gear in C_2 as follows:

$$\begin{cases} x_2 = \left[r_1 + h\cos(z_1\varphi)\right]\sin[\varphi-(1-\mu)\theta_1] + d\sin(\mu\theta_1) \\ y_2 = \left[r_1 + h\cos(z_1\varphi)\right]\cos[\varphi-(1-\mu)\theta_1] + d\cos(\mu\theta_1) \end{cases} \tag{6}$$

where μ is the reciprocal of the gear ratio, $\mu=\theta_2/\theta_1$.

Similarly, the locus of the contact point in the C_f is the meshing line. Therefore, by transforming the coordinates of the contact point from the C_1 to the C_f, the meshing line equation can be obtained as follows:

$$\begin{cases} \mathbf{G}^{(f)} = \mathbf{M}_{f1}\mathbf{G}^{(1)} \\ f(\varphi,\theta_1) = \mathbf{n}_1 \cdot \mathbf{v}_1^{(12)} = 0 \end{cases} \tag{7}$$

where $\mathbf{M}_{f1}$is the transformation matrix of the above coordinate transformation, which can be expressed as:

$$\mathbf{M}_{f1} = \begin{bmatrix} \cos\theta_1 & -\sin\theta_1 & 0 \\ \sin\theta_1 & \cos\theta_1 & -r_1 \\ 0 & 0 & 1 \end{bmatrix} \tag{8}$$

Substitute equations (1), (4), and (8) into equation (7), the meshing line equation is:

$$\begin{cases} x_f = \left[r_1 + h\cos(z_1\varphi)\right]\sin(\varphi-\theta_1) \\ y_f = \left[r_1 + h\cos(z_1\varphi)\right]\cos(\varphi-\theta_1) - r_1 \end{cases} \tag{9}$$

3. Design example and evaluation

3.1. Solid modeling

Assuming that the design parameters are m=1, h^*=1.2, z_1=100, z_2=105. The tooth number difference is $\delta = z_2 - z_1$. First, 2D model of the CG_PTSTD is obtained in Wolfram Mathematica 8.0, as shown in Figure 3. According to [17], the contact ratio of this drive under the given parameters is 1.17, meeting the requirement that the contact ratio should be greater than or equal to 1 to ensure the continuity and stationarity of the transmission. It is assumed that r_{a1} and r_{f1} as well as r_{a2} and r_{f2} represent the addendum and dedendum circles' radii of the cosine gear and its conjugate internal gear, respectively. They can be obtained in the calculation process of the contact ratio. It can be obtained that t r_{a1}=50.84 mm, r_{f1}=48.80 mm, r_{a2}=51.65 mm and r_{f2}=53.70 mm. According to the above parameters, the 2D model is imported into NX 10.0 to establish the 3D model of the CG_ PTSTD, as shown in Figure 4. It can be seen that there is no tooth tip interference between the two engaged gears. Therefore, the cosine gear can be successfully applied in the PTSTD without tooth interference.

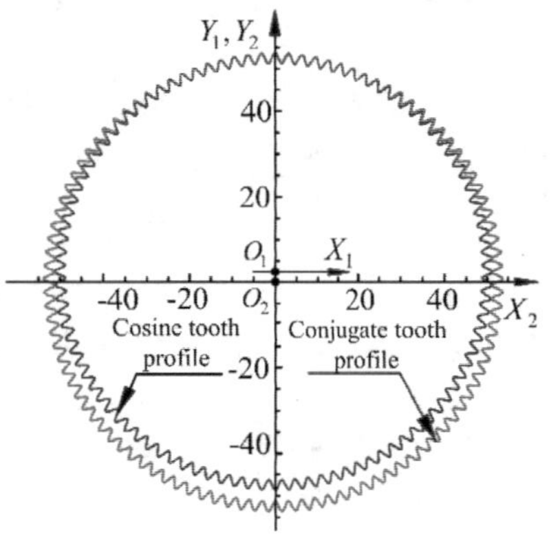

Figure 3. The 2D model

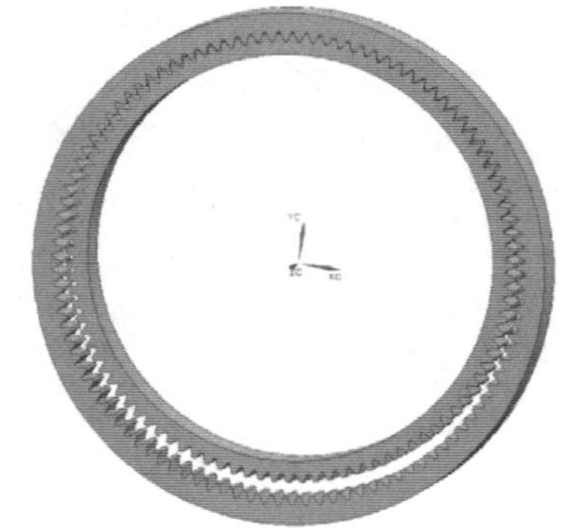

Figure 4. The 3D model

3.2. Evaluation

The analysis of bending and contact stresses of the CG_PTSTD are carried out by the FEA program ABAQUS. The hexahedral solid elements are used to construct the FEA model. The model is partitioned and the mesh density of the contact area was increased. The minimum mesh width is 0.2mm. The properties of materials are: μ =0.3, E = 200GPa, density=7.8e3kg/m3. It is assumed that the sliding coefficient is 0.01. The internal gear was fixed. Except for the rotation about the Z_1 axis, other freedoms of the cosine gear were fixed. The torque applied to the cosine gear was 180 Nm. Figure 5 shows the simulation analysis results. Figure 5 (a) shows that the maximum contact stress is about 199.11 MPa, which is distributed near the tip of the cosine tooth profile. The maximum bending stress is about 164.35 MPa, distributed near the root of the cosine tooth profile. Compared with the applied torque, the stresses are relatively small. Therefore, the cosine gear PTSTD has greater carrying capacity. Figure 5(b) shows that the number of teeth that are simultaneously engaged are 4. However, as mentioned before, the theoretical contact ratio of this gear drive is 1.17. Therefore, there are multi-tooth elastic meshing in the cosine gear PTSTD.

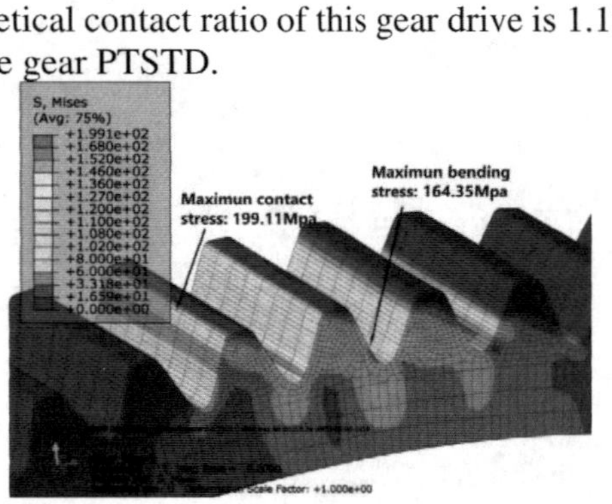

(a) Stress distribution

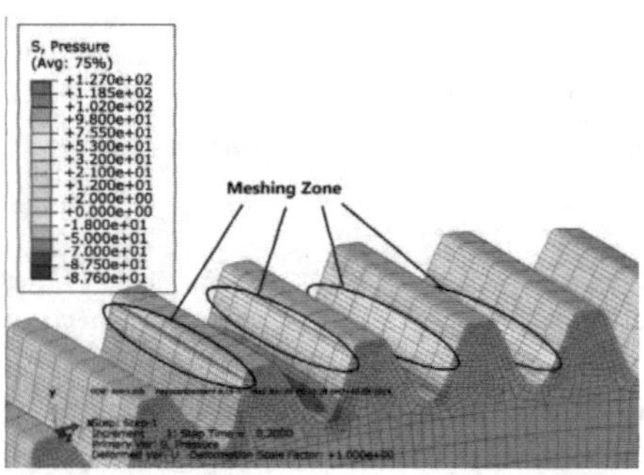

(b) Multi-tooth Meshing

Figure 5. FEM analysis

4. Conclusion

The model of the CG_PTSTD is proposed in this paper. First, the equations of the conjugate internal gear and the meshing line are deduced based on the meshing theory and the equation of the cosine gear. Then a solid model example of the CG_ PTSTD is established. The analysis of the bending and contact stresses of this CG_ PTSTD are carried out by FEA program. The results show that the cosine gear can be successfully applied to the PTSTD without tooth tip interference and there are multi teeth simultaneously participating in the mesh.

Acknowledgments

Authors appreciate the financial support from the China National Science and Technology Major Project (under Grant No. 2018ZX04004001). Besides, authors also thank for the comments from the editors and referees.

References

[1] Li, S. (2014). The latest design technologies for gear devices with great transmission ratios. In: International Gear Conference 2014. Lyon. pp. 427-436.
[2] Chen X, Liu Y, Xing J, Lin S, Xu W. (2014). The parametric design of double-circular-arc tooth profile and its influence on the functional backlash of harmonic drive. Mechanism and Machine Theory, 73: 1-24.
[3] Wang R, Gao F, Lu M, Liu T. (2019). Meshing Efficiency Analysis of Modified Cycloidal Gear Used in the RV Reducer. Tribology Transactions, 62: 337-349.
[4] Chen B, Zhong H, Liu J, Li C, Fang T. (2012). Generation and investigation of a new cycloid drive with double contact. Mechanism and Machine Theory, 49: 270-283.
[5] Litvin F L, Demenego A, Vecchiato D. (2001). Formation by branches of envelope to parametric families of surfaces and curves. Computer Methods in Applied Mechanics and Engineering, 190: 4587-4608.
[6] Li S. (2014). Design and strength analysis methods of the trochoidal gear reducers. Mechanism and Machine Theory, 81: 140-154.
[7] Li X, Li C, Wang Y, Chen B, Lim T. (2017). Analysis of a Cycloid Speed Reducer Considering Tooth Profile Modification and Clearance-Fit Output Mechanism. Journal of Mechanical Design, 139: 33303.
[8] Lai T. (2005). Geometric design of roller drives with cylindrical meshing elements. Mechanism and Machine Theory, 40: 55-67.
[9] Hsieh C. (2015). Traditional versus improved designs for cycloidal speed reducers with a small tooth difference: The effect on dynamics. Mechanism and Machine Theory, 86: 15-35.
[10] Fong Z, Chiang T, Tsay C. (2002). Mathematical model for parametric tooth profile of spur gear using line of action. Mathematical and Computer Modelling, 36: 603-614.
[11] Li S. (2008). Contact problem and numeric method of a planetary drive with small teeth number difference. Mechanism and Machine Theory, 43: 1065-1086.
[12] Maiti R, Roy A K. (1996). Minimum tooth difference in internal-external involute gear pair. Mechanism and Machine Theory, 31: 475-485.
[13] Lin W, Hsiao K, Ke J. (2019). Optimum design of involute tooth profiles for K-H-V planetary drives with small teeth number differences. Journal of Advanced Mechanical Design, Systems and Manufacturing, 13: JAMDSM0007-JAMDSM0007.
[14] Luo S, Wu Y, Wang J. (2008). The generation principle and mathematical models of a novel cosine gear drive. Mechanism and Machine Theory, 43: 1543-1556.
[15] Wang J, Luo S, Xu P. (2014). Mathematical model of slotting manufacture and simulation for cosine gears. Journal of Central South University, 21: 933-941.
[16] Litvin F L, Fuentes A. (2004). Gear Geometry and Applied Theory. Cambridge University Press, Cambridge.

[17] Wang J, Luo S, Chen L, Hu H. (2008). Analysis of transmission characteristics of cosine gear drive. Chinese journal of mechanical engineering: English version, 21: 94-98.

Research on Ship Fault Diagnosis Based on Deep Learning

Zhu Jun[1,2*], Wang Lei[2], Zhang di[3]

[1] NO. 716 Research Institute of China Shipbuilding Industry, Jiang Su 222002, China

[2] CSIC Information Technology Co., Ltd, Jiang Su 222002, China

[3] College of Mechatronics, Beijing University of Technology, Beijing 100041, China

Zhu Jun, e-mail: huzi881210@foxmail.com

Abstract. This paper proposes a deep learning algorithm to diagnose ship faults in order to improve the accuracy and diagnostic efficiency of ship fault diagnosis. 90% of the large number of unlabeled ship operational data samples are selected for model training, and the remaining 10% is used for model testing. We optimize model parameters and improve the accuracy of the deep learning model for fault diagnosis classification. Hidden layer functions are used to extract multi-layer data features and perform feature fusion. Gain values are used to define ship faults, primary faults, secondary faults, and tertiary faults. Finally, we use the soft-max classifier to fault output the fault and get the fault level output. The experimental results on the ship simulation fault dataset show that compared with other traditional algorithms, the accuracy of the fault diagnosis of this method is greatly improved, and the simulation result is 92.5%. Experiments show that the model based on deep learning algorithm for multi-layer feature fusion training can better meet the needs of ship fault diagnosis under complex systems.

1. Introduction

In the event of a major accident in the operation of the ship system, the ship's safe operation will have a serious impact, and improving the efficiency of fault diagnosis is the direction of ship fault diagnosis. The technology of the ship system is constantly improving, the connection between the internal structure of the equipment is complicated, and the coupling work of the ship system is getting higher and higher, making the implementation of effective fault diagnosis more difficult [1]. Ship fault diagnosis is an important part of ship system maintenance decision. Therefore, how to establish an effective ship fault diagnosis technology has become the focus of research [2].

The development of artificial intelligence is currently divided into knowledge-based rules, model-based and data processing-based fault diagnosis methods [3]. The knowledge rule-based method combines artificial intelligence technology with fault diagnosis technology, and uses fault diagnosis and reasoning through knowledge rules, but it requires a lot of knowledge and experience in related fields or expert experience, so it is difficult to achieve [4]. Data-based and model-based methods are driven by data. By analyzing the relationship between device type parameters and fault types, an appropriate machine learning model is selected for fault classification. Related research scholars have published a large number of research results in this field. Lei Yaguo et al. [5] overcome the dependence of fault diagnosis on empirical knowledge, extract features from equipment time domain signals, select sensitive features and pass RBF radial basis. Function neural networks reduce the network structure model. Zhou Hongming et al. [6] combined the extreme learning autoencoder with deep learning to establish a deep limit learning model. Zhao Zhihong et al [7] first used wavelet packet transform and Hilbert-Huang transform to extract fault features, and then used SVM method to classify fault diagnosis types. By analyzing the traditional fault diagnosis classification methods such as SVM, artificial neural network and extreme learning machine ELM, the SVM algorithm can solve the complementary problem of the fault diagnosis classification algorithm and the direct training prediction problem, but the SVM algorithm requires a large number of label data samples. And it is more difficult to deal with the problem of data imbalance [8]. The difference between the observed and actual values of the artificial neural network model is compared with the decision function established in advance to determine whether the equipment fault has occurred. However, the model training process is more difficult, which makes the method have certain limitations. The ELM method is fast, but its sum function is a specified function, and the adaptability of the kernel function in processing different data is not considered, which makes the stability of the method poor. The above methods are traditional machine learning methods, the learning ability has limitations, and the accuracy of fault diagnosis is difficult to further improve [9]. In view of the above shortcomings, a deep learning method is proposed to complete the training model process from a large number of unlabeled data samples, optimize the model parameters and improve the accuracy of the deep learning model for fault diagnosis classification.

In recent years, the deep learning research boom has been rising. The deep learning algorithm is a multi-hidden layer network model structure, which can effectively analyze and extract large amounts of data, and is widely used [10]. Deep learning is mature and has significant effects in the fields of computer vision, speech recognition, image recognition and natural language processing, but deep learning applications are still in their infancy in the field of fault diagnosis. Sun et al. [11] combined the automatic encoder with the deep learning algorithm to propose a deep neural network algorithm implemented by a fine-numbered auto-encoder with denoising coding. Jia et al [12] proposed an intelligent diagnosis method based on deep learning algorithm, which overcomes the defects of traditional methods on expert experience and professional knowledge, and verifies through data sets to achieve better classification results. Lu et al. [13] used the deep learning algorithm to calculate and analyze the frequency domain data of the bearing dataset, and visualized the results to obtain better classification accuracy.

Combined with the above-mentioned application of traditional fault diagnosis technology and deep learning algorithm in the field of fault diagnosis, there are still some problems, such as the increase of data dimension and the increase of network layer. This paper proposes a ship fault diagnosis based on deep learning algorithm. The multi-layer data features are extracted by deep learning; the extracted multi-layer feature vectors are feature-fused; the soft-max classifier is used to classify and output the fault results.

2. Data acquisition and data preprocessing

2.1. Data acquisition

Due to the complexity of the actual working conditions of the ship, it is difficult to collect a large number of actual fault data due to the influence of many external environmental factors. There is no authoritative data set available in the industry. This article was completed on the lab platform. The platform includes a current sensor, a voltage sensor, a rotational speed sensor, and a tension and pressure sensor. The sensor includes four states: normal, complete failure, constant gain, and constant deviation. Faults are classified into: primary fault, secondary fault, and tertiary fault. Definition: When the gain value interval is $\pm$ $(10\% \sim 20\%)$, it is a first-level fault; when the gain is $\pm$ $(20\% \sim 40\%)$, it is a secondary fault; when the gain value interval is $\pm$ $(40\%\ or\ more)$ Three-level failure. The sampling frequency is $40ks/s$, the number of samples per cycle is 4000, the normal state data is collected for $20s$, and the fault state data is collected for $20s$.

2.2. Data preprocessing

The experiment collected 61 cycles for data collection. The data set had 50020 sets of data, 90% of which were used as training sets and 10% as test sets. Because each set of data contains data from four

IOP Conf. Series: Materials Science and Engineering **793** (2020) 012035 doi:10.1088/1757-899X/793/1/012035

different sensors, the data needs to be separately Z-Score standardized to ensure the comparability between the data, defined as:

$$z = \frac{x - \mu}{\sigma} \tag{1}$$

x is the observed value, μ is the population mean mean, and σ is the population standard deviation.

3. Modeling of ship fault diagnosis based on deep learning

The ship fault diagnosis model based on deep learning is divided into four steps: The Inception-v3 model is used to extract the feature vectors in the data set at the front end, and the feature vectors in the last two layers of the pooling layer are used for feature fusion. Vector, used as input to the Softmax classifier, to classify fault levels. The ship fault diagnosis model based on deep learning is shown in Figure 1.

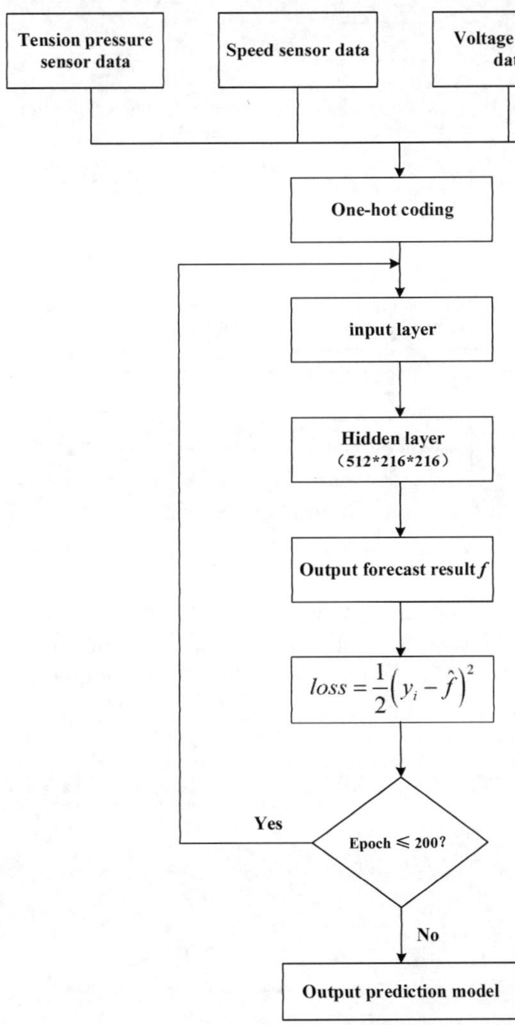

Fig.1 Ship Fault Diagnosis Model Based on Deep Learning

In this paper, $R2$ is used to judge the model fitting degree. The value of the model is used as the model score. The better the model prediction effect is, the closer the $R2$ value is to **1**, the worse the model prediction effect is, and the closer $R2$ is to 0. The definition of $R2$ is as follows:

IOP Conf. Series: Materials Science and Engineering **793** (2020) 012035 doi:10.1088/1757-899X/793/1/012035

$$R^2 = 1 - \frac{\sum_i (y_i - f_i)^2}{\sum_i (y_i - \hat{y})^2} \quad (2)$$

$\boldsymbol{y}$ is the actual value, $\boldsymbol{f}$ is the predicted value, and $\hat{\boldsymbol{y}}$ is the actual value of the training sample average. $\boldsymbol{R2}$ uses the mean as the error reference to calculate whether the prediction error is greater or less than the mean reference error. The algorithm flow is shown in Figure 2.

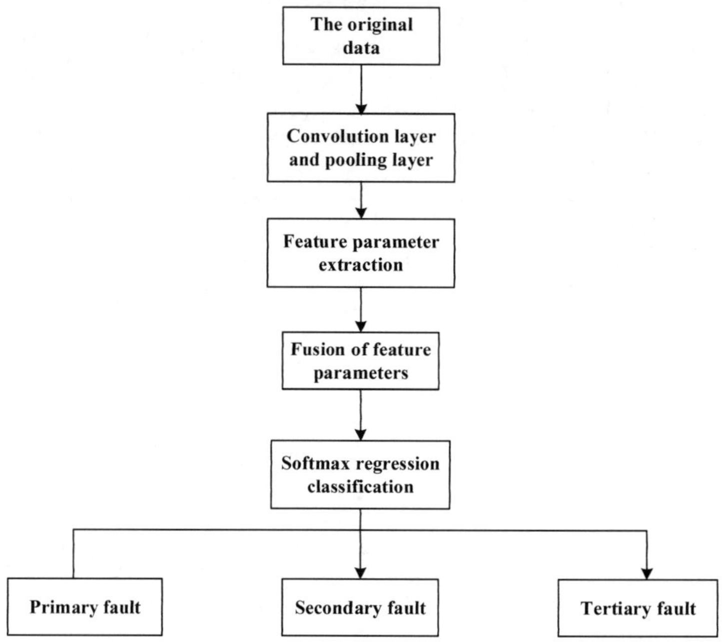

Fig.2 Flow chart of ship fault diagnosis algorithm

4. Experimental results and analysis

In order to evaluate the fault diagnosis performance of ship fault diagnosis model based on deep learning, it is compared with extreme learning machine and support vector machine algorithm. The detection accuracy and detection time of ship fault diagnosis were compared and analyzed. The accuracy comparison analysis is shown in Figure 3.

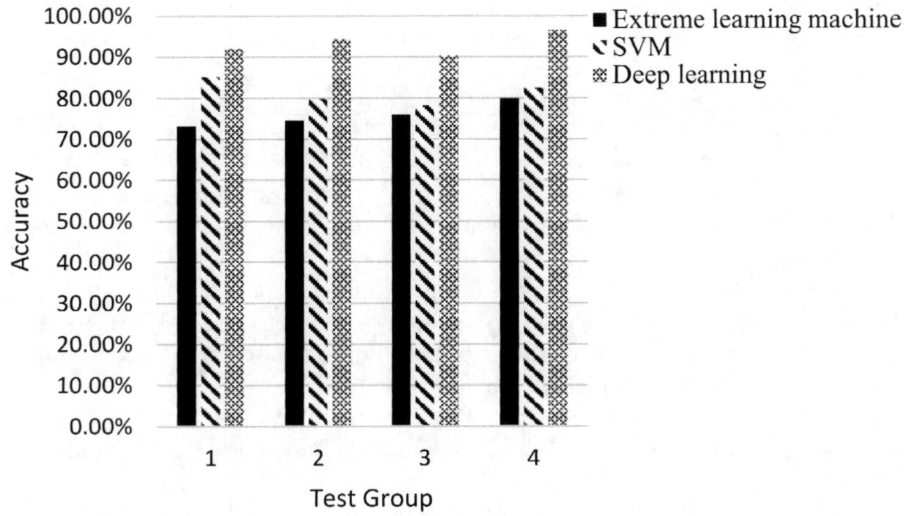

Fig.3 Comparison of Fault Diagnosis Accuracy

As can be seen in Figure 3, compared with other intelligent algorithms, the accuracy of the fault diagnosis of the deep learning algorithm is above 92%, which is much better than the traditional algorithm. The detection time is shown in Figure 4.

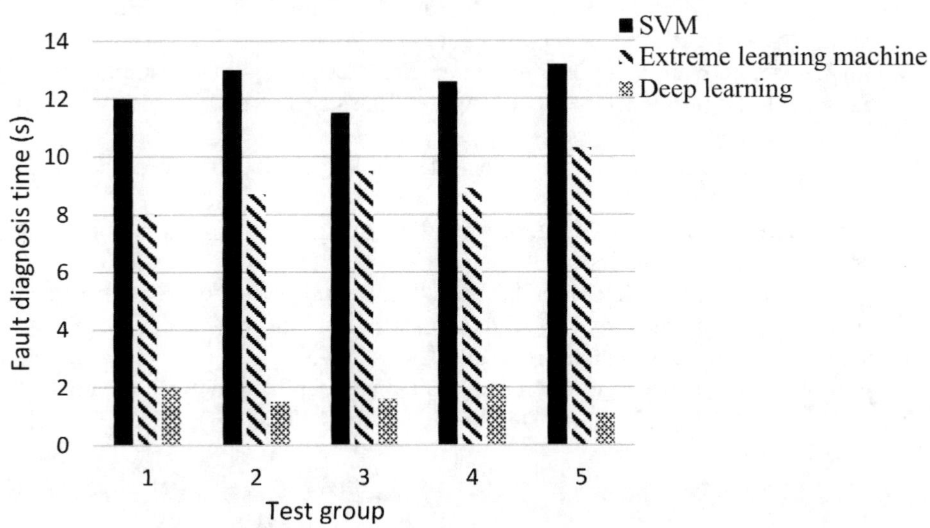

Fig.4 Time comparison of fault diagnosis

5. Conclusion

With the continuous development of intelligent shipbuilding, more and more sensor data has been added to the fault diagnosis. The huge data set and the complex ship operating environment have put forward higher requirements for fault diagnosis. In this paper, a ship fault diagnosis model based on deep learning is proposed. Feature fusion is used to feature the last three layers of the hidden layer. The data set is collected on the experimental bench and the effect of the model is verified. The model parameters are continuously optimized to make it more suitable for ships troubleshooting. The average training precision of the model is 95.67%, and the average test accuracy is 92.52%, which is better than the traditional algorithm. The experimental methods and experimental procedures proposed in this paper can be applied to ship fault diagnosis to achieve better human-computer interaction, high diagnostic efficiency and accuracy.

References

[1] Kano, Y Nakagawa. Data-based process monitoring,process control, and quality improvement: Recent developments and applications in steel induetry[J]. Computers & Chemical Engineering, 2008, 32(1/2): 12-24.

[2] Hua Haide, Ma Ning, Ma Jie, et al. .Robust Intelligent Control Design for Marine Diesel Engine[J]. Journal of Shanghai Jiaotong University(Science), 2013, 18(06):660-666.

[3] Lin Guoyang. Analysis and treatment of vibration faults of circulating water pumps [D]. Nanjing University of Aeronautics and Astronautics, 2016.

[4] Venkata subrarnamiamV, Rengaswamy R, SN. A review of process fault detection and diagnosis:PartII:Qualitative models and search strategies[J]. Computers & Chemical Engineering, 2003, 27(3): 313-326.

[5] Lei Yaguo,Jia Feng,Zhou Wei, et al. A Health Monitoring Method for Mechanical Equipment Big Data Based on Deep Learning Theory[J].Journal of Mechanical Engineering,2015,51(21):49-56

[6] Zhou Hongming, Huang Guangbin. Representational learing with ELMs for big data[J]. IEEE Intelligent Systems,2013, 28(6): 31-34.

[7] Zhao Zhihong. Research on Mechanical Fault Feature Extraction and Diagnosis Based on Vibration Signal[D]. Beijing Jiaotong University, 2012.

[8] Jiang Qiang, Shen Lin, Zhang Wei, et al. Research on Fault Diagnosis Method Based on Deep Learning[J].Computer Simulation,2018,35(07):409-413.

[9] Wu Jianbo, Wang Chunyan, Hong Huajun, et al. Fault Diagnosis of Marine Diesel Engine Based on Extreme Learning Machine[J]. Computer Engineering and Applications, 2019, 55(15): 147-152.

[10] Zhang Shiqiang. Research on fault diagnosis technology based on deep learning [D]. Harbin Institute of Technology, 2018.

[11] Sun Wenzhao. Research on fault diagnosis method of induction motor based on deep learning model [D]. Southeast University, 2017.

[12] Feng Jia,Yaguo Lei,Jing Lin,Xin Zhou,Na Lu. Deep neural networks: A promising tool for fault characteristic mining and intelligent diagnosis of rotating machinery with massive data[J]. Mechanical Systems and Signal Processing,2016,72-73.

[13] Weining Lu. A Novel Feature Extraction Method using Deep Neural Network for Rolling Bearing Fault Diagnosis[C]. Control and Decision Conference (CCDC), 2015 27th Chinese. IEEE, 2015: 2427-2431.

IOP Conf. Series: Materials Science and Engineering **793** (2020) 012036 doi:10.1088/1757-899X/793/1/012036

Wall-climbing obstacle-avoiding full-automatic lampblack cleaning device

Jiarui Qin[1] and Qian Wei[2]

1 Wall-climbing obstacle-avoiding full-automatic lampblack cleaning device, Wuhan University of Technology, Wuhan, Hubei, 430070, China

2 Wall-climbing obstacle-avoiding full-automatic lampblack cleaning device， Wuhan University of Technology, Wuhan, Hubei, 430070,China*Corresponding author's e-mail: 466252181@qq.com.

Abstract. With the development of Chinese society and the progress of productive forces, dust and oil stains are easily produced in kitchens and ceilings when living at home. Dirt deposition in many places cannot be carried out well due to the limitation of space location and cleaning methods, which brings great inconvenience to people's life. At the same time, the existing cleaning robots in the market need to work on a single wall surface, and when encountering obstacles, the positions of the robots need to be adjusted manually to further realize the work and cannot solve the cleaning problems of ceilings and the like. The existing cleaning robots mostly use centrifugal fans to fix on vertical wall surfaces in an adsorption way, thus causing obstacles to be avoided, the connection between different wall surfaces cannot be realized, and the defect that intelligence works on a single wall surface requires manual wall surface replacement, thus reducing automation, intelligence and intelligence. Based on the above background, the members of this group have designed a wall-climbing cleaning machine. The device mainly comprises a walking module, a wall turnover module and a cleaning module [1].

1.Introduction

At present, the scope of most cleaning machines is mostly limited to window glass, which can not be widely applied, thus reducing its efficiency, and correspondingly can not significantly reduce the cleaning cost, which is not easy to popularize and has a poor market prospect. The invention improves the structure of cleaning machines on the market, thereby improving the application range and cleaning efficiency of the cleaning robot for wall cleaning, reducing the potential safety hazard of high-rise glass cleaning, reducing the cleaning cost, and having great significance and economic benefits.

2. Design

2.1 Research content and key technologies

1)The space four-link mechanism controlled by three highly precise steering gears can effectively realize the lateral movement of the range hood on the wall and complete the movement from one wall to the other. And if the recognition module around it recognizes the obstacle, it can complete the obstacle avoidance function by sending instructions to the steering gear [2].

IOP Conf. Series: Materials Science and Engineering **793** (2020) 012036 doi:10.1088/1757-899X/793/1/012036

3. Overall design

3.1 Functional design

1）The device can achieve smooth walking on vertical walls.

2）The device can realize the action from one wall to the other wall.

3）The device can effectively clean the oil stains on the wall.

4）The identification module around the device can realize the identification of obstacles and quickly judge whether the stain is clear so as to execute the next instruction.

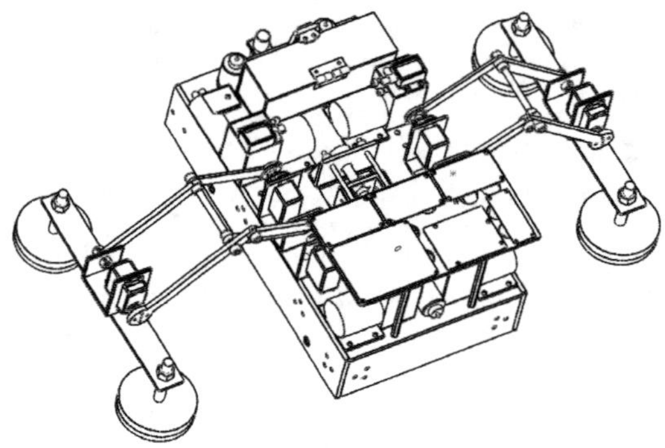

Figure 1 Overall schematic diagram of the device

3.2 Structural design

1）Cleaning section

Vacuum water pump

Disc brush

Roller brush

Water tank with cleaning agent

High-speed motor

Because the oil fume is difficult to clean, in order to better clean the oil fume, we have designed a double cleaning mechanism as shown in the figure. First, the water pump sprays the cleaning liquid inside the water tank into a mist through the spray head and sprays it to the front of the cleaning module. The two front brush heads are driven by the motor to rotate at high speed for the first time cleaning. To clean the remaining cleaning liquid on the wall and the residue after cleaning [3].

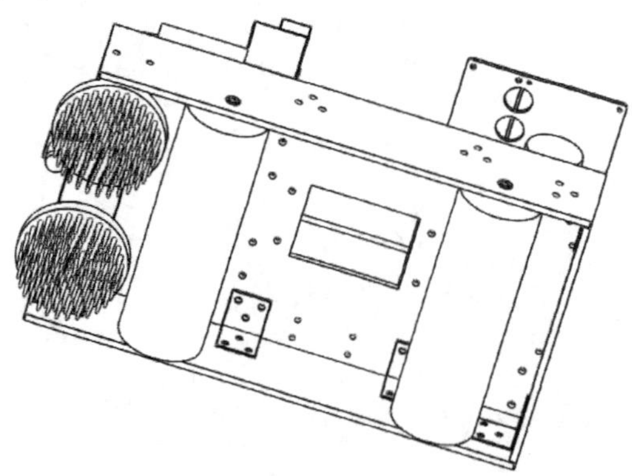

Figure 2 Cleaning part

2) Walking section

In order to realize the transfer of the wall of the mechanism on different planes and the lateral movement on the vertical wall surface, we designed the wall-turning module, as shown in Figure 3. The mechanism is a space four-link linkage mechanism driven by three steering gears. The three steering gears control the forward and backward movement, vertical movement of the connecting rod, and the rotation of the end suction cup. Functions such as lateral movement, over the wall and obstacle avoidance.

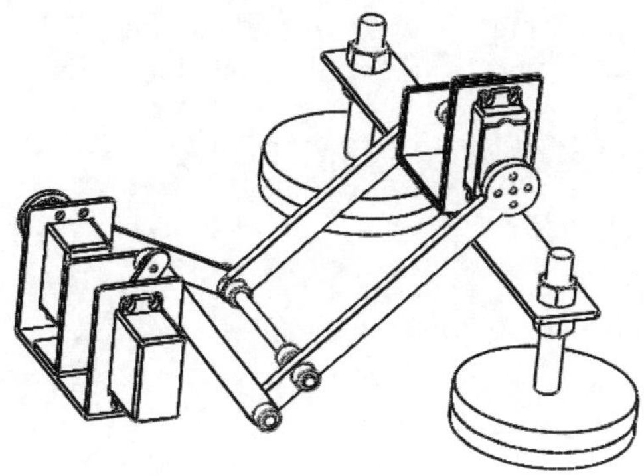

Figure 2 Walking obstacle-crossing module

4. Feasibility analysis

We calculated our space four-linkage according to the calculation formula of the connecting rod in the book "Principle of Machinery and Mechanical Design", and concluded that the machine can take a step up to a stroke of 20cm [4]. The motion path and force situation of the force components in the simulation are simulated, and it is found that the design can achieve our intended purpose. We also used ANSYS FLUENT to perform fluid mechanics simulation of the air pump and suction cup, and concluded that the mutual operation of the suction cup and the air pump can meet our requirements [5].

5. Innovation

1) New structure and ingenious design: Two different adsorption and movement devices are combined, and the suction and release of a plurality of vacuum washing plates are matched with each other, so that the movement is more flexible.

2) Multifunctional and efficient work: The device integrates a cleaning function, a wall climbing function and a barrier surmounting function, and simultaneously combines an ultrasonic detection system, so that the device can complete various motions according to actual environmental requirements, and the purpose of completely cleaning a predetermined area is achieved [3].

6. Conclusion

The project team has designed a wall climbing device which can realize multiple functions such as wall climbing, obstacle avoidance, cleaning, etc. Based on the weakness of existing products, it has good market prospect and broad development space, and can better meet the needs of people's daily life.

Acknowledgments

First and foremost, I would like to show my sincere gratitude to my partner and teachers, who give me support during the research. Besides, I am extremely grateful for my parents for their constant encouragement.

References

[1] SUNG M M, CHANG Y S, JAEMYUNG H, et al. Window cleaning system with water circulation for building facade maintenance robot and its efficiency analysis ［J］. International Journal of Precision Engineering and Manufacturing-Green Technology, 2015 (1): 21-23.
[2] Wang Meiting. Research on the key technology of automatic wall cleaning robot [D]. Shanghai: Shanghai University, 2010.
[3] Du Zhiqiang, Ge Shuqing, Fang Jianfeng, et al. Organization Design Institution Based on MATALBA Language. Shanghai: Shanghai Science and Technology Press, 2011.8
[4] Luo Qingsheng, Korean version Ling. Analysis of the winning works in the extracurricular science and technology innovation competition for college students. Beijing: Mechanical Industry Press, September 2013.
[5] Wang Wenbin et al. Manual of Mechanical Design. Beijing: Mechanical Industry Press, 2004.8.

The Numerical Analysis to Gas-Liquid Separator used in On-orbit Refuelling Mission

Wen Li[1,2], Lei Guo[1], Jintao Liu[1], Yong Li[1] and Lei Chen[1]

[1] Beijing Institute of Control Engineering, Beijing, 100190, China
[2] E-mail: gradylee@126.com

Abstract. In the process of on-orbit refuel, the gas-liquid separator needs to separate the pressurized gas from the propellant, and discharges the gas out of the tank in the process of the tank refilling by propellant. In this paper, the gas-liquid separator has been developed, which could effectively separate the gas-liquid under microgravity conditions. The numerical analysis result shown that the stability of gas-liquid separation was the key characteristics, which was closely related to the geometric parameters of the fluid domain.

1. Introduction

The on-orbit refueling technology of satellite propellant is the key technology to extend the working life of satellite and greatly reduce the cost of satellite. Under microgravity condition, the effective separation the pressurized gas from the propellant is the key technology to achieving the mission of on-orbit refuel of satellite. The process is accomplished by a gas-liquid separator, which needs to separate the pressurized gas from the propellant, and discharges the gas out of the tank in the process of the tank refilling by propellant. The gas-liquid separator involved in this study has been developed based on static gas-liquid separation technology, which was designed based on the surface tension and centrifugal force. The gas-liquid separator has no moving parts, low energy consumption, no maintenance, reliable separation, easy control and other characteristics.

2. The mechanism of gas-liquid separator

2.1. Static gas-liquid separation technology

Static gas-liquid separation technology includes the following three technical approaches [1]: (1) Capillarity gas-liquid separation technology is mainly based on the characteristics of hydrophilic/hydrophobic materials and the capillary effect generated by surface tension. (2) The cyclone gas-liquid separation technology is mainly based on the centrifugal effect of the gas-liquid mixed fluid flowing through the vortex pipeline. (3) Inertial gas-liquid separation technology is mainly based on the principle that the inertia of gas and liquid is very different, and the liquid is not easy to change the direction of motion and is collected by hydrophilic film.

2.2. The mechanism of gas-liquid separator

Based on the above three typical gas-liquid separation technologies, Weislogel [2] proposed a device to solve the problem of static gas-liquid separation on a spacecraft. The operating principle of the device is shown in figure 1. The main body of the device is a spiral tube structure similar to the "ice cream cone section", which consists of three sections, including the inlet section A-A, the diffusion section B-B, the passive capillary separation section and the liquid collection section C-C [3].

Regardless of whether the gas-liquid phase enters the ice cream cone section A-A in any form, the liquid adheres to the internal surface of the device and slows down rapidly, thus being driven downstream by the gas [4]. The function of the tapered section is to guide the large contact angle of the liquid to the bottom sequentially. Once the liquid has accumulated here, the flow will become stable and will not change its flow state even if it enters the straight section again.

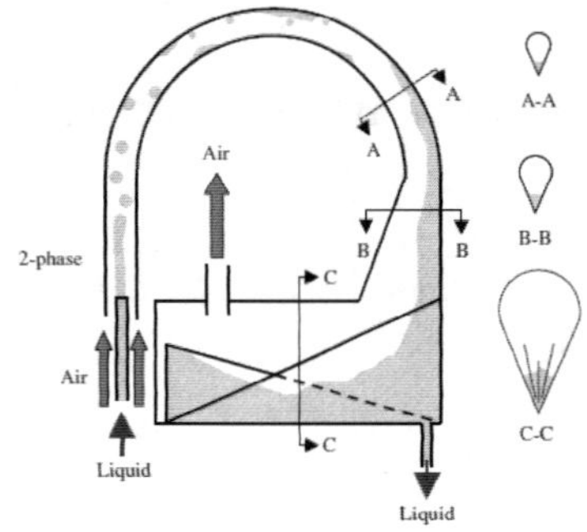

Figure 1. Principle of gas-liquid separator

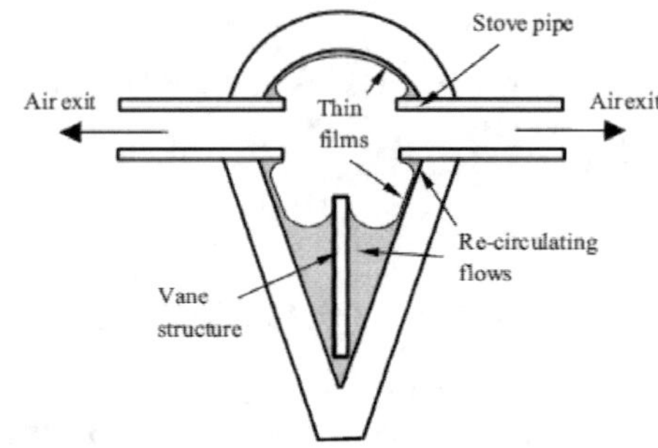

Figure 2. Stovepipe type vent

The centrifugal flow track mainly provides the following functions: (1) It is used to provide a driving force to make free droplets adhere to the dry wall when they hit it or combine with other liquid surfaces. (2) By accumulating the liquid to the apex of the tapered duct, it is easy to drive the air downstream to the liquid collection section C-C without sufficient capillary force due to poor wettability. (3) The results of flight experiments show that when free droplets come into contact with the previously adhered liquid surface, aggregation effect may not always occur. It is possible that free droplets will rebound from the liquid surface, and even aggregation may form free flow. Therefore, the liquid must be accumulated to the bottom of the conical section by a certain centrifugal acceleration. Based on the statistical law, the shortest length of the spiral pipe can be preliminarily determined to realize gas-liquid separation at a given gas velocity.

C-C section guide vane structure mainly provides the following functions: (1) For a liquid with poor wettability, it can be drained to a designated position by the guide vane structure and dragged downstream by the gas in the flow path. In this way, the guide vane provides capillarity and thus gathers the liquid together to facilitate the discharge of the liquid to the outlet. (2) Along the fluid flow path, the effective area and density of the guide vane is changed to provide a better opportunity for the droplet to adhere to the guide vane and to polymerize with a droplet that may already exist at that location. (3) The Angle of the inner guide vane of the guide vane structure decreases continuously along the flow direction, providing a passive way to suck the trapped gas out of the liquid during the downstream movement. The fluid domains between the guide vanes are interconnected, and large liquid substances are removed through a single, larger liquid outlet at the bottom of the conical section, which will not be prone to blockage over time due to pollution.

3. Simulation of gas-liquid separator

3.1. The structure design and fluid domain

According to the function and index requirements of the in-orbit refueling mission, the gas-liquid separator is designed based on the mechanism described above. The index requirements of gas-liquid separator are as follows: flow rate is 10L/min; volume proportion of liquid is 10%. Therefore the main parameters of the gas-liquid separator are as follows: the diameter of the spiral section is 6.35mm, the cone angle is 40 °, the inner diameter of the spiral is 76.2mm, the outer diameter is 101mm; the diameter of the collection section is 14.4mm, the cone angle is 40 ° and the length is 106mm, and the included angle of the three internal guide vane is 15 °. The fluid domain is shown in Figure 3, and the internal guide vanes is shown in Figure 4.

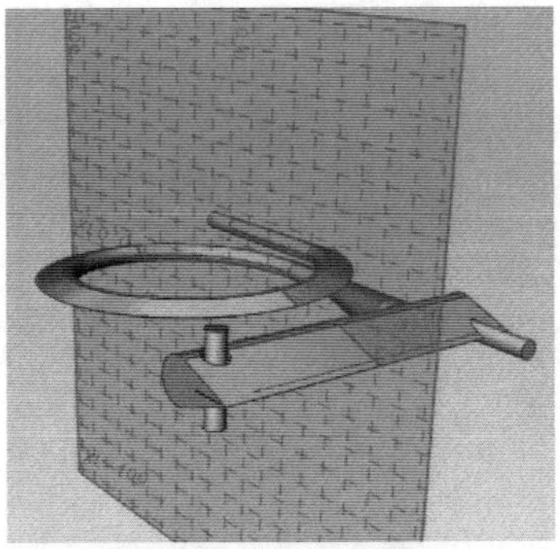

Figure 3. The fluid domain of gas-liquid separator

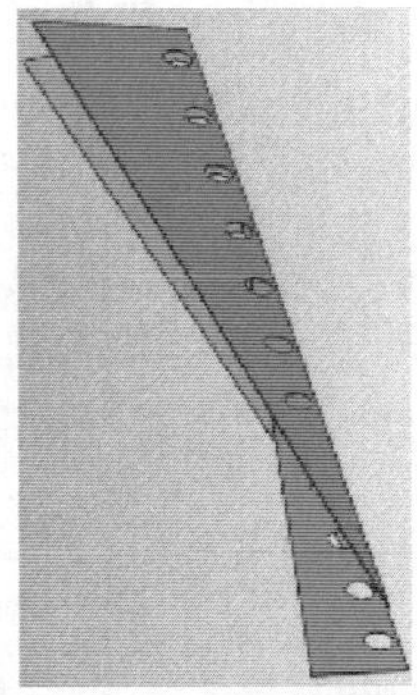

Figure 4. Internal guide vanes

Figure 5. Mesh of spiral separation section

3.2. Calculation model of gas-liquid separator

According to the structure characteristics of gas-liquid separator, the fluid calculation domain is divided into two parts: spiral separation section and collection section. The calculation domain is meshed by ICEM software, in which hexahedral grid is used for spiral separation section as shown in Figure 5, and tetrahedral grid is used for collection section as shown in Figure 6.

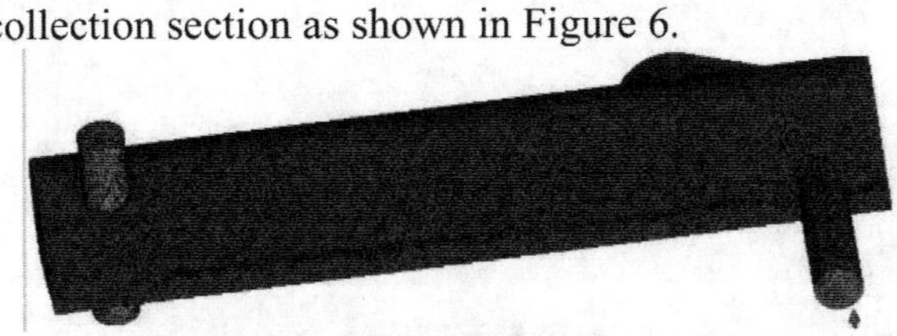

Figure 6. Mesh of collection section

ANSYS-CFX software is selected as the solver [5]. Because the gas-liquid separation process has very strong unsteady characteristics, the unsteady transient multiphase flow simulation is adopted, and the K-ε model is selected as the turbulence model. The gas is ideal helium, which is set as continuous phase, and the liquid is deionized water. As the volume fraction of liquid in this project is less than 10%, it is set as discrete phase, with solid-liquid surface tension coefficient of 0.072n/m and contact angle of 5° .In the initial calculation, the rated flow at the inlet is 10L / min, the volume fraction of liquid is 10%, and the time step is 0.001s.

3.3. Calculation results of gas-liquid separator

In order to evaluate the gas-liquid separation effect of gas-liquid separator, the gas ratio escaping from the liquid outlet and the liquid ratio escaping from the gas outlet are mainly monitored in the calculation process. In CFX, the gas ratio ξ_{air} escaping from the liquid outlet is defined as equation (1), and the liquid ratio ξ_{water} escaping from the gas outlet is defined as equation (2):

$$\xi_{air} = \text{abs(Air.massFlow()@Fluid Outlet/ Air.massFlow()@Inlet)} \quad (1)$$

$$\xi_{water} = \text{abs(Fluid 1.massFlow()@Air Outlet/ Fluid 1.massFlow()@Inlet)} \quad (2)$$

For the initial gas-liquid separator structure designed above, its gas-liquid separation rate is shown in Figure 7. It can be seen that after 2 seconds, the liquid basically fills the cone top area of the collection section and flows out to the gas outlet intermittently. In addition, since the liquid outlet section is well sealed by the liquid, the rate of gas escaping from the liquid outlet after 1.5 seconds is low.

In order to show the two-phase flow pattern in gas-liquid separator more intuitively, the evolution process of gas-liquid two-phase flow in gas-liquid separator structure is shown in Figure 8. It can be clearly seen from the figure that due to the capillary action and centrifugal force of the conical structure, most of the liquid components (red part) accumulate at the top of the conical section. At the same time, due to the drag effect of the gas, when the liquid accumulates at the top of the cone of the collection section, a small amount of liquid accumulates at the left end and then overflows from the gas outlet, which is also the main reason for the fluctuating peak of the liquid separation rate in Figure 7.

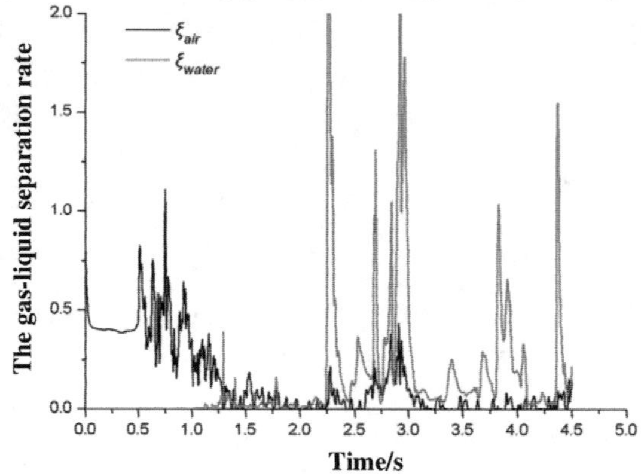

Figure 7. Mesh of collection section

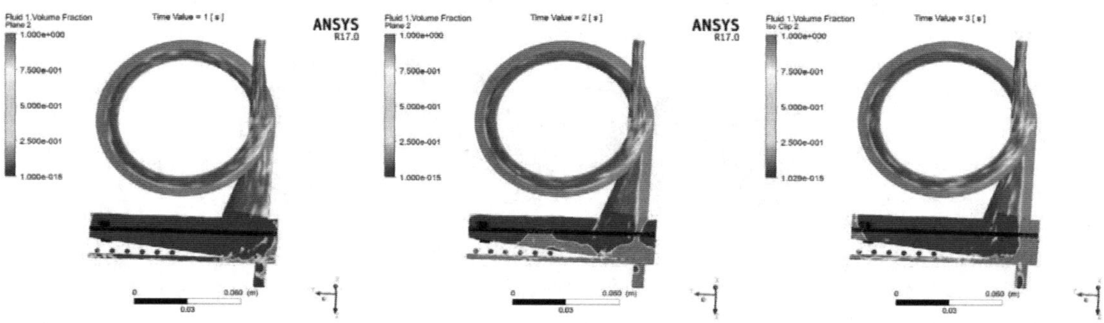

Figure 8. Evolution process of gas-liquid two-phase flow in gas-liquid separator

4. Conclusions

The results of analysis indicates that the gas-liquid separator could effectively separate gas-liquid mixture under microgravity conditions, therefore it would effectively solve the problem of gas-liquid separation of the tank gas-port in the orbital fuel resupply mission. However, the numerical calculation results has shown that the stability of gas-liquid separation was the key characteristics, which was closely related to the geometric parameters of the fluid domain. It is necessary to optimize the parameters of the fluid domain in future.

References

[1] McCraney J M. Analysis of Capillary Flow in Interior Corners : Perturbed Power Law Similarity Solutions[D]. Portland State University, 2015.
[2] Weislogel M M, Thomas E A, Graf J G. A Novel Device Addressing Design Challenges for Passive Fluid Phase Separations Aboard Spacecraft[J]. Microgravity Science and Technology, 2009, 21: 257-268.
[3] Weislogel M M, Thomas E A, Graf J G. Systems and Methods for Separating a Multiphase Fluid[P]. US 7905946 B1, 2011.
[4] Weislogel M M. Capillary flow in interior corners: The infinite column [J]. Physics of Fluids, 2001, 13 (11): 3101-3107.
[5] ANSYS 14.5 Help.

IOP Conf. Series: Materials Science and Engineering **793** (2020) 012038 doi:10.1088/1757-899X/793/1/012038

Side parking device based on automobileautomatic force drive

Jiarui Qin[1], Cancan Wei[2] and Xinhong Xiong[1*]

[1] Side parking device based on automobile automatic force drive, Wuhan University of Technology, Wuhan, Hubei, 430070, China

[2] Side parking device based on automobile automatic force drive， Wuhan University of Technology, Wuhan, Hubei, 430070,China*Corresponding author's e-mail: 466252181@qq.com。

Abstract. With the development of China's economy and the continuous deepening of urbanization, the number of urban private cars has increased dramatically [1]. However, the planning of the community construction was relatively backward at the beginning, this has resulted in many residential areas, especially the old-fashioned ones. The problem of difficult parking is becoming increasingly prominent. Therefore, it is very necessary to design a parking device that can effectively use the existing space and create more parking spaces. In addition, efforts are made to reduce the manufacturing cost of the device, save energy for product operation, avoid complicated maintenance of the device, and reduce potential safety hazards during use. The project team designed a side parking device based on the automatic driving force of the car. The device includes a vehicle-carrying walking module and a motion module. The vehicle-carrying walking module is composed of a laterally-movable vehicle-carrying platform device and a safety baffle mechanism, which realizes the lateral movement of the car parking and prevents accidents during the car parking process. The motion module is composed of a cross-cut telescopic mechanism, a transmission mechanism, a driving mechanism, and a ratchet and pawl mechanism, which realizes the functions of parking the car in the warehouse and preventing the car body from slipping. After many discussions, the members of this project team have fully studied the problems of the existing parking devices and the parking difficulties faced by the old-style residential quarters, the project team decided to start from two aspects: first, to overcome Insufficient; then, to meet the actual needs of parking in the community [3].

1. Introduction

According to a survey, most parking devices or systems in China at this stage have problem of complex mechanisms, poor work continuity, and are often accompanied by high energy consumption and high costs. For example, a three-dimensional garage costs millions of dollars without regarding to site costs and needs to occupy more space [2]. Because of these problems, the usage rate of parking devices in old-fashioned communities with small spaces is very low. The difficulty of parking in the old-fashioned community is relatively prominent, so a parking device with a small size and a low cost is needed to improve the existing parking situation.

2. Design

2.1 Overall design ideas

(1) The car can be safely and smoothly parked in the open space, and the less resources it needs to use, the better.

(2) It can automate the required functions and conform to the development trend of the current era.

(3) Improve the utilization rate of vacant land: Based on the above ideas, the team designed this parking device. The working process is as follows: After the device receives the user's use information, the device is pushed out by the motor. After the car is on the platform, the safety baffle is raised, the car is driven in reverse gear, and the power is transmitted to the crank slider-scissor link mechanism through the transmission mechanism through the electromagnetic clutch, thereby pulling the car into the parking space. Just before reaching the extreme position, the clutch is disengaged and the drive mechanism is turned to idle. When the car needs to be driven out, by virtue of the periodicity of the crank slider, the car continues to drive in reverse gear to complete the vertical ejection. At this time, the locking mechanism is running, and the drum is locked, so that the car can drive out normally.

3. Overall design

3.1 Functional design

1) The device can realize the parking and storage function of the car by using the self-driving device of the car, and the device also has its own power supply, which can also realize the parking function for non-energy vehicles.

2) The device can provide a function of a safety baffle when the car is parked on the platform, effectively preventing safety accidents caused by the car due to inertia and other reasons.

After the car is parked in the warehouse, the device uses a ratchet and pawl mechanism to realize self-locking to ensure the safety of the car parked in.

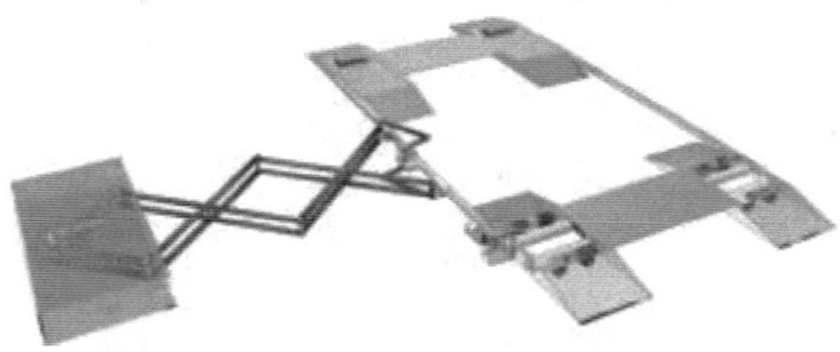

Figure 1 Overall schematic diagram of the device

3.2 Structural design

1) Safety bezel mechanism

Baffle with buckle
bearing shaft
positioning plate
snap-fit movable trigger plate
spring, electromagnet

A circular reed is installed between the baffle and the rotating shaft, so that the baffle can be automatically turned upward without any limit. The buckle of the baffle is connected to the first bearing shaft closest to the buckle. There are two flexible steel wire ropes to limit the rotation angle of the baffle, so as to prevent the vehicle from walking backwards.

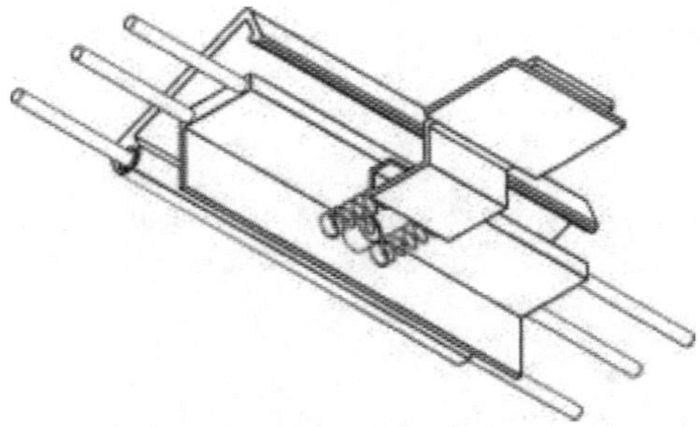

Figure 2 Safety bezel mechanism

2)Ratchet and pawl locking mechanism

The ratchet and pawl mechanism is designed for the car to exit the platform. When the car needs to drive off the loading platform, the drum must be locked to prevent the car from slipping on the drum. The electric signal is fed back to the single chip microcomputer through the travel switch, and then the steering of the steering gear is controlled. The spring is pulled through the wire rope, so that the pawl locks the ratchet and stops the roller.

Figure 3 Ratchet and pawl locking mechanism

4. Feasibility analysis

After we use ADAMS, ANSYS and MATLAB to simulate the machine, we get the following results: The weight of a family car is generally in the range of 1.2t-1.8t, and the device can still maintain various components on the basis of a 2t weight The smooth operation of our machine, and our drive motor uses YS-5D200GN-24 high-power drive motor, which makes the whole machine move more smoothly and efficiently [4] [5].

5. Innovation

1. Use the combination of crank slider and cross-cut link mechanism to realize the two-way operation of the device.

The device is provided with two crank slider mechanisms. The two sliders are respectively connected to both ends of the cross-cut telescopic mechanism. The periodic movement of the crank slider mechanism is used to drive the cross-cut telescopic mechanism, thereby realizing unidirectional driving as a two-way device Operation.

2. Use ratchet and pawl linkage lock to ensure that the car can exit the device normally.

This mechanism is in the motion module. When the car needs to drive out of the platform, it controls a steering gear to steer the wire rope and then the spring through the reel, so that the two pawls can lock the ratchet and lock the drum.

6. Conclusion

The project team developed a side parking device based on the automatic driving force of the car. This device saves the seating space required by the car when it is parked on the normal side, improves the utilization of the plane space of the parking space, and reduces the parking of the owner Technical requirements, and effectively regulates the side parking of the driver, so that it can serve the purpose of serving the community.

Acknowledgments

First and foremost, I would like to show my sincere gratitude to my partner and teachers, who give me support during the research. Besides, I am extremely grateful for my parents for their constant encouragement.

References

[1] Peng Wensheng, Huang Hualiang, Wang Junrong, Li Zhiming. Mechanical Design (Second Edition). Huazhong University of Science and Technology Press, 2003.
[2] Liu Hongwen. Mechanics of Materials. Higher Education Press, 2004.1.
[3] Zheng Wenwei, Wu Kejian. Principles of Machinery (Seventh Edition) [M]. Beijing: Higher Education Press, 2000.2.
[4] Li Zenggang. ADAMS Introduction Details and Examples [M]. Beijing, National Defense Industry Press, 2006.
[5] Li Jun, Xing Junwen, Qin Wenjie. ADAMS Example Tutorial [M]. Beijing, beijing institute of technology press, 2002.

IOP Conf. Series: Materials Science and Engineering **793** (2020) 012039 doi:10.1088/1757-899X/793/1/012039

Virtual Commissioning Modeling and Simulation of Assembly System for Quasi Natural Language Programming

Ya jie Yang[1], Zhi jing Zhang[*] and Ming kuan Shi[2]

[1] School of Mechanical Engineering, Beijing Institute of Technology, 5 South Zhongguancun Street, Haidian District, Beijing, 100081, P.R.China

[*]Corresponding author's e-mail: zhzhj@bit.edu.cn

Abstract. This paper focuses on a kind of quasi natural language programming for automatic assembly system, and studies the modeling and simulation of virtual assembly prototype, including the modeling of mechanical system, control system and the error model of assembly prototype. A multi-domain model is formed based on interfaces between different professional modeling and analysis software. Based on this model, the quasi natural language program is simulated and commissioned. At the same time, the script program is run on the physical prototype to verify the consistency of the assembly system prototype. Tested by experiments, the virtual commissioning modeling and simulation technology of the assembly system can approximately restore the assembly process of the real prototype, perform lifecycle management on the development of the equipment, and improve the commissioning speed and production quality.

1. Introduction

In the process of continuously promoting the "Industry 4.0" and "Made in China 2025" strategies, traditional manufacturing is gradually transforming and upgrading to intelligent manufacturing[1]. However, in the process of transitioning to intelligent manufacturing, virtual commissioning technology for product lifecycle management is particularly important.

The virtual commissioning technology refers to the creation of digital models in the physical environment through virtual reality technology, which are used to test and verify the performance of a series of devices[2]. Products are often designed in the order of "design-debug-verify" during the design phase, which makes the period longer. Virtual commissioning allows the designer to make any modification and optimization before production. During the program verification process, the running state of the prototype can be simulated by 3D simulation software, meanwhile, the automatic assembly system can be improved in time.

This paper proposes a method for commissioning assembly systems virtually based on quasi natural language, and performs modeling and simulation verification for this purpose.

2. Quasi Natural Language Programming

Quasi natural language programming is a reconfigurable programming method for automatic assembly systems, acting on a high-precision pulse sequence multi-axes controller (G-MAC[3]). This programming method divides the automatic assembly process into the smallest action units, which can be configured in servo mode or step mode. Different assembly processes are realized through serial and parallel connection of different action instructions. The assembly system has certain generality and reconfiguration to complete the automatic assembly process for various assembly processes of different parts. The programming instruction sample is shown in Fig. 1.

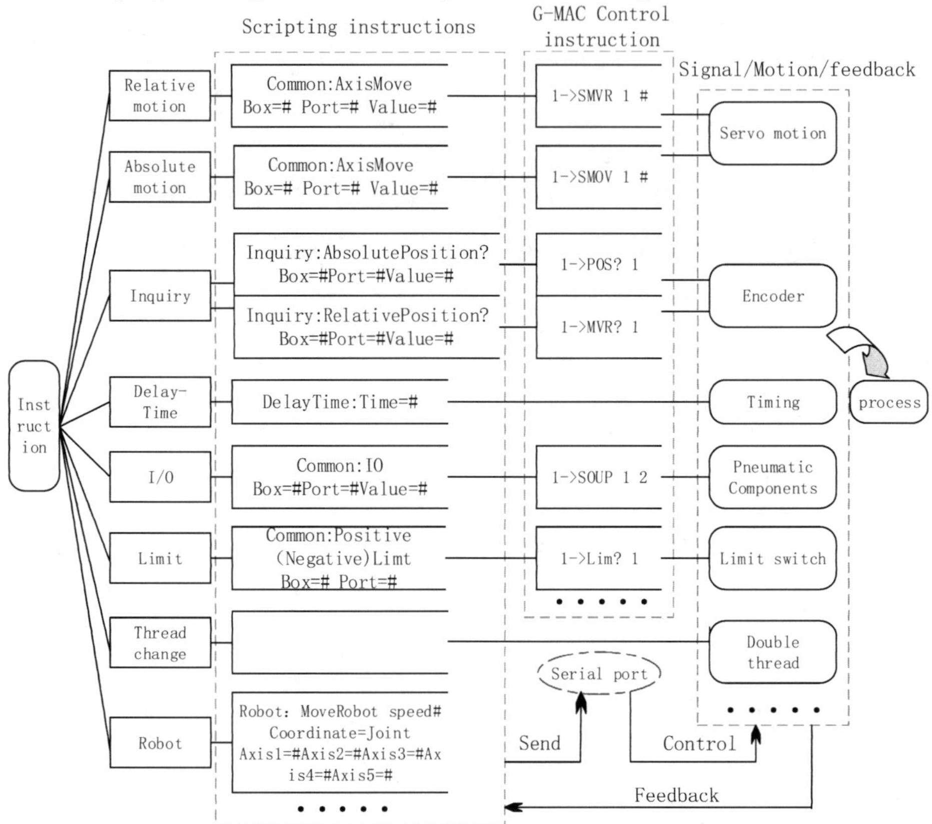

Figure 1. Quasi natural language programming instructions.

In the virtual layer, this paper builds a virtual motion control system using NI Softmotion. It reads and parses quasi natural language programming script files, and uses motion functions to drive virtual prototypes in 3D software for automated assembly. In the real hardware layer, the edited control instructions are sent to the G-MAC through the serial port, which can automatically identify the instructions and complete the control of the multi-axes motion, I/O signals and sensors, limit switches and other signals . And then complete the assembly process of the automatic assembly system.

3. Data Parsing and Logical Controlling

Before performing virtual commissioning, the script files need to be parsed into the information required in NI Softmotion, including motion resources (axes numbers), enable, stop, speed, acceleration and position information. There is a certain relationship between these information, such as sequential structure, nested relationship, etc. The basic structure of data storage and query selects the Hash table structure for fast query of nodes. Hash table is a data structure that is accessed directly according to the key value. The bottom layer is generally an array structure [4].

When we store the data value in the hash table, we will get the key value corresponding to the input information. After the operation, we will get the slot in the corresponding array, which denoted by K; According to the original address value, the corresponding physical address under each different K value is obtained. Then, the data is obtained by querying the address. The address structure is:

$$\text{address } x = address0 + K * M \tag{1}$$

Where address is the address; K is the slot; M is the number of bytes; Taking the absolute motion instruction as an example, the expression structure in the hash table is shown in Fig 2:

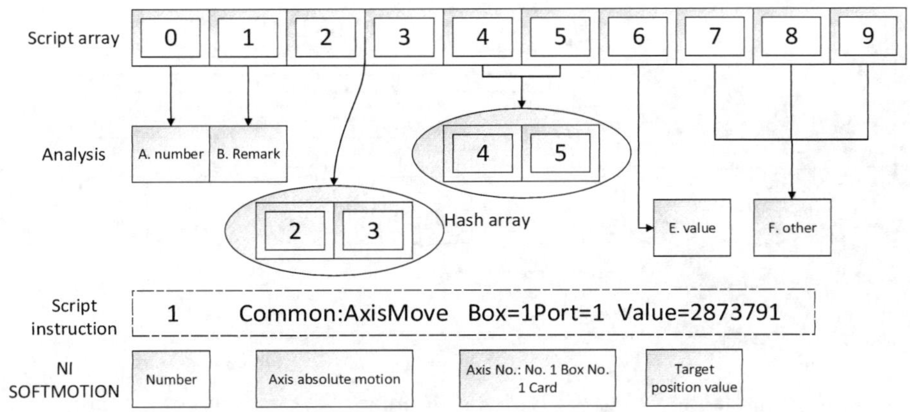

Figure 2.Script instruction expression format.

By reading the script file through LabVIEW, the axis information required for NI Softmotion can be obtained after parsing. The implementation method in LabVIEW is shown in Figure 3.

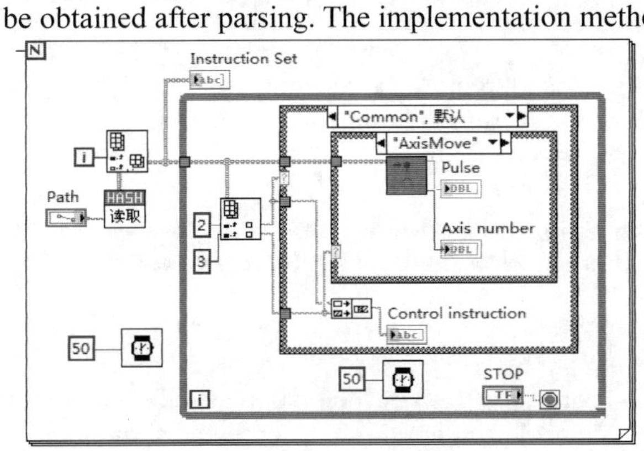

Figure 3. Implementation method in LabVIEW.

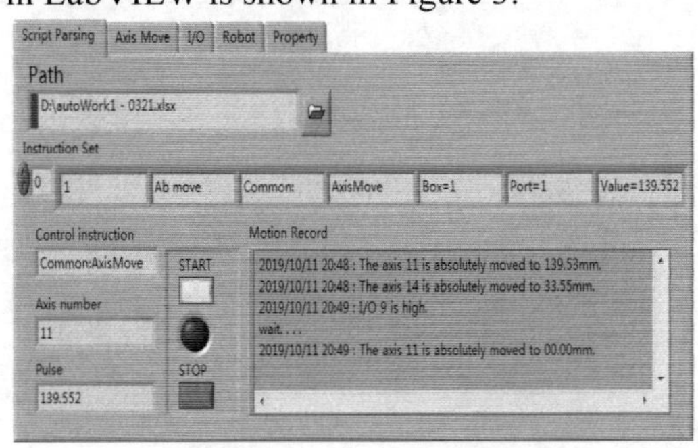

Figure 4. LabVIEW interface.

The script instructions are read cyclically and configured as input information to the control unit. Different axes movements are controlled in parallel. After the simulation is completed, the current cycle of all axes is ended. There are three end conditions: fault stop, manual stop and automatic motion end. When the end condition is a fault stop or manual stop, the simulation is over. Restart the simulation by repairing the fault and checking the model. Until the programming script file is completely read, the operation ends. The interface display is shown in Fig 4.

4. Mechanical System and Control System of Virtual Prototype

4.1. Mechanical System

The basis of automation systems virtual commissioning is the establishment of a "mechanical-control" simulation environment. The mechanical model of the assembly system is built in Solidworks. The whole system builds 5 linear axes, 2 rotary axes and 5-axis robotic arms, as shown in Fig 5.

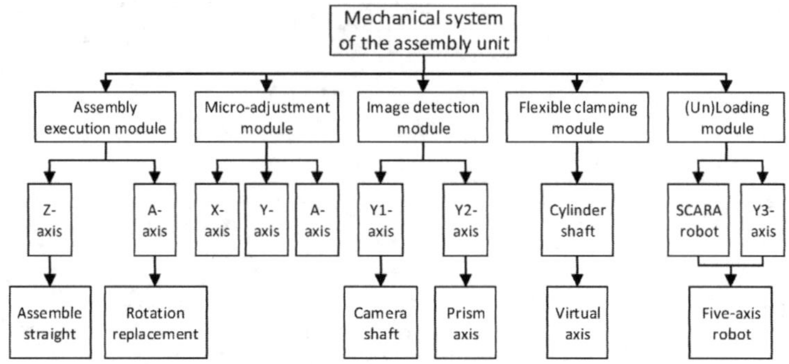

Figure 5. Virtual prototype model. Figure 6. Modular block diagram of assembly system.

In order to achieve the precise assembly of microminiature parts, the assembly system needs to be able to independently complete at least one or more assembly processes. Therefore, the assembly system modeling must be reconfigurable and extensible. Finally, a two-level modular design method oriented to system adaptability is adopted.

According to the classification of assembly objects and requirements, the assembly system is divided into 5 modules to complete the entire process of "loading and unloading, flexible clamping, image detecting, micro-adjustment and assembly". The modular structure of the assembly system is shown in Fig 6.

In order to make the prototype closer to the real, some physical field attributes are added to the model, including attributes such as mass, acceleration, friction, and spring force, which make the virtual prototype closer to the real physical prototype.

4.2. Control System

The control model of the assembly system is built in LabVIEW [5-6], which simulates the input and output of the real motion controller, and optimizes the control program through combined simulation.

The advantage of using a script programming file as the control program is that it already includes path planning and sequential logic design for motion control. We only need to build the smallest motion control unit—axis motion model in NI Softmotion.

Each axis unit constitutes the entire system, and the specific functions include: linear / rotary motion, absolute / relative motion, position reading and feedback of limit signals, etc. The LabVIEW model is shown in Fig 7.

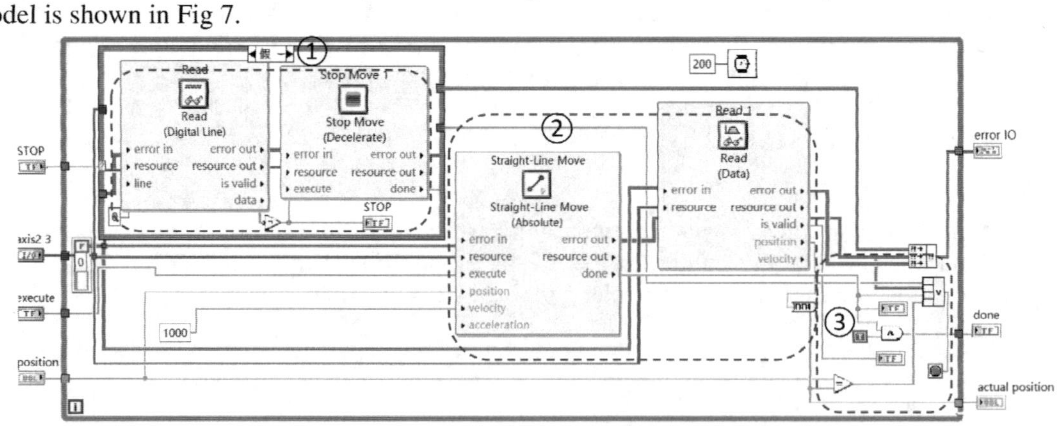

Figure 7. LabVIEW implementation method

Part 1: First, configure the positive and negative limit switches of the motion axis. Then, read the sensor data from Solidworks through the read function. In the end, it is used to inform the end of the travel of the motion system with high / low level signals.

Part 2: Configure the absolute rotary motion of the axis and read the current position in real time.

When the feedback value matches the input, the next position value is read.

Part 3: Configure the end conditions of the cycle: Fault alarm, End of motion and Manual stop.

In addition, for the automatic assembly system, the main function of the IO signal is to control the cylinder movement. The output signal controls the movement of the cylinder through the control model, and the input signal detects the current position fed back by the magnetic switch. In NI Softmotion, the movement of the cylinder uses a point-to-point distance method.

5. Error Model of Virtual Prototype

The error between the assembly parts and the based parts during the assembly process is studied, and the topology of the assembly system is described using a low-order array [7], as shown in Table 1.

Table 1. Assembly system low-order-body array.

	1	2	3	4	5	6	7	8	9
L^0(j)	1	2	3	4	5	6	7	8	9
L^1(j)	0	1	2	3	4	0	6	7	8
L^2(j)	0	0	1	2	3	0	0	6	7
L^3(j)	0	0	0	1	2	0	0	0	6
L^4(j)	0	0	0	0	1	0	0	0	0
L^5(j)	0	0	0	0	0	0	0	0	

0- Main body; 1-X axis; 2-Y axis; 3-Micro-adjustment turntable; 4-Fixture; 5-Based parts; 6-Z axis; 7-A. turntable; 8-A.fixture; 9- Assembly parts;

For the assembly system, the two branches are the assembly branch of the assembly and the micro-adjustment branch of the based part.

According to the multi-domain system kinematics theory and homogeneous coordinate transformation, the feature change matrix between the relative motion structures is:

$$T_{SV} = T_{SV,\mathrm{p}} T_{SV,pe} T_{SV,s} T_{SV,se} \tag{2}$$

The feature change matrix between relatively stationary structures is:

$$T_{SV} = T_{SV,\mathrm{p}} T_{SV,pe} \tag{3}$$

For the assembly unit system, adjacent moving bodies include 12, 23, and 67, and adjacent non-moving bodies include 01, 34, 45, 06, 78, and 89.

The actual space position of the based part in the assembly unit system can be determined.

$$T_{05} = T_{01}T_{12}T_{23}T_{34}T_{45} = T_{01,p}T_{01,pe}T_{12,p}T_{12,pe}T_{12,s}T_{12,se}T_{23,p}T_{23,pe}T_{23,s}T_{23,se}T_{34,p}T_{34,pe}T_{45,p}T_{45,pe} \tag{4}$$

Similarly, the actual space pose of the assembly part is:

$$T_{09} = T_{06}T_{67}T_{78}T_{89} = T_{06,p}T_{06,pe}T_{67,p}T_{67,pe}T_{67,s}T_{67,se}T_{78,p}T_{78,pe}T_{89,p}T_{89,pe} \tag{5}$$

The actual position of the based and assembly part in the coordinate system can be expressed as:

$$P_5 = \left[P_x, P_y, P_z, 1\right]^T = T_{05}\left[0,0,0,1\right]^T \tag{6}$$

$$P_9 = \left[P_x, P_y, P_z, 1\right]^T = T_{09}\left[0,0,0,1\right]^T \tag{7}$$

The position deviation is:

$$\Delta P_5 = \left[\Delta P_{5x}, \Delta P_{5y}, \Delta P_{5z}, 1\right]^T = P_5 - P_5^{'} \tag{8}$$

$$\Delta P_9 = \left[\Delta P_{9x}, \Delta P_{9y}, \Delta P_{9z}, 1\right]^T = P_9 - P_9^{'} \tag{9}$$

During the assembly process, the component errors of each motion axis are brought into the formula, and the maximum coordinate displacement deviation that can be caused in the three coordinate axis directions.

$$\left|\Delta X_{59}\right|_{\max} = \left|\Delta X_5\right|_{\max} + \left|\Delta X_9\right|_{\max} = 0.0264mm \tag{10}$$

$$\left|\Delta X_{59}\right|_{\max} = \left|\Delta X_5\right|_{\max} + \left|\Delta X_9\right|_{\max} = 0.0264mm \tag{11}$$

$$\left|\Delta Z_{59}\right|_{\max} = \left|\Delta Z_5\right|_{\max} + \left|\Delta Z_9\right|_{\max} = 0.0174mm \tag{12}$$

6. Simulation verification of virtual prototype

In this paper, a visual virtual prototype trajectory, logic sequence and collision detection are obtained through simulation. Compensate the movement error of each axis, check the deviation values of the X, Y and Z directions between the assembly and the based parts after operation, and compare with the actual operation. The results are shown in Figure 8-9.

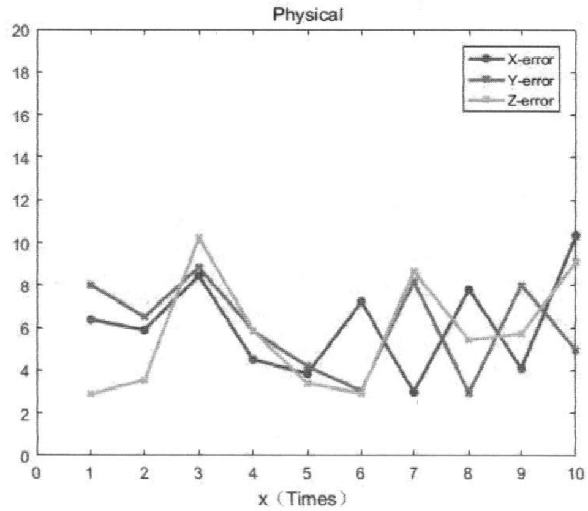

Figure 8. Error on real prototype.

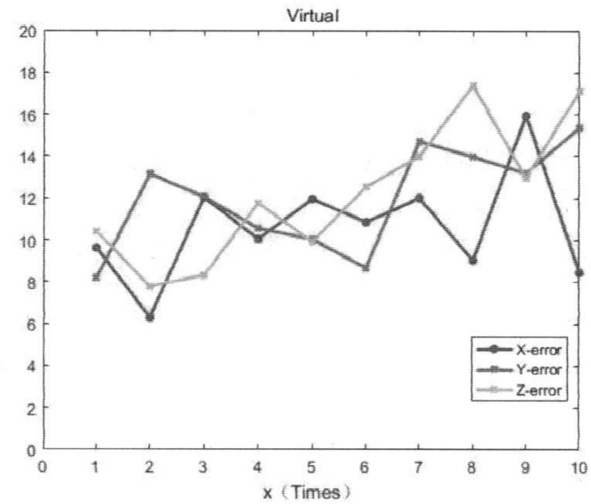

Figure 9. Error on virtual prototype.

7. Conclusion

For the natural language-like programming of the automatic assembly system, this research studies the modeling and simulation of the virtual assembly prototype, and simulates the virtual operation of the program.Based on the kinematic error model of the multi-domain system, the running error of the assembly system is compensated to the virtual prototype, and the movement track, logical sequence and collision detection of the assembly system motion are obtained, which verifies the consistency of the assembly prototype operation.

References

[1] Jianzhong Fu. Development Status and Trend of Intelligent Manufacturing Equipment [J]. Mechanical and Electrical Engineering, 2014, 31 (8): 959-962.

[2] Wang Chunxiao, Luo Weichao, Liu Riliang, Hu Tianliang. Multi-domain Modeling and Virtual Debugging of NC Machine Tools Based on Modelica [J] .Modular Machine Tool & Automatic Manufacturing Technology, 2018 (10): 102-105 + 110.

[3] Gaoyang Zhang. Research on autonomous control system technology of non-silicon MEMS flexible micro-assembly line [D]. Beijing Institute of Technology, 2014.

[4] Longzhen Li.Research on the Data Structure of the Special Spreadsheet Software [J] .Electronic

Production, 2014 (13): 62-63 .

[5] Gadzhanov,S.D,Nafalski,A, Nedic. Z. An application of NI SoftMotion RT system in a motion control workbench [C]. /REV.2013.6502909.

[6] Ke-qing Wang. Motion simulation of robot for nuclear reactor pressure vessel non-destructive detection based on SoftMotion[A]. Technology of China:IEEE BEIJING SECTION,2015:8.

[7] Yan Li, Dapeng Fan. Modeling and Analysis of Assembly Error of Three-axis Turntable Based on Kinematics Theory of Multibody System [J] .Acta Armamentarii, 2007 (08): 981-987.

IOP Conf. Series: Materials Science and Engineering **793** (2020) 012040 doi:10.1088/1757-899X/793/1/012040

Optimization of extended range electric vehicle energy management strategy via driving cycle identification

Chen Yong[12], Zhang Yue[12], Wei Changyin[12], Li Guangxin[12] and Li Congcong[12]

[1] School of Mechanical Engineering, Hebei University of Technology, Tianjin, 300130, China

[2] Tianjin key laboratory of new energy vehicle power transmission and safety technology, Tianjin, 300130, China

[*]Corresponding author's e-mail: weiag2@163.com

Abstract. There are ambitious fuel consumption targets for the manufacturers of heavy-duty vehicles. For this reason, extended range electric vehicle (EREV) is a promising powertrain technology. However, the energy management strategy (EMS) is still an obstacle to the improvement of fuel economy. This paper introduces an energy management strategy for driving cycle identification. Twenty-two typical driving cycles are divided into five categories through Q-type clustering. Euclidean distance is used to identify the driving cycle. For each type of driving cycle, the energy management strategy parameters are optimized through genetic algorithms with fuel consumption and emissions as the goals. The results show that the EMS via driving cycle identification is more effective than a strategy that does not identify it. For comprehensive test cycles, the former's fuel consumption is optimized by 6%, SOC consumption is optimized by 1%, and there is a slight improvement in emissions.

1. Introduction

At present, pure electric vehicles are faced with some disadvantages, such as short driving range, inconvenient charge service, long charging time, short life and high battery pack prices. To solve these disadvantages, extended range electric vehicle (EREV) is designed with APU and small capacity battery pack [1]. EREV has the advantages of both the pure electric drive model and the hybrid power drive model, and it also can extend driving ranges and reduce cost. Thus, this model is gradually becoming a development trend of electric vehicle products [2]. Optimizing Energy management strategy (EMS) is an effective way to improve fuel consumption and emissions. Research shows that the uncertainty in vehicle driving cycles has an important impact on fuel consumption. EMS control parameters were adjusted in real time in a study [3] in which six representative standard cycles and 24 characteristic parameters were selected and used to identify comprehensive drive cycles. In another study [4], six urban conditions typical of China and the United States were selected as objectives of off-line optimization, and eight characteristic parameters including maximum vehicle speed were used to driving cycle identification [5].

In this paper, a variety of road conditions is considered, 22 typical driving cycles are selected as the standard mode library, and cluster analysis is conducted on them. The driving cycles are identified during the running of the vehicle, and then the control parameters are adjusted to minimize fuel consumption and emissions. The important control parameters are optimized offline by a genetic algorithm. Finally, the proposed vehicle EMS is simulated using driving cycle identification, and the results are analyzed

IOP Conf. Series: Materials Science and Engineering **793** (2020) 012040 doi:10.1088/1757-899X/793/1/012040

in a MATLAB/Simulink environment.

2. Classification and identification of driving cycle

In this section, the EREV architecture is illustrated in figure 1. The parameter configuration of the whole vehicle is shown in table 1.

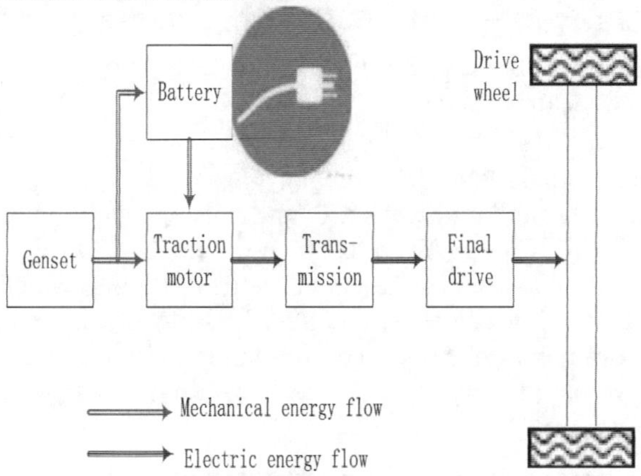

Figure 1. System configuration of the EREV.

Table 1. Vehicle parameters.

Name	Parameter
Curing quality (kg)	6000
FC max power(kw)	128
Motor max power(kw)	105
Generator max power(kw)	44
Wheelbase(m)	3.36
Wind resistance	0.44
Windward area A (m^2)	4.8
Acceleration time(0-50km/h) (s)	22
Maximum grade	30%
Pure electric range (km)	60

2.1. Classification of driving cycle

In this paper, the maximum speed of urban logistics vehicle is 90km/h is studied, so the driving cycles are selected from advisor as follows:

Table 2. Driving cycles.

1.1015	2.5PEAK	3.ARTERLAL	4.BUSRTE	5.CBD14	6.COAST
7.COMMUTER	8.CONSTANT	9.CSHVR	10.ECE	11.UDDS	12.MANHATTAN
13.NYCC	14.NYCCOMP	15.NYCTRUCK	16.NYGTC	17.NewYorkBus	18.SC03
19.STEP	20.WVUCIFY	21.FTP	22.IM240		

According to reference [6], 10 characteristic parameters are selected for identifying the driving cycles: time, distance, maximum speed, average speed, maximum acceleration, maximum deceleration, average acceleration, average deceleration, idle time, and number of stops.

To ensure the accuracy of classification, the 22 typical driving modes are divided into five categories according to the cluster analysis. In this paper, Q-type clustering is selected because it better suits the characteristics of the research objects. The distance is used to measure the similarity between different driving cycles. Euclidean distance [7] is used in this paper because it is the simplest and most accurate.

$$\begin{cases} \| y_i - y_j \| = \sqrt{\left(y_{i1} - y_{j1}\right)^2 + \cdots + \left(y_{im} - y_{jm}\right)^2} = \sqrt{\sum_{m-1}^{7} \left(y_{im} - y_{jm}\right)^2} \\ i \neq j \cap i,j \in Z^+ \cap i,j \in [1,22]\,. \end{cases} \tag{1}$$

The classification results are shown in figure 2.

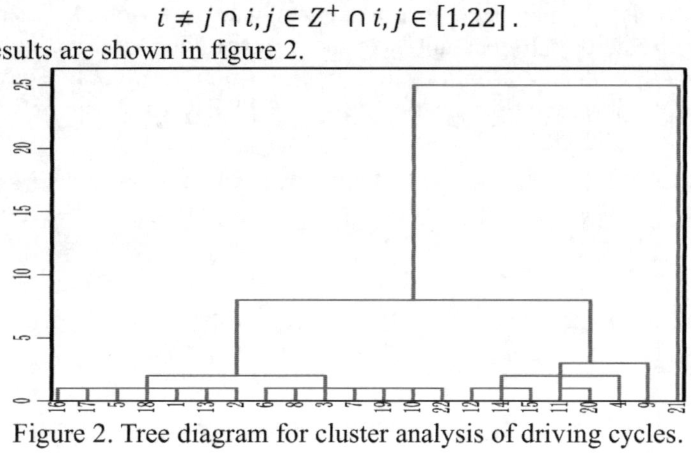

Figure 2. Tree diagram for cluster analysis of driving cycles.

2.2. Identification of driving cycle

To measure the similarity between driving cycles, a parameter is introduced to represent the progress [8]. If the five typical driving cycles in the library are the sample A_n (n= 1–5), and the actual driving cycle sample is X, then the distance $\sigma(A_n, \mathrm{X})$ is defined as

$$\sigma(A_n, \mathrm{X}) = 1 - \frac{1}{\sqrt{m}}\left(\sum_k^m \left(A_n(k) - X(k)\right)^2\right)^{\frac{1}{2}} \tag{2}$$

where m=10 is the number of characteristic parameters of the driving cycles, and k=1. The distance between A_n and X is calculated according to the Euclidean formula

$$\sigma(A_i, \mathrm{X}) = \max\{\sigma(A_1, X), \sigma(A_2, X), \ldots \sigma(A_5, X)\} \tag{3}$$

The actual driving cycle X of the vehicle is considered to belong to the typical cycle sample A_i.

Five types of driving cycles in different cities, suburbs, and highways are randomly selected in this paper: The comprehensive test cycles (CYC_CBDTRUCK+ CYC_CHTC_HT+ CYC_CWTVC+ CYC_WVUSUB+ CYC_MANHATTAN) are obtained and serve as the actual vehicle driving cycles. In order to ensure the accuracy of the drive cycle identification, select the smallest possible time interval to identify, a 200 s segment is selected to extract the parameter information of the standard driving cycle. The Euclidean distance is used to calculate the similarity to identify the drive cycles of each time period.

The driving cycle identification result is shown in figure 3.

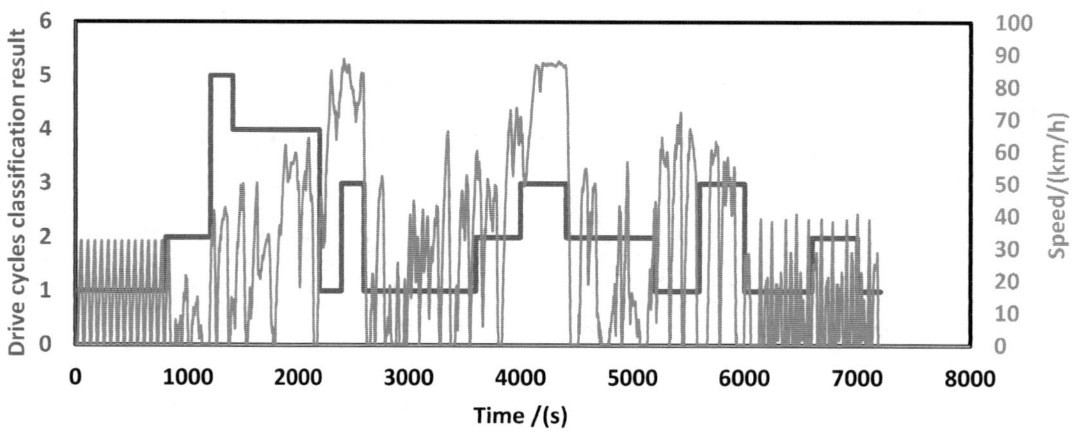

Figure 3. Speed-time curve of comprehensive driving cycle and result of driving cycle identification

3. EREV energy management strategy optimizations

The rule-based energy management strategy is applied in this paper, as shown in Figure 4. where A is the minimum power output from the engine. In addition, limit the engine speed at 2000-3500r/min.

A genetic algorithm is a randomized search method that draws on the mechanisms of natural selection and genetics in biology. It does not have the same problem and has powerful global search ability, which is especially suitable for optimization in highly nonlinear complex systems. This method is more suitable for multi-objective optimization design of EREV. The basic optimization steps are in figure 5.

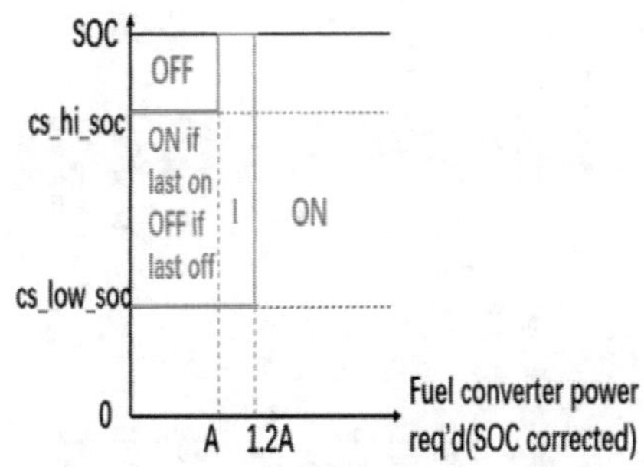

Figure 4. Energy management strategy for EREV.

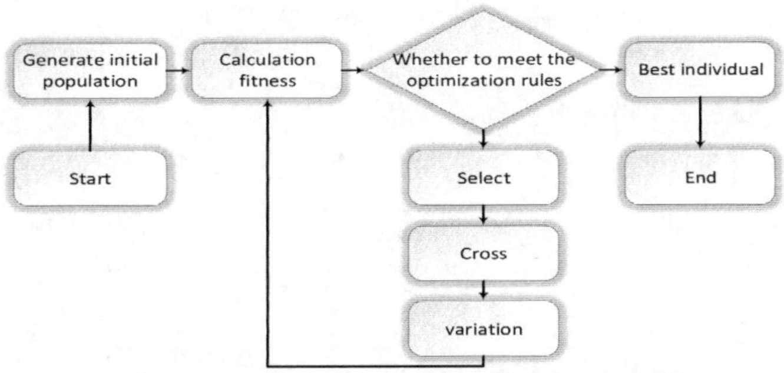

Figure 5. Step diagram for genetic algorithm optimization.

The system parameters selected for optimization are the allowed SOC upper limit Hi_{soc}, allowed SOC lower limit Low_{soc}, maximum rate of increase of genset power $Max_{pwr\ rise\ rate}$, and maximum rate of decrease of genset power $Max_{pwr\ fall\ rate}$.

Genetic operations can be expressed as

$$G_A = (C, J, P_0, M, \Phi, \Gamma, \Psi, \mathrm{T}) \tag{4}$$

where C is the individual coding method, J is the individual fitness function, P_0 is the initial population, M is the population size, Φ is the genetic operator, Γ is the crossover operator, Ψ is the mutation operator, and T is the termination condition of the genetic operation.
The objective function is the fitness function defined as

$$\mathrm{J(x)} = F_c + E_c = \omega_1 \frac{F_c}{F_{cmax}} + \omega_2 \frac{E_{HC}}{E_{HCmax}} + \omega_3 \frac{E_{co}}{E_{comax}} + \omega_4 \frac{E_{NO_\chi}}{E_{NO_\chi max}} \tag{5}$$

where ω_i (i = 1–4) represents the weight of each optimization target, F_c is the cumulative value of each instantaneous fuel consumption $F_c(t)$, E_c is the cumulative value of other pollutant emissions. The weight ω_i is initially taken to be 0.25 to consider each sub-objective function in the optimization as being equally important. Taking the fitness function of equation (5), the optimized results are shown in table 3.

Table 3. Optimal values of control parameters for five categories of driving cycles.

Driving cycle category	Hi_{soc}	Lo_{soc}	$Max_{pwr\ rise\ rate}$	$Max_{pwr\ fall\ rate}$
Category one	0.81	0.49	4920	-3120
Category two	0.77	0.43	5124	-2991
Category three	0.72	0.47	5207	-3080
Category four	0.79	0.40	4980	-3233
Category five	0.88	0.37	5119	-2895

4. Results and discussion

The simulation model is built with the MATLAB/SIMULINK platform. The simulation analysis is carried out under the comprehensive test cycle (CYC_CBDTRUCK+ CYC_CHTC_HT+ CYC_CWTVC+ CYC_WVUSUB+ CYC_MANHATTAN). The initial battery SOC is 0.7. During operation of the vehicle, the battery SOC is always changing in the charging and discharging high efficiency zone.

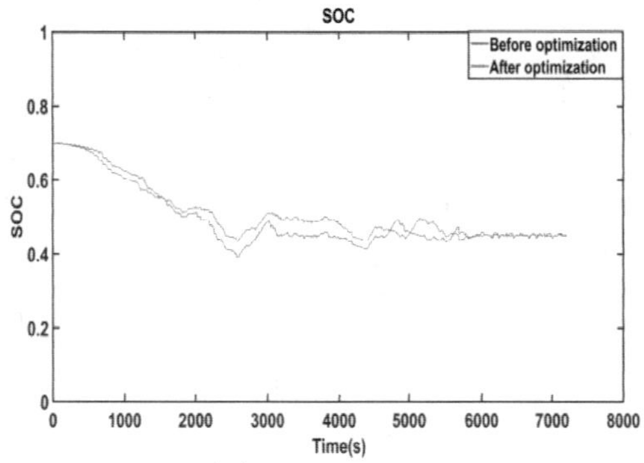

(a) SOC curve

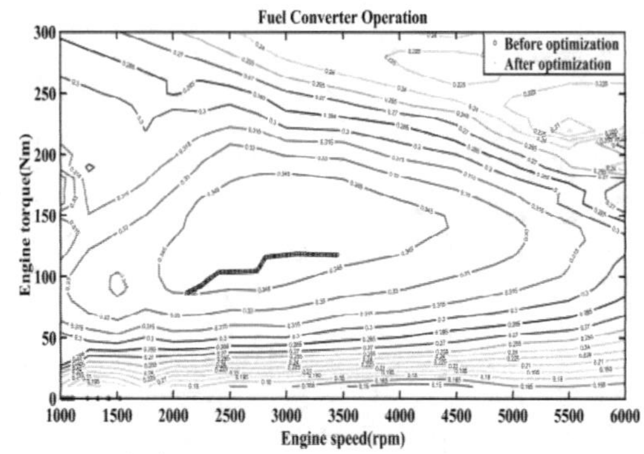

(b) Fule convertur operation

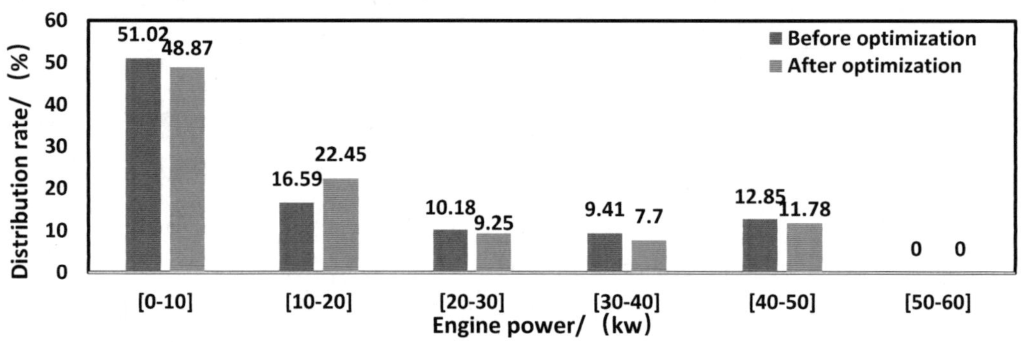

(c) Engine power distribution

Figure 6. Simulation result.

As seen in figure 6, the operating points of the engine are concentrated in the high-efficiency area. It can be seen from the graphs b and c that the engine operating point is on the optimal working curve, but the engine power distribution is different, and the optimized engine power distribution is more reasonable. The results for EMSs with and without cycle identification are shown in table 4 for comprehensive test driving cycles.

Table 4. Simulation results.

Aims	No driving cycle identification	Driving cycle identification	improvement
Fuel Consumption(L/100km)	14.1	13.25	6%
△SOC	0.249	0.246	1%
HC(g/km)	1.416	1.410	0.4%
CO(g/km)	1.616	1.602	0.8%
NOx(g/km)	0.660	0.654	0.9%

5. Conclusion

In this paper, an EMS based on the driving cycle identification has been proposed for power distribution of a hybrid power source. Cluster analysis is carried out on typical driving cycles, and the closeness of driving cycle is calculated to identify driving cycle. On this basis, the genetic algorithm is used to optimize the control parameters, which ultimately improves the economy and emission rate of the vehicle.

The main ideas and findings are:

1. Because of the randomness of hybrid vehicle driving cycles, driving cycles are identified to optimize the key parameters of vehicle control strategies. It is effective to optimize the fuel consumption and emissions of the EREV through the method of drive cycle identification.

2. The simulation and optimization results show that the fuel economy and emission performance of the vehicle are significantly improved by the proposed method. The fuel consumption is reduced by 6%, the battery consumption is improved by 1% and there is a slight improvement in emissions.

Acknowledgements
This research was funded by the National Key R&D Program of China (grant number 2018YFB0106403) and Hebei full-time top talents project (NO. 20181228).

References

[1] Li, J., Jin, X., & Xiong, R. (2017) Multi-objective optimization study of energy management strategy and economic analysis for a range-extended electric bus. Applied energy, 194:798-807.
[2] Chen, Z., Xia, B., You, C., & Mi, C. C. (2015) A novel energy management method for series plug-in hybrid electric vehicles. Applied Energy, 145:172-179.
[3] Jeon, S. I., Jo, S. T., Park, Y. I., & Lee, J. M. (2000) Multi-mode driving control of a parallel hybrid electric vehicle using driving pattern recognition. J. Dyn. Sys., Meas., Control, 124(1): 141-149.
[4] Guoqing, H., Xiaodong, W., Fengying, M., & Qingsheng, Z. (2003) Application of dynamometer card identification in diagnosis of working condition for suck rod pump. Oil Drilling & Production Technology, 5.
[5] Hu, X., Murgovski, N., Johannesson, L. M., & Egardt, B. (2014) Comparison of three electrochemical energy buffers applied to a hybrid bus powertrain with simultaneous optimal sizing and energy management. IEEE transactions on intelligent transportation systems, 15(3): 1193-1205.
[6] Fang, L., Qin, S., Xu, G., Li, T., & Zhu, K. (2011) Simultaneous optimization for hybrid electric vehicle parameters based on multi-objective genetic algorithms. Energies, 4(3): 532-544.
[7] Lei, Z., Cheng, D., Liu, Y., Qin, D., Zhang, Y., & Xie, Q. (2017) A dynamic control strategy for hybrid electric vehicles based on parameter optimization for multiple driving cycles and driving pattern recognition. Energies, 10(1): 54.
[8] Lei, Z., Cheng, D., Liu, Y., Qin, D., Zhang, Y., & Xie, Q. (2017) A dynamic control strategy for hybrid electric vehicles based on parameter optimization for multiple driving cycles and driving pattern recognition. Energies, 10(1): 54.

IOP Conf. Series: Materials Science and Engineering **793** (2020) 012041 doi:10.1088/1757-899X/793/1/012041

Fault diagnosis of mechanical spindle drive chain based on time-frequency domain characteristics

Zhao Guobo[1], Chen Xuezhen[1],*, Li Ying[1], Pan Shilu[1], Dong Guangliang[1], He Yi[1]

([1]AVIC Chengdu Aircraft Industrial (Group) CO., LTD., 610092, Chengdu, Sichuan, China. *Corresponding author: Dr. Chen Xuezhen, Senior Engineer, email: 532015901@qq.com.)

Abstract: As a key component of CNC (computer numerical control) machines, spindle's performance directly affects the performance of machines. The mechanical spindle, which is often used in the processing of complex aeronautical structures such as titanium alloys, has the characteristics of large torque and impact resistance. However, its long drive chain and complex structure make it difficult to effectively forewarn and diagnose its faults. In this paper, a fault forewarning and diagnosis method is designed, which is based on the analysis of characteristics of vibration time-frequency domain of mechanical spindle. Firstly, a method for calculating the frequency characteristics of key components in the drive chain of mechanical spindle is proposed, and the signal characteristics of vibration are obtained. Secondly, according to the position and layout of each component in the spindle drive chain, the layout scheme of signal acquisition points is designed based on the principle of point-to-surface integration. This scheme ensures that the signals of each point can not only reflect the state of the spindle, but also reflect the vibration characteristics of different parts. Variable speed test is carried out on different equipment of the same type, and the signals processed by Fourier transform, and the frequency domain signals at different speeds are obtained. The time-frequency signal of the spindle is analyzed to find its vibration characteristics. Combining with the results of theoretical calculation, the working state of the spindle and the working performance of each component, as well as the specific fault parts and types of fault are obtained. Through experimental verification, the method proposed in this paper effectively improves the ability of working state monitoring and fault diagnosis of the spindle system.

1. Introduction

CNC machines play an important role in supporting and safeguarding in the development of manufacturing industry. As the core component of CNC machine tool, the spindle's performance is directly related to the machining performance and accuracy of machine tool [1]. A spindle with good performance can effectively guarantee the quality of part processing. However, in actual processing, failures of the spindle system can be caused by many factors, such as aging and wear of components, unreasonable machining parameters, overload operation, and insufficient cooling and lubrication. Although it does not necessarily stop the machine, there are major safety and processing quality hazards. Therefore, monitoring the performance and status of the spindle is particularly important.

At present, monitoring its vibration, noise and other characteristics is the method commonly used of spindle monitoring by equipment maintenance personnel [2], but there are two defects in this method. First of all, the mechanical spindle is a complex mechanical transmission system, there are many possible parts for failure. It's hard to locate the specific failure point. Secondly, if significant noise and vibration have occurred, the failure of the spindle is already very serious, and many components may be failures already. Therefore, the methods and measures which can precisely and effectively monitor spindle's performance are needed.

In this paper, a fault forewarning and diagnosis method based on the vibration time-frequency signal of spindle system is proposed. By designing a reasonable signal monitoring point, the vibration signal of the whole spindle and each key component is collected. Set the comparison experiment of variable speed and variable equipment, the data is processed by Fourier transform, and the time-frequency signal of the spindle system is analyzed synthetically to find its typical vibration characteristics. Combined with the theoretical calculation and analysis of key component's vibration, the problem can be quickly and effectively found.

2. The analysis of mechanical spindle's vibration characteristic

A domestic 5-axis CNC machine uses a mechanical spindle, and its drive chain is long, as shown in Figure 1. It mainly includes the motor, gearbox, drive shaft, bevel gear pair, helical gear pair, coupling and spindle, and every part may fail after long-term use. Due to the long spindle drive chain and the large number of components, faults are difficult to be located. In order to improve the efficiency of maintenance and the utilization of equipment, and avoid removing the precision installed parts in conditions of uncertainty. It's very important to locate the faults quickly and accurately. In order to determine the faulty components, we must know the vibration signal of each component.

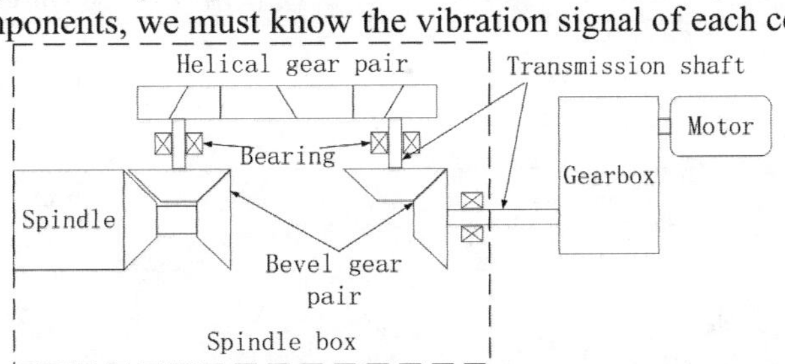

Fig.1. diagram of mechanical spindle drive chain of a 5-axis machine tool

2.1 Gear vibration analysis

There are two sets of bevel gear pairs and helical gear pairs in the spindle drive chain, which will produce a variety of vibration signals while spindle is running. For the gear transmission system, the calculation method of characteristic frequency is already exist. By analyzing the vibration signal of the gear pair, various faults of the gear transmission can be found out, including wear, overload, eccentricity, backlash migration, teeth rupture, teeth broken and chasing teeth [3]. The GMF (Gear-mesh Frequency) is an important parameter for studying the vibration of the gear. All the gears will produce a Gear-mesh Frequency when meshing. Generally, from whether there is a certain amplitude of side band in the meshing frequency, we can analyze whether the gear transmission effect is good or not [4].

When the gear system is running, the frequency with large amplitude is more related to the GMF, For example, there will be some peak amplitude at GMF, 2GMF, 3GMF and other frequencies [5]. The vibration signal characteristics of gear in common fault state have been studied in detail, For example, when the gear wear is obvious, the sideband of the GMF increases, and the amplitude is always higher at 2GMF and 3GMF. When the gear bear a large load, the amplitude of the GMF is generally large; when the gear is eccentric and there is a backlash, a significant sideband is generated at the gear GMF; when the gear is cracked or broken, a large amplitude will be at the GMF [6].

2.2 Bearing analysis

There are many bearings in the spindle drive chain, which play an important role in the stability of mechanical transmission. However, bearing failure takes a large proportion in the traditional mechanical failure, bearing wear, corrosion, breakdown and other faults have a large impact on the transmission of the mechanical system, and always accompanied by varying degrees of vibration.

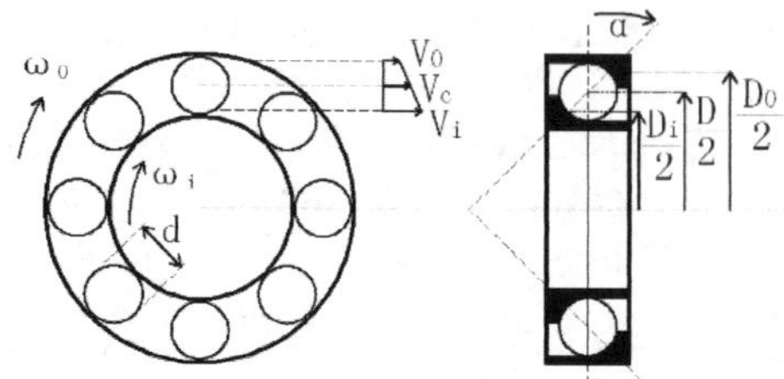

Fig.2. diagram of rolling bearing

The "characteristic frequency" of bearing has been deeply studied by scholars, so this paper will not elaborated it in detail. The calculation formula of characteristic frequency of bearing fault is shown in Table 1 [7]. Where, D is the bearing diameter corresponding to the roller center; d is the diameter of the bearing roller; Z is the number of rollers in the bearing; α is the contact angle; f_i is the rotation frequency of the inner race of the bearing.

Table.1. Calculation formula of bearing characteristic frequency

Bearing parts	Calculation formula
Inner race raceway	$f_{bpf\,i} = \dfrac{-Zf_i(1+(d/D)\cos\alpha)}{2}$
Outer race raceway	$f_{bpf\,o} = \dfrac{-Zf_i(1-(d/D)\cos\alpha)}{2}$
Cage	$f_c = \dfrac{f_i(1-(d/D)\cos\alpha)}{2}$
Roller	$f_{bsf} = -\dfrac{f_i}{2}\dfrac{D}{d}(1-(\dfrac{D}{d}\cos\alpha)^2)$

3. Vibration analysis experiment of key parts of spindle

Since the mechanical spindle has a long drive chain and many components, it is difficult to diagnose and locate faults. At present, the common method is to check the bearing, gear and other parts in turn after removal, and find the fault point by means of troubleshooting. This method is time-consuming and labor-intensive, and for some precision-mounted components, it will seriously affect the transmission accuracy and effect after disassembly. In this paper, a new method is proposed. First of all, the vibration detection experiment of each component is needed.

In order to accurately collect the vibration signals of the various components in spindle system, the layout of the collection points should be close to each measured component, and the collecting direction of components such as helical gears should be concentrated in the axial direction of the gears. However, the spindle drive chain is mainly distributed inside the spindle box and cannot be directly measured. Therefore, according to the main position layout of each key component in the spindle drive chain and the allowable conditions of the experimental equipment, three measuring points are selected for measuring the vibration outside of the spindle box by the comprehensive action of each part. The position of three measuring points on the spindle is shown in Fig. 3.

IOP Conf. Series: Materials Science and Engineering **793** (2020) 012041 doi:10.1088/1757-899X/793/1/012041

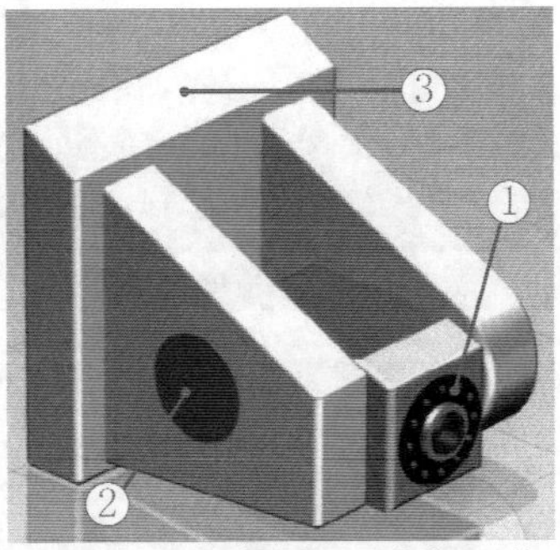

Fig.3. diagram of the installation position of the vibration measuring point

Point 1 is set at the end of the spindle, which mainly monitors the actual vibration of the spindle head. Point 2 is set on the side of the spindle box, because there are many gears running here. Point 3 is set at the back-end of the spindle box, which mainly reflect the influence of the vibration of the motor and gearbox on the spindle system.

By comparing the vibration signals of two machines at the same acquisition point and the same speed, the state performance of two machines can be diagnosed, as well as the main contribution parts of each machine's vibration. Comparing the vibration signals of the same collection point at different speeds of a single machine, it can be known whether the vibration mainly comes from the front or the back of the spindle. And the performance at high and low speeds.

3.1 Vibration experiment design

In order to ensure the authenticity and reliability of the measured vibration data, equipment comparison and speed comparison experiments are implemented. In equipment comparison experiment, we will collect and compare the vibration signals of two machines which belong to the same type, installed at the same time, under the same working condition and experimental conditions the speed comparison experiment was carried out by using a same input motor speed and changing the output speed for comparison.

The spindle drive chain is shifted by gearbox. In order to make sure whether the main vibration of the spindle system comes from the front spindle or the motor and gearbox, in the experiment, the motor will keep a same speed and spindle work in high and low gear respectively. The transmission ratio of the machine in the high and low gear position is i_1=0.2266 and i_2=1.323, when the speed is lower than 800rpm and the low gear is used. When the speed is higher than 1200rpm, the high gear is used, when the speed is between 800~1200rpm, the shift is not adopted.

In order to distinguish the main contribution points of vibration signals, the test is carried out by switching to high and low gears respectively under the condition of ensuring the same motor speed. After calculation, take the motor speed N_0 = 2000 rpm, at this time, the output speed of the spindle c is 453 rpm and 2646 rpm when working in high and low gear.

Table. 2. Machine vibration experiment parameters

machine	**Rotating speed/rpm**	
Machine 1	453	2646
machine 2	453	2646

3.2 Vibration data analysis

According to the calculation of the rotation speed of the parts in spindle system, and in order to ensure the accuracy of data acquisition, the sampling frequency f = 1000Hz is taken, and the vibration signals of three points are collected by the vibration sensors. Finally, Fourier transform is carried out to analyze the vibration signals.

3.2.1 Variable machine experiment

The experiment set at a speed of 453 rpm. The vibration characteristic frequency of the bearing and each gear in the spindle drive chain is calculated, as shown in Table 3.

Table.3. vibration characteristic frequency of bearings and gears in spindle system

Spindle components	Parameter characteristics	Vibration frequency/Hz
Bevel gear	Z_1=36 n=453rpm	f_{m1}=271.8
Small helical gear	Z_2=36 n=453rpm	f_{m2}=271.8
Large helical gear	Z_3=46 n=354.5rpm	f_{m3}=271.8
bearing outer ring	D=78mm d=15mm α=25° Z=11	f_{bpfi}=48.76
bearing inner ring	D=78mm d=15mm α=25° Z=11	f_{bpfo}=34.28
Cage	D=78mm d=15mm α=25° Z=11	f_c=3.12
Bearing roller	D=78mm d=15mm α=25° Z=11	f_{bsf}=691.15

Fourier transform is performed on the vibration signals of the three points of two machines, in order to get the time-frequency characteristics, as shown in Fig. 4.

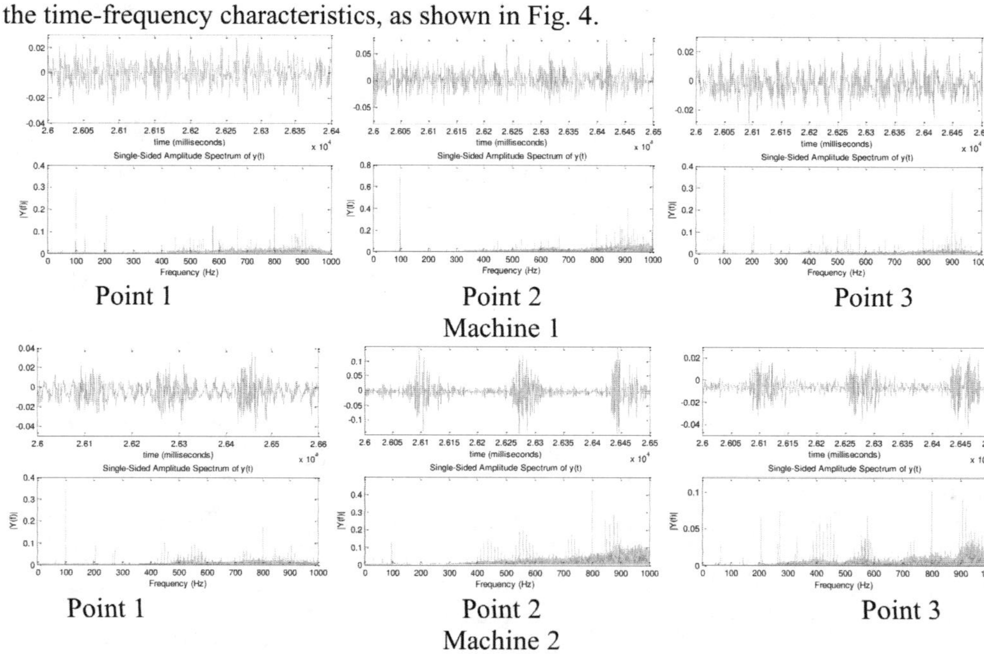

Fig.4. Time-frequency characteristics of vibration signals at three points on two machines at 453 rpm

A comprehensive analysis of Figure 4 shows that the vibration characteristics of the two machines have many in common. For example, the vibration of the point 2 in the two machines is more obvious than the other two points, it means the vibration of the gear transmission on the side of the spindle box is more pronounced than the spindle head and the rear drive. There is a certain commonality between the two, that is, a high amplitude at 100 Hz and 800 Hz.

In addition, the vibration signals of the two machines also have a certain difference, which represents the difference in performance and state of the spindle system of the two machines. Firstly, the overall amplitude of machine1 is smaller and more uniform than that of machine2; the overall amplitude of machine 2 is larger and more regular. Through time domain graph analysis, its period $T\approx$170ms, and its vibration frequency is 353Hz, which is close to the rotating frequency of the large helical gear, so it could be speculated that the periodic vibration is mainly generated by the rotation of large helical gear. Among the three groups of vibration signals of machine 1, there is a large amplitude at 100Hz, which is exactly 2 times of the vibration frequency of the outer ring of angular contact ball bearing. Therefore, several groups of angular contact ball bearings in machine 1 are likely to have problems such as wear, which need to be checked in time. However, the amplitude of the machine 2 has a large amplitude at 100 Hz, but the amplitude at the point 2 and point 3 is relatively small, indicating that there is a problem with the front bearing of the machine 2 and the rear bearing is in a good state. The amplitudes of the two machines are large at 800Hz, and it can be seen from table 2 that the frequency is just at 3GMF of the gear transmission system, indicating that the gear transmission system has certain wear. In addition, point 3 of machine 2 has a large amplitude at 200Hz, 270Hz, 900Hz, etc. Indicating that the rear end of machine 2 has a more obvious impact on the vibration of the spindle system, and the parts such as the motor and the gearbox should be checked in time.

3.3.2 Variable speed experiment

According to the experimental scheme, the two machines are analyzed separately, and the experiments were carried out at 453 rpm and 3646 rpm respectively. Comparing the vibration signals of different rotating speeds of a machine can effectively analyze the main contribution sources of vibration, and the performance of the spindle at high and low speed. Fourier transform is performed on the measured experimental data to obtain its frequency domain features, as shown in Fig. 5 and Fig. 6.

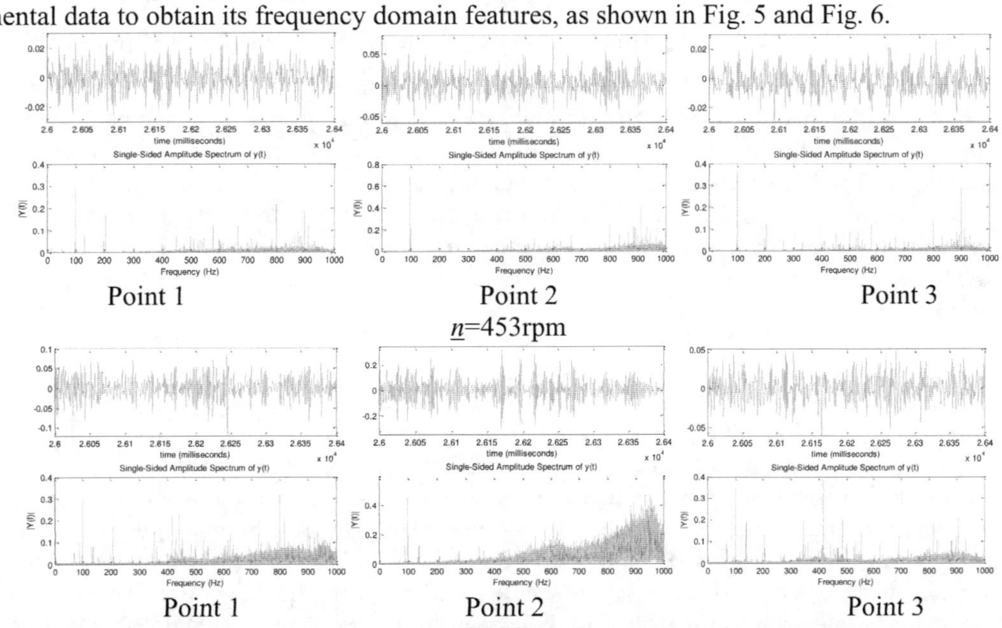

Fig. 5 Time-frequency characteristics of vibration signals of machine 1 at 453 rpm and 2646 rpm

By comparing and analyzing the time-frequency characteristics of the vibration of the machine 1 at two speeds, the high amplitudes at 100 Hz and 800 Hz are found when n=453 rpm. However, when n = 2646rpm, the higher amplitude value is also obtained at 410Hz, especially at points 1 and 3. It is calculated that the frequency is exactly 2 times of the vibration frequency of the inner ring of the bearing, so the bearing of the front and rea[r bevel gears has strong vibration when running at high speed, and the inner ring of the bearing may have problems such as wear and even sintering. The vibration signal of point 2 has a high amplitude in the high frequency range. Due to the large number of components at point 2, and the vibration frequency of each component is different. Therefore, when the vibration of many components is obvious, the phenomenon of concentrated high amplitude appears as shown in Figure 5.

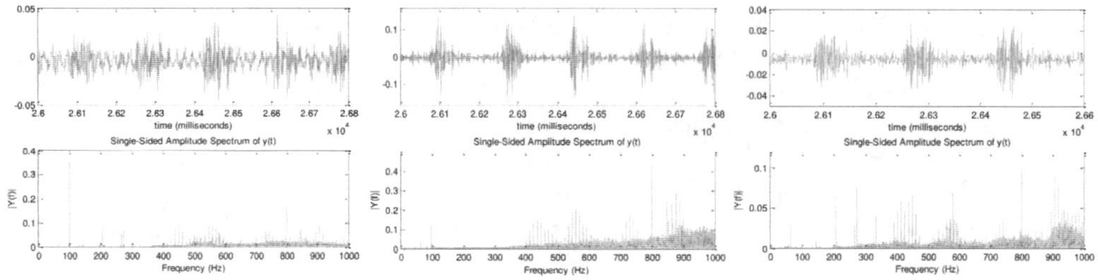

Point 1 Point 2 Point 3

n=453rpm

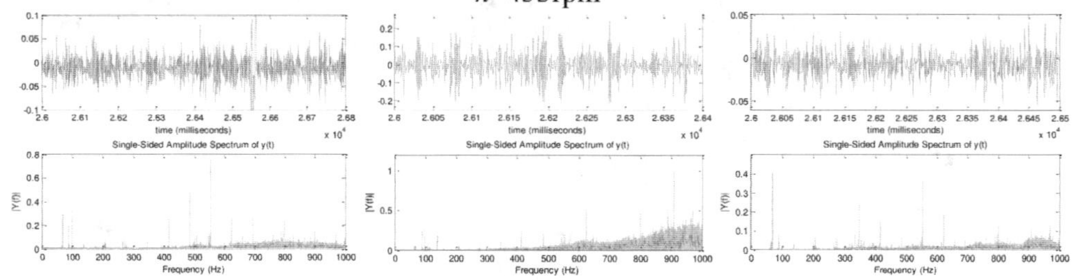

Point 1 Point 2 Point 3

n=2646rpm

Fig.6 Time-frequency characteristics of vibration signals of machine 2 at 453 rpm and 2646 rpm

The time-frequency characteristics at different rotating speeds of machine 2 are compared and analyzed. The difference between high and low speed is similar to that of machine 1, and there are concentrated high amplitude values at point 2. However, the vibration characteristics of point 1 and point 3 have changed. The high amplitude appears at about 560Hz. It is twice the vibration frequency of the outer ring of the bearing. So, it can be inferred that the front and rear bearings of equipment 2 have abnormal vibration, and the failure point lies in the outer ring of the bearing.

4. Conclusion

The vibration signals of different parts of the spindle are collected by setting the speed and machine comparison experiments. The Fourier transform is performed on the vibration data, and the time-frequency characteristics are analyzed to find the main vibration frequency. Compare the results with the theoretical calculation values of the vibration frequency of each component. And the state of the spindle system and the parts with wear and other faults can be effectively judged.

The vibration of the side gear transmission part is more violent than the front and rear ends of the spindle system. From the analysis of the frequency domain diagram, the main vibration is caused by bearings regardless of the equipment and speed, especially the installation position of the bevel gears. Therefore, as the most vulnerable part of the whole spindle system, the bevel gear’s bearings should be maintained to ensure good lubrication. The helical gears on the side of spindle box also need to be attention because of the significant vibration at low speed.

The results show that the spindle system fault warning and diagnosis method based on the mechanical spindle time-frequency domain analysis can effectively detect the working performance and state of the spindle, and it is also helpful for locating specific fault parts. It has a great value in engineering application

Funding information

This work is supported by the Major Project of National Science and Technology (Grant No.2017ZX04002001).

References

[1]Zhigang Wang,Shogo Nakashima,Mark Larson. Energy Efficient Machining of Titanium Alloys by

IOP Conf. Series: Materials Science and Engineering **793** (2020) 012041 doi:10.1088/1757-899X/793/1/012041

Controlling Cutting Temperature and Vibration[J]. Procedia CIRP,2014,17.
[2]Pengju Kang, Birtwhistle, D.. Analysis of vibration signals for condition monitoring of power switching equipment using wavelet transform[P]. Time-Frequency and Time-Scale Analysis, 1998. Proceedings of the IEEE-SP International Symposium on,1998.
[3]Feng Li,Xinyu Pang,Zhaojian Yang. Motor current signal analysis using deep neural networks for planetary gear fault diagnosis[J]. Measurement,2019,145.
[4]Jian Cheng,Yu Yang,Xin Li,Haiyang Pan,Junsheng Cheng. An early fault diagnosis method of gear based on improved symplectic geometry mode decomposition[J]. Measurement,2019.
[5]LI Yongbo,SI Shubin,LIU Zhiliang,LIANG Xihui.Review of local mean decomposition and its application in fault diagnosis of rotating machinery[J].Journal of Systems Engineering and Electronics,2019,30(04):799-814.
[6]Hu Shengyang,Fang Zongde,Liu Chao,Xiang Long. Measurements and theoretical analysis of a helical gear meshing impact signal[J]. Proceedings of the Institution of Mechanical Engineers,2019,233(4).
[7]Rui Yang,Yulin Jin,Lei Hou,Yushu Chen. Study for ball bearing outer race characteristic defect frequency based on nonlinear dynamics analysis[J]. Nonlinear Dynamics,2017,90(2).

IOP Conf. Series: Materials Science and Engineering **793** (2020) 012042 doi:10.1088/1757-899X/793/1/012042

The wind load prediction and comparative study of asmmetric semi-submersible lifting platform based on CFD

Zhenqiu Yao[1], Xiangyu Tang[2*] and Hongjie Ling[1]

[1] Marine equipment and technology institute, Jiangsu university of science and technology, Zhenjiang, Jiangsu, 212000, China

[2] School of naval architecture and ocean engineering, Jiangsu university of science and technology, Zhenjiang, Jiangsu, 212000, China

Fund projiect: Research and development of semi-submersible lifting platform and collaborative innovation of supporting industry chain, a model city construction project of innovation and development of Marine economy in Nantong, Jiangsu,China

*Corresponding author's e-mail: tangxiangyu0625@163.com

Abstract. In order to accurately predict the wind load of asymmetric offshore structures, this paper studied the wind load of an asymmetric semi-submersible lifting platform based on the CFD software FINE/ MARINE,and calculated the wind load of operation state, lifting state and survival state respectively, and compared the results with those of China Classification Society(CCS) specification. The results show that under the same wind speed, the change of wind direction angle has a great influence on the distribution of wind field around the platform. The specification calculation can not fully consider the influence of the shielding effect between the components of the superstructure, therefore, the results are larger than the numerical simulation value.In addition, the wind field distribution near the platform can be used to analyze the variation of wind load, which provides an effective method for wind load prediction and wind field analysis of asymmetric offshore structures.

1. Introduction

Semi-submersible lifting platform has been widely used in offshore lifting,submarine pipeline laying, offshore platform demolition and other aspects due to its advantages of large lifting capacity, strong ability to wiithstand wind and waves, and wide range of working water depth[1-2]. In the process of offshore platform design, wind load is one of the important loads to be considered. In case of severe wind conditions, wind load will seriously threaten the safety of the platform. Therefore, it is particularly important to accurately calculate the wind load of the platform. At present, there are four main methods of wind load research in the world: field measurement, wind tunnel test, numerical simulation, and the specification calculation. The field measurement is easy to be interfered by the surrounding environment, and the cost is high. Wind tunnel test, with its high reliability, is often used as an important means of wind load research[3], but the test period is long. Therefore, the research on wind load of offshore platform is still based on the specification calculation and numerical simulation. At home and abroad, many scholars have studied the wind load of offshore platforms. Boonstra H[4] calculated and compared the wind force of a semi-submersible offshore platform by two methods of specification calculation and field measurement. The results show that the field measurement value is only half of the specification calculation. Egon T D[5] found that the wind load of the test is less than the specification value through wind tunnel test over a semi-submersible platform. Furnes G.K[6] calculated the wind load of a drilling platform at different wind directions and compared it with the wind tunnel test data. The results are in good agreement. Lin Yi[7] studied the platform load through wind tunnel test, specification calculation and numerical simulation. He sumarized that the numerical simulation value is smaller than the specification calculation value, but similar to the wind tunnel test value. Cao Mingqiang[8] justified that the reliability of the numerical simulation to calculate the wind load by using numerical calculation and model test to study the wind load of an ultra deep water drilling semi-submersible. Chen Gang[9] used wind tunnel test to study the average wind static force of the platform, and compared the calculation results with the specification calculation values. Through wind tunnel test, Zhu Hang[10] conducted the wind resistance test of HYSY-981 semi-submersible platform at the ratio of 1:100 under the steady-state gradient, and compared the results with the numerical analysis results. The results indicate that the two results are relatively consistent, and the reliability of the numerical method is explained.

In this paper, on the basis of existing research, based on the FINE / MARINE software, the wind loads of an asymmetric semi-submersible liftint platform under three states of operation, lifting and survival are numerically simulated, with the emphasis on the load size of the platform with different wind directions and the distribution of the wind field near the platform, and the calculation results are compared with the China Classification Society(CCS) specification calculation.

2. Calculation model

2.1. Model establishment

SolidWorks software is used to complete the geometric modeling of the platform. In order to improve the calculation efficiency, this paper only establishes the parts above the waterline of the platform, and simplifies or deletes the parts of the platform that have little impact on the wind load. The complete model for wind load calculation is shown in Figure 1.

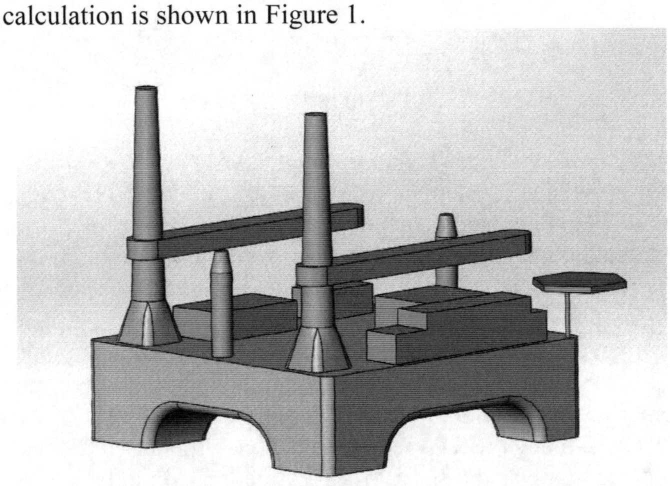

Figure 1. Three dimensional model of the semi-submersible lifting platform

2.2. Computing domain and grid division

The SolidWorks model was saved as an x_t format and imported into the Hexpress software to establish the calculation domain. In order to minimize the influence of the boundary on the calculation results, the calculation domain was set as a cuboid of sufficient size. The model was placed at the bottom of the calculation domain, with the boundary 3 times of the length from the bow, 6 times of the length from the tail, 2 times of the length from both sides of the ship's side, and 3 times of the length from the bottom to the upper boundary. In order to accurately and conveniently simulate the wind loads of the platform under different wind directions, the cylindrical domain envelope platform was built around the platform, which is defined as the calculation inner domain and the outer domain as the Boolean operation region. In order to complete the grid division of different wind direction angles, the internal domain of spin mounting calculation was adopted. Figure 2. shows the mesh division of the calculation domain and platform surface.

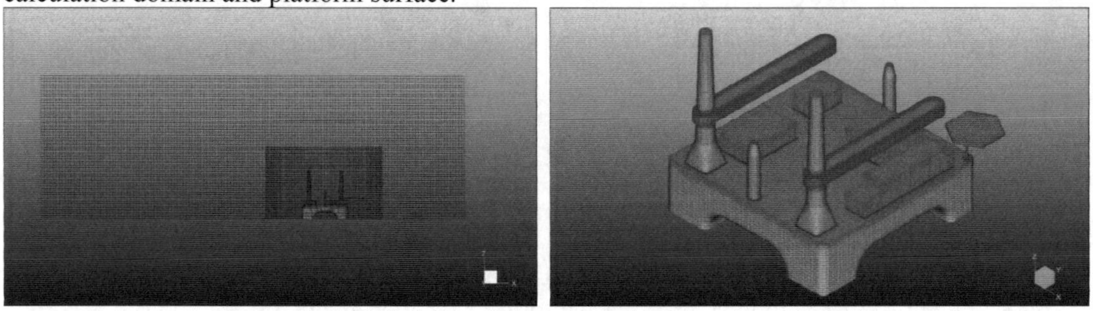

Figure 2. Grid division of the calculation domain and platform surface

3. Numerical simulation calculation

3.1. Governing equation

Most of the local structures of offshore platforms are blunt bodies. The numerical simulation of the surrounding flow field is similar to the flow of real air. The Reynolds time averaged Navier stocks (RANS) equation is used as the control equation of the whole flow field:

$$\frac{\partial \rho}{\partial t}+\frac{\partial}{\partial z_i}(\rho u_i)=0 \tag{1}$$

$$\frac{\partial \rho}{\partial t}(\rho u_i)+\frac{\partial}{\partial x_i}(\rho u_i u_j)=-\frac{\partial \rho}{\partial x_i}\left(\mu\frac{\partial u_i}{\partial z_j}-\rho\overline{u_i' u_j'}\right)+S_i \tag{2}$$

In the formula, I, j=1, 2, 3; Air density ρ=1.218Kg/m^3; Dynamic viscosity μ=1.785×10^{-5}Pa · s.

3.2. Calculation conditions and numerical calculation methods

The origin of the model coordinate system is located at the intersection of the middle longitudinal section, the middle transverse section and the waterline plane of the platform. The x-axis parallel waterline plane points to the bow, and the y-axis points to the left chord. The wind direction angle is defined as the angle between the wind direction and the positive direction of x-axis. As shown in Table 1, the wind load of the platform under the three states of operation, lifting and survival is calculated respectively, the wind direction angle is 0 ° to 360 °, the step size is 15 °, and the total wind load is 72 working conditions. Based on FINE / MARINE software, the parameters of the wind load prediction of the asymmetric semi-submersible lifting platform are set as follows: three-dimensional unsteady single-phase flow; k-omega (SST-menter) turbulence model is selected, the values of K and ω are related to the Reynolds number; the six degrees of freedom of the fixed platform, the inlet, outlet and left and right boundary velocities of the calculation domain are given, the upper boundary is set as the specified pressure condition, and the lower boundary is set as the velocity far field; The turbulence and momentum equations are discretized by AVLSMART scheme.

Table 1. Calculation condition

Condition	Draft（m）	Wind speed（m/s）	Wind direction (°)
Operation	20	20	0~360°， 15°interval
Lifting	26.4	10.7	
Survival	17	28	

4. Calculation results and analysis

4.1. Expression form of calculation results

In order to facilitate the comparison between different platform draught, wind speed and other change factors, the dimensionless coefficient is adopted in this paper. The dimensionless coefficient is defined as:

$$(a)\ C_x = \frac{F_x}{0.5\rho V^2 A} \qquad (b)\ C_y = \frac{F_y}{0.5\rho V^2 A} \qquad (c)\ C_{M_z} = \frac{M_z}{0.5\rho V^2 AL} \tag{3}$$

In the formula: C_x, C_y, C_{M_z} respectively represent longitudinal wind load coefficient, the transverse wind load coefficient and yaw moment coefficient, F_x, F_y, M_z respectively represent longitudinal wind load force, transverse wind load force and yaw moment respectively; ρ is air density, take saturated moist air at 15℃, and its value is 1.218 kg / m3; V is the wind speed; A i represents the positive wind area of the platform; L is the total length of the platform.

4.2. Comparative analysis of loads under different wind directions

The wind load of the platform under the wind direction of 0 ° ~ 360 ° is calculated based on FINE / MARINE software. Figure 3(a) - (c) shows the comparison of the longitudinal and transverse load coefficients and the yaw moment coefficients of the platform under three conditions respectively. Some rules of the load on the platform can be summarized from Figure 3.

1) Longitudinal load coefficients C_x: The longitudinal load coefficient reaches the maximum at the wind direction angle of 0° and 180°, and approaches 0 at 90° and 270°, decrease gradually from 0° to 90° and increase gradually from 90° to 180°. The change trend between 180° and 360° is basically the same as that between 0° and 180°.

2) Transverse load coefficients C_y: The transverse load coefficients is close to 0 when the wind direction angle is 0° and 180° and approaches the biggest when the wind direction angle is 90° and 270°, it increases gradually from 0° to 90° and decreases gradually from 90° to 180°. The change trend between 180° and 360° is basically the same as that between 0° and 180°.

3) Yaw moment coefficients C_{M_z}: The peak value of yaw moment coefficient appears near 30°, 120°, 195° and 345° respectively. It is found that the moment coefficients at 345° is the largest.

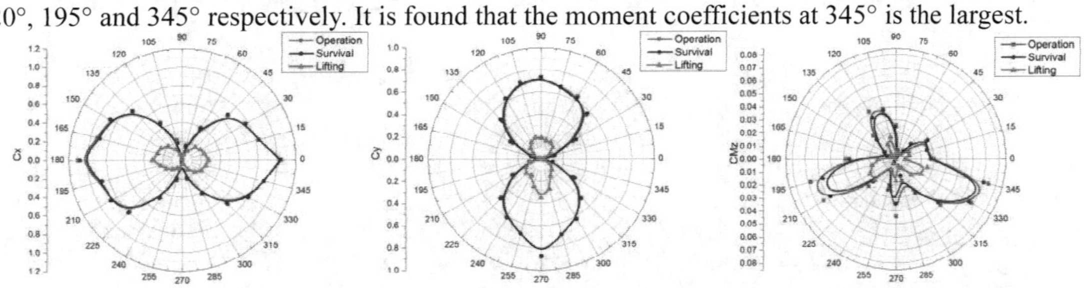

(a) Comparison of C_x (b) Comparison of C_y (c) Comparison of C_{M_z}

Figure 3. Comparison of dimensionless coefficients in three states

4.3. Comparative analysis of CFD calculation and CCS specification calculation

Figure 4 shows the comparison between the CFD calculation value of wind load and the CCS specification calculation value under three conditions of the platform, from which we can see:

1) The results of CCS and CFD numerical simulation have the same trend with the wind direction angle. The results of CCS and CFD numerical simulation basically show symmetrical distribution with the wind direction angle, and the numerical simulation can reflect the subtle differences of each angle.

2) The results of CCS are larger than those of the numerical simulation, which shows that the specification calculation method can not fully consider the influence of the front and back shielding effect between the components of the superstructure.

IOP Conf. Series: Materials Science and Engineering **793** (2020) 012042 doi:10.1088/1757-899X/793/1/012042

3）The wind load is closely related to the platform draft. Among the three states, the survival state has the largest load, the second is the operation state, and the smallest is the lifting state.

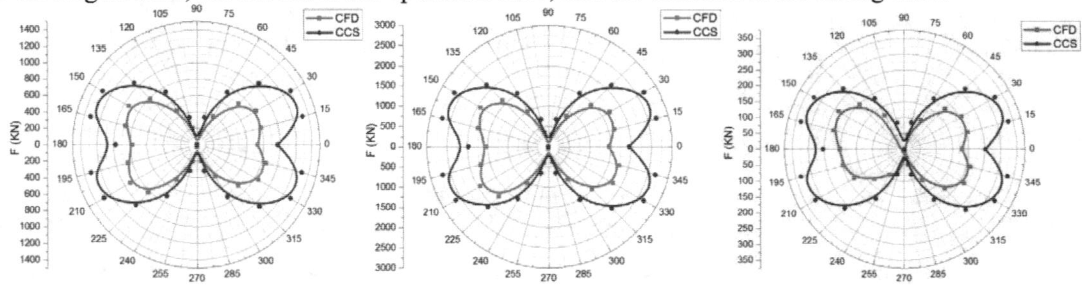

（a）Operation （b）Survival （c）Lifting

Figure 4. Stress comparison of three states

4.4. *Detailed analysis of wind field*

Taking the wind direction angle of 180° as an example, Figure 5 shows the flow field distribution near the platform surface. The flow field near the platform surface reflects the characteristics of velocity, vortex and circumfluence. It can be seen from Figure 5 that vortices and backflows are very easy to occur at platform corners and between structures. It can be seen from Figure 5(a) that the high pressure area is located at the windward side; from Figure 5(b) and Figure 5(c), when the fluid encounters the vertical windward side of the superstructure, the velocity will drop sharply and flow around, and at the same time, the flow back behind the superstructure will generate a relatively large vortex, as shown in Figure 5(c), which is due to the effect of air viscosity, resulting in the phenomenon of flow separation on the surface of the object, and the flow on the leeward side is caused by the low flow rate and the pressure difference caused the backflow of the fluid.

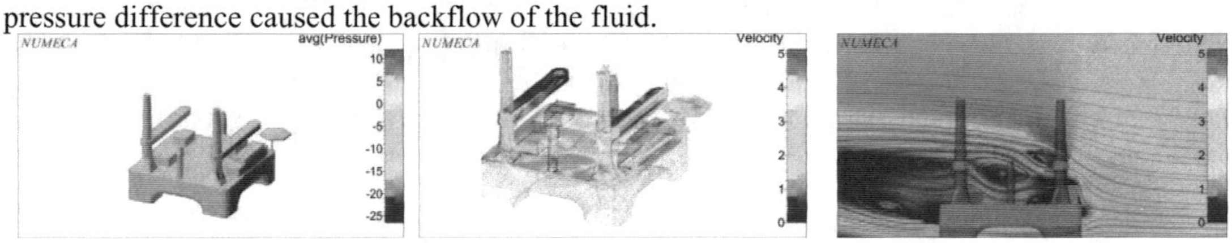

(a) Pressure cloud chart (b) Velocity vector diagram (c) Streamline diagram

Figure 5. Wind field distribution near the platform surface with 180°

5. Results and discussion

In this paper, an asymmetric semi-submersible lifting platform is taken as the main research object. Based on FINE / MARINE software, the wind loads of platform with different wind directions are studied, and the conclusions are as follows:

1）The CCS specification calculats that the longitudinal force of the platform is the largest when the wind direction angle is 165°;the transverse force is the largest when the wind direction angle is 60°. The longitudinal force of the platform calculated by CFD is the largest when the wind direction angle is 30°;the transverse force calculated by CFD is the largest when the wind direction angle is 315°.

2）Due to the asymmetry of the platform, although the wind loads of 0°~180° and 180°~360° are basically the same, but the values are not the same.

3）The results of CCS specification and CFD numerical simulation are consistent with the wind direction. However the influence of the front and back shielding effect between the components of superstructure can not be fully considered in the specification calculation method, so the results are larger than those of numerical simulation.

4）Due to the viscous effect of the fluid, when the air meets the vertical windward side of the superstructure, the velocity will suddenly drop and flow around, and at the same time, there will be a relatively large vortex in the back of the superstructure.

Reference

[1] Yuan Ruhua, Wang Minglun, Li Litao, yuan Shangchen, Yin Hao, Wang Yufeng. Development status and application of SSCV for semi submersible crane pipe laying vessel [J]. China water transport (second half), 2017,17 (08): 1-6 + 64

[2] Min Bing, Shu Shiyong. Application of semi submersible crane engineering ship in marine engineering [J]. China shipbuilding, 2010,51 (01): 156-161

[3] Wang Zhongchang, Yi Cong, Li Da, Bai Xueping, Zheng Wentao, Peng Chao. Study on wind load characteristics of FPSO based on wind tunnel test [J]. Offshore engineering, 2016,34 (05): 125-130

[4] Boonstra H, Marcon I.Analysis of full scale wind force on a semi-submersible platform using operator data[C].OTC-3628, Houston, USA, 1979.

[5] Egon T D, Svenn V. Wind overturing effect on a semi-submersible[C].OTC-3036, Houston, USA, 1978.

[6] Furnes G.K. Numerical simulations of wind forces on Troll B[J]. Marine Structures, 11(1998):273-289.

[7] Lin Yi, Hu Ankang, Xiong Fei. Numerical simulation and experimental study of wind load on jack up platform [J]. Hydrodynamic research and progress Series A, 2012,27 (2): 208-215

[8] Cao Mingqiang, Wang Lei, Zhou Li. Wind load test analysis of deepwater semi submersible [J]. Laboratory research and exploration, 2009,28 (09): 17-19 + 28

[9] Chen Gang, Wang Yi, Huang Zhuyu. Wind load test of jack up drilling platform [J]. Ship engineering, 2012,34 (02): 74-75 + 83

[10] Zhu Hang, Ma Zhe, Zhai Gangjun, Xie bin, Fu Yingjun, Ou Jinping. Wind load numerical simulation and wind tunnel experiment of HYSY-981 semi submersible [J]. Naval engineering, 2009,38 (05): 149-152

Investigation into convective heat transfer of nanofluids at the thermal boundary layer development region in micro-channels

Shuxiang Wang[1], Li Xu[2],* and Junjie Tong[1]

[1]GuangZhou Maritime University, Guangzhou 510725, PR China

[2]Guangdong Provincial Institute of Metrology, Guangzhou 510405, PR China

*Correspondence: li_xu_scm@163.com

Abstract. The convective heat transfer at the thermal boundary layer development region in microchannels was theoretical and experimently investigated with de-ionized water and Al_2O_3-H_2O nanofluids as the working fluids. It was found that nanofluids could effectively decreased the wall temperature and improve the heat transfer coefficient compared with the de-ionized water at the thermal boundary layer development zone. When the serious agglomeration of nanoparticles didn't appear, increasing the concentration of Al_2O_3 nanoparticles could effectively increase the heat transfer enhancement effect. And the local heat transfer coefficient increase amplitude by using the nanofluids decreased along the direction of refrigerant flow. At the thermal boundary layer development zone, higher inlet velocity would get the larger increase amplitude of the heat transfer coefficient. At the thermal boundary layer development region, lower heat flux could achieve more pronounced heat transfer enhancement by nanofluids. When the thermal boundary layer developed, the heat transfer enhancement by nanofluids decreased and affected by heat flow become weak. When the mass flow rate increases, the better improvement can be achieved by nanofluids in the micro-channels. The nanoparticles have greater impact on local heat transfer coefficient in thinner thermal boundary layer, leading to the heat transfer enhancement is better at the thermal boundary layer development region in micro-channels.

1. Intruduction

Nanofluids have attracted much attention recently because of their extraordinary fluid transport properties and flow features. As a new working fluids, it is liquid suspension containing metallic and non-metallic nanoparticles with high thermal conductivity. Unlike the milliscale or microscale particles in liquid suspension, the nanoscale particles have excellent stability, litter penalty in pressure and wall abrasion. As a consequence, many experimental studies investigated the characteristics of nanofluids concerning thermal conductivity, such as the species, size, and dispersion of the nanoparticles and the concentration, temperature and pH value of the nanofluids [1,2].

Compared to the plentiful studies on thermal conductivity of nanofluids, there have been relatively few literatures on the nanofluids flow and heat transfer characteristics, and even in microchannel. Chein and Chuang [3] presented a theoretical analysis on the performance of microchannel heat sink using nanofluids. Wu et al. [4], and Zhang et al. [5] used the Al_2O_3-water nanofluid as the coolant in the parallel rectangular microchannels. Liu and Yu [6] and Azizi Z. et al. [7] studied the convective heat transfer of nanofluids through the microchannel by experiment.

Akbari et al. [8] investigated the laminar forced convection heat transfer of water–Al_2O_3 nanofluids through a horizontal rib-microchannel by the computational fluid dynamics methods. Shamsi et al. [9] numerically studied the increasing heat transfer of non-Newtonian nanofluid in rectangular microchannel with triangular ribs. Heydari A et al. [10] and Alipour et al. [11] also numerically studied the heat transfer of water/Ag nanofluid in rectangular microchannel with triangular ribs or in trapezoidal microchannel with T-semi attached ribs.

Although some of these researches [3,5] referred to the different enhancement of nanofluids at the entrance region, very few experiments were focus on nanofluids' heat transfer at the thermal boundary layer development region. Because the microchannel heat sink is short, in most case the entrance effect greatly impacts microchannel heat sink performance. It is important to research the entrance effect for nanofluids in microchannels.

In present work, theoretical analysis and an experimental study on convective heat transfer behavior of nanofluids at the entrance region under laminar flow in microchannels were carried out. Factors effecting the heat transfer enhancement of nanofluids were investigated.

2. Analysis into enhancement of the nanofluids for laminar flow in the entrance regime

Most researchers carried out experimental or numerical studies base on the thermal boundary fully developed condition. As the microchannel heat sink is always very short, the entrance effect can't be neglected. This point is illustrated with the aid of a simple example to analyze the convective heat transfer behavior of nanofluids under laminar flow in the entrance regime.

Consider a fluid being heated uniformly as it flows inside a tube of diameter D. According to Shah [12] equation for laminar flow, the Nu at the entrance regime is

$$Nu(z) = 1.953\ (RePrD/z)^{1/3} \tag{1}$$

The axial mean temperature distribution of the fluid is determined by way of a simple energy balance. That is

$$q''D\pi x = C_p\dot{m}(T_f(z) - T_{in}) \tag{2}$$

For pure water: (3-a)

$$Tw(z)\big|_{pw} - T_{in} = q''\left(\frac{D}{1.953(\mathrm{Re}\,\mathrm{Pr}\frac{D}{z})^{1/3}k_{pw}} + \frac{D\pi z}{C_{p.pw}\dot{m}}\right)$$

For nanofluids: (3-b)

$$Tw(z)\big|_{nf} - T_{in} = q''\left(\frac{D}{1.953(\mathrm{Re}\,\mathrm{Pr}\frac{D}{z})^{1/3}k_{nf}} + \frac{D\pi z}{C_{p.nf}\dot{m}}\right)$$

When the tube parameters is shown in Table 1, the length of entrance region is as follows:

$$L/D = 0.07RePr = 0.07Pe = 0.07Du\rho C_p/k$$

$$L = 0.07D^2u\rho C_p/k = 0.018573\text{m} > 0.5\text{mm}$$

So the whole tube is in the entrance region.

Table 1. Parameters of the tube.

D	L	Q	q_{flim}
0.5mm	5mm	2.945243063×10^{-8}m^3/s	50kW/m^2

According to the Hamilton–Crosser model [13], effective thermal conductivity of different concertration Al_2O_3-H_2O nanofluids at 30℃ is,

k_{pw}= 0.613285 W/mK (φ=0); k_{nf}=0.631004 W/mK(φ=1%); k_{nf}=0.649067194 W/mK(φ=2%)

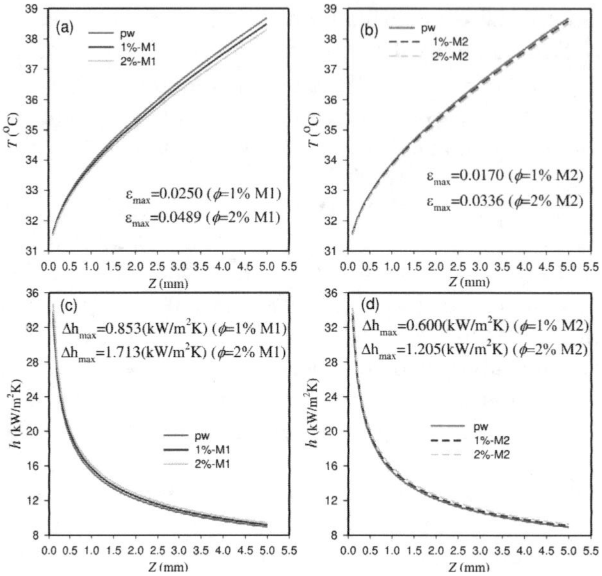

Figure 1. Wall temperature and heat transfer coefficient in the development of thermal boundary layer (H-C model).

According to Bergman [14] research, the dimensionless nanofluids effectiveness ε is

$$\varepsilon = 1 - \frac{T_w(z)|_{nf} - T_i}{T_w(z)|_{pw} - T_i} \tag{4}$$

In this literature, the properties of nanofluids can be got as follows:

Viscosity [15] $$\mu_{nf} = (1+2.5\phi)\mu_{pw} \tag{5}$$

Density [16] $$\rho_{nf} = \phi\rho_{np} + (1-\phi)\rho_{pw} \tag{6}$$

There are two models about the specific heat of nanofluids. Model 1 based on analogous to the mixing theory for ideal gas mixtures. Model 2 is based on classical and statistical mechanism, assuming that the base fluid and the nanoparticles are in thermal equilibrium.

Specific heat:

Model 1: $$C_{p,nf}\rho_{nf} = \rho_{nf}[\phi C_{p,s} + (1-\phi)C_{p,pw}] = \rho_{nf}\phi C_{p,s} + \rho_{nf}(1-\phi)C_{p,pw} \tag{7}$$

Model 2: $$C_{p,nf}\rho_{nf} = \phi(\rho C_p)_s + (1-\phi)(\rho C_p)_{pw} \tag{8}$$

Fig.1 is the wall temperature and heat transfer coefficient along the flow direction in the development of thermal boundary layer by different specific heat with models.

Increasing the concentration of nanoparticles , the wall temperature would decrease, indicating that the heat transfer coefficient of the microchannel heat sink increased with the use of nanofluids as working fluids.The prediction carried out by model 1 shows better heat transfer enhancement than by model 2. However, the temperature decreased by using nanofluids is less than 1℃ no matter whether Model 1 or Model 2 prediction . And the drop is quite miniscule especially based on model 2 (Fig 1(a,b)). The maximum dimensionless nanofluids effectiveness ε is only 0.017 by Model 2 at 1% volume concentration nanofluids. The increase of heat transfer coefficient is also very miniscule, and the maximum enhancement is only 1.713kW/m^{2}K by model 1 at 2% volume concentration nanofluids.

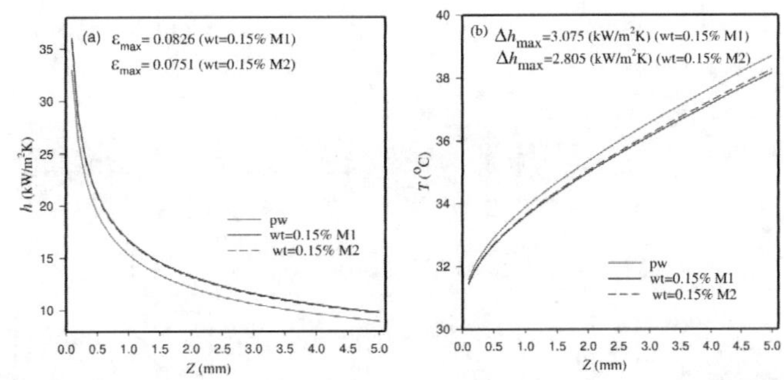

Figure 2. Wall temperature and heat transfer coefficient in the development of thermal boundary layer.

3. Description of the experiment

3.1. Preparation of nanofluids

In the paper, the Al_2O_3-H_2O nanofluid was prepared by two-step methods. The average diameter of the quasi-spherical Al_2O_3 nanoparticles is 13nm. Nanoparticles with SDBS (one tenth of the mass of nanoparticles) were dispersed into DI water. Then the mixture was vibrated in an ultrasonic bath for 3h.

3.2. Test loop and operating procedure

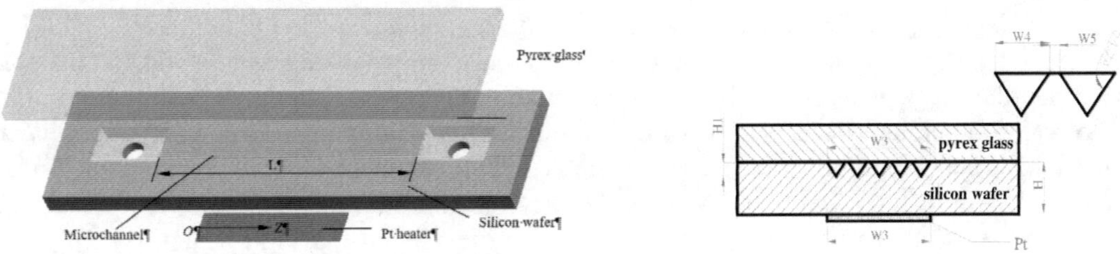

(a) **(b)**

Figure 3. (a) The encapsulated model of the test section; (b) Schematic of the microchannel heat sink.

As shown in Fig.3(a), the test section consisted of a top 7740 glass cover bonded with a bottom silicon substrate. At the back surface of the silicon substrate, a platinum film heater was deposited centrally, acting as the heating source for the heat transfer system. Schematic of the microchannel heat sink is shown in Fig 3(b). Geometric parameters are listed in table 2. The hydraulic diameter of a miarochannel is 130μm.

Fig.4 shows the experiment setup. The working fluid was pushed by the syringe pump (HARVARD, PUMP33), flowed through a 7μm filter, constant temperature bath, test section, and finally was collected in a container. The fluid temperature at the inlet and outlet of the microchannels were measured by two K-type thermocouples (OMEGA, 5TC-GG-K) with a 0.1 mm diameter installed as close to the microsystem as possible.

Table 2. Dimentions of microchannel.

L(μm)	H(μm)	$H1$(μm)	W_3(μm)	W_4(μm)	W_5(μm)
12000	500	177.6	1.65559×10^3	251.12	100

The red line in the Fig.4 was the schematic of the experimental heating circuit. It consisted of the DC power supplier, the thin film heater (4500×1655.59 μm), the given resistance and the voltmeters. The DC power supplier provided a constant heat flux boundary condition at the precisely selected area under the microchannel heat sink. When the voltage applied on the heating area was V_1 and the voltage applied on the given resistance was V_2, the heating power on the heating area was computed as V_1V_2/R, where R is the given resistance.

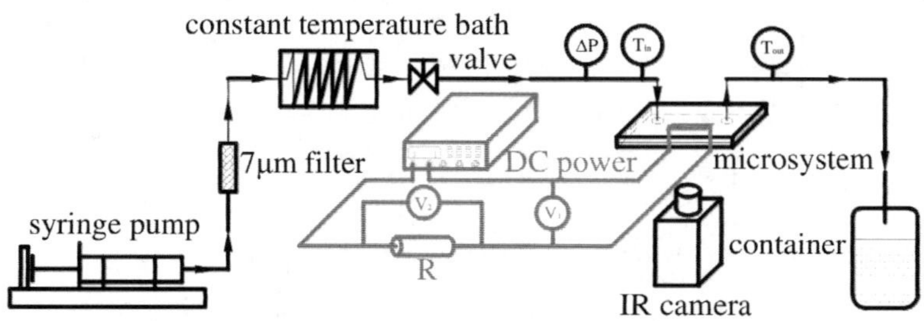

Figure 4. Schematic of experimental systems.

At the back surface of the silicon substrate, the thin film heating area was measured by a high resolution, high accuracy infrared radiator imaging system (FLIR ThermaCAM SC3000IR).The temperature dependence of emissivity in the present experimental data range can be neglected. The IR imaging system was calibrated against a set of standard temperatures with an accuracy of 0.5℃ [17].

The mass flux *G*, liquid Reynolds number *Re*, heat flux at the main heater surface q_{film}, heating power applied on the total inside wall surface of microchannels q_w are similar to Xu et al[17].

4. Results and discussion

4.1. Average wall temperature by using pure water and different concentration nanofluids

Figure 5 is the average wall temperature of the heating area with increasing heating flux at different mass flux. When the mass flux increases, the average wall temperatures lift slowly with heating flux. The average wall temperature decreased obviously by using nanofluids instead of pure water. While the nanoparticles' concentration increased, the lower wall temperature achieved. That means enhancement of heat transfer is improved by using higher concentration nanofluids. At a certain mass flux, the decrease of wall temperature by nanofluids is more obvious in high heat flux. These are agreed with the analysis in Fig.2.

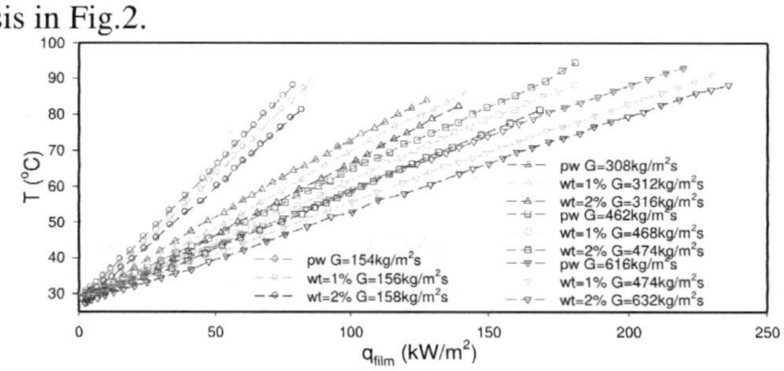

Figure 5. Curve of average wall temperature versus heat flux at different mass flux.

4.2. Effect of the flow distance for the enhancement of heat transfer coefficient by nanofluids

To verify the wall temperature distribution along flow direction as theoretical prediction in Fig.2, the heater wall temperature distribution by infrared radiator imaging system is shown in Fig.6(a) ($G=624\pm8kg/m^2s$, $q_{film}=35kW/m^2$). And Fig.6(b) is the corresponding distribution of wall temperature. The measured wall temperatures in Fig6(b) show that the Al_2O_3-H_2O nanofluids reduce the wall temperature in the entrance region. This is agreed with the theoretical prediction. For the Al_2O_3-nanoparticle concentration studied, it was found that the wall temperatures decrease with the increase in Al_2O_3-nanoparticle concentration. Fig6(c) shows the heat transfer coefficient along the flow direction with different concentration nanofluids. The heat transfer coefficient is higher by using nanofluids than pure water. And the increasemeant is also enhanced further by using higher concentration nanofluids. From Fig6(c), it is show that the enhancement is higher near the inlet region.

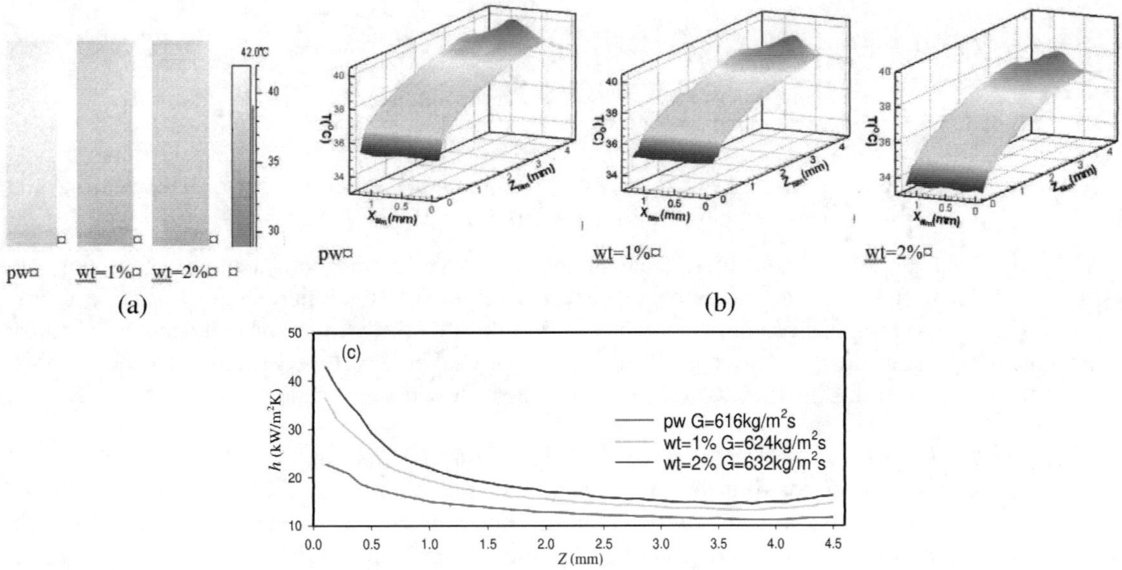

(c)

Figure 6. (a) IR camera pictures on the back heating surface with pure water and nanofluids; (b) Temperature distribution of back heating surface with pure water and nanofluids; (c) Heating transfer coefficient at the central line of back heating surface with pure water and nanofluids (q_{film}=35kw/m^2).

Fig.7 shows the enhancement of heat transfer coefficient along flow direction clearly. It implies that heat transfer could be enhanced further by using nanofluids with higher particle concentration once more. The enhancement decreases along the flow direction. Near the inlet region the enhancement is largest (zone I), then droped sharply in this region. At the area between zone I and II, the enhancement is tended to constant. In zone II, the variation of enhancement is almost negligible. This means that the enhancement of heat transfer coefficient by nanofluids reduces with the development of thermal boundary layer.

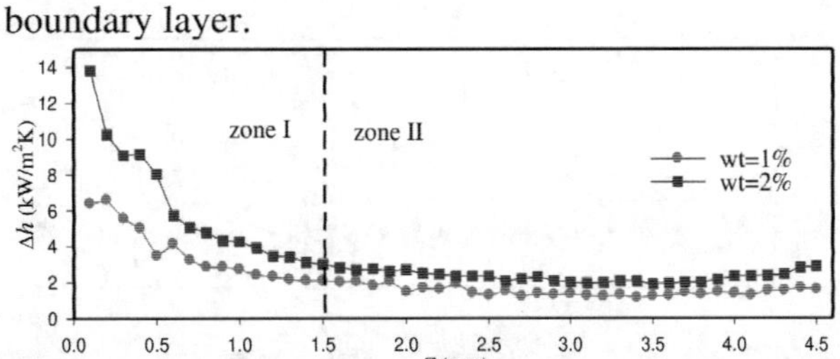

Figure 7. Enhancement of heating transfer coefficient along the flow direction by different concentration nanofluids(u=0.6m/s, q_{film}=350±10kw/m^2).

4.3. Effect of the heat flux for the enhancement of heat transfer coefficient by nanofluids

The effect of heat flux for the enhancement of heat transfer coefficient by nanofluids is shown in Figure 8. As in Figure 7, heat transfer could be enhanced further by using nanofluids with higher particle concentration and the enhancement decreases along the flow direction at various heat flux. However, the enhancement curves are overlapped for different heat fluxes, especially in zone II. The region near the inlet, heat transfer enhancement decreases as heat fluxes grow up. While the working fluids flow into zone II, there is no obvious difference between various heat flux.

IOP Conf. Series: Materials Science and Engineering **793** (2020) 012043 doi:10.1088/1757-899X/793/1/012043

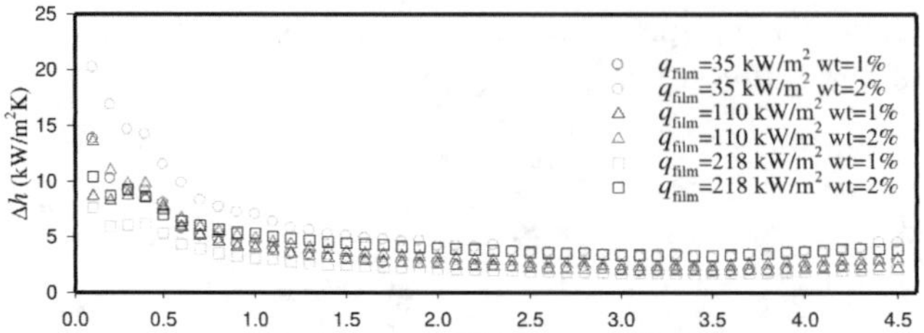

Figure 8. Enhancement of the heating transfer coefficient at the central line of back heating surface with different heat flux (G=624±8 kg/m^{2}s).

The experimental results show that effect of the heat flux for the enhancement of heat transfer coefficient by nanofluids is not the same in different regions of the microchannel. At the thermal boundary layer development region, lower heat flux could achieve more pronounced heat transfer enhancement by nanofluids. But at the thermal boundary layer developed region, the heat transfer enhancement by nanofluids decreased and affected by heat flow density become weak.

4.4. Effect of the flow rate for the enhancement of heat transfer coefficient by nanofluids

Fig.9 is the enhancement of heat transfer at different flow rate. Appreciable enhancement can be seen at higher flow rate. At a certain nanoparticle weight friction, the enhancement becomes large by higher flow rate, especially in the entrance region. And the area where nanofluids can improve heat transfer obviously will become wider when mass flow rate increases. In other words, zone I is broaden by increasing flow rate. There is is quite miniscule enhancement at low mass flow rate, especially in zone II. For example, when the mass flow is 156 kg/m^{2}s (G=156 kg/m^{2}s), the enhancement by weight friction 1% (wt=1%) nanofluids is only 1kW/m^{2}K. However, the enhancement grow up rapidly to 14 kW/m^{2}K at the mass flow 624kg/m^{2}s near the inlet region. In zone II, the wt=1% nanofluids hardly make any enhancement of heat transfer coefficient at the mass flow 156 kg/m^{2}s. But at this mass flow, the enhancement can reach 3 kW/m^{2}K in this region. Despite the enhancement is much smaller than in zone I, it is much higher than low mass flow rate.

From Fig.9, there may be close relation between the mass flow rate and heat transfer enhancement by nanofluids. When the mass flow rate is higher, the better improvement can be achieved by nanofluids.

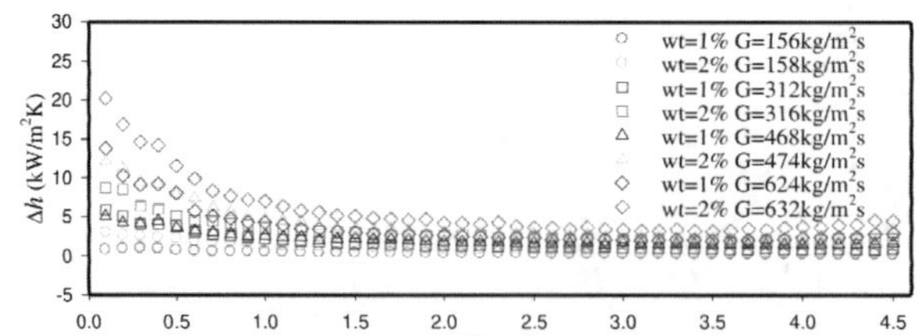

Figure 9. Enhancement of the heating transfer coefficient at the central line of back heating surface with different mass flux (q_{film}=35 kw/m^2).

4.5. Reasons for these experiment results

There may be four mechanisms for heat transfer enhancement with nanofluids, including the interfacial layering, Brownian motion, ballistic transport of energy carriers, and nanoparitles clustering[18]. For the experiment in this paper, we know that the enhancement of heat transfer by nanofluids is effected by the distance of flow and the flow rate, and the heating power seems no obvious effect for the enhancement.

In the analysis, assuming the nanoparticles didn't effect the thermal boundary layer thickness, that is the thermal boundary layer of nanofluids is the same as pure water. Fig.10 shows the development of thermal boundary layer in a tube with constant heat flux at the entrance region. According to Shah's equation[12],

The local heat transfer $$h(z) \propto u^{1/3}k^{2/3}z^{-1/3} \tag{9}$$

And we can get $h(z) = \zeta u^{1/3}k^{2/3}z^{-1/3}$, where ζ is a plus.

At the point of A, the heat transfer of pure water is as follows:

$$h_{A,pw} = \zeta u^{1/3}k^{2/3}z_A^{-1/3} \tag{10}$$

When the effective thermal conductivity of nanofluids increases Δk, that is $k_{nf}=k+\Delta k$.

The local heat transfer of nanofluids at A is: $$h_{A,nf} = \zeta u^{1/3}(k+\Delta k)^{2/3}z_A^{-1/3} \tag{11}$$

(11)-(10): $$\Delta h_A = h_{A,nf} - h_{A,pw} = \zeta u^{1/3}z_A^{-1/3}[(k+\Delta k)^{2/3} - k^{2/3}] \tag{12}$$

This is the enhancement of heat transfer at point A.

At point B $z_B=27z_A$

The local heat transfer of pure water and nanofluids is as follows:

$$\Delta h_B = h_{B,nf} - h_{B,pw} = \zeta u^{1/3}(27z_A)^{-1/3}[(k+\Delta k)^{2/3} - k^{2/3}] = \frac{1}{3}\Delta h_A \tag{13}$$

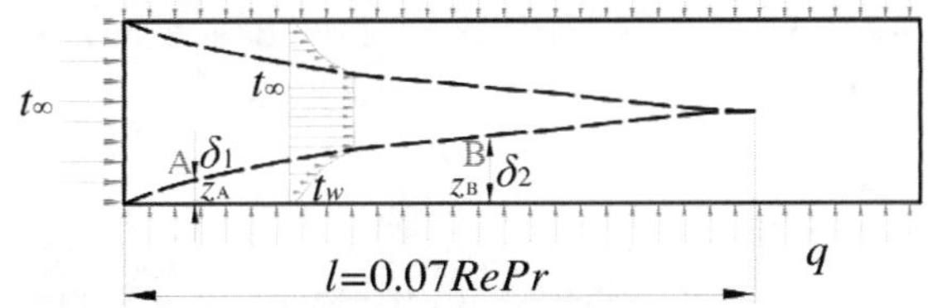

Figure 10. Development of thermal boundary layer.

The enhancement of heat transfer at point B is only 1/3 times as at the point A. And the enhancement is much smaller at point B than point A.

When the effective thermal conductivity of nanofluids k_{nf} is contant, the enhancement is the same situation as the above analysis by changing the flow rate.

From the analysis, the enhancement of heat transfer is more obvious when the thermal boundary layer is thin. Decreasing the flow distance (z) and increasing the flow rate (u) can improve the enhancement of heat transfer by nanofluids.

This result may be used to explained the heat flux effect for the heat transfer enhancement by nanofluids. When the heating power increased, the thermal boundary layer become thicker. According to the previous analysis, the enhancement heat transfer effect of nanofluids should decrease. However, the heating power increase would lead to the temperature of the nanofluids increased. And the effective thermal conductivity of nanofluids also increased with the increase of temperature. This would lead to an increase enhancement heat transfer effect of nanofluids.

So the two contradictory factors conflict and decide the enhancement heat transfer effect of nanofluids when the heating power increased. In zone I, the thermal boundary layer is so thin that plays the major role for the enhancement. When heat power increases, the layer becomes thicker, and the enhancement decreases. In zone II, with the development of thermal boundary layer, its affect is no longer as strong as in zone I. The effective thermal conductivity and thermal boundary layer thickness determines the enhancement in this zone. So there is no obvious relationship about the heat flux and heat transfer enhancement at this region.

5. Conclusions

This paper is an assessment of effectiveness of Al_2O_3 nanoparticles with the average diameter 13nm at enhancing single-phase convective heat transfer in silicon microchannels. Analysis and experiments were carried out at the entrance region under laminar flow conditions. The following conclusions were obtained:

(1) The high thermal conductivity of nanoparticles relative to common pure fluids can significantly enhance the convective heat transfer at the entrance region under laminar flow conditions, and increasing nanoparticle concentration can increase the enhancement. At a certain concentration, the enhancement becomes more obvious with larger effective thermal conductivity.

(2) The enhancement is particularly significant in the entrance region, and decreases with thermal boundary layer developing.

(3) Increasing the mass flow rate can make the enhancement become more obvious. The heat flux effect for enhancement can be divided into two zones. Near the inlet zone (zone I), the enhancement decreases with the heat flux going up. In the other zone (zone II), there seems to be no sensitive affects about the heat flux for enhancement.

(4) The thickness of the thermal boundary layer may be one of the main factors which affect the heat transfer enhancement by nanofluids. The nanofluids can play a better performance when the thermal boundary layer is thinner.

Acknowledgements

The authors thank for the financial support by the National Natural Science Foundation of China with the contract numbers of 51806041, the Natural Science Foundation of Guangdong Province with the contract numbers of 2018A0303130013, the Guangdong Science and Technology Project (2014A040401044).

References

[1] Kakaç S, Pramuanjaroenkij A. Single-phase and two-phase treatments of convective heat transfer enhancement with nanofluids–A state-of-the-art review[J]. International journal of thermal sciences, 2016, 100: 75-97.

[2] Azmi W H, Sharma K V, Mamat R, et al. The enhancement of effective thermal conductivity and effective dynamic viscosity of nanofluids–a review[J]. Renewable and Sustainable Energy Reviews, 2016, 53: 1046-1058.

[3] Chein R, Huang G. Analysis of microchannel heat sink performance using nanofluids[J]. Applied thermal engineering, 2005, 25(17-18): 3104-3114.

[4] Wu X, Wu H, Cheng P. Pressure drop and heat transfer of Al2O3-H2O nanofluids through silicon microchannels[J]. Journal of micromechanics and microengineering, 2009, 19(10): 105020.

[5] Zhang H, Shao S, Xu H, et al. Heat transfer and flow features of Al2O3–water nanofluids flowing through a circular microchannel–Experimental results and correlations[J]. Applied Thermal Engineering, 2013, 61(2): 86-92.

[6] Liu D, Yu L. Single-phase thermal transport of nanofluids in a minichannel[J]. Journal of Heat Transfer, 2011, 133(3): 031009.

[7] Azizi Z, Alamdari A, Malayeri M R. Thermal performance and friction factor of a cylindrical microchannel heat sink cooled by Cu-water nanofluid[J]. Applied Thermal Engineering, 2016, 99: 970-978.

[8] Akbari O A, Toghraie D, Karimipour A, et al. Investigation of rib's height effect on heat transfer and flow parameters of laminar water–Al2O3 nanofluid in a rib-microchannel[J]. Applied Mathematics and Computation, 2016, 290: 135-153.

[9] Shamsi M R, Akbari O A, Marzban A, et al. Increasing heat transfer of non-Newtonian nanofluid in rectangular microchannel with triangular ribs[J]. Physica E: Low-Dimensional Systems and Nanostructures, 2017, 93: 167-178.

[10] Heydari A, Akbari O A, Safaei M R, et al. The effect of attack angle of triangular ribs on heat transfer of nanofluids in a microchannel[J]. Journal of Thermal Analysis and Calorimetry, 2018, 131(3): 2893-2912.

[11] Alipour H, Karimipour A, Safaei M R, et al. Influence of T-semi attached rib on turbulent flow and heat transfer parameters of a silver-water nanofluid with different volume fractions in a

three-dimensional trapezoidal microchannel[J]. Physica E: Low-Dimensional Systems and Nanostructures, 2017, 88: 60-76.
[12] Shah R K. Thermal entry length solutions for the circular tube and parallel plates[C]//Third national heat mass transfer conference, Indian Institute of Technology, Bombay, India. 1975, 1: 11-75.
[13] Hamilton R L, Crosser O K. Thermal conductivity of heterogeneous two-component systems[J]. Industrial & Engineering chemistry fundamentals, 1962, 1(3): 187-191.
[14] Bergman T L. Effect of reduced specific heats of nanofluids on single phase, laminar internal forced convection[J]. International Journal of heat and mass Transfer, 2009, 52(5-6): 1240-1244.
[15] Einstein A. Investigations on the Theory of the Brownian Movement[M]. Courier Corporation, 1956.
[16] Drew D A, Passman S L. Theory of multicomponent fluids[M]. Springer Science & Business Media, 2006.
[17] Xu J, Liu G, Zhang W, et al. Seed bubbles stabilize flow and heat transfer in parallel microchannels[J]. International Journal of Multiphase Flow, 2009, 35(8): 773-790.
[18] Keblinski P, Phillpot S R, Choi S U S, et al. Mechanisms of heat flow in suspensions of nano-sized particles (nanofluids)[J]. International journal of heat and mass transfer, 2002, 45(4): 855-863.

Research on multipath suppression performance of multiple BOC signals of navigation satellite

Yan Wang[1], Wei Zhao[2], YangHe Shen[1] and Su Yu[1]

[1]104 Youyi Road, Haidian District, Beijing,100094, China,Beijing Institute of Spacecraft System Engineering

[2]22 Beiqing Road,Haidian District,Beijing,100094,China

wangyan198567@126.com

Abstract. BOC(Binary Offset Carrier) signal is a kind of navigation satellite downlink signal. Compared with BPSK(Binary Phase Shift Keying) modulation signal, it can improve the multipath suppression performance of navigation system. BOC signal modulation includes a variety of extended signal modulation modes, including TMBOC(time-multiplexed BOC), AltBOC(Alternative BOC) signals, etc. It is an important subject to study the performance of multipath suppression of various BOC signals. In this paper, the multipath error model is constructed by using the noncoherent early-late processing track loop, the performance of multipath suppression of TD-AltBOC(Time Division Alternative BOC) and AltBOC signals for different front-end bandwidth of the receiver is analyzed. By studying the average multipath error and envelope value of AltBOC (15,10), BOC (1,1) and TMBOC (6,1,4/33) signals, the performance comparison results of three kinds of BOC modulation are given. In order to verify the research results of multipath suppression performance, the receiver is used to receive the signals of different BOC modulation systems transmitted by navigation satellite, The multipath error is obtained by the combination of dual frequency pseudo range and carrier phase.The experimental results show that AltBOC (15,10) has the best multi-path suppression performance, followed by TMBOC (6,1,4/33), and BOC (1,1) has the worst multi-path suppression performance compared with the others, which proves the correctness of the research results of multi-path suppression performance of various BOC signals.

1. Introduction

With the development of navigation satellite system, the navigation and positioning technology become more and more significant in national economy development and national defence day by day, and various powerful countries have engaged in the development of their own navigation satellite system. At present, there are four main established/establishing navigation satellite system: GPS(Global Positioning System), GLONASS(Global Navigation Satellite System), Galileo(Galileo Positioning System) and BDS(Beidou navigation satellite system). The more wide-spreading of the navigation and positioning application, the higher requirement of its positioning accuracy, anti-interference performance and multipath mitigation capability. With the increasing of satellite number in space, the navigation frequency is more and more crowd. Using the traditional BPSK-R modulation method to improve navigation performance costs too much. Therefore, seeking a new signal modulation method may be the best way to solve the problem. BOC is a new kind of modulation method as the development of navigation technology, whose definition was first introduced in GPS modernization reformation. It was initially suggested to be the modulation method for GPS military signal, but now it has been wide applied in modern GPS civil signal design[1]. In 2001, CNES(Centre National d'Etudes Spatiales) proposed AltBOC modulation method with four pseudocode constant-envelope, which was adopted by Galileo E5a and E5b[2].

During the establishment of Beidou global navigation satellite system, the selection of BOC is conducted as an important job. Since 2009, a plentiful of domestic scientific agencies have worked over BOC for Beidou global signal system. Tang zuping promoted TD-AltBOC modulation method, which was a time-division AltBOC modulation technology to alternately propagate 4 signal components in two time slot and each time slot propagated tow constant-envelope AltBOC signals[3]. Though this method could avoid patent issue, it also brings other problems. As indicated in reference [4], this method may reduce the affective length of the pseudocode and weaken cross-correlation between different pseudocodes. What's more, in order to proceed time-division signal, it would force receivers to change correlator structure and add time slot switching circuit. To solve the above AltBOC shortcoming, Yao Zheng promoted ACED(Asymmetric Constant Envelope Double-sideband) multiplexing technology to improve AltBOC, increase signal power in pilot component and keep the current receiver status with good interoperability[5].

Hence, the current research in Beidou BOC modulation method is focused on compatibility, interoperability, receiver's complexity and signal multiplexing method and efficiency, and the capability of mitigating multipath is rarely lucubrated. This paper engaged in the capability of mitigating multipath for BOC modulation signal and analyzed with simulation and actual data in Beidou navigation satellite system.

2. Multipath Error Characteristic of Beidou navigation satellite System

Multipath means the receiving of unexpected signal after being reflected and scattered. As the difference of the propagation path, there are relative phase, relative delay and scope characteristic between direct signal and reflected and scattered signal, which result in the receiving loop signal distortion and zero crossing point excursion of correlate peak discriminating curve, affecting the measurement accuracy. Normally, multipath signal is supposed to be caused by reflection and the dispersion signal model is conducted as a noise term, which has little effect to code tracking.

In Beidou regional system, there are 5 GEO(Geosynchronous Earth Orbit) satellites to cover the most area of China. According to the research of satellite based augmentation system abroad, there was "inherent multipath error" in GEO pseudorange measurement, meaning GEO trajectory characteristic had remarkable affection to pseudorange multipath error. The moving trajectory satellite, such as IGSO(Inclined GeoSynchronous Orbit) and MEO(Medium Earth Orbit), does not have obvious and inherent multipath error and their multipath error may only increase for low observation elevation during inbound and outbound[6].

According the current Beidou global blueprint, the Beidou global satellites will all be moving trajectory satellites, which composes of IGSO and MEO, and trajectory characteristic will not be the major factor to pseudorange multipath as the signal modulation structure is the major factor. During the experiment segment of Beidou global system construction, downlink signal system will be one of the key chain-link. In order to realize compatibility and interoperability with GPS and Galileo,in Beidou navigation satellite system, TMBOC, AltBOC and TD-AltBOC had be chosen. Therefore, the research of the three modulation methods for pseudorange multipath is an urgent and essential work.

3. Simulation and Analysis on Multipath-mitigating Capability of BOC Singal

3.1. Model Choosing and Estimating Criterion of Multipath Mitigation

As mentioned in chapter 2, the major factor affecting user positioning accuracy is reflected signal. The multipath mathematical model is normally expressed as:

$$r(t) = D(t) \cdot C(t) \cdot \sin(\omega_0 t) + \sum_{i=0}^{N-1} \alpha_i \cdot D(t) \cdot C(t+\delta_i) \cdot \sin(\omega_0 t + \theta_i) \tag{1}$$

Where D(t) is base-band modulation signal, C(t) is satellite spreading code, $\sin(\omega_0 t)$ is the intermediate frequency after down-conversion, α_i is multipath signal gain compared to direct signal. Usually, the amplitude of multipath signal is less than the direct signa. Hence, $\alpha_i < 1$. δ_i is the phase delay of spreading code, θ_i is the effect of multipath to carrier phase. The multipath signal in base-band of receiver is equivalent discribed as following.

$$r(t) = \alpha_0 e^{j\varphi_0} x(t-\tau_0) + \sum_{n=1}^{N} \alpha_n e^{j\varphi_n} x(t-\tau_n) \tag{2}$$

Where α_0 is amplitude of direct signal, φ_0 is the phase of direct signal, x(t) is the complexing envelope of receiving signal, τ_0 is the propagation delay of direct signal, N is the path number of reflecting multipath singals; α_n, φ_n and τ_n repectely mean the amplitude, phase and propagation delay of reflected multipath signal.

The receiving of the reflected multipath signal leads the balance point offseting the zero point of receiver discriminating curve and produces pseudorange tracking deviation which means pseudorange multipath error. Therefore, the multipath error analysis should be engaged on the ouput function of discriminator. In order to realise convenient annalysis and possessing generality, the analysis model with single reflected path is widely applyed during the design of navigation signal. Under the condition of one reflected path considered, the envelope model of multipath error for loop discriminator is established. Then, the non-coherent early-late processing track loop is chosen, whose loop discriminator is shown as following.

$$D(\varepsilon_\tau) = \alpha_0 \left[R\left(\varepsilon_\tau - \frac{d}{2}\right) - R\left(\varepsilon_\tau + \frac{d}{2}\right) \right] \pm \alpha_1 \left[R\left(\varepsilon_\tau - \tau - \frac{d}{2}\right) - R\left(\varepsilon_\tau - \tau + \frac{d}{2}\right) \right] \tag{3}$$

Where, R(τ) is the correlation function, ετ is delay error estimation of direct signal, τ is multipath delay, and d is the interval of early-late correlator.

The answer of the equation which makes the $D(\varepsilon_\tau) = 0$ is the multipath error. With the signal bandwidth being considered, the final output of multipath error is described as following.

$$\varepsilon = \frac{\pm\alpha \int_{-\beta_r/2}^{\beta_r/2} S(f)\sin(2\pi f\tau)\sin(\pi f d)\,df}{2\pi \int_{-\beta_r/2}^{\beta_r/2} f S(f)\sin(\pi f d)\left[1 \pm \alpha\cos(2\pi f\tau)\right] df} \tag{4}$$

Where, $\alpha = \alpha_1 / \alpha_0$ is MDR, meaning the signal amplitude rate of multipath signal and direct signal, β_r is signal bandwidth, d is the interval of early-late correlator, and τ is multipath delay. When the phase difference between multipath signal and direct signal is 0°, + is selected in above equation; when the phase difference between multipath signal and direct signal is 180°, - is selected in above equation, which means minimum and maximum influence to multipath signal respectively. Hence, the above equation shows the envelope of multipath error. Besides, the average of the multipath error can be obtained with the following equation.

$$\varepsilon(\tau_1) = \frac{1}{\tau_1} \int_0^{\tau_1} \left[\frac{abs(\varepsilon(\tau)|_{\varphi=0}) + abs(\varepsilon(\tau)|_{\varphi=180})}{2} \right] d\tau \tag{5}$$

Where φ is relative phase of multipath signal, respectively corresponding with the + and – condition. When the multipath delay τ1 is known, the average of the multipath error can be obtained.

TD-AltBOC(15,10) is one of the modulation methods used in Beidou navigation satellite system, whose power spectrum density can be described as following.

$$G_{TD-AltBOC}(f) = \frac{10.23}{\pi^2 f^2} \cos^2\left(\frac{\pi f}{10.23}\right) \frac{\sin^2\left(\frac{\pi f}{6\times 10.23}\right)}{\cos^2\left(\frac{\pi f}{3\times 10.23}\right)} \left(2 - \cos\left(\frac{\pi f}{3\times 10.23}\right)\right) \tag{6}$$

AltBOC(15,10) is another modulation methods used in Beidou navigation satellite system, whose power spectrum density can be described as following.

$$G_{AltBOC}(f)=\frac{4\times10.23}{\pi^2 f^2}\frac{\cos^2(\frac{\pi f}{10.23})}{\cos^2(\frac{\pi f}{3\times10.23})}\left\{\begin{array}{l}\cos^2(\frac{\pi f}{2\times10.23})-\cos(\frac{\pi f}{2\times10.23})-\\ 2\cos(\frac{\pi f}{2\times10.23})\cos(\frac{\pi f}{4\times10.23})+2\end{array}\right\} \quad (7)$$

Putting equation (6) and (7) into (4) and (5) respectively, the multipath error envelope and its mean value can be gained with two different BOC modulation methods.

3.2. Capability Comparison of Mitigating Multipath between AltBOC and TD-AltBOC modulation method

Set MDR= -6 dB, d = 24.4 ns, and the frontend signal bandwidth β_r =30MHz. According to equation (4), the multipath error envelope with TD-AltBOC and AltBOC modulation method can be gained and the comparison result can be seen in figure 1. According to equation (5), the mean value of multipath error with TD-AltBOC and AltBOC modulation method can be gained and the comparison result can be seen in figure 2.

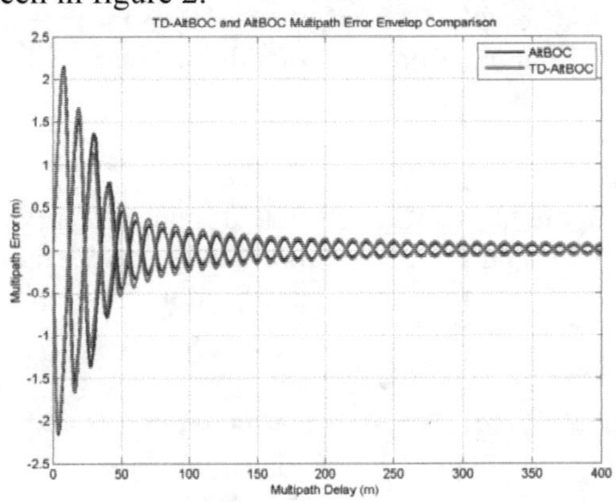

Figure 1. AltBOC and TD-AltBOC multipath error envelop comparison (frontend bandwidth 30MHz)

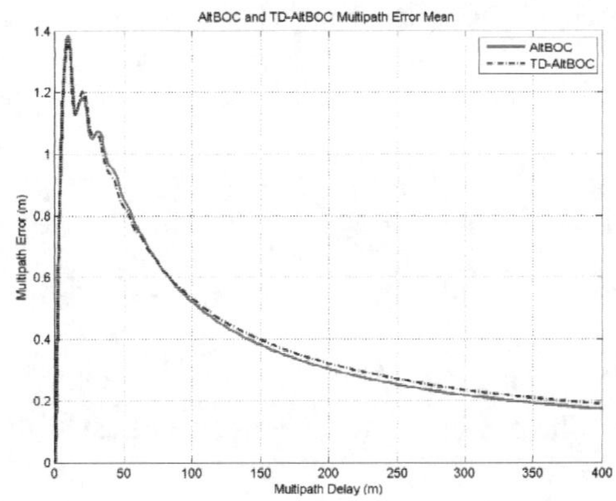

Figure 2. AltBOC and TD-AltBOC mean multipath error comparison(frontend bandwidth 30MHz)

It can be seen that the capability of mitigating multipath with AltBOC and TD-AltBOC was approximate when the front bandwidth of receiver was 30MHz.

Keep other parameters unchanged and increases the signal bandwidth to 40MHz, the multipath error envelope could be seen in figure 3 with mean value showed in figure 4. Seen from the figure 3 and 4, the capability of mitigating multipath of TD-AltBOC was better than AltBOC.

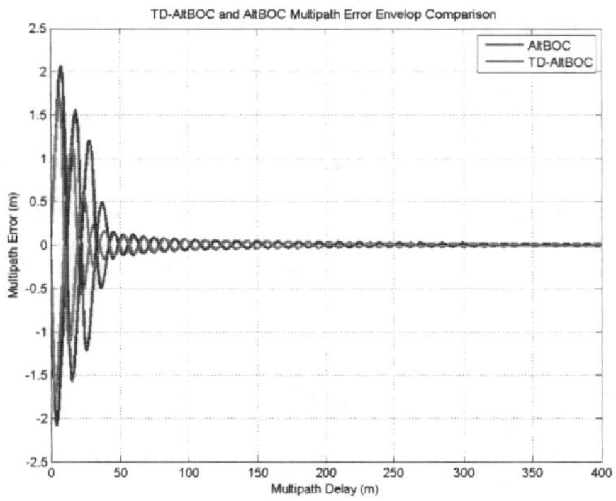

Figure 3. AltBOC and TD-AltBOC multipath error envelop comparison (frontend bandwidth 40MHz)

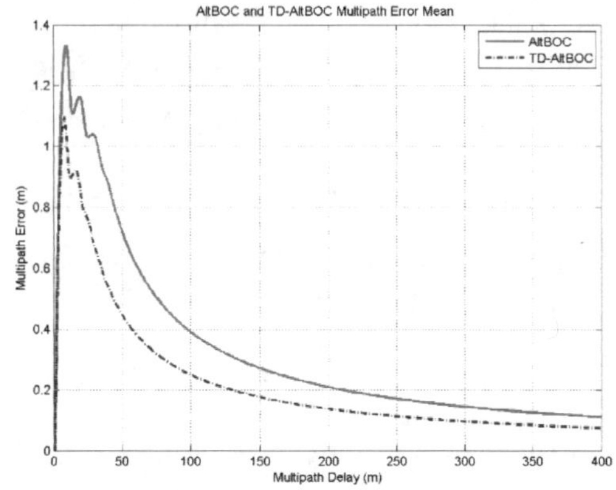

Figure 4. AltBOC and TD-AltBOC mean multipath error comparison(frontend bandwidth 40MHz)

In conclusion, the front bandwidth of receiver could affect the capability of mitigating multipath with two modulation methods. Under the condition of larger front bandwidth, the capability of mitigating multipath with TD-AltBOC was superior to AltBOC.

3.3. Capability Comparison of Mitigating Multipath with BOC Modulation Method

For Beidou navigation satellite system, data component BOC (1,1) and pilot component TMBOC(6,1,4/33) are combined with MOC(6,1,4/33), realizing the compatibility and interoperability with GPS L1. The power spectrum density of the pilot component is showed as equation (8).

$$G_1(f)=\frac{29}{33}G_{BOC(1,1)}(f)+\frac{4}{33}G_{BOC(6,1)}(f) \tag{8}$$

The power spectrum density with BOC (1,1) and BOC (6,1) are shown as equation (9) and (10).

$$G_{BOC(1,1)}=1.023\times\left[\frac{\tan(\frac{\pi f}{2\times1.023})\sin(\frac{\pi f}{1.023})}{\pi f}\right]^2 \tag{9}$$

$$G_{BOC(6,1)}=1.023\times\left[\frac{\tan(\frac{\pi f}{2\times6\times1.023})\sin(\frac{\pi f}{1.023})}{\pi f}\right]^2 \tag{10}$$

Putting the equation (9) and (10) into (8), the power spectrum density of pilot component can be gained and shown as equation (11).

$$G_1(f)=\frac{\sin^2(\frac{\pi f}{1.023})}{(\pi f)^2}\left[\frac{29\times1.023}{33}\tan^2(\frac{\pi f}{2\times1.023})+\frac{4\times1.023}{33}\tan^2(\frac{\pi f}{2\times6\times1.023})\right] \tag{11}$$

The power spectrum density of data component is shown in equation (9). Putting the equation (11) and (9) into (4) and (5) respectively, the multipath error envelope and its average value can be obtained in data component and pilot component.

Set MDR= -6 dB, d = 24.4 ns, and the signal bandwidth β_r =40MHz. The figure 5 showed the comparison result of pseudorange multipath error envelope with AltBOC (15,10), BOC (1,1) and TMBOC (6,1,4/33). According to equation (5), the mean value of multipath error with AltBOC (15,10), BOC (1,1) and TMBOC (6,1,4/33) could be gained. The comparison result was shown in figure 6.

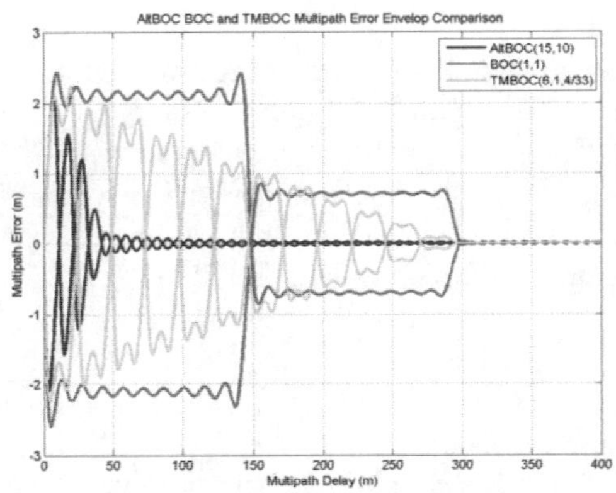

Figure 5. multipath error envelop comparison of three BOC signals (frontend bandwidth 40 MHz)

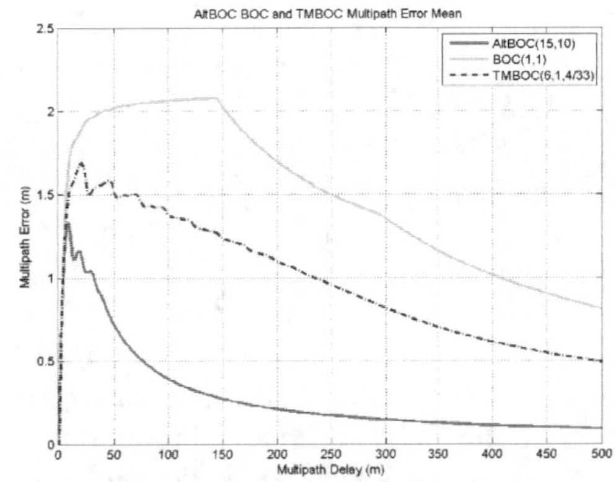

Figure 6. mean multipath error comparison of three BOC signals (frontend bandwidth 40 MHz)

In conclusion, when the delay of multipath signal was small, the average of multipath error with three modulation methods was rather approach; when the delay of multipath signal was large enough, the average of multipath error with AltBOC (15,10) was lower than MBOC (6,1,4/33) and the average value with BOC (1,1) was largest.

4. Analysis of Test Data

The RMS(Root Mean Square) of pseudorange multipath error was gained as receivers in two ground monitoring stations received down-link signal of Beidou global system and proceeded pseudorange measurement when satellites could be observed within 24hours.The result of pseudorange multipath error in BOC (1,1) andTMBOC(6,1,4/33) was obtained with two ground station observing two satellites S1 and S2.

The RMS statistics results of the two satellites was shown in Table 1.

Table 1. Multipath error statistics results of two monitoring stations

satellite	*Monitoring station*	*TMBOC*	*BOC*	*AltBOC*
S1	Sta1	0.187	0.243	0.123
	Sta2	0.335	0.408	0.156
S2	Sta1	0.186	0.214	0.118
	Sta2	0.171	0.215	0.095

Seen from the Table 1, the statistics results of pseudorange multipath error were consistent with the simulation results in chapter 3 as the comparison result is AltBOC> TMBOC>BOC.

5. Conclusion

Global navigation satellite system adopts various BOC modulation signals. Based on noncoherent early-late processing track loop, the paper simulated the capability of mitigating multipath with TD-AltBOC and AltBOC modulation methods under different front bandwidth conditions, compared the capability of AltBOC and TMBOC under the same front bandwidth, and analyzed pseudorange measurements with two ground monitoring stations observing two Beidou global satellites, calculating the RMS of the pseudorange multipath error. The simulation results showed: when the front bandwidth was 30MHz, the capability of pseudorange multipath error with TD-AltBOC (15,10) and AltBOC (15,10) was similar; when the front bandwidth was 40MHz, the capability of pseudorange multipath error with TD-AltBOC (15,10) was better than AltBOC(15,10); The capability of pseudorange multipath error with AltBOC(15,10) is superior to TMBOC(6,1,4/33) and the capability of TMBOC(6,1,4/33) is better than BOC(1,1). The analysis results with actual data was consistent with the simulation results.

References

[1] Betz J W. The offset carrier modulation for GPS modulation. In: Proceedings of ION NTM'99 and 19th Biennial Guidance Test Symposium. Washington: Institute of Navigation, 1999. 639—648

[2] Laurent Lestarquit, Géraldine Artaud, Jean-Luc Issler, AltBOC for Dummies or Everything You Always Wanted to Know About AltBOC[C], ION GNSS 21st. International Technical Meeting of the Satellite Division, 16-19, September 2008, Savannah, GA, P961-970.

[3] TANG Z，ZHOU H，WEI J，et al. TD-AltBOC: a new COMPASS B2 modulation ［J］. Science China: Physics，Mechanics and Astronomy，2011，54(6) : 1014-1021.

[4] Betz JW, Cahn CR, Dafesh PA et al (2006) L1C signal design options. In Technical national meeting of the institute of navigation. CA, Monterey, pp 685-697.

[5] Kharisov V, Povalyaev A(2011) Optimal aligning of GNSS navigation signals sum. In: The 24 th international technical meeting of the satellite division of the institute of navigation. Portland, Oregon, pp 3141-3155.

[6] Wei Zhao, Na Zhao. Analysis of the Pseudorange Multipath Impact on Dual-Frequency Ionospheri Delay Correction in BeiDou System. [C]. China Satellite Navigation Conference(CSNC) 2013 Proceedings, LNEE 243, p.355 ff.

The State Analysis of Five-axis NC Machine with Large Torque Mechanical Spindle Based on Hilbert Transform

Fu Yaohuimei [1], Xia Yuanmeng[1,2], Chen Xuezhen [1,*], Zhao Guobo [1], Zhang Yun [1]

[1]CNC processing plant, Chengdu Aircraft Industry Co., Ltd. Chengdu, Sichuan, 610091, China.

[2]University of Electronic Science and Technology of China, Chengdu, 611731, China

[*]Corresponding author's e-mail: 532015901@qq.com

Abstract. Only five-axis machine with large torque mechanical spindle is capable to complete the process of titanium alloy aerospace structure in general. But mechanical spindle is too complex to locate the fault rapidly when the part shows anomalies or damage, and this reduces high-value machine's utilization rate as well as brings difficulties in solving part's quality problems. Pointing at surface quality problems and spindle component's wearing down that emerge frequently, this paper simulates various parameters in NC(numerically controlled) machining process and monitors spindle states of a five-axis machine with large torque mechanical spindle, then amplitude, time-frequency, and frequency curvature characteristics of the tested signals are obtained by Hilbert Transform and fast identification of key component's wearing down is realized. The results show that the state analysis method of five-axis NC machine with large torque mechanical spindle based on Hilbert transform presented in this paper has strong engineering application value.

1. Introduction

Five-axis NC(numerically controlled) machine with large torque mechanical spindle is the main source of titanium alloy parts manufacturing, the spindle performance directly determines the processing performance, which in turn affects the final product's quality [1]. However, in the actual production process, because such machine is equipped with mechanical spindle with complex structure, there is difficulty in diagnosing faults for wearing down and overload, and the final results are reflected on the part surface, which are manifested as tool flapping and ripples. Therefore, timely monitoring of the spindle operating state and effective fault diagnosis will greatly reduce the downtime caused by equipment failure, thereby improving the production efficiency and accuracy of aviation structural components.

The most convenient way to monitor the spindle is to monitor its vibration [2]. The vibration signal contains real-time status information of the power system. Based on vibration data, the most common method is spectrum detection based on Fourier transform. But in reality, the mechanical spindle is a nonlinear time-varying system, which is constantly changing due to the influence of load and equipment state, and since the Fourier integral is limited to the entire time domain, even if a certain frequency component of the signal changes, it cannot find the transformation time and the duration of a certain frequency, the most intuitive result is the frequency aliasing phenomenon after the transformation and the reasonable frequency component of the signal cannot be identified. The essence is that the

IOP Conf. Series: Materials Science and Engineering **793** (2020) 012045 doi:10.1088/1757-899X/793/1/012045

fundamental wave selected by the transform is an infinitely long cosine signal, and there is a great limitation on the localization problem of time domain and the frequency domain for the spindle vibration's unstable signal[3].

Considering the limitations of the Fourier transform in analyzing the spindle state signal, Hilbert transform, is introduced. Based on clear mathematical principles, by converting the measured real-number domain signal into a complex-domain signal, the envelope and instantaneous frequency of the signal can be presented, and the analytic function thus constructed does not contain the negative frequency portion, so it is suitable for analysing spindle status signals[4]. In this paper, the vibration of the large torque mechanical spindle of the five-axis machine is tested for its titanium alloy parts' surface ripples. The Hilbert transform method is used to analyse the cause of the fault and realize rapid identification of the internal structure's wearing down of the spindle.

2. Hilbert transform principle

For continuous time signals $x(t)$, the Hilbert transform is defined as follows[5]:

$$\hat{x}(t) = \frac{1}{\pi}\int_{-\infty}^{\infty}\frac{x(t)}{t-\tau}dt = \frac{1}{\pi}\int_{-\infty}^{\infty}\frac{x(t-\tau)}{t}d\tau = x(t)\frac{1}{\pi t} \tag{1}$$

Considering Fourier transform of $\frac{1}{\pi t}$:

$$F(j\omega) = -jsign(w) = \begin{cases} -j\ (\omega > 0) \\ j \quad (\omega < 0) \end{cases} \tag{2}$$

For $X(j\omega)$ is $x(t)$'s Fourier transform, according to convolution theorem, then:

$$\hat{x}(j\omega) = F(j\omega)X(j\omega) = -jsign(jw)X(jw) \tag{3}$$

Now define the analytical signal for initial $x(t)$ to $z(t)$:

$$z(t) = x(t) + j\hat{x}(t) \tag{4}$$

Thus,

$$Z(j\omega) = X(j\omega) + j\hat{x}(j\omega) = X(j\omega) + jF(j\omega)X(j\omega) = \begin{cases} 2X(j\omega) & \omega > 0 \\ 0 & \omega < 0 \end{cases} \tag{5}$$

This shows that the analytical signal consisited of Hilbert transform has only positive frequency, which is very suitable for the analysis of the spindle's vibration signal.

Then write the analytical signal as an exponential form:

$$z(t) = A\exp(-j\phi) \tag{6}$$

Let A be the amplitude, and ϕ the phase angle, then the original signal can be written as:

$$x(t) = A\cos\phi \tag{7}$$

Where,

$$\phi = \arctan(\hat{x} / x) \quad \text{and} \quad A = \sqrt{x^2 + \hat{x}^2} \tag{8}$$

Therefore, the Hilbert transform of the signal is actually a convolution of amplitude A and frequency modulated signal $\cos\phi$.

3. Experiment analysis

3.1 Experimental preparation

Part surface of Zhongjie 1# NC machine has been frequently corrugated, fortunately, the quality of final product has not been affected. In order to ensure the efficiency and stability of the subsequent processing, the vibration detection test of the spindle head state of the machine is performed to analyse the cause of the problem and identify the fault location. The Zhongjie 1# five-axis NC machine for machining titanium alloy parts was selected as the experimental object, and the SmartBalancer instrument was used for vibration detection. The instrument is equipped with a three-phase acceleration sensor, which can quickly collect and analyse the vibration signal without disassembling the spindle. In order to facilitate the observation and comparative analysis, and taking into account the structural characteristics of the five-axis mechanical spindle of the machine tool's long-drive chain, in each experiment, observe from the forward direction of X-axis, and test three parts respectively: directly below the spindle head, left side of the spindle head, right side of the spindle head[6].

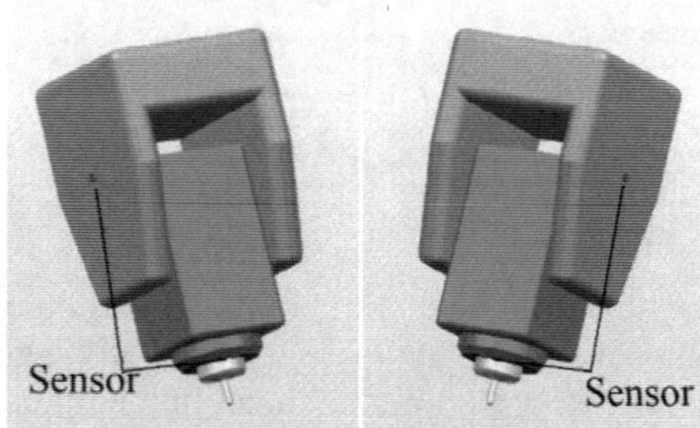

Figure1. Sensor installation diagram

In order to ensure the validity of the measured data and simulate the normal processing state of the machine, each measurement time is consistent, which is 8:30 on Friday morning, and the spindle is preheated for 15 minutes before each experiment. In addition, the spindle speed and the equipped tool were consistent for each experiment, respectively 1500 rpm, and ø20-70R3 end mill.

3.2 Experimental data collection and analysis

Since the spindle speed is 1500rpmin, the sampling frequency is set at 1000HZ. After many experiments, according to the information measured by the vibration sensor, the Hilbert transform principle is used to obtain the analytical signal of the original signal, then the frequency and amplitude of the spectrum line with time. After that, solve the curvature change's average value of the analytical signal.

3.2.1 Time-frequency signal

Immediately directly below the spindle head, the left side of the spindle head, and the right side of the spindle head, the frequency and amplitude of each part's analysing signals in four experiments are changed over time, as shown below:

IOP Conf. Series: Materials Science and Engineering **793** (2020) 012045 doi:10.1088/1757-899X/793/1/012045

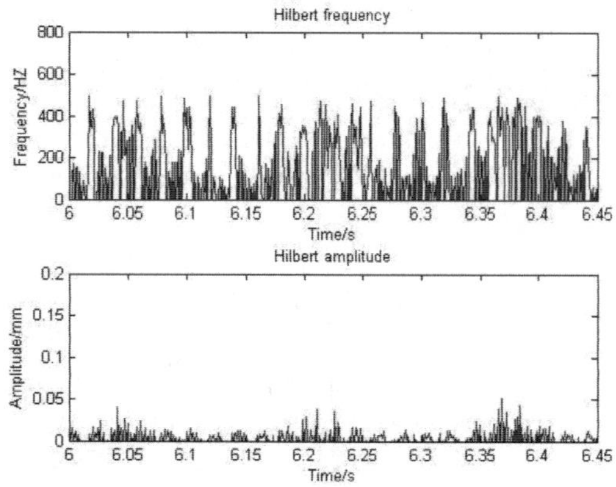

(a)

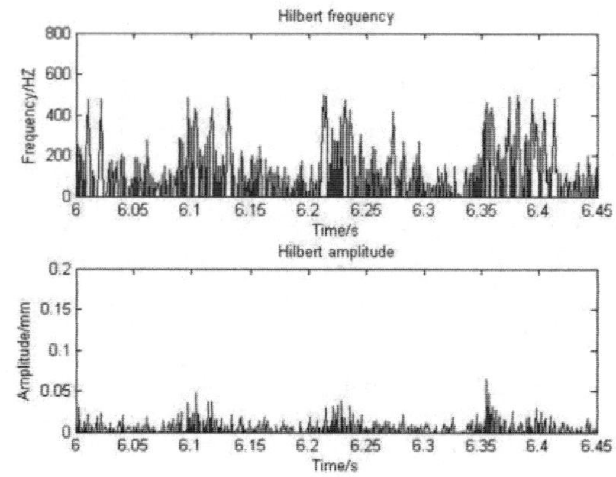

(b)

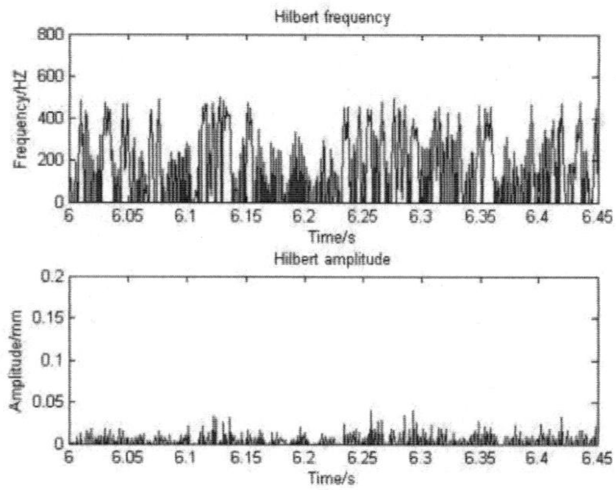

(c)

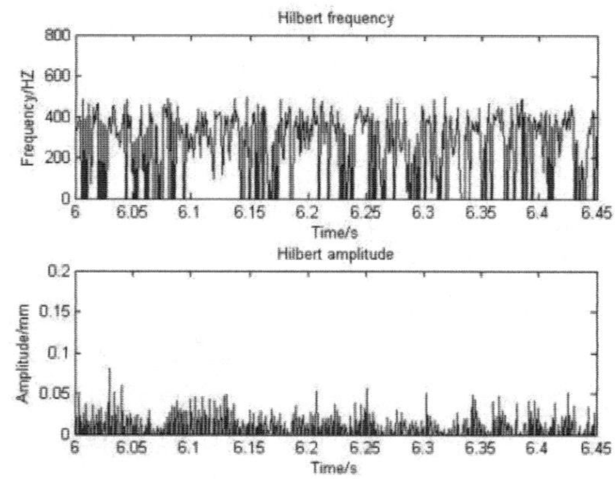

(d)

Figure2. Spectrum distribution directly below the spindle head

Analyse the spectrogram directly below the spindle head, in the intervals of figure 2(a)'s 6.35s~6.4s, figure 2(b)'s 6.35s~6.4s, figure 2(c)'s 6.25~6.3s, the proportions that the corresponding frequency exceed 200HZ are 94.7%, 95.9%, and 94.3% respectively, and this proportion in figure 2(d) reaches to 97.6% on the whole. It can be found that in these intervals, high frequency concentrate, the amplitudes have obvious high points and even the highest points.

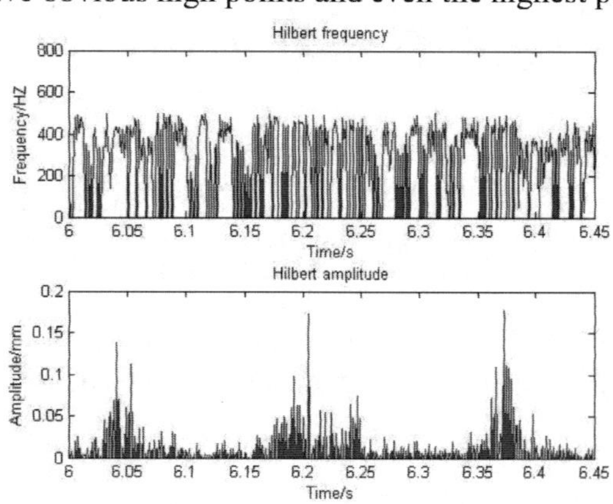

(a)

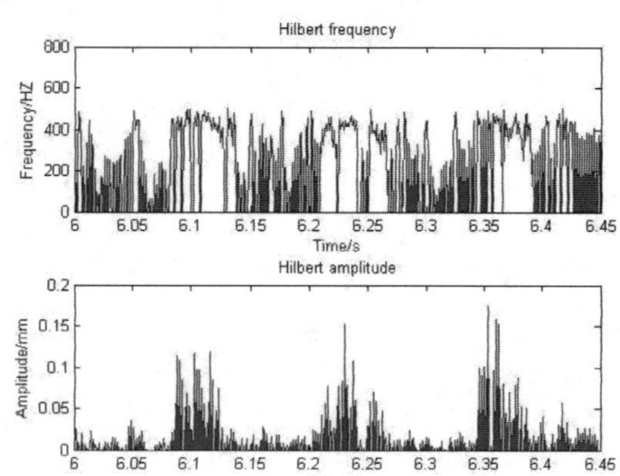

(b)

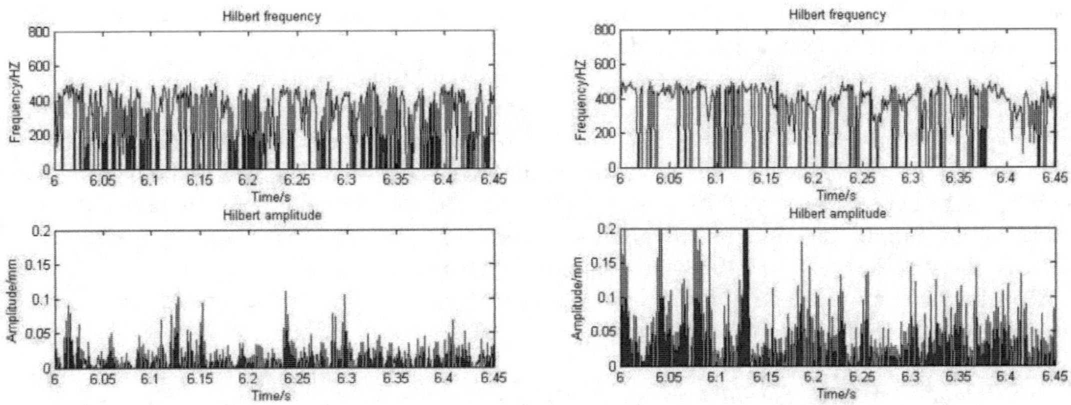

(c) (d)

Figure3. Spectrum distribution on the left side of the spindle head

Analyse the spectrum on the left side of the spindle head, the proportions that frequency exceed 200HZ are 98.6%,96.4%,97.7%,99.3% respectively, with an overall high frequency in each experiment, so more amplitude peak points appeared in the four experiments.

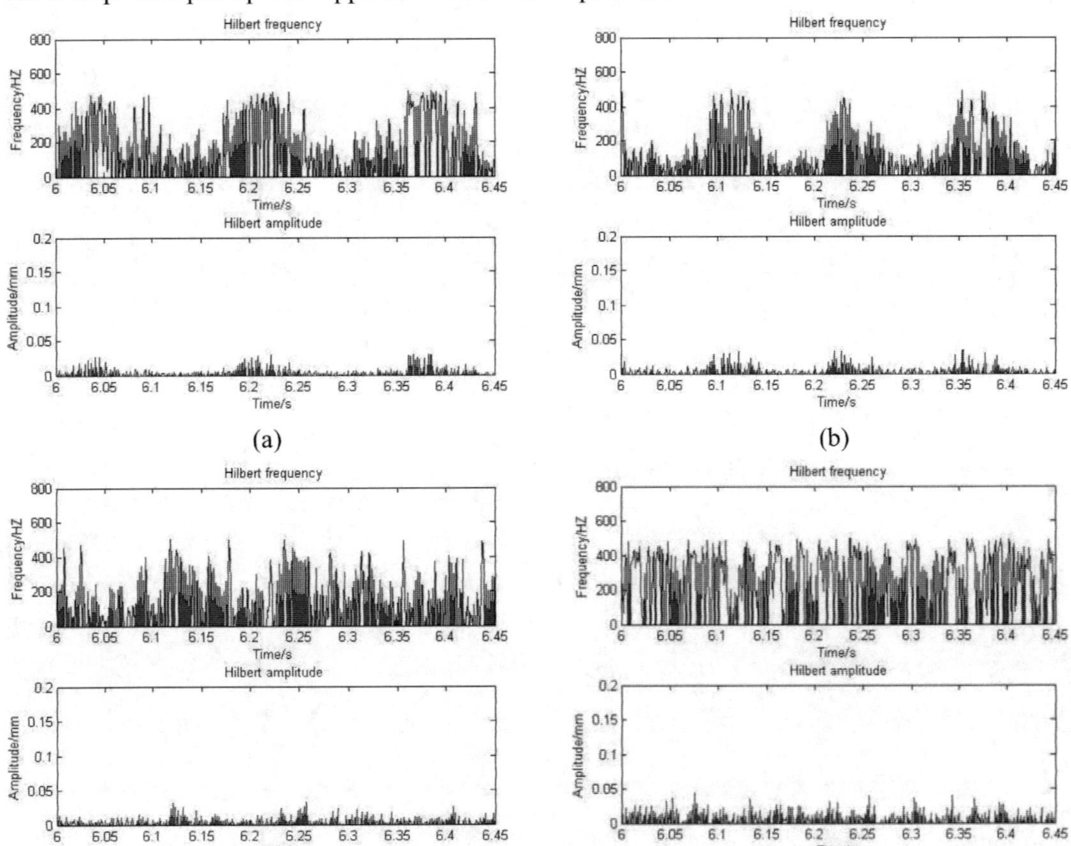

(c) (d)

Figure4. Spectrum distribution on the right side of the spindle head

Analyse the spectrogram on the right side of the spindle head, the intervals that the frequency concentratedly exceed 200HZ are 6.2s~6.25s and 6.35s~6.45s in figure 4(a), and their proportions are 93.4% and 97.5%. Figure 4(b) owns three intervals that high frequency concentrate, and this

IOP Conf. Series: Materials Science and Engineering **793** (2020) 012045 doi:10.1088/1757-899X/793/1/012045

phenomenon is more obvious in figure 4(c) and figure 4(d). In the interval where these high frequency occurrences are relatively high, the amplitude also presents corresponding cusps.

According to the results of the above four experiments, in the same period of time, where the high-frequency concentrated, the appearancing frequency of strong amplitudes increases of the part directly below the spindle head, indicating that the accumulation of high-frequency has an adverse effect on the spindle's operating state. Under the same monitoring state, in general, the interval of the high frequency on the left side of the spindle head is obviously higher than that on the directly below part and the right side part, and the amplitude value of this portion reaches 0.2 mm, which is obviously higher than the right side of the spindle head's 0.04mm and 0.06mm directly below the spindle head.

Obviously, the amplitude and frequency information of the vibration on the left side of the spindle head is more prominent than the other two parts.

3.2.2 Time-frequency signal curvature

In order to verify the results of the time-frequency signal analysis, according to the above measurement data, the average curvature of the time-frequency discrete signal changes of the four experiments in each part is solved, and the results are as follows:

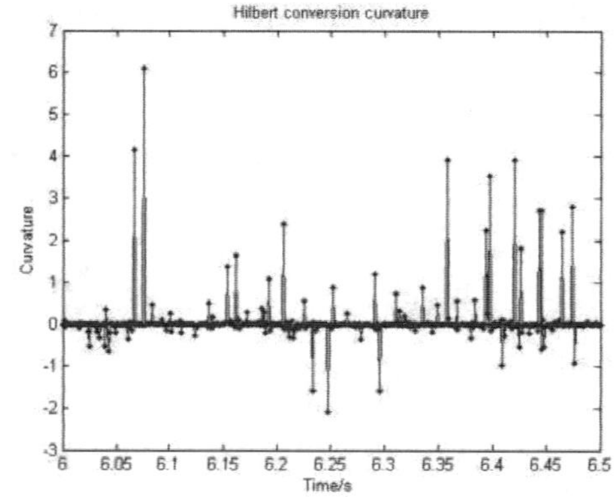

(a)

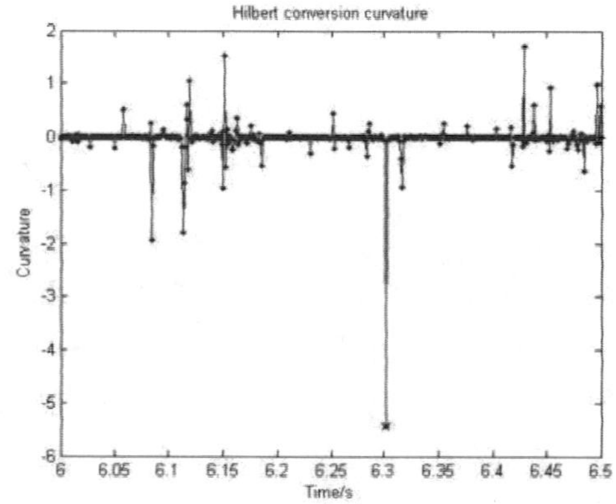

(b)

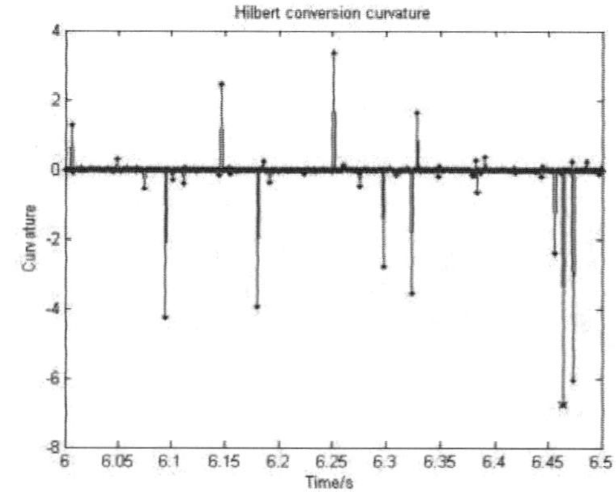

(c)

Figure5. Curvature diagram of the three parts' frequency change

According to the curvature diagrams of the above three discrete signals, it is obvious that the order of the singularity's number of the three curvatures is: directly below the spindle head, the right side of the spindle head, and the left side of the spindle head, which is consistent with the result of that high frequency on the left side is more common and its frequency change is less in time-frequency distribution, which verifies the conclusion of the previous section. Since the number of singular points directly below the spindle head is the largest, it indicates that this part is subjected to the greatest impact,

IOP Conf. Series: Materials Science and Engineering **793** (2020) 012045 doi:10.1088/1757-899X/793/1/012045

which is consistent with the general machine phenomenon, and attention should be paid to the maintenance of the connecting parts.

3.3 Experimental result

According to the spindle state detection experiment, after the amplitude, time-frequency signal and its curvature test analysis, it can be judged that the vibration on the left side of the spindle head is abnormal. Considering the mechanical structure of the Zhongjie 1# long-drive chain (as shown in Figure 6), it is possible to locate the mechanical parts directly related to the left side of the spindle head is the bevel gear and their bearing parts. So, it can be judged that under the same processing conditions, the wearing down of the bevel gear and its bearing will be stronger, which is also the main cause of frequent ripples on the surface of parts. It should be paid attention to in daily maintenance.

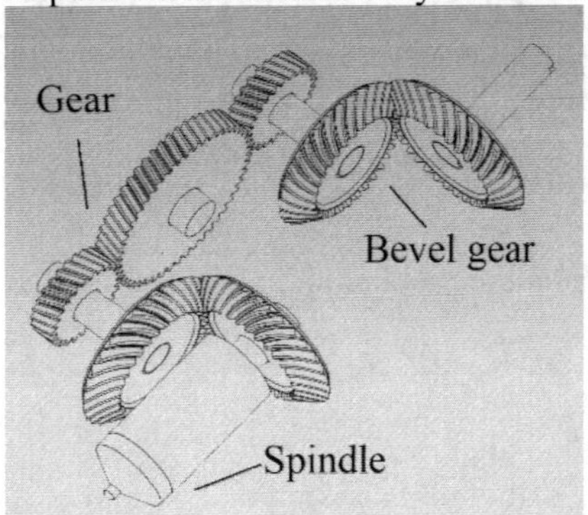

Figure6. Spindle structure diagram for Zhongjie 1#

4. Conclusion

The five-axis NC machine with large torque mechanical spindle has complex mechanical structure, which makes fault detection get hard. Under the simulated actual machining conditions, the real number signal of the three parts' vibration signal of the faulty machine's spindle head is converted into analytic signal in the complex domain by Hilbert transform, the signal of one dimension is presented as a signal in the two-dimensional complex plane, thereby extracting the frequency and amplitude of the signal. The method overcomes the problem that the traditional Fourier transform cannot perform the horizontal comparison of the frequency of various parts under the same time condition. Through the presentation of the frequency in the time domain, the frequency of each part and the aggregation and distribution can be clearly judged, combining the amplitude change, the mechanical state of the three parts of the spindle head can be analysed and diagnosed.

Further, by solving the variation curvature of the discrete frequency signal, the frequency variation of the state of the mechanical spindle head's three parts can be quantitatively analysed, and the characteristics of the three parts' frequency can be clearly compared according to the number, distribution and numerical size of the singular points. This verified the results of the spectrum analysis, and then infer the mechanical state of each part to find the cause of the surface ripple problem.

The results show that the large torque mechanical spindle state analysis method of the five-axis NC machine based on Hilbert transform can quickly identify the wearing down condition of the transmission components in the spindle, thereby improving the quality of parts and improving production efficiency, and has strong engineering application value.

Funding information

This work is supported by the Major Project of National Science and Technology (Grant No.2017ZX04002001).

References

[1]Mark,A.(2018)Machine built for full five-axis precision.J.Modern Machine Shop,91(3).

[2]Jedd,C.(2018)Five-axis VMC increases accuracy by reducing vibration.J.Modern Machine Shop,90(10).

[3]Kowalski,E.(2011)A Note on fourier coefficients of poincare series.J.DOI: 10.1112/S0025579310001804.57(1):31-40.

[4] Cao,H.Xi,S.Chen, X.(2019)Dynamic modeling of spindle bearing system and vibration response investigation.J.Mechanical Systems & Signal Processing,114:486-511.

[5]Zhang, Z.Huang, Zhen.Li,Cheng.(2019)Vibration measurement based on multiple Hilbert transform for self-mixing interferometry.J.Optics Communications: A Journal Devoted to the Rapid Publication of Short Contributions in the Field of Optics and Interaction of Light with Matter,436:192-196.

[6]Kong,X.Zou,H. et al.(2019)Acceleration and Deceleration Motion Control and Linear Interpolation based on Five-axis Linkage NC Machine.J.Materials Science and Engineering.612(3):032088.

Dynamics analysis of planar linear array deployable structures with multiple clearance joints

Bo Li[1*], Hong zhao Liu[1] and Jia hua Zhang[1]

[1]School of Mechanical and Precision Instrument Engineering, Xi'an University of Technology, Xi'an 710048, China

[*]Corresponding author's e-mail: yunfenglibo@126.com

Abstract. The purpose of this paper is to investigate the dynamic characteristic of the scissor linear array deployable structure under the coupling action of clearance joints, where the kinematic model of the deployable structure with clearance is established according to the principle of the motion between joint elements. The normal and tangential contact forces between contact components are respectively described using the Flores contact force model and modified Column friction model. In the sequel of this process, the impact forces between the connected components are converted to the mass centers of links connected by this clearance joint, and the converted contact force and additional torque were integrated to the generalized force in motion equations of the deployable structure, and then the effect of joint clearance is introduced to deployable syetem. The simulation results show that the dynamic characteristics of the mechanism are very sensitive to the clearance position, and the movement of a joint clearance will affect the movement of the other joint clearance, which shows that clearance joints in deployable structure have a strong dynamic interaction and all the joints should be modeled as real joints.

1. Introduction

The deployable structure has the characteristics of small volume and large space. Moreover, this structure can be expanded from the contraction state to the preset expansion state and keeps the stable configuration during the transportation, so it has been widely used in aviation, aerospace and construction fields [1-5]. Li et al [2] proposed a new inter-joint force model to predict the dynamics of planar linear deployable structure.

In the movement of the real mechanism, the joint clearance is inevitable. And it greatly affects the motion precision and stability of the structure. Yan et al [4] and Flores [5] investigated the influence of clearance on the structural dynamics, but the dynamic interaction between different clearance joints is not clearly stated. In reality, each joint in a multibody system has a gap, so the interaction of the clearances at different location must be considered.

Therefore, the clearances are introduced into the kinematic pairs at different locations for modeling the dynamics of deployable structure consisting of SLE. Also, the intra-joint forces between the contact bodies are evaluated using the Flores contact model and modified Column law. In the sequel of this process, the resulting general forces are included into the motion equation of the system to model the contact impact process between the journal and bearing.

IOP Conf. Series: Materials Science and Engineering **793** (2020) 012046 doi:10.1088/1757-899X/793/1/012046

2. Equations of motion of planar deployable structure

The deployable structure consisting of two SLEs is presented in Figure 1, where each SLE is composed of two bars and can be rotated around the midpoint. Joints A and D are real joint, other joints are ideal joints. This structure has only one degree of freedom and is subject to a horizontal load at point F, 100 N.

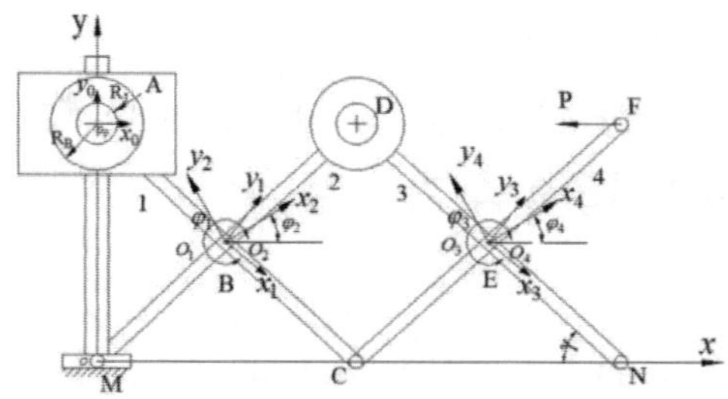

Figure 1 Scissor linear array deployable mechanism with clearance

The motion equation of the deployable structure can be expressed as:

$$\begin{bmatrix} \mathbf{M} & \mathbf{\Phi}_q^{\mathrm{T}} \\ \mathbf{\Phi}_q & \mathbf{0} \end{bmatrix} \begin{bmatrix} \ddot{\mathbf{q}} \\ \boldsymbol{\lambda} \end{bmatrix} = \begin{bmatrix} \mathbf{Q} \\ -(\mathbf{\Phi}_{\mathbf{q}}\dot{\mathbf{q}})_{\mathbf{q}}\dot{\mathbf{q}} - 2\mathbf{\Phi}_{\mathbf{q}t}\dot{\mathbf{q}} - \mathbf{\Phi}_{tt} \end{bmatrix}, \quad (1)$$

where $\mathbf{M}$ and $\mathbf{\Phi}_{\mathbf{q}}$ are respectively the mass matrix and Jacobian matrix, $\mathbf{q}$ is the generalized coordinate, the vectors $\dot{\mathbf{q}}$ and $\ddot{\mathbf{q}}$ separately the first and second derivatives of $\mathbf{q}$ with respect to time.$\mathbf{Q}$ and λ are the generalized force vector and Lagrange multiplier vector. In order to avoid the constraint violation, Baumgarte method is used where the violation parameters α and β are taken as 5.

3. Modeling the revolute clearance joint

During the analysis, it is assumed that the revolute joint consists of two components, journal and bearing, shown in Figure 2.

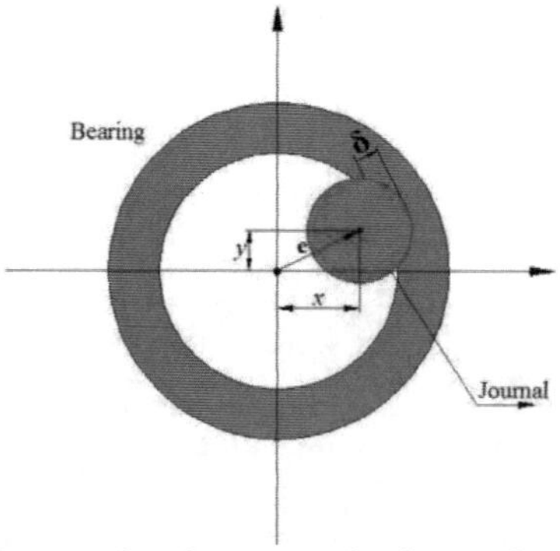

Figure 2 Penetration between the journal and bearing

The penetration δ shown in Figure 2 can be obtained : $\delta = e - c$, where c is the radial clearance size. e is the eccentricity between the centers of journal and bearing. Also, δ can be used to determine the positional relationship between the connecting bodies. When $\delta < 0$, it indicates that the contact bodies are in a free movement state. On the contrary, the contact deformation takes place between the connected components.

Compared with the Lankarani-Nikravesh model commonly used, Flores model broadens the application range of the Hertz law and has more stable dynamic solution, it has obtained good numerical results in the dynamic solution of soft materials and hard materials.

$$F_N = K\delta^n[1+\frac{8(1-c_r)}{5c_r}\frac{\dot{\delta}}{\dot{\delta}^{(-)}}], \tag{2}$$

where δ and $\dot{\delta}$ are respectively the indentation depth and relative normal contact velocity of the contacting bodies. $\dot{\delta}^{(-)}$ is the initial impact velocity. K and n are respectively the contact stiffness parameter.

On the other hand, when different bodies where there are contacting surfaces have relative sliding motion the friction may likely appear in joints. In order to describe the friction phenomenon, the modified Coulomb friction model is utilized which effectively prevents the unstable numerical integration operation caused by the change of friction direction when the tangential velocity is near 0.

$$F_t = -\mu_d c_d F_N \boldsymbol{v}_t, \tag{3}$$

where F_N is the normal intra-joint force obtained from Equation (2), μ_d and c_d denote the slide friction coefficient and dynamic modification coefficient, and $\boldsymbol{v}_t$ is relative tangential velocity.

4. Simulation and discussion

Figure 1 is used as an example to verify how the joint clearance affects the dynamic behavior of the deployable structure, where the bars in this structure have the same length and mass, 566 mm and 3.7 kg, and the endpoint F is subject to the horizontal load, 100 N. At the beginning of the simulation, the bearing and journal coincide, and the parameters required to solve the dynamic model are shown in Table 1.

A typical planar deployable structure as shown in Figure 1 is used as a demonstrative example to study the parametric effect of a revolute clearance joint on the dynamic response of the mechanical system. The bars in the SLE are of the same length and mass which are respectively 566mm and 3.7kg. The end F of bar 4 bears the horizontal constant force which is taken as 100 N, and the journal and bearing centers coincide. The parameters used for the numerical calculations required to solve the mechanism dynamics are listed in Table 1.

Table 1. Parameters required to solve the dynamic model of the deployable structure.

Bearing radius	10.0mm	Poisson's ratio	0.29
Restitution coefficient	0.7	Density	7.8g/cm^3
Young's modulus	206.8GPa	Nonlinear power exponent	3/2

According to the initial parameters, Figures 3 and 4 show when the revolute joints A and D have different clearance sizes, the dynamic response of the center of mass of bar 1 with respect to the ideal mechanism. The clearance values of joints A and D in Figure 3 are 0.2mm and 0.5mm respectively, the situation in Figure 4 is just the opposite.

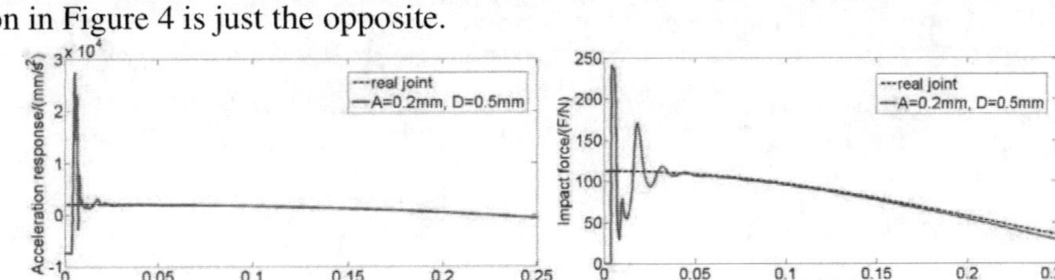

Figure 3 (1) Acceleration response (2) Impact force at the clearance joint. A=0.2mm, D=0.5mm

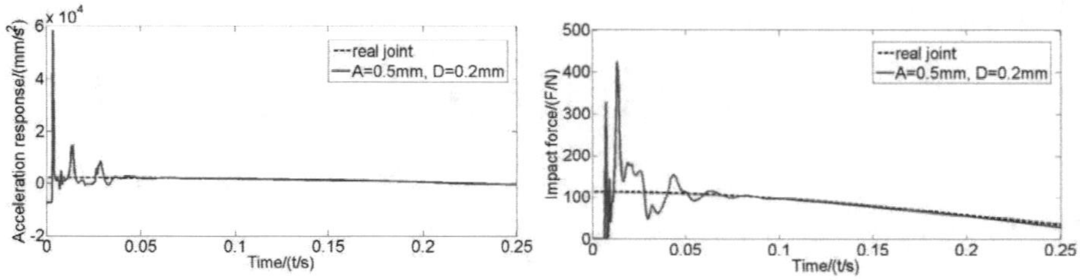

Figure 4 (1) Acceleration response (2) Impact force at the clearance joint. A=0.5mm, D=0.2mm

It can be seen from Figures 3 and 4 that as the clearance value increases in joint A, the dynamic characteristics of the center of mass of bar 1 change more obviously. Compared with Figure 3, the acceleration response and impact force amplitude in Figure 4 show a more pronounced high frequency oscillation, which indicates that the influence of the clearance coupling between the different joints on the dynamic performance of mechanical system cannot be ignored.

5. Conclusions

According to the simulations, it can be concluded that the coupling between the joint clearance at different positions is complicated, and the dynamic behavior of the mechanical system become irregular. For the deployable structure, when the revolute joints have different clearance values, the increase or decrease of the structural dynamic behavior is consistent with those of the clearance size at joint A. On the contrary, when the clearance size at joint D is increased, the dynamic behavior of this structure is suppressed, which indicates that the clearance in joint D is the main factor causing the nonlinearity of the deployable structure, and it should be analyzed in more detail in order to optimize the structure.

Acknowledgments

The authors gratefully acknowledge the financial support of the Natural Science Basic Research Plan in Shaanxi Province of China (Program No. 2019JQ-753) and National Natural Science Foundation of China (Grant No.51175422).

Conflict of Interests

The authors declare that there is no conflict of interests regarding the publication of this paper.

References

[1] Li B, Wang S M, Zhi C J, et al. (2016) Analytical and numerical study of the buckling of planar linear array deployable structures based on scissor-like element under its own weight. Mechanical Systems & Signal Processing, 83:474-488.

[2] Li B, Wang S. M, Yuan R, et al. (2016) Dynamic characteristics of planar linear array deployable structure based on scissor-like element with joint clearance using a new mixed contact force model.P I MECH ENG C-J MEC, 230(18): 3161-3174.

[3] Chen Y, Peng R, You Z. Origami of thick panels. Science, 349(6246):396-400.

[4] Yan S, Shen Y, Chen H. (2003) Dynamic performance of deployable structures with flexible members and clearance connections. Journal of Tsinghua University, 43(2): 145-148.

[5] Flores P, Ambrósio J. (2014) Revolute joints with clearance in multibody systems. Computers & structures, 82(17): 1359-1369.

IOP Conf. Series: Materials Science and Engineering **793** (2020) 012047 doi:10.1088/1757-899X/793/1/012047

Gas drainage radius evaluation and well pattern optimization of multilayer gas reservoirs

Zhijun Liu[1,2] Zhenglin Mao[3] Haobo Zhang[1,2] Yongbin Zhang [4] Qian Liu[1,2] Zhonglin Wen [4]

[1]Petroleum Exploration and Development Research Institute of PetroChina Changqing Oilfield Company;

[2]China National Engineering Laboratory for Exploration and Development of Low Permeability Oil and Gas Fields

[3]State Key Laboratory of Oil and Gas Reservoir Geology and Exploitation, Southwest Petroleum University, Chengdu 610500, China

[4]Qinghai Oilfield Company, PetroChina, Dunhuang 736200, China

Abstract. In the development process of shenmu gas field, there are "few wells and low production of reserves; Many wells, high development investment and easy cross-well interference ", it is particularly important to optimize the well spacing of development well pattern. In view of this problem, on the basis of deepening geological understanding, it is necessary to clarify the gas drainage radius of each well, and combine various methods to demonstrate and optimize the well spacing of gas field well pattern.

1. Introduction

Shenmu gas field is a multilayer gas reservoir with no prominent longitudinal predominance.At present, there is a lack of productivity evaluation method for single well of multi-layer system, and the scope of discharge and control factors of small layer of gas well are not clear.

Based on the results of controlling reserves and reservoir physical characteristics of gas Wells of different layers, the gas drainage radius of each layer in typical block is clarified, and the reasonable well spacing is calculated and analyzed.The current method of calculating radius of gas discharge mainly has four kinds[1]: the pressure drop funnel method, well control reserves of backstepping method, frustrated radius analysis method and economic limit well spacing method.

In order to solve the corresponding key technical problems, this paper calculates the deflating radius of shenmu gas field by combining various methods, so as to determine the contribution degree of stratified productivity of multilayer gas reservoirs, and puts forward the optimization scheme of well pattern of multilayer gas reservoirs.

2. Discharge range evaluation method

2.1. pressure drop funnel method

The pressure square variance (P_e^2-P_{wf}^2) was calculated by using the output formula of the radial flow gas well in the form of pressure square:

$$q_{sd} = \frac{774.6Kh(p_e^2 - p_{wf}^2)}{T\bar{\mu}\bar{z}\ln(\frac{r_e}{r_w})} \quad (1)$$

Where q_{sd} is the single well productivity of the perforating unit,$10^4 m^3/d$; K is the permeability of the gas layer,10^{-3} m^2; h is the thickness of the gas layer,m; P_e is the supply boundary pressure,MPa; P_{wf} is bottom hole pressure (MPa); T is the gas layer temperature,K; $\bar{\mu}$ is the gas viscosity under average pressure and temperature,mPa·s; $\bar{z}$ Is the gas deviation factor under average pressure and temperature; r_e is the supply boundary radius,m; r_w is the gas well radius,m.

The square distribution expression of stable seepage pressure expressed in the form of pressure square is as equation (2).

$$p^2 = p_e^2 - \frac{p_e^2 - p_{wf}^2}{\ln\frac{r_e}{r_w}}\ln\frac{r_e}{r} \quad (2)$$

Take the value between the bottom hole flow pressure and the formation pressure, calculate the corresponding r value according to equation (2), and substitute it into the formula to calculate the pressure gradient dp/dr at any point:

$$\frac{dp}{dr} = \frac{p_e^2 - p_{wf}^2}{\ln\frac{r_e}{r_w}}\frac{1}{2pr} \quad (3)$$

Where : dp/dr is the pressure gradient at any point in the gas layer.

When the pressure gradient[2] is less than the set value, it can be determined that the change of formation pressure tends to slow down, and the corresponding radius r is the deflating radius d.

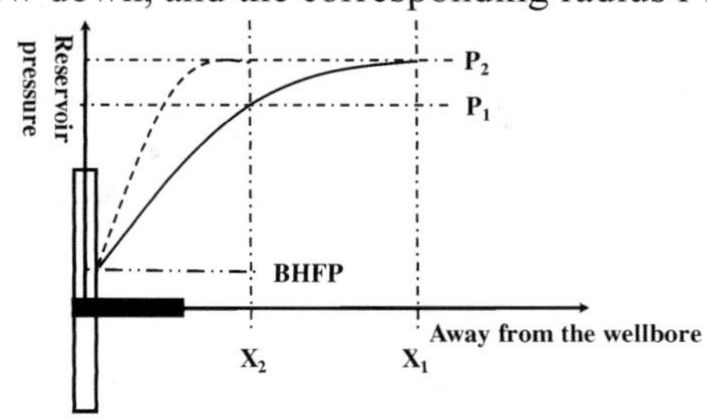

Figure 1 schematic diagram of pressure drop funnel

2.2. the method of well control dynamic reserve backdating

In this method, the dynamic reserves of a single well are calculated by pressure drop method, and then the control area of a single well is deduced reversely according to the reserves calculation formula of volume method, and the control radius, that is, the control range of a single well, is obtained.

$$A = \frac{G_0}{0.01\phi S_{gi}E_{gi}h} \quad (4)$$

Where :G_0 is well-controlled dynamic reserves,10^8 m^3; A is the controlled gas-bearing area,km^2; h is the mean effective thickness,m; φ is the average effective porosity, decimal; S_{gi} is the average original gas saturation, decimal; E_i is the expansion coefficient of natural gas.

2.3. analysis of gas radius

According to the theory of unstable seepage flow in natural gas, when gas well unstable seepage, radius of gas well discourage had semilog relationship with permeability[3]. When reach the quasi steady seepage, experience in statistical analysis of low permeability gas reservoir, frustrated radius with reservoir permeability is shown in Figure 2 semilog relationship. According to the shenmu gas permeability distribution histogram (Figure 3), you can see, shenmu gas permeability of the reservoirs are mainly concentrated in the mD (0.2 ~ 0.5), accounting for 78% of the total percentage.

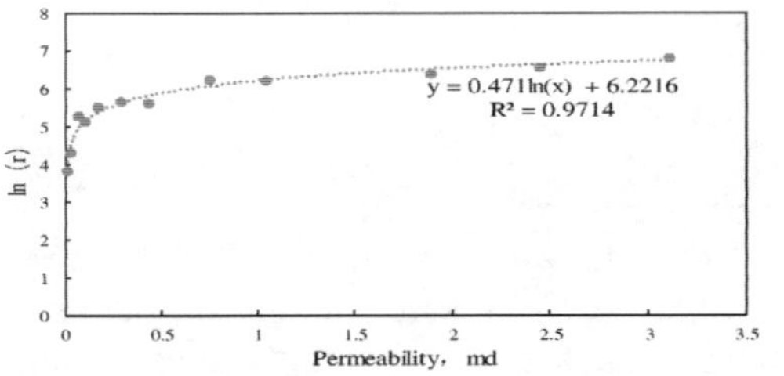

Figure 2 relationship between discharge radius and reservoir permeability

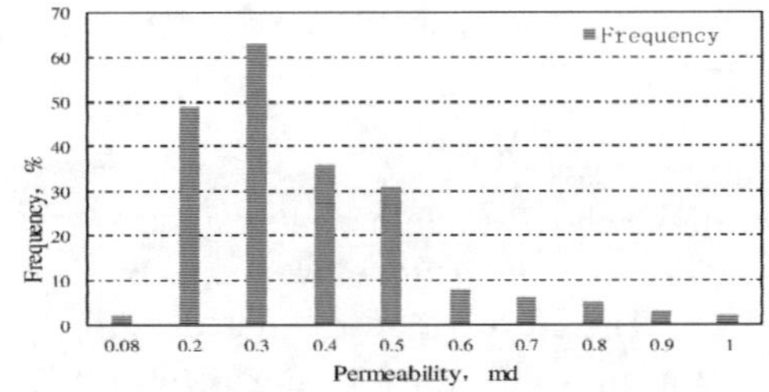

Figure 3.cumulative frequency distribution of reservoir permeability in shenmu gas field

2.4. economic limit well spacing

According to the balance principle of input and output, the economic limit control reserves of a single well are estimated. Based on the reserves of each development group and the control area of well pattern, the maximum economic limit Wells, well spacing and economic limit well pattern density are calculated.

2.4.1. Economic limit well control reserves

It means that after all the recoverable reserves controlled by a single well have been produced and all of them have been sold, the revenue generated will be used to exactly cover all the investment needed for development and all the taxes for development. The formula is as follows:

$$G_{sg} = \frac{C + tP}{A_G \times E_R} \quad (5)$$

Where :A_G is the selling price of natural gas,0.711 yuan /m^3; C is the total cost of single well drilling and oil construction; E_R is the natural gas recovery factor, estimated at 0.54; P is the average annual operating cost of gas recovery per well (648,181 yuan/year · well); t is the mining life,25 years; G_g is the economic limit well controlled reserves,10^{8}m^3.

2.4.2. Economic limit well spacing

Economic limit well spacing refers to the average well spacing when the value of recoverable reserves controlled by a single well is equal to the input cost of gas field construction and single well of gas extraction under certain development and geological conditions.

Formula for calculating economic limit well spacing:

$$d = 2\left(G_{sg} \times \frac{A \times S_p}{G \times R_{sg}} / \pi\right)^{1/2} \quad (6)$$

Where, A is the distribution area,m^2; S_p is the distribution area/gas-bearing area, and the value of weak edge water reservoir is 0.5. G is proved geological reserves of natural gas,m^3; d is the economic limit well spacing,m.

Economic limit well pattern density and reasonable well pattern density should be involved in the process of drilling infiling Wells. The economic limit and economic optimal well pattern density of different types under different recovery rates are calculated by gas reservoir engineering method, and the reasonable well pattern density is calculated by adding 1/3 difference method[4].

Table 1. results of reasonable well pattern density and well spacing in vertical/directional Wells

Recovery / %	Pattern density/ A well • km^{-2}			Reasonable spacing
	Economic limit method	Economic best practice	Plus one third difference	
30	2.69	2.54	2.59	538×717m
35	3.14	2.93	3.00	500×666m
40	3.59	3.32	3.41	469×625m
45	4.04	3.69	3.81	444×592m
50	4.49	4.06	4.20	422×563m

The calculated results show that the reasonable well spacing at this stage is (500×600)m for enhanced recovery.

3. Case analysis

Through the above methods, an example of shenmu gas field is analyzed.

The pressure drop funnel method was used to analyze 18 Wells with an average deflating radius of 245.79m. Therefore, the calculated well spacing was about 491.58m.According to the reverse calculation of well-controlled dynamic reserves, it is found that the overall gas drainage radius is 170.8m-254.9m, so the approximate range of reasonable well spacing is 340m-510m.Combined with the empirical method, the relationship between the discharge radius and the reservoir permeability curve, the gas well deflation radius 210m-350m and the reasonable well spacing range 420m-700m were finally determined.The economic limit well spacing is calculated and analyzed to obtain the well spacing under different recovery rates. In order to improve the recovery rate, the reasonable well spacing at the present stage is 500×600m.

Various methods were used to analyze the discharge range and reasonable well spacing in the target block.The results show that the gas deflection radius is about 180m-300m and the reasonable well spacing is about 360m-600m.The research results can be used to guide the well pattern adjustment analysis of the target block, facilitate the fine management of small layers, and improve the overall development benefit.

References

[1] Kai Zhang, Yuxue Chen and Liming Zhang 2015 Well pattern optimization using NEWUOA algorithm *Journal of Petroleum Science and Engineering*. **134** 257-272

[2] Tang LG, Ding GS, Sun CH and Wang JM 2018 Mathematical description of gas drainage radius for underground gas storage *Chemistry and Technology of Fuels and Oils*. **54** 500-508

[3] Quanyou Liu, Zhijun lin and Qingqiang Meng 2015 Genetic types of natural gas and filling patterns in Daniudi gas field, Ordos Basin, China *Journal of Asian Earth Sciences*. **107** 1-11

[4] Dongbo He, Alba Jia and Guang Ji 2013 Well type and pattern optimization technology for large scale tight sand gas, Sulige gas field, NW China *Petroleum Exploration and Development*.**40** 84-95

IOP Conf. Series: Materials Science and Engineering **793** (2020) 012048 doi:10.1088/1757-899X/793/1/012048

Mechanism design and kinematics simulation of 3-DOF parallel mechanism with low coupling degree

J Xie[1], F F Zuo[1], Y P Li[1] and Q Z Yang[2]

[1] School of Mechanical Engineering, Jiangsu University, Zhenjiang, 212013, China

[2] School of Agricultural Equipment Engineering, Jiangsu University, Zhenjiang, 212013, China

Abstract. According to the topological design theory of parallel mechanism(PM) based on the position and orientation characteristic(POC), a new PM with two translational degree of freedom(DOF) and one rotational DOF (2T1R) was proposed. The topological characteristics of the PM was analyzed, including POC set, DOF and coupling degree. Based on analytic method, the solving modeling was established, and the forward and inverse solutions were obtained. Then the correctness of the position solution was verified by Matlab. Velocity and acceleration formulas of PM were calculated by Jacobian matri, and simulation curves of velocity and acceleration was put forword. Kinematic analysis shows that the 2T1R PM has both translational and rotational characteristics, moreover, the moving platform has stable output and good dynamic characteristics.

1. Introduction

The 2T1R PM with has the characteristics of both movement and rotation. Due to large stiffness, small inertia and high precision, it has attracted the attention and research of many scholars. Mao[1]proposed the RRC+RRRR+SSR type of 2T1R parallel mechanism;

Wang[2]gave the positive and negative solution equations and mechanism velocity formulas of 2-CU+1PU parallel mechanism; Xie[3]studied the application of configuration synthesis of 2T1R parallel mechanism in manufacturing; Ni[4]proposed a new 2T1R parallel mechanism and obtained the error Jacobian matrix. Mazare[5] calculated the global dexterity index of the 3-[P2(US)] PM. Qu[6]and Liu[7]carried out configuration synthesis of 2T1R mechanism. Deng[8]studied(RPa ∥ 3R)-R+RSS type parasitic motion parallel mechanism.

Most of the existing 2T1R parallel mechanisms have high coupling degree, which brings difficulties to the mechanism design and kinematic analysis. According to the problem mentioned above, a new type of 2T1R parallel mechanism was proposed, which can realize the translation along the x and y axes and rotation about y axes in space. The PM is designed with weak coupling, and the forward and inverse solutions can be obtained by one-dimensional search. In addition, the kinematics characteristics of the PM are analyzed respectively.

2. topology analysis

According to parallel mechanism topology design theory, and based on position and orientation characteristic[9], the 2T1R PM including the moving platform 1, static platform 0 and single open chain I, II, III was put forward.

IOP Conf. Series: Materials Science and Engineering **793** (2020) 012048 doi:10.1088/1757-899X/793/1/012048

As shown in figure 1, the topology of PM is PRR+2-PRRU. And the single open chain(SOC) I($P_1 \| R_1 \| R_2$-U_1) structure is equivalent to ($P_1 \| R_1 \| R_2 \| R_{11} \perp R_{12}$). Similarly, SOC II($P_2 \| R_3 \| R_4$-$U_2$) structure is equivalent to ($P_2 \| R_3 \| R_4 \| R_{21} \perp R_{22}$). And SOC III structure is ($P_3 \| R_5 \| R_6$). The relationship of axis layout in each prismatic pair on the static platform 0 is as follows: the prismatic pairs P_1, P_2 and P_3 are arranged along the x, z and y axis respectively. In addition, the axes of the prismatic pairs P_1, P_2 and P_3 are mutual perpendicular.

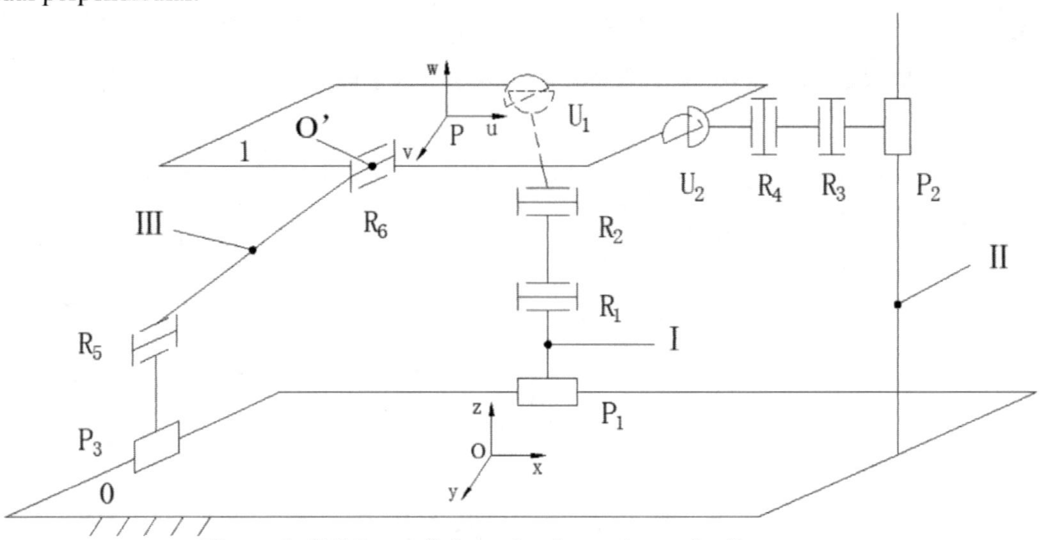

Figure 1. 2T1R parallel mechanism schematic diagram

The POC equation of parallel mechanism[9] are as follows:

$$M_{bi} = \bigcup_{j=1}^{k} M_{sj} \tag{1}$$

$$M_{pa} = \bigcap_{i=1}^{n} M_{bi} \tag{2}$$

where: M_{pa} ——POC set of mechanism moving platform

M_{bi} ——The POC set at the end of the i th chain

M_{sj} ——The POC set of the j th SOC in the branched chain

(1) Topology of the mechanism

As shown in figure 1, the topology structure of SOC_j (j=1,2,3) composed of the PM is:

$SOC_1\{-P_1 \| R_1 \| R_2 - U_1 -\}$

$SOC_2\{-P_2 \| R_3 \| R_4 - U_2 -\}$

$SOC_3\{-P_3 \| R_5 \| R_6 -\}$

On the static platform 0, kinematic pairs P_1, P_2 and P_3 are perpendicular each other; and choose the end point O 'of SOC III as the basis for analysis.

(2) Determine each chain POC set

Equation (1) shows that:

$$M_{b1} = \begin{bmatrix} t^1(\| P_1) \\ r^0 \end{bmatrix} \cup \begin{bmatrix} t^1(\perp R_1) \cup t^1(\perp R_2) \\ r^1(\| R_1) \end{bmatrix} \cup \begin{bmatrix} t^2(\perp \rho) \\ r^2(\| \lozenge P_1P_3) \end{bmatrix}$$

$$= \begin{bmatrix} t^3 \\ r^2(\| \lozenge P_1P_3) \end{bmatrix}$$

$$M_{b2}=\begin{bmatrix} t^1(\| \mathrm{P}_2) \\ r^0 \end{bmatrix} \cup \begin{bmatrix} t^1(\perp \mathrm{R}_3) \cup t^1(\perp \mathrm{R}_4) \\ r^1(\| \mathrm{R}_3) \end{bmatrix} \cup \begin{bmatrix} t^2(\perp \rho) \\ r^2(\| \Diamond \mathrm{P}_2\mathrm{P}_3) \end{bmatrix}$$
$$=\begin{bmatrix} t^3 \\ r^2(\| \Diamond \mathrm{P}_2\mathrm{P}_3) \end{bmatrix}$$
$$M_{b3}=\begin{bmatrix} t^1(\| \mathrm{P}_3) \\ r^0 \end{bmatrix} \cup \begin{bmatrix} t^1(\perp \mathrm{R}_5) \\ r^1(\| \mathrm{R}_5) \end{bmatrix} \cup \begin{bmatrix} t^1(\perp \mathrm{R}_6) \\ r^1(\| \mathrm{R}_6) \end{bmatrix}$$
$$=\begin{bmatrix} t^2 \\ r^1(\| \mathrm{P}_3) \end{bmatrix}$$

Freedom degree formula[9] of PM is :

$$F=\sum_{i=1}^{m} f_i - \sum_{j=1}^{v} \xi_{Lj} \tag{3}$$

Among them:

$$\sum_{j=1}^{v} \xi_{Lj} = \dim\left\{\left(\bigcap_{i=1}^{j} M_{bi}\right) \cup M_{b(j+1)}\right\} \tag{4}$$

where:

F——freedom in mechanism

f_i——the i th degree of kinematic pair

m——number of kinematic pair

n——number of component

v——independent loop number, where $v=m-n+1$

ξ_{Lj}——the number of independent equations of the j th loop

$\bigcap_{i=1}^{j} M_{bi}$——the subordinative parallel mechanism POC set composed of the j th chain

$M_{b(j+1)}$——the POC set of the terminal component of $j+1$ chain

(1) Determine the ξ_{Lj} of each chain

The mechanism can be decomposed into two independent loops :

$\mathrm{SOC}_1\{-\mathrm{P}_1 \| \mathrm{R}_1 \| \mathrm{R}_2 - \mathrm{U}_1 - \mathrm{P}_3 \| \mathrm{R}_5 \| \mathrm{R}_6 -\}$

$\mathrm{SOC}_2\{-\mathrm{P}_2 \| \mathrm{R}_3 \| \mathrm{R}_4 - \mathrm{U}_2 -\}$

subordinative PM consisted by SOC I and SOC III is the first independent loop, It can be obtained from equation(4):

$$\xi_{L1} = \dim.\{M_{b1} \cup M_{b3}\}$$
$$=\dim.\left\{\begin{bmatrix} t^3 \\ r^2(\| \Diamond \mathrm{P}_1\mathrm{P}_3) \end{bmatrix} \cup \begin{bmatrix} t^2 \\ r^1(\| \mathrm{P}_3) \end{bmatrix}\right\}=5$$

The POC set of the subordinative mechanism is:

$$M_{pa(1-3)} = M_{b1} \cap M_{b3}$$
$$=\begin{bmatrix} t^3 \\ r^2(\| \Diamond \mathrm{P}_1\mathrm{P}_3) \end{bmatrix} \cap \begin{bmatrix} t^2 \\ r^1(\| \mathrm{P}_3) \end{bmatrix}$$
$$=\begin{bmatrix} t^2 \\ r^1(\| \mathrm{P}_3) \end{bmatrix}$$

The subordinative PM and the SOC II constitute the second independent loop, It can be obtained from equation(4):

$$\xi_{L2} = \dim.\{M_{pa(1-3)} \cup M_{b2}\}$$

$$= \dim.\left\{\begin{bmatrix} t^2 \\ r^1(\| P_3) \end{bmatrix} \cup \begin{bmatrix} t^3 \\ r^2(\| \lozenge P_2 P_3) \end{bmatrix}\right\} = 5$$

(2) Determine the DOF of the PM

The DOF of the parallel mechanism can be obtained from equation (4):

$$F = \sum_{i=1}^{11} f_i - \sum_{j=1}^{2} \xi_{Lj} = 13 - 10 = 3$$

Therefore, the DOF of the PM is 3, and kinematic pairs P_1, P_2 and P_3 on the static platform 0 are selected as the driving pair according to the criteria for driving pair subexistence[9].

3. Coupling degree κ

According to the mechanism composition principle based on SOC, arbitrary components can be decomposed into several basic kinematics chains (BKC). The BKC of v independent loops can be further decomposed into $\mathrm{SOC}(\Delta_j)$ ($j=1,2,\ldots,v$), and the constraint degree of the j th SOC_j is defined as:

$$\Delta_j = \sum_{i}^{m_j} f_i - I_j - \xi_{Lj} \tag{5}$$

Among them:

$$\Delta_j = \begin{Bmatrix} \Delta_j^+ = 1,2,\ldots \\ \Delta_j^0 = 0 \\ \Delta_j^- = -1,-2,\ldots \end{Bmatrix}$$

where: m_j——the prismatic pairs of the jth SOC_j

I_j——the number of driving pairs of the jth SOC_j

Δ_j has positive, zero and negative forms, and must meets $\sum \Delta_j = 0$. Therefore, the coupling degree of BKC is defined as:

$$\kappa = \frac{1}{2} \min \sum_{j=1}^{v} |\Delta_j| \tag{6}$$

Where, BKC can be decomposed into several allocation schemes, $\min \sum_{j=1}^{v} |\Delta_j|$ should be taken as the smallest.

Coupling degree[9] κ reflects the correlation and dependence degree between variables in BKC. The higher value of κ is, the more complex kinematics and dynamics problem is. when $\kappa=0$, kinematics and dynamics analysis of the BKC can be solved separately. when $\kappa>0$, the kinematics and dynamics analysis of the BKC requires simultaneous solution of multiple loops.

According to section 1, the number of independent equations of the two loops is $\xi_{L1}=\xi_{L2}=5$, and the number of driving pairs of each loop is $I_1=2$, $I_2=1$, so the constraint degree Δ_1, Δ_2 are:

$$\Delta_1 = \sum_{i=1}^{7} f_i - I_1 - \xi_{L1} = 8 - 2 - 5 = 1$$

$$\Delta_2 = \sum_{i=1}^{4} f_i - I_2 - \xi_{L2} = 5 - 1 - 5 = -1$$

$$\Delta_1 + \Delta_2 = 0 \rightarrow BKC = 1$$

Therefore, the institution only contains one BKC. According to equation (6), the coupling degree κ is 1; So the parallel mechanism can obtain all real solutions of the position problem by one-dimensional search method.

IOP Conf. Series: Materials Science and Engineering **793** (2020) 012048 doi:10.1088/1757-899X/793/1/012048

4. kinematic analysis

As shown in figure 2, both static platform 0 and moving platform 1 are rectangular. They are arranged in parallel. The moving platform 1 is located directly above the static platform 0, and the distance between them is L_0; Moving coordinate system P-{uvw} is established with center P of moving platform 1 as the origin. The u axis is parallel to the length direction of moving platform and points to A_2. The w axis is vertically upward and perpendicular to the moving platform. And the v axis is determined by the right hand rule. Similarly, the fixed coordinate system O-{ xyz } is established with center O of the static platform 0 as the origin.

The size parameters of the parallel mechanism are as follows: the static platform 0 is $2l_1$ in length and $2l_2$ in width; the moving platform 1 is $2l_3$ in length and $2l_4$ in width. Other dimension parameters are: $B_3C_3=l_5$, $A_3B_3=l_6$, $A_2B_2=l_7$, $B_2C_2=l_8$, $C_2D_2=l_9$.

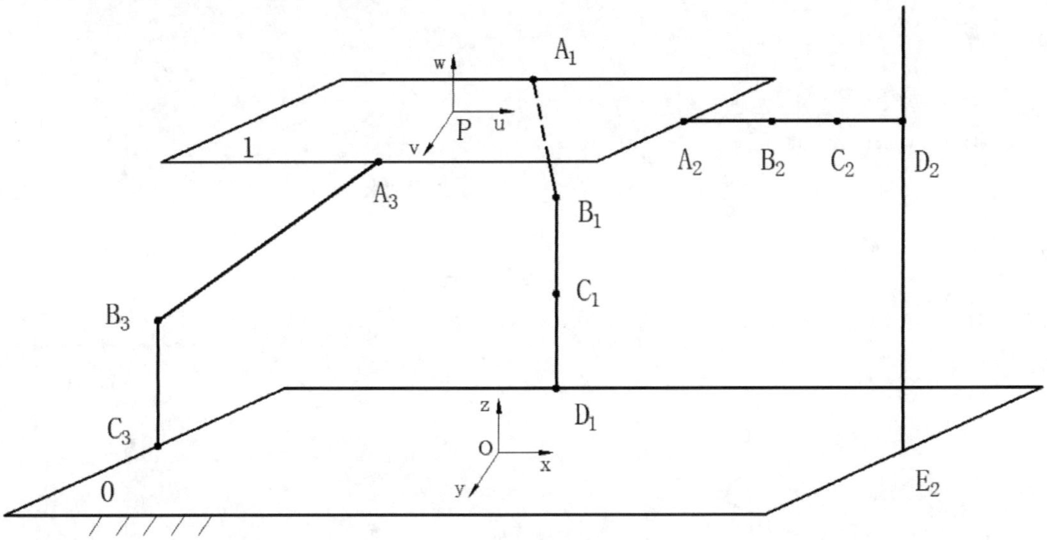

Figure 2. kinematics model of 2T1R parallel mechanism

4.1 Forward kinematics

The forward kinematics of PM means that get the input parameters (x_1, y_1, z_1) to solve the position and pose parameters (x, y, β).

(1) solition of x and y

As shown in figure 1, during the movement of PM, the displacement of the prismatic pair P_1 and P_3 is the displacement of moving platform along the u and v axis, then:

$$\begin{cases} x = x_1 \\ y = y_1 \end{cases} \tag{7}$$

(2) solition of β

As shown in figure 3, According to the closed loop[10]consisted by SOC II and SOC III, the following equations can be established:

$$\begin{cases} l_6 \cos\theta + l_3 \cos\beta + l_7 + l_8 + l_9 = 2l_1 \\ l_6 \sin\theta + l_3 \sin\beta + l_5 = L_0 + z_1 \end{cases} \tag{8}$$

Let $a_1 = 2l_1 - l_7 - l_8 - l_9$, add the sum of squares of equation (8):

$$l_3^2 + a_1^2 + (L_0 + z_1 - l_5)^2 - l_6^2 - 2a_1 l_3 \cos\beta - 2l_3 (L_0 + z_1 - l_5)\sin\beta = 0 \tag{9}$$

Further, equation (9) can be obtained:

$$A\cos\beta + B\sin\beta + C = 0 \tag{10}$$

Where:

$$\begin{cases} A = -2l_3 a_1 \\ B = -2l_3 (L_0 + z_1 - l_5) \\ C = l_3^2 + a_1^2 + (L_0 + z_1 - l_5)^2 - l_6^2 \end{cases}$$

Arranged equation (10):

$$B^2 (1-\cos^2 \beta) = A^2 \cos^2 \beta + C^2 + 2AC \cos \beta \tag{11}$$

Further, equation (11) can be obtained:

$$a \cos^2 \beta + b \cos \beta + c = 0 \tag{12}$$

Where:

$$\begin{cases} a = A^2 + B^2 \\ b = 2AC \\ c = C^2 - B^2 \end{cases}$$

Solution of equation (12) can be obtained as follows:

$$u = \cos \beta = \frac{-b + \sqrt{b^2 - 4ac}}{2a} \tag{13}$$

That is:

$$\beta = \arccos u \tag{14}$$

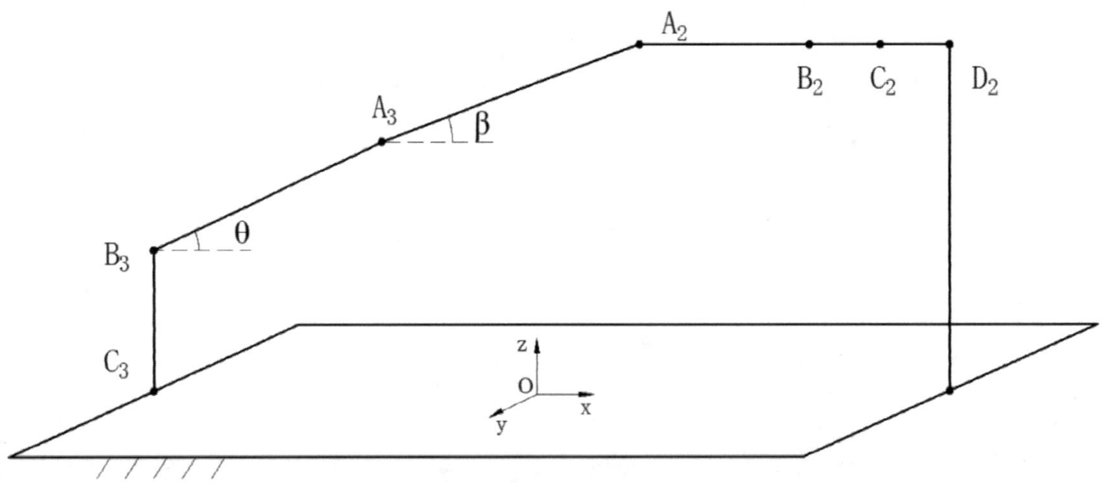

Figure 3. kinematics model of closed loop

4.2 Inverse kinematics

The inverse kinematics of PM means that get the position and pose parameters (x, y, β) to solve the input parameters (x_1, y_1, z_1).

(1) solition of x_1 and y_1

$$\begin{cases} x_1 = x \\ y_1 = y \end{cases} \tag{15}$$

(2) solition of z_1

According to equation (9), it can be known that:

$$(L_0 + z_1 - l_5)^2 + d(L_0 + z_1 - l_5) + e = 0 \tag{16}$$

Where:

$$\begin{cases} d = -2l_3 \sin\beta \\ e = l_3^2 + a_1^2 - l_6^2 - 2a_1 l_3 \cos\beta \end{cases}$$

Solution of equation (16) can be obtained as follows:

$$(L_0 + z_1 - l_5) = \frac{-d + \sqrt{d^2 - 4e}}{2} = u \tag{17}$$

That is:

$$z_1 = u + l_5 - L_0 \tag{18}$$

4.3 Numerical example

The structure parameters of PM are set as follows： l_1=400mm, l_2=370mm, l_3=150mm, l_4=75mm, l_5=185mm, l_6=350mm, l_7=165mm, l_8=60mm, l_9=210mm, L_0=415mm. Matlab software was used to verify the correctness of the forward and inverse kinematics equations. And the simulation steps were as follows:

(1) set input parameters (x_1, y_1, z_1)

(2) substitute the input parameters (x_1, y_1, z_1) into the forward solution model and calculate the forward solution parameters

(3) substitute the forward solution parameters into the inverse solution model and calculate the inverse solution parameters (x_1', y_1', z_1')

(4) calculate the difference between the inverse solution parameters (x_1', y_1', z_1') and the input parameters (x_1, y_1, z_1)

(5) according to the difference judge the correctness of kinematic analysis

After above simulation steps, the numerical example of kinematic analysis can be obtained as shown in table 1.

Table 1. example of kinematic analysis

serial number	input parameters (x_1, y_1, z_1)	inverse solution parameters (x_1', y_1', z_1')	Difference
1	(165,90,55)	(165,90,49.9999)	(0,0,0.0001)
2	(170,95,55)	(170,95,55)	0
3	(175,100,60)	(175,100,60.0001)	(0,0,0.0001)
4	(180,105,65)	(180,105,64.9999)	(0,0,0.0001)
5	(185,110,70)	(185,110,70.0001)	(0,0,0.0001)

As shown in table 1, It can be found that the inverse solution parameters (x_1', y_1', z_1') are approximately equal to the input parameters (x_1, y_1, z_1), which verifies the correctness of the position analysis.

5. velocity and acceleration analysis

5.1 Solution of velocity formula

By taking the first derivative of equation (7) and equation (9) by time, the velocity formulas of the three driving sliders can be obtained:

$$\begin{cases} \dot{x} = \dot{x}_1 \\ \dot{y} = \dot{y}_1 \\ D\dot{\beta} = E\dot{z}_1 \end{cases} \tag{19}$$

Where:

$$\begin{cases} D = a_1 l_3 \sin\beta - l_3 \cos\beta (L_0 + z_1 - l_5) \\ E = l_3 \sin\beta - (L_0 + z_1 - l_5) \end{cases}$$

According to equation (19), the formula based on Jacobian matrix can be obtained:

$$J_P \dot{p} = J_q \dot{q} \tag{20}$$

Where $\dot{p} = (\dot{x}, \dot{y}, \dot{\beta})^T$ is the output velocity vector, $\dot{q} = (\dot{x}_1, \dot{y}_1, \dot{z}_1)^T$ is the input velocity vector, J_p and J_q are the forward and inverse Jacobian matrix of PM respectively.

Among them:

$$J_p = \begin{bmatrix} 1 & 0 & 0 \\ 0 & 1 & 0 \\ 0 & 0 & D \end{bmatrix}$$

$$J_q = \begin{bmatrix} 1 & 0 & 0 \\ 0 & 1 & 0 \\ 0 & 0 & E \end{bmatrix}$$

When there is no singular position of the mechanism, namely J_p reversible, the velocity solution of PM is:

$$\dot{p} = J_p^{-1} J_q \dot{q} \tag{21}$$

5.2 Solution of acceleration formula

The first derivative of each variable in equation (20) by time, the acceleration formulas can be obtained as follows:

$$\dot{J}_p \dot{p} + J_P \ddot{p} = \dot{J}_q \dot{q} + J_q \ddot{q} \tag{22}$$

Where:

$$\begin{cases} \ddot{p} = (\ddot{x}, \ddot{y}, \ddot{\beta})^T \\ \ddot{q} = (\ddot{x}_1, \ddot{y}_1, \ddot{z}_1)^T \\ \dot{J}_p = \begin{bmatrix} 0 & 0 & 0 \\ 0 & 0 & 0 \\ 0 & 0 & \dot{D} \end{bmatrix} \\ \dot{J}_q = \begin{bmatrix} 0 & 0 & 0 \\ 0 & 0 & 0 \\ 0 & 0 & \dot{E} \end{bmatrix} \end{cases}$$

$$\begin{cases} \dot{D} = a_1 l_3 \dot{\beta} \cos\beta + l_3 \left(L_0 + z_1 - l_5\right) \dot{\beta} \sin\beta - l_3 \dot{z}_1 \cos\beta \\ \dot{E} = l_3 \dot{\beta} \cos\beta - \dot{z}_1 \end{cases}$$

When there is no singular position of the mechanism, namely J_p reversible, the acceleration solution of PM can be obtained:

$$\ddot{p} = J_p^{-1} \left(\dot{J}_q \dot{q} + J_q \ddot{q} - \dot{J}_p \dot{p} \right) \tag{23}$$

5.3 Simulated analysis

In order to further understand the output characteristic of moving platform, the 3D model of PM built by Ug software imported into Adams software.

According to the simulation, the translational velocity $\dot{x}$, $\dot{y}$ and angular velocity $\dot{\beta}$ were obtained as shown in figure 4. As shown in figure 5, translational acceleration $\ddot{x}$, $\ddot{y}$ and angular acceleration $\ddot{\beta}$ were also obtained.

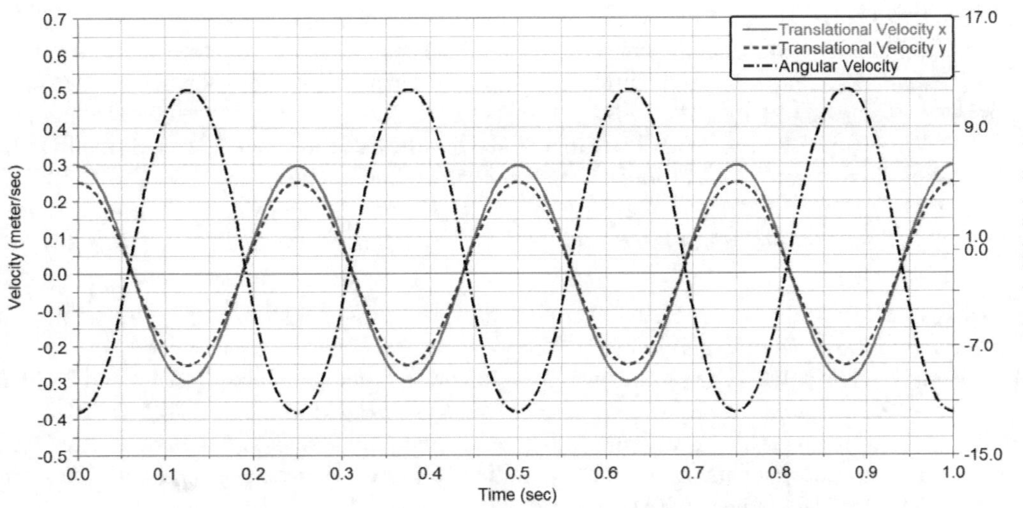

Figure 4. velocity of moving platform verified by Adams

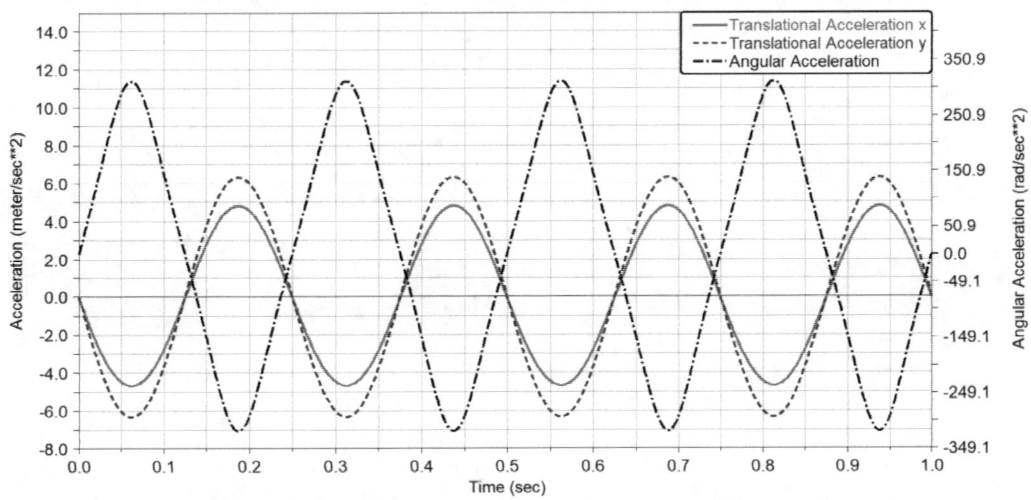

Figure 5. acceleration of moving platform verified by Adams

It can be found that the velocity and acceleration curves of the PM change steadily without sharp points and have good dynamic performance.

6.Conclusion

To simplify kinematics analysis of PM, a new 2T1R weakly coupled PM is designed, which can realize the translation along the x and y axes and rotation along the y axes. The weakly coupled design of the PM can reduce the number of simultaneous kinematic branches required for the analysis of the forward and inverse kinematics solutions, and it simplify the complexity of kinematics analysis.

Moreover, kinematics modeling of the PM was carried out to obtain the forward and inverse solutions. The correctness of kinematic analysis of the mechanism was verified by Matlab software. The simulation analysis of velocity and acceleration shows that the PM has good dynamic performance.

References

[1] Mao J H, Gao G Q, Kuai S S and Zhao D A 2007 The research on the 3-DOF parallel mechanism robot *Machine Tool & Hydraulics*. **2007**(12) 48-51.

[2] Wang H D, Bi Y Q, Yang B H and Huang K 2010 Kinematics analysis of parallel mechanisms with two translations and one rotation *Science Technology and Engineering*. **10**(35) 826--

8829+8833.

[3] Xie F G, Liu X J, You Z and Wang J S 2014 Type synthesis of 2T1R-type parallel kinematic mechanisms and the application in manufacturing *Robotics & Computer Integrated Manufacturing*. **30**(1) 1-10.

[4] Ni Y B, Wu N and Zhong X Y 2015 Dimensional synthesis of a 3-DOF parallel manipulator with full circle rotation *Chinese Journal of Mechanical Engineering*. **28**(4) 830-840.

[5] Mazare M, Taghizadeh M and Najafi R M 2017 Kinematic analysis and design of a 3-DOF translational parallel robot *International Journal of Automation and Computing*. **2017**(4) 432--441.

[6] Qu S W, Li R Q and Guo Z H 2017 Type synthesis of a 3-DOFs decouple parallel mechanisms *Machine Design & Research*. **2017**(1) 38-41.

[7] Liu W and Liu H Z 2018 Type synthesis of 3-DOF parallel mechanism with 2T1R and 2R1T motion mode. *Transactions of the Chinese Society for Agricultural Machinery*. **49**(7) 401-409.

[8] Deng J M, Xu K and Zhao Y C 2018 Design and kinematics analysis of asymmetric 2T1R-type spatial parallel mechanism without parasitic motion *Transactions of the Chinese Society for Agricultural Machinery*. **49**(6) 408-417.

[9] Yang T L, Liu A X and Luo Y F 2012 *Topological Structure Design of Robot Echanisms* Beijing:Science Press.

[10] Xie J, Zhang X Q and Wang C J 2012 Tea sieving based on three translation parallel mechanism *Transactions of the Chinese Society for Agricultural Machinery.* **43**(10):101-104.

Research on Air Pressure Cycle Test of Liquefied Petroleum Gas Cylinder

Songqing Jia[1], Bing Han[2], Xingtao Dai[2], Ke Bo[3], Zhihao Wei[1] and Hongxin You[1*]

[1]Department of Chemical Engineering, Dalian University of Technology, Dalian, Liaoning, 116024, China

[2]Dalian Boiler and Pressure Vessel Inspection & Testing Institute Co., Ltd, Dalian, Liaoning, 116013, China

[3]China Special Equipment Inspection and Research Institute, Beijing, 100029, China

[*]E-mail: youhx@sina.com

Abstract. Air pressure cycle tests were performed on liquefied petroleum gas (LPG) high-density polyethylene liner glass fiber fully wound cylinders. During the repeated filling and discharging of the gas cylinder, the changes of the internal and external surface temperature and internal pressure of the gas cylinder with time were obtained. Through experimental analysis and numerical simulation, the characteristics of the flow field in the cylinder during the pressure cycle test at different pressure cycle rates were studied. The results show that the temperature in the gas cylinder rose steadily during the test. The temperature rise trend of the gas cylinder in Fluent simulation was the same as that in the test. In the air pressure cycle test, the temperature change of the cylinder was larger than that of the cylinder when it was filled with LPG, and the structure and material properties of the cylinder were unchanged. This test can check whether the LPG high-density polyethylene liner glass fiber fully wound cylinders are safe.

1. Introduction

Liquefied petroleum gas (LPG), as a clean energy, has been in a steady growth stage in China [1]. With the laying of domestic natural gas pipelines, the modernization of cities and the increase in natural gas imports, civilian LPG consumption has been impacted. However, due to the decline in international oil prices at the end of 2014 and the strict implementation of safety and environmental protection requirements, dimethyl ether, an alternative to LPG, has gradually lost its original market, instead promoting the recovery of LPG in the civilian sector. In the first-tier and second-tier cities in China, the floating population is relatively large, and LPG cylinders have a rigid demand. The demand for liquefied petroleum gas in the field of civil combustion has maintained a good trend. With the development of ethylene production and processes such as aromatization, alkylation, methyl tert-butyl ether (MTBE) and propane dehydrogenation (PDH), LPG has developed rapidly in the field of industrial raw materials, and LPG consumption space has ushered in huge increase.

There is a good prospect for LPG high-density polyethylene liner glass fiber fully wound cylinder. Since the supply of carbon fiber in 1998 increased [2], its price has started to fall, making the cost of manufacturing carbon fiber composite cylinders even more attractive. Compared with LPG steel cylinders, these new gas cylinders have the following advantages [3]: (1) Light weight. The specific strength of the fiber material is very high, which is about 50% to 75% of the titanium alloy cylinder.

And at the same time, the bearing capacity of the cylinder is improved. (2) Long working life under load. (3) High safety, fatigue failure mode with leakage before blasting. (4) Low production cost and short cycle. The cylinder is wound with a winding machine, which has high precision, high production efficiency and low labor intensity.

Gas cylinder pressure cycle test [4] is of great significance for verifying the reliability of repeated filling of gas cylinders. And it is an important means for testing the safety margin, structural rationality and reliability of composite gas cylinders. During the air pressure cycle test, effective control of pressure and temperature should be implemented. However, there is little research on the law of pressure increase and temperature change of composite gas cylinders during the air pressure cycle test. There is currently no method to accurately control the boost rate and the temperature of the cylinders [5]. Therefore, it is necessary to study the flow field characteristics in the air pressure cycle test of LPG cylinders.

2. Experiment methods

The test cylinders are LPG high-density polyethylene liner glass fiber fully wound cylinders. The outer diameter of the cylinder is 545 mm and the volume is 27.5 L. The temperature measurement uses a Pt100 temperature sensor with a temperature measurement range of -60 to 200 ℃ and an accuracy of 0.1 ℃. The pressure measurement uses a pressure sensor with a range of 5 MPa and a level of accuracy of 2. The tested temperature and pressure were recorded and saved by a paperless recorder, and the relevant temperature and pressure were recorded every 1 s. The armored thermocouple model is WZP 231, which is used to measure the internal temperature of the gas cylinder. The position of the internal temperature measurement points of the gas cylinder is shown in Figure 1. The cylinder port uses a three-way valve.

The cylinder is placed horizontally. Adhesive Pt100 platinum thermal resistance measures the temperature of three measuring points on the outer wall surface. The measurement points are in the middle of the bottle, near the mouth of the bottle, and at the flat bottom of the bottle, as shown in Figure 2.

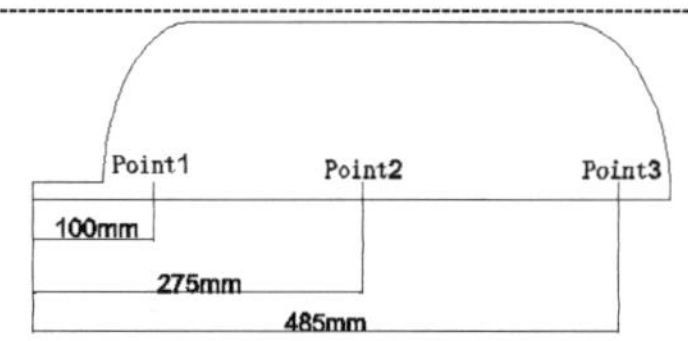

Figure 1. Position of the internal temperature measurement points of the gas cylinder.

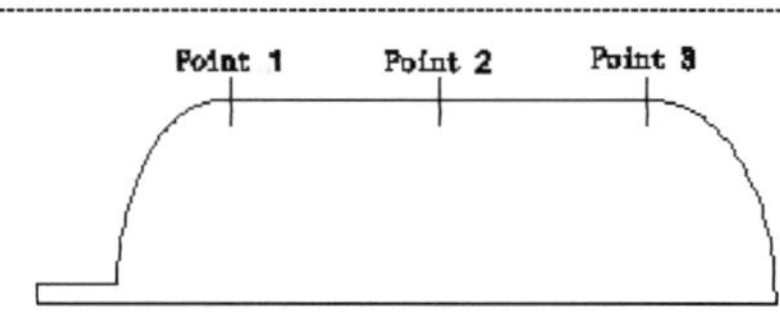

Figure 2. Position of temperature measurement points on the outer wall of the cylinder.

Pressure cycling tests require repeated filling of the cylinder. During the pressure cycle test, the internal pressure of the gas cylinder first rose to the upper limit of the specified cycle pressure to maintain the pressure, and then decreased to the lower limit of the specified cycle pressure to continue the pressure retention until the end of the test. The pressure cycling rate was changed to do two sets of experiments to explore the effect of the pressure cycling rate on the performance of the gas cylinders to ensure that the structure and material properties of the gas cylinders remained unchanged after the test.

According to the above test process, after the test data is collected, the collected data needs to be sorted by relevant software, and the pressure cycle test is numerically analyzed using Fluent. The heat transfer coefficient of each interface in Fluent is 0.1 $W/m^2 \cdot K$; the environment and initial temperature are 295 K; the total filling process takes 108 s, and the termination pressure is 3.2 MPa. The gas inlet takes the pressure inlet conditions. The relationship between pressure and time is shown in the following formula (1):

$$P= 0.0288\times t^4 -8.0815\times t^3 +807.18\times t^2 +1018.4\times t^1 +2613.7 \quad (1)$$

The changes of inlet pressure with time are consistent in the test and simulation, as shown in Figure 3. In Figure 3, P_{test} is the cylinder inlet pressure during the test, MPa; P_{si} is the cylinder inlet pressure during the simulation, MPa.

3. Test results and discussion

The data collected in the test has been completed, and the collected data was organized using the origin software. The temperature rise during air filling is shown in Figure 4. During the test air filling process, the measured temperature change in the bottom of the cylinder is only 2 ℃, because the flow field was hindered by the penetration of the armored thermocouple into the metal rod in the cylinder. In Figure 4, $T_{M1.S}$, $T_{M2.S}$, and $T_{B.S}$ are respectively the temperature of the mouth, middle and bottom of the cylinder during simulation, K; $T_{M1.T}$, $T_{M2.T}$, $T_{B.T}$ are respectively the temperature of the mouth, middle and bottom of the cylinder during the test, K; P is the cylinder inlet pressure during the test and simulation, MPa.

Before Fluent numerically simulated the pressure cycle test, the correctness of the model was verified. The temperature rise trend of the air filling process simulated by Fluent basically accords with the actual filling process (see Figure 4), and this model could be used. In the air filling process simulation, the temperature changes of the three measurement positions at the bottom, middle and mouth of the cylinder with time is consistent with that during the test and the maximum temperature difference is 8.63 ℃. Due to the lagging thermal response of platinum resistance, there is an error between the variation trends of temperature in simulation and test. The temperature rise trend at the bottom of the bottle is very different from the test.

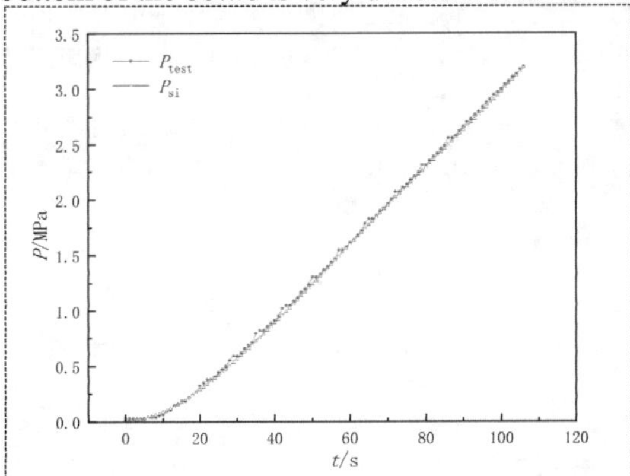

Figure 3. Changes in inlet pressure over time in test and simulation.

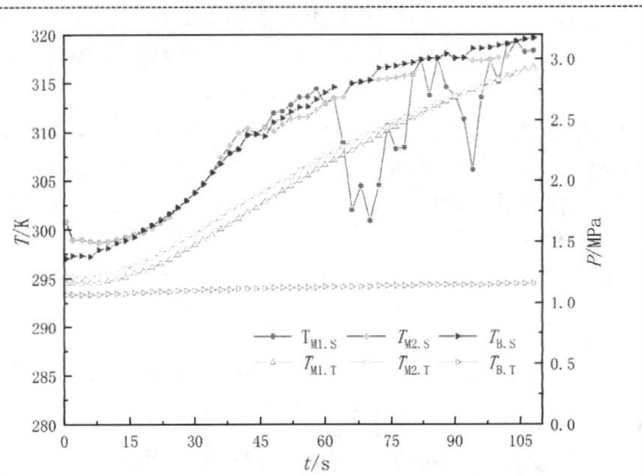

Figure 4. Changes in temperature of cylinder over time in test and simulation.

The pressure cycle test data obtained by changing the pressure cycle rate is shown in Figure 5. In Figure 5, T_{M1}, T_{M2}, and T_B are respectively the temperature of the mouth, middle and bottom of the cylinder during the test, K; P is the cylinder inlet pressure during the test, MPa.

The throttling expansion of the filled air produces the Joule-Thomson effect. The temperature will decrease and the cooling effect will occur. Therefore, the temperature of the cylinder started to fall within 15 seconds. After the air entered the cylinder from the pipe, the speed decreased, and the kinetic energy changed to internal energy, which caused the temperature of the gas in the cylinder to rise. During the filling process, the gas in the bottle was compressed to do work, which caused the gas temperature to rise. A metal rod with a thermocouple in the gas cylinder hinders the flow of air in the bottle, making it turbulent and unable to flow to the bottom of the gas cylinder. After the pressure was released, the gas in the container and the container wall, and the container wall and the environment performed heat transfer, so that the temperature of the gas in the container decreased.

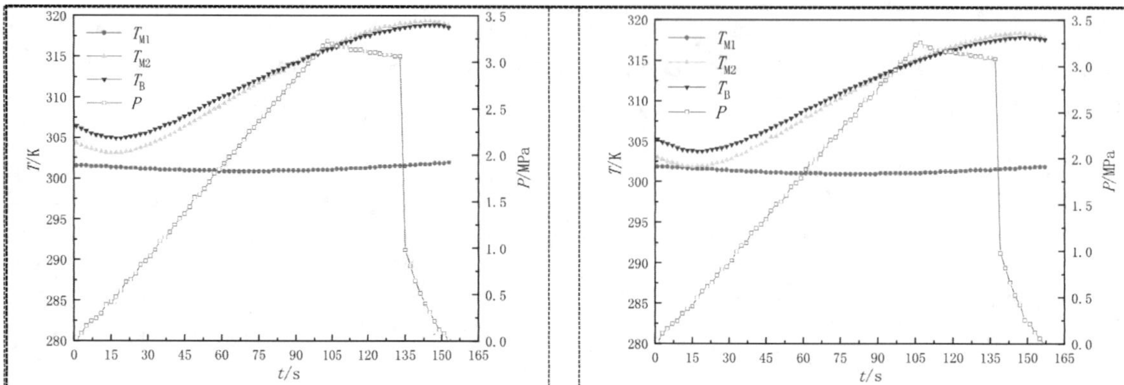

(**a**) 0.46 times / minute. (**b**) 0.44 times / minute.

Figure 5. Changes in temperature and pressure of cylinder over time in test.

4. Conclusion

An air pressure cycle test was performed on a 27.5 L LPG high-density polyethylene liner glass fiber fully wound cylinder, and the safety margin, structural design, and reliability of the cylinder were tested. During the test, the temperature variation in the cylinder was 17 ℃, which was greater than the temperature variation of the cylinder when it was filled with LPG, and the structure and material properties of the cylinder were unchanged. Therefore, this test can test the safety performance of this cylinder. In Fluent simulation, the temperature rise trend of the cylinder was the same as that in the test, and the maximum temperature difference was within the acceptable range. This model can be used to simulate the internal flow field of a gas cylinder in an air pressure cycle test, which is helpful to explore the method of controlling the pressure increase rate and temperature of a composite material cylinder in the test.

Acknowledgments

This work was supported by National Key R&D Program of China(No. 2017YFC0805600).

References

[1] Hu, Q.Q. (2019) Feasibility Study of LPG Supplementary LNG Application. Chemical Enterprise Management., 29: 94–98.

[2] Xie, Y.M., Tan, Y.Q. (2003) Insect Diversity and the Trophic Structure of Communities. In: 2003 Western Clean Car Technology Symposium. Chengdu. pp. 58-61.

[3] Fei, C., Dong, P. (2014) Characteristic and Application of Composite Gas Cylinder. Fiber Composites., 31: 53–55.

[4] Li, Y., Fu, D.L. (2019) Analysis of Medium Selection for Cylinder Pressure Cycling Test. Chemical Equipment Technology., 40: 14–16.

[5] Chen, H.G. (2014) Research on Hydraulic Fatigue Test System for 70MPa Composite Hydrogen Cylinder and Control Method of Pressure and Temperature. Zhejiang University., 31: 53–55.

Safety performance optimization of front bumper system based on pedestrian lower leg protection

Jiabin Zhong, Baijun Shi*, Yihuai Zhang

South China University of Technology, Guangzhou, China

Corresponding author e-mail: bjshi@scut.edu.cn

Abstract. Taking the front bumper system of a passenger car as the research object, a finite element model of pedestrian lower leg collision was first established, and the tibial acceleration at Y400 was found to be high in the performance analysis. Second, the sensitivity analysis was performed on the main parameters of the bumper system to determine the skin Thickness and foam density are the main influencing variables. Finally, a performance improvement scheme is proposed, and the lower leg tibia acceleration is reduced by 47.71% after the improvement, indicating that the scheme is conducive to lower leg protection.

1. Introduction

In traffic accidents, pedestrians are the most vulnerable groups, and pedestrians' lower leg are the most vulnerable parts. The front bumper system first contacted the pedestrian's leg, causing tibia fractures, ligament tears, and knee injuries. The current research method is to reduce the damage to the lower leg by changing the thickness of the relevant materials, but less consideration is given to the overall collision performance, and it does not involve the development and design of new components. The optimization scheme proposed in this article satisfies various performances, provides a reference for the development of the front bumper system, saves development costs, and has certain research significance.

2. Pedestrian lower leg collision performance analysis

2.1. Establishment of finite element model for lower leg collision

The front bumper system mainly includes the front bumper skin, energy absorbing foam, and beams. The front bumper skin includes parts such as skin, upper grille, lower grille, left and right brackets, etc. In Hypermesh, it is divided according to a grid size of 5mm, foam and beams are divided according to a grid size of 8mm, and the skin and beam use thin shells. Foam uses solid units, as shown in Figure 1. In terms of material selection, the skin and beams were simulated using MTL24, and the energy-absorbing foam was MTL57[1]. A shell unit is covered on the outer surface of the energy-absorbing foam to prevent the occurrence of negative volume. The material parameters are shown in Table 1.

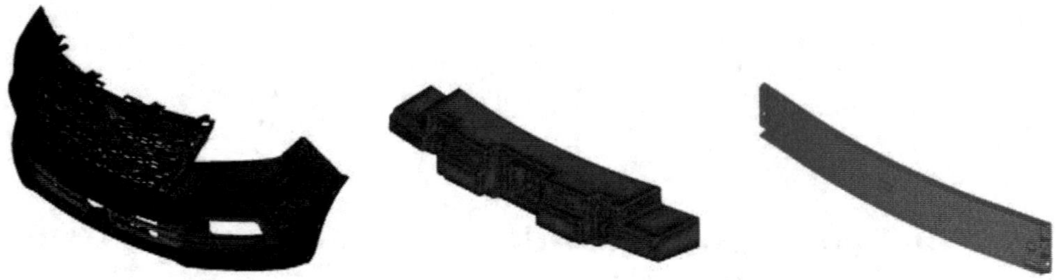

(a) Front bumper skin (b) Energy absorbing foam (c) Beams

Figure 1 The finite element model of front bumper system

Table 1 Material parameters

Name	Density/ $t \cdot mm^{-3}$	Elastic modulus/ MPa	Poisson's ratio	Yield Strength/ MPa	Failure strain/ %
Beam	7.86×10^{-9}	200000	0.28	414	7
skin	9.7×10^{-10}	1400	0.45	14	—
foam	9×10^{-11}	6.5	0	—	—

In terms of connection settings, this study uses rigidbody to simulate the connection between deformable locks; by creating a nodal rigid body, the bolt connection between parts is achieved[2]. Set the center of mass coordinates and give the vehicle a mass of 1.595t.

This article uses a pedestrian protection test leg impactor developed by LSTC. This model is based on the physical structure of the lower limb module of EEVC, as shown in Figure 2.

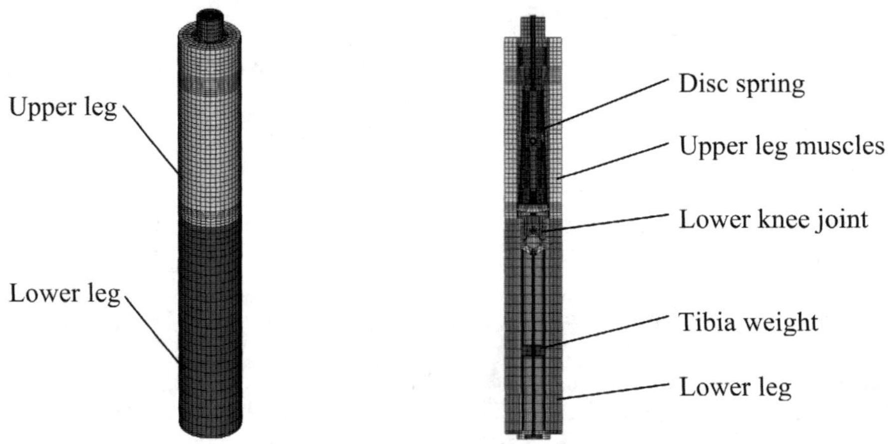

Leg bumper module Section

Figure 2 Finite element model of leg impactor

The model contains 29 components, divided into upper leg and lower leg parts, and they are connected by ligaments. The pedestrian lower leg finite element model has passed the static and dynamic calibration tests required by European regulations (EC631 / 2009) [3].

The six degrees of freedom in the x, y, and z directions must be constrained on the bumper system. In addition, according to the requirements of pedestrian collision regulations, a leg impactor of 40km/h in the x direction is given. Bumper self contact, leg bumper self contact, and bumper and calf bumper contact must be set. The calculation time of the pedestrian leg collision model set in this paper is 30ms. In terms of the collision area, the collision points Y0, Y100, Y200, Y300, and Y400 with distances from the center of symmetry of 0mm, 100mm, 200mm, 300mm, and 400mm were selected. According to EEVC regulations, pedestrian evaluation indicators are shown in Table 2 below.

Table 2 Pedestrian lower leg evaluation index target value

Evaluation index	Tibia acceleration/g	Knee joint bending angle/°	Knee shear displacement/mm
Target limit	≤ 150	≤ 15	≤ 6

Taking the Y0 collision point as an example. The total energy is basically unchanged, and the initial total energy is 825.25J. The hourglass energy is 12.02J, and the maximum slip energy is 23.39J. Both the hourglass energy and the slip energy are less than 5% of the total energy, and the curve is smooth and there is no abrupt phenomenon. Therefore, it can be shown that the calculation result of this model is more accurate.

2.2. Crash performance analysis

Simulate on each collision point, and finally get the tibial acceleration, knee bending angle, and knee shear displacement of the five collision points of Y0, Y100, Y200, Y300, Y400[4]. The lower leg index statistics of each collision point are shown in Table 3.

Table 3 Lower leg injury index statistics

Collision point	Peak tibia acceleration/g	Peak knee joint bending angle/°	Peak knee shear displacement/mm
Y0	103.92	10.59	1.31
Y100	88.72	4.08	1.08
Y200	75.72	4.28	0.96
Y300	87.19	3.52	0.98
Y400	189.60	4.25	1.30

As can be seen from the table 3, the tibia acceleration at the collision point of Y400 exceeded the target value, and there was a huge risk to the protection of the lower leg, which needs to be optimized. Because Y400 is located in the collision angle area, there are components with high rigidity such as headlight mounting brackets. Compared with the middle position, the overall rigidity of the collision point is greater. In addition, the knee joint bending angle and knee shear displacement of the five collision points meet the requirements.

3. Parameter sensitivity analysis of front bumper system

The parameters that affect the lower leg injury value are skin thickness, upper grille thickness, foam density, beam material, beam thickness, and beam height. In order to effectively optimize the front bumper system of an automobile, it is necessary to analyze the degree of influence of six parameters [5]. Taking the above six parameters as input variables, each variable is set to 3 levels, and the table of input variable levels is shown in Table 4.

Table 4 Bumper system parameter input variable level table

	Input variable	Level 1	Level 2	Level 3
A	Skin thickness/mm	2	2.6	3
B	Upper grid thickness/mm	2	2.5	3
C	Foam density/kg m^{-3}	50	90	110
D	Beam material	Steel M414	Steel M1000	Aluminum alloy
E	Beam thickness/mm	2	2.5	3
F	Beam height/ mm	145.6	155.6	165.6

In the pedestrian lower leg collision at the Y400 collision point, the maximum tibial acceleration, the maximum knee bending angle, the maximum knee shear displacement, and the front bumper system mass were used as output variables. The test results were imported into Hyperstudy for DOE analysis, and the results were analyzed and processed from the experimental index average and range respectively. Recorded as the average of the corresponding levels of each impact factor, R is the range of the mean. The magnitude of the range reflects the influence of each factor on the output response. The larger R is, the greater the influence of this factor on the output response, and the more critical the factor [6, 7].

Table 5 shows the corresponding range of the influence factors of the peak acceleration of tibia. Among them, the degree of influence of the factors on the peak acceleration of the tibia is ranked as A >C>F>E>D>B, which shows the thickness of the skin and the density of the foam have a greater impact on the acceleration of the tibia. Figure 3 shows the main effect of the peak acceleration of the tibia. It can be seen that as the thickness of the skin increases, the acceleration of the tibia decreases sharply; as the density of the energy-absorbing foam increases, the acceleration of the tibia also decreases sharply. Therefore, consider increasing the thickness of the skin and the density of the foam, which is conducive to reducing the peak acceleration of the tibia.

Table 5 Analysis of peak range of tibia acceleration

Evaluation index		Impact factor					
		A	B	C	D	E	F
Peak tibia acceleration a_{max} / g	K_1	172.13	145.73	172.77	144.38	140.93	164.37
	K_2	158.16	142.26	148.90	159.09	138.28	133.20
	K_3	111.43	153.72	120.04	138.25	162.51	144.14
	R	60.70	11.46	52.73	20.84	24.23	31.17

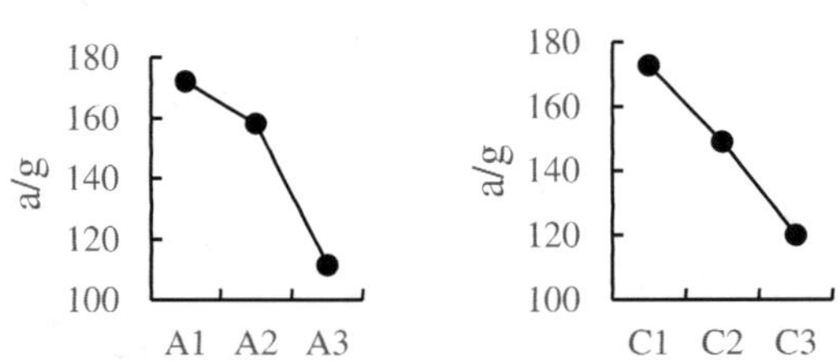

Figure 3 The main effect of tibia acceleration

4. Optimization of front bumper safety performance parameters

According to the sensitivity analysis, among the six parameter variables, the most sensitive variables to the collision point Y400 are the skin thickness and foam density. The optimization scheme is proposed. The design variable values of the optimization scheme and the original scheme are shown in Table 6.

Table 6 Optimal and original design variables

Scheme	A/mm	B/mm	C/kg mm^{-3}	D	E/mm	F/mm
Original	2.6	2.5	90	Steel M414	2.5	145.6
Optimal	3.0	2.5	110	Steel M414	2.5	145.6

The five collision points of Y0, Y100, Y200, Y300, and Y400 in the lower leg collision area are analyzed. Table 7 shows the comparison of the tibia acceleration before and after the optimization of each collision point, which below the legal limit of 150g and meets the design requirements. Among them, the simulation values at the three collision points of Y0, Y100, and Y200 have increased. This is due to the increase in the thickness of the skin and the density of the foam, resulting in an increase in the overall stiffness. The acceleration front end is reduced by 47.71%, which is due to the skin and foam can not play a good role in energy absorption. Increasing the stiffness can reduce the energy transmitted to the beam and other parts of the stiffness height, and reduce peak tibia acceleration.

Table 7 Comparison of peak initial value and optimal value of tibial acceleration

Collision point	Y0	Y100	Y200	Y300	Y400
initial tibia acceleration/g	103.92	88.72	75.72	87.19	189.60
optimal tibia acceleration/g	120.34	103.65	95.30	83.91	99.14
Rate of change/%	15.80	16.83	25.86	-3.76	-47.71

5. Conclusion

In this article, the front bumper of a passenger car is taken as the research object to establish an accurate and reliable crash simulation model. Through safety performance analysis, it is found that the acceleration of the tibia at the collision point Y400 is high, which is not conducive to pedestrian lower leg protection. Through the sensitivity analysis, it can be seen that the thickness of the skin and the density of the energy-absorbing foam have a greater impact on the tibia acceleration, and its parameters need to be optimized. The skin thickness of the initial scheme was changed from 2.6mm to 3mm, and the foam density was changed from 90 $kg \cdot mm^{-3}$ to 110 $kg \cdot mm^{-3}$. The tibia acceleration at the collision point Y400 was reduced by 47.71% after optimization, and all other collision points could meet the limit conditions. The optimization methods have certain guiding significance for future research.

References

[1] Zhao. LS_DYNA Dynamic Analysis Guide [M]. Tianjin: Ordnance Industry Press, 2003: 12-148.
[2] Xu. Research on front bumper system based on low-speed collision and pedestrian leg protection [D]. Changsha: Hunan University, 2011.
[3] Xu. Optimization analysis of calf protection of a car colliding with a pedestrian [D]. Changchun: Jilin University, 2017.
[4] Micuno K, Ueyama T, Nakane, et al. Comparison of reponses of the Flex-PLI and TRL legform impactors in pedestrian tests. SAE International Journal, 2012, 5 (1): 203-213.
[5] Lim J.M, Ju Y.P, Shin D.W. Development of an optimized structure for meeting pedestrian protection requirements [J]. SAE World Congress, 2011-01-0770.
[6] Mu. Optimal design of front bumper system based on pedestrian lower limb protection and low speed collision [D]. Changsha: Hunan University, 2013.
[7] Cheng. Design and Performance Analysis of Passenger Car Exhaust Muffler [D]. Guangzhou: South China University of Technology, 2016.

IOP Conf. Series: Materials Science and Engineering **793** (2020) 012051 doi:10.1088/1757-899X/793/1/012051

Study on the Atmospheric Environmental Efficiency of the Silk Road Economic Belt of China Based on Energy conservation and Emission Reduction

Nini Yin[1*], Pengcheng Yan[2]

[1]School of economics and management, Anhui university of science and technology, Huainan, Anhui, 232001, China

[2]School of electrical and information engineering, Anhui university of science and technology, Huainan,Anhui, 232001, China

[*]Corresponding author's e-mail: nnYIN306@163.com

Abstract. Based on SBM-Undesirable model, this paper evaluated air environmental efficiency of 8 provinces of the Silk Road Economic Belt of China during the period of 2010—2017, and investigated the temporal-variation and energy conservation and emission reduction potential. Empirical analysis shows that the overall atmospheric environmental efficiency of Silk Road Economic Belt is still at low point but presents a mild rising trend over the period 2010—2017, regional differences are remarkable, and it has great potential of energy conservation and emission reduction in the future.

1. Introduction

In 2013, Chinese government proposed a major initiative to jointly build the "Silk Road Economic Belt" and the "21st century Maritime Silk Road", which put the green concept of ecological civilization construction on priority. However, due to the high consumption and low utilization of energy and the high-intensity emission of air pollutants such as SO2 and NOX, the air pollution problem along the line has become increasingly serious. Therefore, the scientific measurement and analysis of the Silk Road Economic Belt's atmospheric environmental efficiency can provide data reference and theoretical basis for the sustainable economic development and ecological civilization construction in China.

DEA (date envelopment analysis) method is widely used by scholars at home and abroad. Since it was first proposed by A Charnes and W W Cooper in the late 1970s [1,2], it has been widely used and developed in the research of energy and environmental efficiency. J L Hu et al. first used DEA method to calculate the TFEE index of 29 provinces in China, and verified the hypothesis that the energy efficiency is directly proportional to the economic growth of China [3]. With the development of energy and environmental efficiency research, scholars at home and abroad begin to consider the impact of input and output slack variables in the process of energy efficiency measurement. Y Choi et al. measured China's energy efficiency and CO2 emission efficiency based on the SBM model of slack measure, and estimated its potential emission reduction and marginal emission reduction cost [4].In recent years, using DEA to deal with unexpected output has become a hot topic of domestic and foreign scholars, H. George et al.summarized four treatment methods of unexpected output in DEA [5].The above researched proved that DEA method is feasible in the measurement of energy

IOP Conf. Series: Materials Science and Engineering **793** (2020) 012051 doi:10.1088/1757-899X/793/1/012051

environmental efficiency, and also provides reference and model support for other scholars in the measurement of atmospheric environmental efficiency index.At present, there are few researches on the atmospheric environmental efficiency of the Silk Road Economic Belt. This paper evaluated air environmental efficiency of 8 provinces of the Silk Road Economic Belt of China during the period of 2010—2017 Based on SBM-Undesirable model.

2. Methods and models

2.1. SBM-Undesirable model

This model is improved by K Tone based on non radial and non angular SBM model with slack variable, which introduces "unexpected output" by splitting output variable [6]. According to "unexpected output", build a possible production set. Suppose that there is N (n = 1, … , N) decision cells, including M (m =1, … , M) factor inputs and P (p = 1, … , P) factor outputs, which are divided into P_1 expected outputs and P_2 unexpected outputs. Define matrix input indicators $X \in R^{m\times n}$, Expected output index $Y^g \in R^{p1\times n}$, Unexpected output index $Y^b \in R^{p2\times n}$, In the case of constant returns to scale (CRS), the set of production possibilities:

$$G = \{(x, y^g, y^b) | x \geq X\lambda,\ y^g \leq Y^g\lambda,\ y^b \geq Y^b\lambda,\ \lambda \geq 0\} \tag{1}$$

$$\min \rho = \frac{1 - \frac{1}{M}\sum_{m=1}^{M}\frac{s_m^-}{x_{m0}}}{1 + \frac{1}{P_1 + P_2}\left(\sum_{i=1}^{p1}\frac{s_i^{g+}}{y_{i0}} + \sum_{i=1}^{p2}\frac{s_i^{b+}}{y_{i0}}\right)} \tag{2}$$

$$S.T.\ x_0 = X\lambda + s^-, y^g = Y\lambda + s^{g+},\ y^b = Y\lambda + s^{b+},\ \lambda \geq 0,\ s^- \geq 0,\ s^{g+} \geq 0,\ s^{b+} \geq 0$$

ρ is the value of atmospheric environmental efficiency, s^-、s^{g+}、s^{b+} is the slack value of factor input, expected output and unexpected output, λ is the weight matrix. ρ On s^-、s^{g+}、s^{b+} strictly monotone decreasing, and $0 \leq \rho \leq 1$, if and only if $\rho = 1$,current DMU is valid, Otherwise, it is invalid.

3. sample variables and data

Considering the availability of data, the sample investigation period is 2010—2017, and the research object is the main 8 provinces along the Silk Road Economic Belt. Data of the above indicators can be obtained directly from China Statistical Yearbook, The descriptive statistical characteristics of input-output variable data are shown in Table 1.

Table 1. Descriptive statistics of sample's input and output indicators during 2010—2017.

Variable		Units	Min	Max	Mean	Std.Dev
Input factors	Labor	Million workers	307.650	2998.890	1596.595	940.068
	Capital	RMB Billion	2786.836	25064.281	12804.812	6470.016
	Energy	Million tons of coal	2814.570	17391.700	8129.471	3376.812
Desirable outputs	GDP	RMB Billion	1070.381	16343.242	7417.766	4498.344
Undesirable outputs	SO2	104tons	9.240	91.680	48.563	22.739
	NOX	104tons	7.230	88.690	42.029	20.490
	Dust	104tons	8.330	81.390	31.697	16.789

4. Empirical measurement and result analysis

4.1. Temporal-variation of atmospheric environmental efficiency

According to formula (1) and formula (2), the atmospheric environmental efficiency of 8 provinces along the Silk Road Economic Belt in 2010-2017 is calculated based on CRS, the results are shown in Table 2.

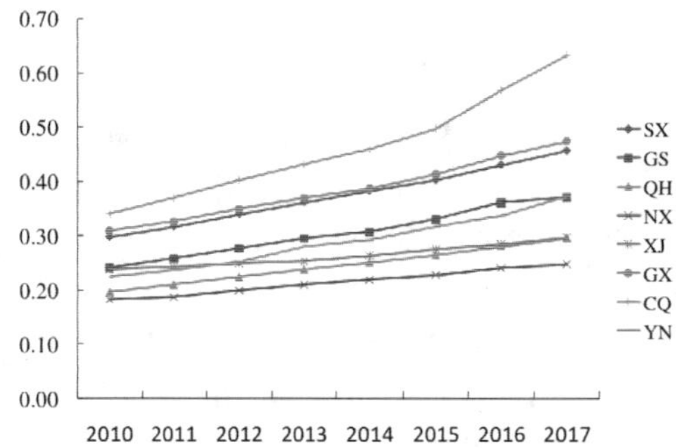

Figure 1. Atmospheric environmental efficiency of 8 provinces of the Silk Road Economic Belt during 2010—2017

According to figure 1, The mean value of atmospheric environmental efficiency of SX, GS and other provinces in Northwest China and GX, CQ and YN provinces in Southwest China is between 0.463 and 0.214. Because the northwest and southwest Regions are located in the inland of China, the ecological and natural conditions are relatively complex, and the formation and accumulation of air pollutants are difficult to spread. The economic foundation of some provinces and cities is weak, the level of industrial structure is low, and the level of technology is backward, which makes the atmospheric environmental efficiency at a low level for a long time.

4.2. Redundancy analysis of atmospheric environmental efficiency

According to formula (1), calculate the redundancy rate of atmospheric environmental efficiency of 8 provinces along the Silk Road Economic Belt during 2010-2017. In order to provide scientific and reasonable improvement direction for atmospheric environmental efficiency, the potential of energy conservation and emission reduction of input-output variables in each province is analyzed. The results are shown in Figure 2.

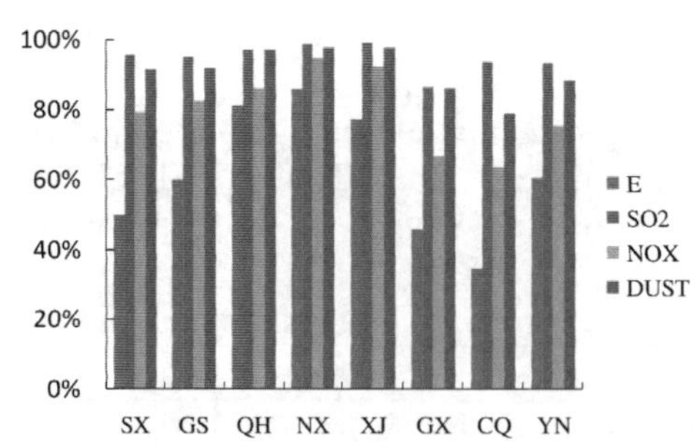

Figure 2. Redundancy ratio of input and output variables of 8 provinces of the Silk Road Economic Belt during 2010—2017

According to figure 2, in addition to energy input, the redundancy rates of the other three unexpected output variables are presented SO2＞DUST＞NOX. From the dynamic angle, The redundancies of energy input and unexpected output are decreasing with each passing year, this is consistent with the time trend of annual increase of atmospheric environmental efficiency given above.

IOP Conf. Series: Materials Science and Engineering **793** (2020) 012051 doi:10.1088/1757-899X/793/1/012051

5. Conclusion

According to the research results, it can be found that in the six years from the initiation to the implementation of the initiative, although the atmospheric environmental efficiency of the areas along the line in China is still low, it shows a positive trend of increasing year by year and has great potential for energy conservation and emission reduction.In order to better promote the construction of ecological civilization, we should take green as the base color of the Silk Road Economic Belt. The provinces should further improve energy utilization rate and industrial structure, introduce and promote the green technology focusing on energy conservation and emission reduction, so as to prevent and control air pollution from the source.

References

[1]Charnes A, Cooper W W, Rhodes E. Measuring the efficiency of decision making units [J]. European Journal of Operational Research, 1978, 2(6): 429-444.
[2]Banker R D, Charnes A, Cooper W W. Some models for estimating technical and scale inefficiencies in data envelopment analysis[J]. Management Science, 1984, 30(9): 1078-1092.
[3]Hu J L, Wang S C. Total-factor Energy Efficiency of Regions in China [J]. Energy Policy, 2006,(34):3206-3217.
[4]Choi Y , Zhang N , Zhou P . Efficiency and abatement costs of energy-related CO2 emissions in China: A slacks-based efficiency measure[J]. Applied Energy, 2012, 98:198-208.
[5]George Halkos, Kleoniki Natalia Petrou. Treating undesirable outputs in DEA: A critical review[J].Economic Analysis and Policy,2019,(62):97-104.
[6]Tone K . A slacks-based measure of efficiency in data envelopment analysis[J]. European Journal of Operational Research, 2001, 130(3):498-509.

IOP Conf. Series: Materials Science and Engineering **793** (2020) 012052 doi:10.1088/1757-899X/793/1/012052

The research on the management strategy of exclusive parts for satellite integration

Chen Zou[1]

[1] Institute of Telecommunication Satellite, China Academy of Space Technology, Peking, 100094, China

Abstract. As an important component of satellites, the demand characteristics of exclusive parts for integration are short cycle, large quantity and low cost. It is urgent to solve the problem of meeting the demand with the current production capacity without affecting the quality level . Based on the analysis of the management strategy of the exclusive parts, it is indicated that the root cause of the problems such as the left backup and the structural shortage is the separated management system. It is also indicated the generality of some exclusive parts. In this paper, the general strategy is proposed to form a unified management system for general exclusive parts for integration, which is conducive to the full use of resources, and cost reduction.

1. Introduction

As one of the most important parts of satellites, exclusive parts for integration indicate connectors for satellite integration. With the rapid growth of the quantity of satellites, the demand for exclusive parts is growing rapidly. However, in the case of relatively stable production capacity, it is necessary to make more reasonable arrangements for resources to meet the rapid growth of demand and reduce the cost without affecting the quality level. The management of the exclusive parts for integration is relatively extensive, and the research on the management mode is rare. This paper analyzes the management of the exclusive parts for integration, studies the cost characteristics and demand characteristics of the exclusive parts for integration to discover the existing problems, and puts forward new suggestions for the problems .

2. The current management strategy of exclusive parts for satellite integration

Satellites are highly customized products. Each satellite has its own independent design-management system. The exclusive parts for integration belong to the separated management system of the satellite who uses them. The management mode of "separated design, separated manufacture, separated storage and separated management" is adopted.

Because the actual use of some exclusive parts is not clear at the beginning, in order to avoid shortage, there will be a lot of backup when making production plan. These unused backups remain in the warehouse of the satellites launched. Because the scope of use is limited to the satellite which the parts belong to, they will be kept in the warehouse all the time and accumulate with the increase of the number of satellites. As shown in Table 1, the remaining unused exclusive parts in the warehouse of 5 launched satellites are counted.

Table 1. The exclusive parts left

Satellite	Exclusive parts	Amount

Satellite 1	Washer	13
	Grounding line	41
Satellite 2	Grounding line	168
	Washer	71
Satellite 3	Grounding line	458
Satellite 4	Washer	4
Satellite 5	Grounding line	220
Total		975

There are 975 exclusive parts for 5 satellites remaining in the warehouse in Table 1, with an average of about 195 for each satellite. Washers and grounding lines are exclusive parts commonly used on satellites. Taking the washers as an example, the washers in Table 1 are all of fimilar structures. Taking the washers of Satellites 1 and 4 in Table 1 as an example, it is indicated that there are two kinds of washers in satellite 1, one of which is the same with the ones in Satellite 4 in material and size. Washers of the same configuration are also used in other satellites. It is indicated the washers are widely used in satellites. The same conclusion can be get in the analysis of the ground lines.More analysis shows these exclusive parts are interchangeable.

2.1. The cost characteristics of exclusive parts for satellite integration

The cost C of exclusive parts for satellite integration consists of four parts.

$$C = C_1 + C_2 + C_3 + C_4 \tag{1}$$

C_1: Design and manufactured cost. The design of satellites is complex, and the cost of exclusive parts is considerable. Manufactured cost is the entity cost. Due to the particularity of aerospace materials and devices, the cost of exclusive parts is quiet high. Since the case of backup is normal, there are 975 useless exclusive parts left in warehouse at present, because of the limitation of the current management mode, resulting in waste of resources and the increasing cost.

C_2: Inventory cost means holding cost. Due to the high requirements of environment for satellite product storage, requirements on the warehouse is quiet strict. There are a large number of non reusable exclusive parts in the warehouse due to the problems of the management system. They occupy the inventory space, resulting in increased cost. Furthermore, the inventory of exclusive parts is predicted to grow with the rapid growth of the number of satellites, resulting in the rapid increasing of inventory costs.

C_3: Quality cost. Quality is the most important factor for satellite products, and the quality investment is also astonishing.[1]

C_4: Shortage cost refers to the punitive cost induced by the failure for exclusive parts meeting the demand during integration. The project progress is delayed because of the shortage of exclusive parts, that induces the shortage cost.

2.2. The demand characteristics of exclusive parts for satellite integration

During the integration of a specific satellite, the requirements of exclusive parts are distributed discretely in time (t), usually corresponding to a specific integration stage; in the parallel integration of multiple satellites, the demand distribution of exclusive parts for a specific satellite with time is independent with others. Suppose Part 1, Part 2 required for Satellite A, Satellite B are completely interchangeable. In general, when satellites integrate parallelly, due to the different progress, the demand time nodes for the parts are different, that is, the possibility of superposition referred to multi-satellite integration at a specific time point is relatively small, and the total demand usually maintains a relatively stable state in unit time.

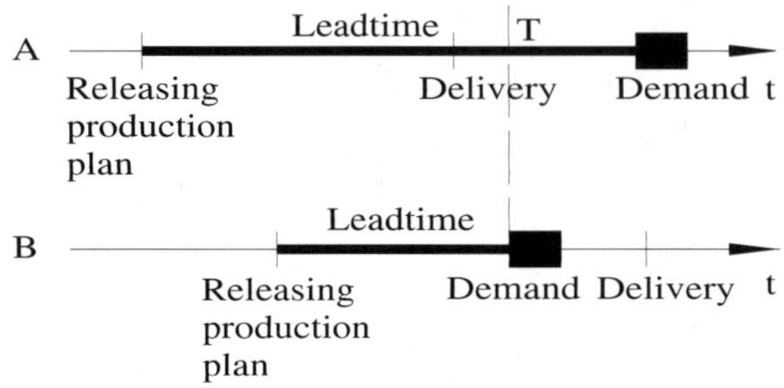

Figure 1. Structural shortage

The time from the release of production plan to the demanded time node of integration is defined as the leadtime, which indicates the time for producing. As satellite projects implement the production plan of exclusive parts separately, it may exist time point T, as shown in Figure 1, at which exclusive parts have been delivered to Satellite A, but it is not necessary to install Part 1, while Part 2 has not been delivered, but Satellite B urgently needs Part 2. Due to the independence of satellite project management systems, Satellite B can not use Part 1, resulting in structural shortage, and a substantial increase in the shortage cost.

3. The solution strategy

It is indicated that a considerable number of exclusive parts can be used interchangeably between satellites. However, with the current strategy, the root cause of problems such as backup legacy and structural shortage is that the exclusive parts are managed by separated management systems. In order to solve the problems above, it is necessary to break the barriers between satellite projects and manage the parts as a whole.

The requirements of satellites are considerably diversified. The research [2] believes the use of general parts can well adapt to the diversity needs. Based on the generality of the exclusive parts, the general strategy is adopted, which is to separate the exclusive parts with generality from the separated whole-satellite management system to form a unified general management system of the exclusive parts. The general parts have a unified code system. The general part managers unify the production plan as a whole, and form the mode of "unified production, unified inventory".

A unified warehouse is used to avoid structural shortage, and because the parts are no longer belonged to the separated systems, the backups are cyclically utilized. Furthermore the general parts strategy can reduce inventory cost through risk sharing[3]. It is also indicated [4]the cost of development and design, the cost of manufacturing are reduced by generalization. The general parts have successful flight experiences, which are of higher maturity and lower quality cost.

The specific flow chart of the production planning process is shown in Figure 2. According to the total demand Q and the existing inventory level W, the net gap quantity L can be obtained.In the actual manufacture, the parts delivered at a certain time point may not match the demand, inducing the risk of shortage. Under the original management mode, such risks cannot be avoided. In the general mode, in order to avoid shortage, dynamic supervision is carried out on the inventory, and a certain safety stock S is set up. The sum of the inventory of the exclusive parts W and the ones to be produced in the plan P is greater than or equal to the sum of the safety stock S and the net gap quantity L.

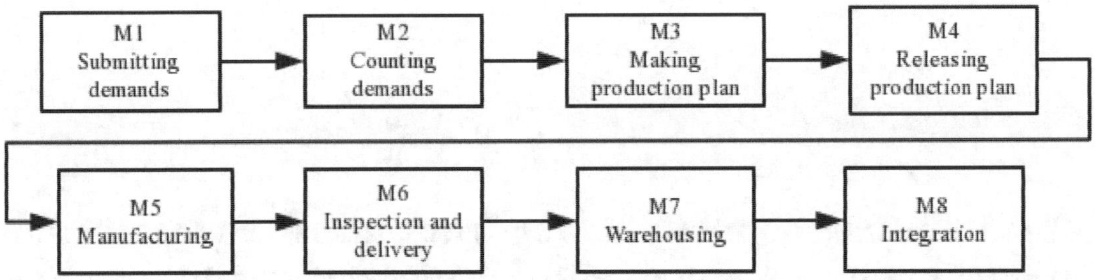

Figure 2. Production planning process

$$Q=\sum_{i=1}^{n} Q_i \tag{2}$$

$$L=Q-W \tag{3}$$

$$P+W \geqslant S+L \tag{4}$$

At any point in time, the inventory W should be maintained higher than the safety stock level.

$$W \geqslant S \tag{5}$$

4. Conclusion

Some parts are universal, while in the current strategy of satellite integration, a separated management system is used for the exclusive parts. The analysis indicates the problems, such as backup residual, structural shortage and so on, are induced by the separated management. In this paper, the general parts strategy is proposed to form a unified general management system for exclusive parts, which is conducive to the full utilization of resources and cost reduction.

References

[1] Tan, W., Hu, J.(2009) SPACECRAFT SYSTEMS ENGINEERING.Science and technology of China press, Peking;

[2] Chen, G. Zhang. X., JIN, S.(2013) Product line design based on commonality of development-sensitivity. Forecasting, 32:76-80.

[3] Lin, Y., Li, L., Liao, L.(2009) A Study on Component Commonality Inventory Model with Guaranteed Delivery Time and Price. Industrial Engineering and Management, 14:14-19;

[4] Zhou.L., Wan.X.,(2002) Grey Decision Making in Selection and Substitution of Component Conlmonality, JOURNAL OF NANCHANG UNIVERSITY(ENGINEERING & TECHNOLOGY, 24:66-69.

IOP Conf. Series: Materials Science and Engineering **793** (2020) 012053 doi:10.1088/1757-899X/793/1/012053

A momentum-based method for the mass concentration measurement of pneumatically conveyed solid: design optimization

Guanqing Liu[1*]

[1]Huadian Electric Power Research Institute Co., LTD. Beijing, 100170, China

[*]Corresponding author's e-mail: liugqthu@gmail.com

Abstract. Mass concentration measurement of pneumatically conveyed solid is still challeging in industry. In our previous work (Liu G.Q., *The 5th International Conference on Advances in Energy Resources and Environment Engineering*, December 6-8, 2019, Chongqing, China.), a novel momentum-based method with more reliable measurement mechanism and simpler structure was developed and computationally validated with a one-end closed straight tube. In this paper, optimization on the structure of the metering tube is conducted to improve the measurement sensitivity represented by the conversion factor k. Tubes with a conical inlet are found to possess higher k than that of the straight tube. Computational fluid dynamics (CFD) computations show that k generally increases with decreasing half-cone angle at fixed radius and decreases with decreasing restitution coefficient of collision between particles and the tube walls.

1. Introduction

Though pneumatic transportation of bulk solids widely exists in industry, the measurement of the concentration of solid particles is still challenging. Various measurement techniques that have been developed in past decades can not work well in situ as in laboratory environment, due to reasons such as[1-4]: (1) non-uniform flow in large-diameter pipes, (2) disturbances like pipe vibration and particle contamination and (3) high sensitivity to secondary factors such as moisture content of solid particles that varies in wide ranges. In a previous work[5], we have proposed a momentum-based measurement mechanism that only relies on the essential flow parameters and is easy to implement. The mechanism was computationally verified with a straight metering tube. In current work, we seek to make optimization on the structure design of the metering tube to improve the measurement sensitivity which is represented by the conversion vector k. k is enlarged by 'collecting' more particles into the tube through a conical opening, the diameter of which is larger than that of the straight part of the tube. Effects of the half-cone angle and the restitution coefficient of collision on k are investigated and an optimized design of the tube structure is proposed.

2. Mechanism of measurement and computational setup

2.1. Measurement mechanism

As reported previously[5], a one-end closed tube with its opening facing towards a uniform gas-solid flow will see an increase in the fluid static pressure at its inner bottom end. The increment is proportional to the mass concentration of solid particles in the flow, as given by:

$$\Delta p = \frac{1}{2} k c u_0^2 \tag{1}$$

where Δp is the difference between the fluid static pressure at the inner bottom of the tube and the fluid total pressure of the main flow, k is the conversion factor, c is the mass concentration of particles and u_0 is the velocity of the main flow.

In our previous computational fluid dynamics (CFD) computations, the maximum value of k for a straight tube was found to be around 1.5, smaller than the theoretical upper limit of 2. Since Δp originates from the drag force exerted by particles (especially those having entered the tube) on the fluid inside the tube, Δp and hence k can be enlarged if more particles 'collide into' the tube. A tube with a conical opening is a straightforward option. The conical opening is characterized by diameter D_o and half-cone angle θ, as shown in figure 1.

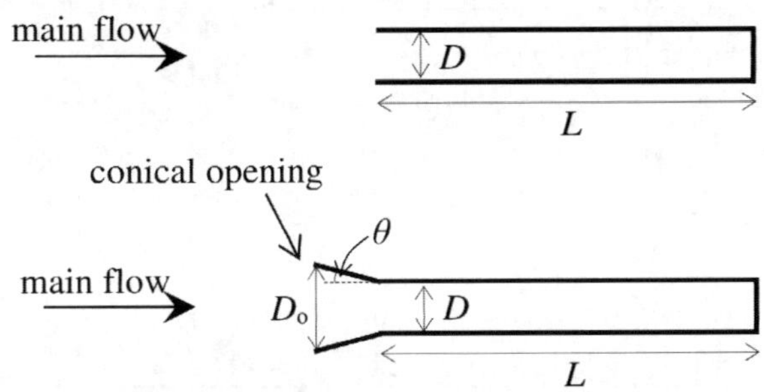

Figure 1. Schematic of the metering tube

In this work, D_o is fixed to $\sqrt{2}D$ and hence k is expected to be enlarged by up to two times. On the other hand, particles colliding with the conical wall will lose certain amount of momentum. Therefore, e, the restitution coefficient of collision, is expected to play an important role in affecting particle motion and hence k.

2.2. Computational setup

The CFD computational setup follows that of our previous work[5]. Simulations are carried out with Fluent in two-dimensional with axisymmetric boundary condition set on the axis of the tube. SST k-ω model was adopted to model turbulence. The main flow is with air at 25°C and u_0=15 m/s. Structured grid is used and refined in the vicinity of the tube. Details of the computation setup is given in table 1. Particle motion and interaction with the fluid phase are computed with DPM model. Transient computations with time step of 0.0001s are conducted.

Coefficient of restitution (both tangential and normal) for the impact of particles with the tube wall are set to 0.9 for standard cases and varied in other cases to investigate the effect of wall collisions on k. Trap condition is applied to the bottom of the tube.

Table 1. Computational parameters.

Variable / Parameter	Unit	Default value
computation domain	-	1150mm × 100mm
tube length, L	mm	200
tube diameter, D	mm	5
fluid velocity, u_0	m/s	15
fluid temperature, t_f	°C	25
fluid density, ρ_f	kg/m^3	1.225
fluid dynamic viscosity, μ_f	kg/m·s	1.789×10^{-5} kg/m·s
particle density, ρ_p	kg/m^3	1550

particle mass concentration, *c*	kg/m^3	0.5
half-cone angle, θ	°	6, 15, 30, 90

3. Results and discussion

The conical wall affects the motion of particles through two basic mechanisms. The first is that fluid motion around the tube opening is changed due to the existence of the conical opening and motion of particles are thereby affected. The second mechanism is that particles colliding with the conical wall will not only lose certain amount of momentum, but also are 'redirected' either into or away from the tube depending on the half-cone angle θ. To reveal how these mechanisms quantitatively affect *k*, computations under various *e* and θ are conducted, with results shown in table 2.

Table 2. Computation results.

case	d_p (μm)	θ=15°			θ=6°	θ=30°	θ=90°	straight tube[5]
		e=0.9	*e*=0.5	*e*=0.2	*e*=0.9	*e*=0.9	*e*=0.9	*e*=0.5
1	1	1.01	1.01	1.01	1.03	1.00	1.01	1.01
2	20	1.78	1.50	1.33	1.62	1.44	1.47	1.39
3	40	2.60	1.83	1.59	2.76	1.95	1.67	1.5
4	60	2.83	1.90	1.49	3.01	2.06	1.68	1.53
5	90	2.71	1.77	1.46	3.02	1.85	1.69	1.48
6	105	2.49	1.63	1.37	2.74	1.66	1.54	1.36
7	120	2.24	1.46	1.25	2.45	1.43	1.37	1.21
8	150	1.85	1.18	1.04	1.96	1.07	1.10	0.98
9	200	1.33	0.87	0.78	1.43	0.69	0.76	0.73

3.1. Effect of the restitution coefficient

Figure 2 shows variation of *k* versus d_p at *e*=0.9, 0.5 and 0.2. θ is fixed at 15°, which is a moderate value, ensuring that particles colliding with the conical wall bounce and keep moving inwards (i.e. are 'redirected') into the tube. The general trends under varied *e* are similar, i.e. *k* firstly increases with d_p due to increment of particle inertia, and then decreases since the stop distance of particles of higher inertia is longer than the tube length. Particles with d_p as low as 1μm have negligible inertia and behave as part of the fluid phase itself such that *k* is 1, as theoretically predicted.

Figure 2 shows *e* has significant influence on *k*, especially for d_p in the range of 40 to 105μm. The maximum value of *k* is increased by approximately 75% when *e* increases from 0.2 to 0.9. Therefore, the material of the tube especially the conical part must be carefully chosen to maximize *e* and hence *k*. The material should be hard enough as well to resist corrosion of particles, and the inner surface of the cone should also be as smooth as possible.

It is seen that the curve for d_p in between 40 and 105μm at *e*=0.2 is rather flat, similar as when a straight tube is used[5]. This phenomenon becomes less distinct when *e* is increased, i.e. *k* is more sensitive to particle inertia at higher *e*. The percentage of increment of *k* at *e*=0.9 comparing to *e*=0.2, however, lies in a narrow range of 1.8 to 1.9 for d_p in between 60 and 150μm.

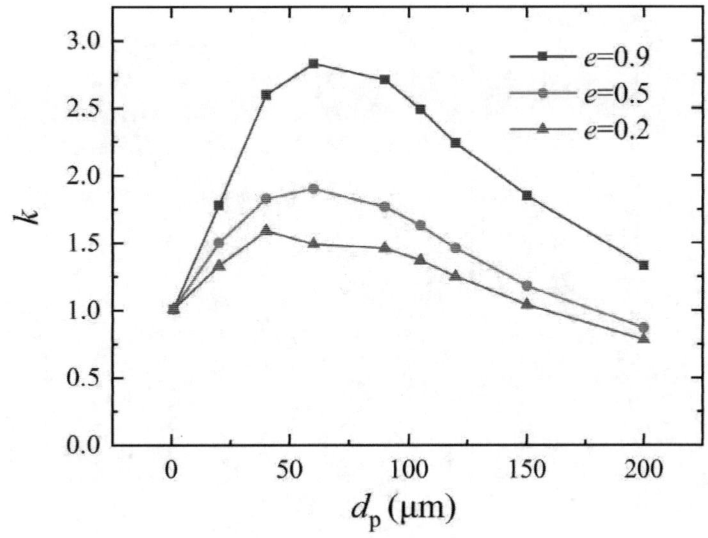

Figure 2. Variation of the conversion factor k with e at θ =15°.

3.2. Effect of the half-cone angle

Figure 3 shows k under different half-cone angles. One extreme condition is θ at 90°, in which particles colliding with the 'conical' wall will directly bounce outwards and fly away from the tube. The conical opening was designed to 'collect' more particles into the tube to enlarge k. However, such a design will take minor effects when θ is too big and hence the second mechanism as mentioned at the beginning of this section fails. On the other hand, k at θ=90° for all investigated particle sizes are indeed higher than that of a straight tube (smaller e of the later case is believed to have negligible effect on k, as shown in [5]). Therefore, the 'conical' wall at larger θ does help enlarge k, though not significantly. Such an effect is probably due to the first mechanism mentioned previously, as the 'conical' wall at larger θ helps creating an enlarged stagnation zone of fluid in front of the tube opening.

k at θ=90° and 30° does not differ remarkably, especially for $d_p \leq 20\mu m$ and $d_p \geq 120\mu m$. However, when θ is further decreased to 15° and 6°, k is significantly enlarged, especially for $d_p \geq 40\mu m$. k at θ=6° is even larger than the 15° case for $d_p \geq 40\mu m$. At small θ, particles still possess a large velocity component along the axial direction of the tube after colliding with the conical wall. Particles with moderate particle size have higher k. For smaller d_p, the forementioned second mechanism plays a less important role since fewer particles will collide with the conical wall due to their short stop distances.

Compared to the straight tube, k of the tube with θ=6° and e=0.9 is increased by nearly 100% for d_p>40μm. As previously mentioned, the expected of increment of k is one time for $D_o = \sqrt{2}D$. Consequently, the newly proposed design of the metering tube is effective.

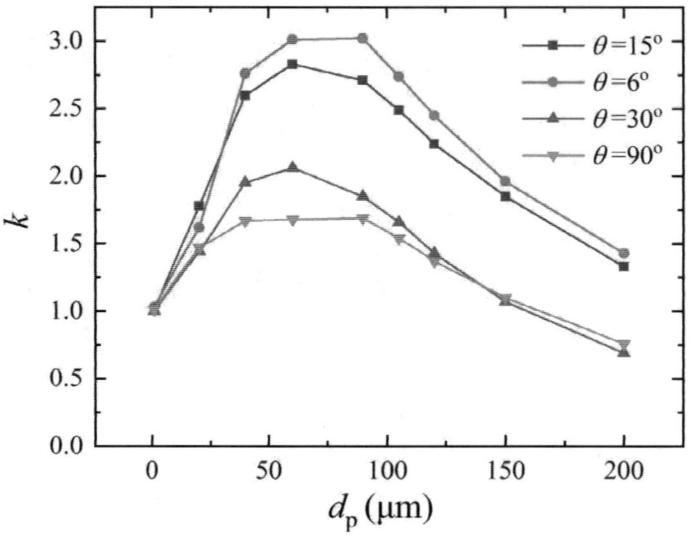

Figure 3. Variation of the conversion factor k with θ at e =0.9.

4. Conclusions

Optimization on the structure of the metering tube of a momentum-based measurement technique of solid concentration is conducted in this work. Tubes with a conical inlet are found to possess a higher conversion factor (k) than that of the straight tube, leading to improved measurement sensitivity. k is increased by nearly 100% under optimized design. k increases with decreasing half-cone angle and decreases with decreasing restitution coefficient of collision. θ around 6°~15° and higher e are recommended for the optimization of design.

Acknowledgments

This work has been supported by the CHDER R&D Project No.CHDERKJ19-02-08.

References

[1] Zhang, W.B., Wang, C., Yang, W.Q., Wang, C.H. (2014) Application of electrical capacitance tomography in particulate process measurement - a review. Advanced Powder Technology, 25: 174-188.
[2] Zheng, Y.N., Liu, Q. (2011) Review of techniques for the mass flow rate measurement of pneumatically conveyed solids. Measurement, 44: 589-604.
[3] Black, D.L., McQuay, M.Q., Bonin, M.P. (1996) Laser-based techniques for particle-size measurement: a review of sizing methods and their industrial applications. Progress in Energy and Combustion Science, 22: 267-306.
[4] Yan, Y. (1996) Mass flow measurement of bulk solids in pneumatic pipelines. Measurement Science and Technology, 7: 1687–1706.
[5] Liu G.Q. (2019) A momentum-based method for the mass concentration measurement of pneumatically conveyed solid: computational verification. In: The 5th International Conference on Advances in Energy Resources and Environment Engineering. Chongqing, China.

Task maintainability evaluation of complex system based on simulation

Song Mingchang, Shi Quan[a] and Wang Yadong

Shijiazhuang campus, army engineering university, 050003 Shijiazhuang City in Hebei Province, China

Shi Quan E-mail 1143190903@qq.com

Abstract: In view of the problem that the randomness of maintenance time of functional units cannot be considered in traditional system maintainability evaluation, this paper firstly analyzes the general process of system task maintainability evaluation and determines the basic method of task maintainability evaluation. Then, aiming at the problem that it is difficult to solve the problem analytically when several functional units fail at the same time, a task maintainability evaluation algorithm based on simulation is proposed. Finally, an example is given to illustrate the effectiveness and practicability of the proposed method.

1. Introduction

Maintainability refers to the ability to maintain or restore the product to the specified state when it is repaired according to the specified procedure and method under the specified conditions and within the specified time [1]. As a special case of maintainability, mission maintainability mainly reflects the equipment's ability to recover to the specified function after maintenance due to failure during the mission. Its measurement index mainly includes: task maintenance degree, task time to restore function, average downtime, etc. Reasonable evaluation of equipment task maintainability is the basic premise of implementing task-based maintenance decision.

Mission maintenance degree refers to the probability that the failure of mission function unit occurs in the mission profile, and the time consumed to restore to the specified function after maintenance is less than the specified time [2]. As with the recovery task time, the essence is to evaluate the maintenance time of the equipment. Literature [3] simulated the maintainability of basic mission of missile battalion through MATLAB simulation, and obtained the task time of mission maintenance degree and recovery function. In literature [4], Petri net technology was applied to establish a maintainability simulation model for mobile air defense mission of polymer-level equipment combat unit under ExSpect simulation environment. Literature [5] constructed a maintenance process model based on colored random time Petri nets, and on this basis, monte carlo algorithm was used to realize maintenance time simulation. In literature [6], the maintainability model block diagram is used to analyze the maintenance time of the hybrid system when the unit maintenance time is known. Literature [7] proposed the idea of considering parallel functional units as a whole to conduct task maintainability evaluation on the basis of related component state evaluation. In general, the current research focuses on the maintenance time estimation of a single unit, and rarely studies the overall maintenance time estimation required to maintain the equipment in normal state in the specific mission profile. Although literature [7] has analyzed the maintenance time of the system, it is assumed that the maintenance time of each unit is known in the analysis process. However, the maintenance time of

IOP Conf. Series: Materials Science and Engineering **793** (2020) 012054 doi:10.1088/1757-899X/793/1/012054

each unit is not an accurate value, so it is impossible to accurately estimate the system maintenance time. Therefore, this paper mainly establishes a quantitative evaluation model of mission maintainability which is applicable to the system failure caused by multiple failures, and evaluates the mission maintainability of the system by means of simulation and fitting.

2. System task maintainability evaluation modeling

For the modeling process of task maintainability evaluation of complex systems, the main steps are described as follows:

2.1 Draw a task reliability block diagram

The logical relationship between functional units, such as series mode, parallel mode and mixed relationship between series and parallel mode, is indicated by the task reliability block diagram, so as to prepare for the task maintainability evaluation of complex systems.

2.2 Distinguish repairable units from unrepairable units

Once the unrepairable unit fails, it is unable to implement any repair measures, and its task maintenance degree can be considered as 0. For repairable units, only by repairing all the repairable units that have failed in the task can the task continue to execute. Therefore, the distinction between repairable units and unrepairable units in the mission profile is the basis for assessing the maintainability of the mission [8]. The distinction is based on the possible failure modes of the functional unit during the mission, the support resources and forces in the mission profile, and the allowable downtime for maintenance in the mission profile.

2.3 Determine the probability of fault occurrence and repair time of repairable unit

In the process of a task, if several units fail at the same time, the task can be affected. It can be considered that there is a functional "parallel relationship" amone these units [9 10]. In order to ensure the timeliness of the task, only the unit with the minimum maintenance time consumption is repaired after the overall failure of the unit (all units fail) occurs during the task. Therefore, for repairable units with parallel connection, the failure probability is

$$P(a_k)=\prod_{j=1}^{m}P(a_j) \tag{1}$$

Where $P(a_k)$ represents the probability of failure of the JTH parallel element group, $P(a_{kj})$ represents the probability of failure of the JTH element in the group of parallel elements.

The overall fault removal time of the parallel part is:

$$t_k = \min(t_{kj}) \tag{2}$$

Where t_k Represents the maintenance time of the KTH parallel part, Represents the maintenance time of the JTH functional unit of the KTH parallel part, whose value is determined by the maintenance time probability density function of the functional unit.

2.4 Determine the mission maintenance of the system

In order to facilitate calculation and description, when constructing the combination of repairable units, the functional units with "parallel relationship" are regarded as a whole, so that the combination of repairable units becomes a pure series system.

Suppose an equipment performs a task, and there are n repairable functional units associated with the task, by a_1, a_2, a_3, $\cdots a_n$, respectively. The failure probability of each functional unit is represented by $p(a_1)$, $p(a_2)$, $p(a_3)$, $\cdots p(a_n)$ respectively. The probability density function of maintenance time of each unit is $m_{a1}(t)$, $m_{a2}(t)$, $m_{a3}(t)$ and $m_{an}(t)$ respectively. A represents the maintenance task and T represents the maintenance time allowed to troubleshoot this fault. If any of n repairable parts is damaged, maintenance task A will occur. It can be seen that the probability of maintenance task A is as follows.

$$P(A)=\sum_{i=1}^{2^n-1}P(A_i) \tag{3}$$

Where $P(A)$ represents the probability of maintenance task A. $P(A_i)$ represents the probability of maintenance events A_i. The events A_i is fault combination of functional units. There are n combinations, Its occurrence probability can be expressed by

$$P(A_i)=\prod_{x\in U}p(a_x)\prod_{y\notin U}(1-p(a_y)) \tag{4}$$

Where U represents the set of repairable units; a_x represents the failed unit; a_y represents the unit that did not fail;

The probability of exclusion within time T after the occurrence of fault event A_i is as follows:

$$M(A_i)=P(T_s\le T) \tag{5}$$

T_s represents the sum of repair time consumed by repairable unit. Its value is related to the probability density function of maintenance time of each failure unit. Therefore, the maintenance degree of maintenance task A can be obtained according to the posterior probability calculation principle.

$$M(A)=\frac{\sum_{i=1}^{2^n-1}P(A_i)M(A_i)}{P(A)} \tag{6}$$

3. System task maintenance degree calculation method

When multiple functional units fail, it is difficult to solve the maintenance time by means of analytical calculation. Therefore, monte carlo simulation can be used to solve the elimination probability. Specific calculation steps are as follows:

1) Based on the probability density function of repair time of n repairable units, repair time data was extracted. The maintenance time of parallel part units can be obtained through the method of section 1.4. Get the time data set $t_s=(t_s^{a_1},t_s^{a_2},\cdots,t_s^{a_n})$, $s=1,2\cdots,r$;

2) Analyze the repairable units with faults in fault events E_j, extract corresponding data from each group of data, and the repair time data of group r for repairable unit in E_j are obtained.

3) According to the set of allowable maintenance time of all possible task types of the equipment, the possible value interval $[T_{min},T_{max}]$ of allowable maintenance time when the equipment performs a certain task is determined.

Select m time parameters $T_1,T_2,\cdots T_m$ at equal intervals within this interval. Take the parameter T_k. Sum the data of group r obtained in step 3 respectively, count the number of groups whose summation result is not greater than T_k, and calculate the ratio of this number to the total number of data, q_k. Similarly, m two-dimensional vectors are constructed: $(T^1,q_1),(T^2,q_2),\cdots,(T^m,q_m)$.

4) The two-dimensional vector obtained in step 4 is fitted to obtain the functional expression with time variable T as the independent variable and ratio q as the dependent variable.In the function expression, T is replaced by t and q is replaced by $M_j(t)$, and the analytic expression can be obtained.

By using the above steps, the probability distribution function of elimination of each fault event can be obtained, and then the probability $M(A_1),M(A_2),\cdots,M(A_{2^n-1})$ of elimination of each fault event within a given maintenance time can be calculated.

According to the failure probability $p(a_1),p(a_2),\cdots p(a_n)$ of each repairable unit, use formula (4) to calculate the probability $P(A_1),P(A_2),\cdots,P(A_{2^n-1})$ of each failure event. Finally, the maintenance degree $M(A)$ of system task is obtained through formula (6).

4. Analysis of examples

When a certain type of equipment performs a task, there are 5 functional units related to the task.

Firstly, the task reliability block diagram is determined to clarify the relationship between each functional unit, as shown in figure 1.

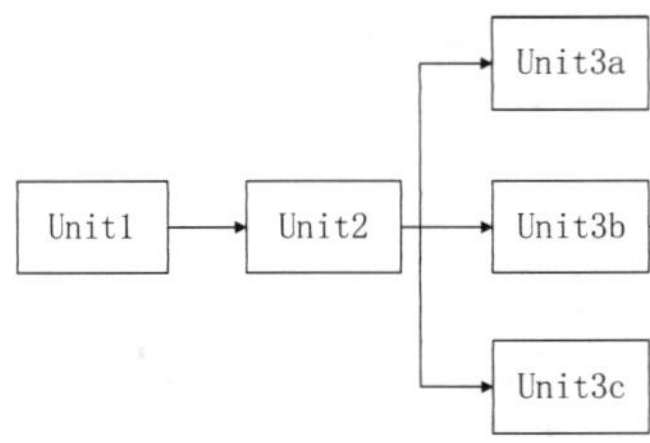

Figure 1. Task reliability block diagram.

According to the maintenance support force configuration and the design characteristics of the equipment, it can be concluded that the repairable unit has 4 functional units (2, 3a, 3b and 5), and the rest are unrepairable units. Therefore, the evaluation is mainly carried out around functional unit 2, functional unit 3a, functional unit 3b and functional unit 5.

Assume that the failure probability of task-related functional units during the task is shown in table 1.

Table 1 Failure probability of each functional unit during the task

Functional units	2	3a	3b	5
Failure probability	0.5	0.1	0.3	0.2

The probability density function of maintenance time of functional unit 2 is a positive distribution with mean value of 30.9 and standard deviation of 5.5. Functional unit 3a obeys the exponential distribution with a mean of 45, and functional unit 3b obeys the exponential distribution with a mean of 30. The functional unit obeys logarithmic normal distribution with mean value of 4.8 and standard deviation of 0.1.

Since there is a functional parallel relationship between functional unit 3a and functional unit 3b, it is regarded as a functional unit and treated as a functional unit 3 during evaluation. According to formula (1), the failure probability can be obtained. $p(a_3) = 0.03$.

Take the minimum time generated by the two units as the maintenance time data group element. It was simulated for several times and the obtained data was fitted. The maintenance time of the parallel part was exponential distribution with a mean value of 18.17, which was denoted as $M_2(t)$.

If the parallel part is regarded as a whole, the fault events can be divided into 7 categories, which are denoted as fault events $A_1 \sim A_7$.

Failure event A_1 indicates failure of function unit 2, normal of function unit 3 and function unit 5. According to formula (4), $P(A_1) = 0.3880$. The probability of elimination of the fault event can be expressed as $M_1(\mathrm{t})$, which obeys the probability distribution function of positive distribution with a mean value of 30.9 and a standard deviation of 5.5.

Failure event A_2 indicates failure of function unit 3, normal of function unit 2 and function unit 5. Its occurrence probability is $P(A_2) = 0.012$. The probability of fault event elimination can be expressed as $M_2(\mathrm{t})$, which obeys the exponential distribution probability distribution function with a mean value of 18.17.

Failure event A_3 indicates failure of function unit 5, normal of function unit 2 and function unit 3. Its occurrence probability is $P(A_3) = 0.097$. The probability of fault event elimination can be expressed as $M_3(\mathrm{t})$, which obeys the logarithmic positive distribution probability distribution function with mean value of 4.8 and standard deviation of 0.1.

Failure event A_4 represents failure of function unit 2 and function unit 3, and normal function unit 5. Its occurrence probability is $P(A_4)=0.012$. The probability of troubleshooting is expressed as $M_4(\mathrm{t})$.

Failure event A_5 represents failure of function unit 2 and function unit 5, and normal function unit 3. Its occurrence probability is $P(A_5)=0.097$. The probability of troubleshooting is expressed as $M_5(\mathrm{t})$.

Failure event A_6 represents failure of function unit 3 and function unit 5, and normal function unit 2. Its occurrence probability is $P(A_6)=0.003$. The probability of troubleshooting is expressed as $M_6(\mathrm{t})$.

Failure event A_7 represents failure of function unit 2 and function unit 3, and normal function unit 5. Its occurrence probability is $P(A_7)=0.003$. The probability of troubleshooting is expressed as $M_7(\mathrm{t})$.

The fault time probability functions of A_4, A_5, A_6 and A_7 are simulated and fitted by the calculation method of fault event elimination probability in section 2. And set $r=10^5$, $m=44$, the curve result can be obtained as shown in figure 3.

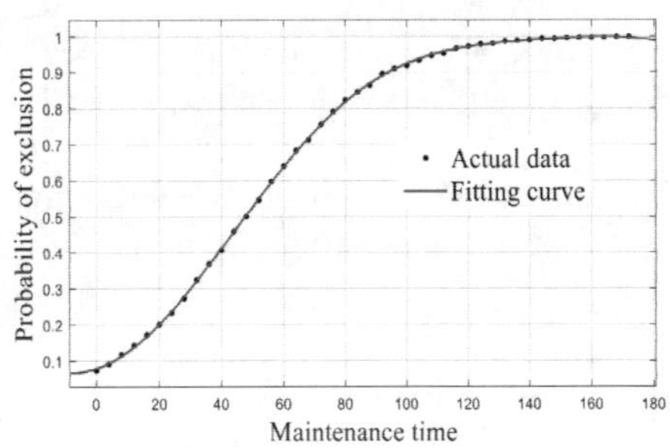

(a) $M_4(\mathrm{t})$ Fitting curve

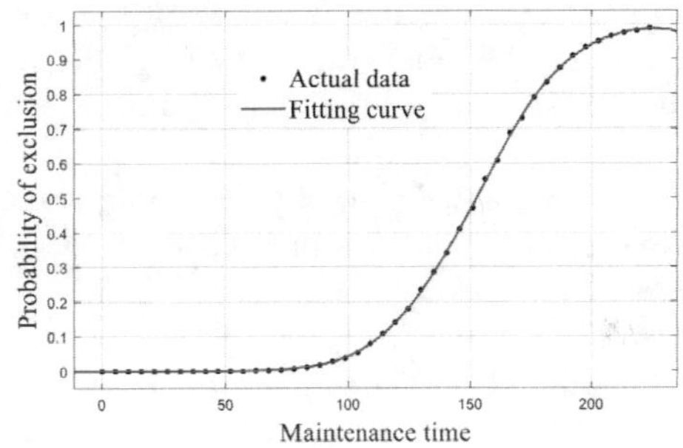

(b) $M_5(\mathrm{t})$ Fitting curve

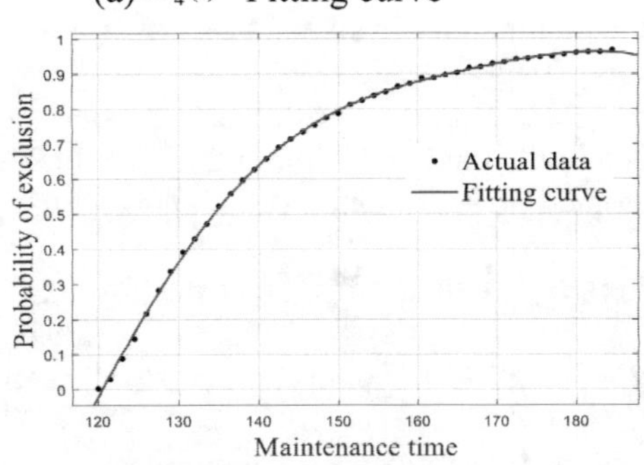

(c) $M_6(\mathrm{t})$ Fitting curve

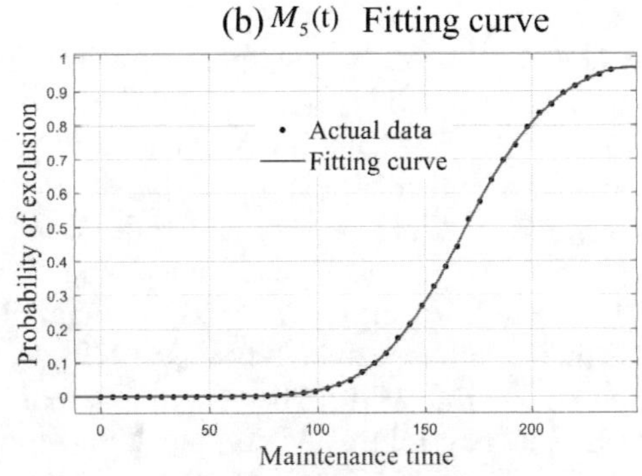

(d) $M_7(\mathrm{t})$ Fitting curve

Figure 3. Fitting curve of probability of fault event elimination

Fit the function curve, Where, $M_4(\mathrm{t})$, $M_5(\mathrm{t})$ and $M_7(\mathrm{t})$ adopt gaussian fitting to have the highest accuracy, thus the probability of elimination of each fault event can be obtained as follows:

$$M_4(\mathrm{t})=0.9193e^{-[\frac{(t-142.7)}{70.06}]^2}+0.425e^{-[\frac{(t-68.88)}{49.44}]^2} \tag{7}$$

$$M_5(\mathrm{t})=0.7025e^{-[\frac{(t-235.5)}{43.41}]^2}+0.6808e^{-[\frac{(t-181.2)}{48.81}]^2} \tag{8}$$

$$M_7(\mathrm{t}) = 0.8613e^{-[\frac{(t-252.7)}{63.42}]^2} + 0.3904e^{-[\frac{(t-187.8)}{49.05}]^2} \tag{9}$$

$M_6(\mathrm{t})$ adopts the highest accuracy of Fourier fitting, and then:

$$M_6(t) = 0.1005+1.069\cos(0.0364t)+(-0.0233)\sin(0.0364t) + (-0.2277)\cos(0.0727t)+ 0.0670\sin(0.0727t) \tag{10}$$

Finally, the task maintenance degree of the system was evaluated by formula (6), and the following results were obtained:

$$M(\mathrm{t}) = \frac{\sum_{i=1}^{7} P(\mathrm{A}_i)\mathrm{M}_i(\mathrm{t})}{\sum_{i=1}^{7} P(\mathrm{A}_i)} \tag{10}$$

If the maximum allowable maintenance time of the task is 140min, the probability of each fault event being eliminated is $M(\mathrm{A}_1) = 0.9998$, $M(\mathrm{A}_2) = 0.9995$, $M(\mathrm{A}_3) = 1$, $M(\mathrm{A}_4) = 0.9716$, $M(\mathrm{A}_6) = 0.6370$, $M(\mathrm{A}_7) = 0.1877$. The task maintenance degree of the system is obtained $M(\mathrm{A}) = 0.8889$.

5. Conclusion

This paper analyzes the maintainability of the complex system and gives the general process of the maintainability evaluation. The task maintainability of complex system is evaluated by drawing the task reliability block diagram, distinguishing repairable units from unrepairable units, and determining the failure probability and repair time of each repairable unit, especially the parallel part. Then aiming at the problem that multi-function convolution is difficult to calculate when multiple fault events occur simultaneously, a probabilistic calculation method of maintenance event elimination based on simulation is proposed. Finally by an example analysis shows that the proposed evaluation method can make up for the deficiency of the traditional task maintainability evaluation method.

References

[1]Qin Y X. Reliability. Maintainability. Guarantee management [M]. 2003.
[2]Zhu Z, Sun Q. Study on system maintenance degree based on fuzzy Markov process [J]. Arms automation, 2008(12):13-14.
[3]Feng H J, Liu X K, tang yanfeng. Basic mission maintainability simulation of missile battalion based on Matlab [J]. Tactical missile technology, 2012(4):48-52.
[4]Feng H J, Yu Y L, Zhang liu. Maintainability simulation of mobile air defense mission aggregation level equipment combat unit [J]. Computer technology and development, 2009(11):236-239.
[5]Lu Z, Sun Y C, Wu H Q. Maintainability modeling method based on colored random time Petri nets [J]. Chinese journal of mechanical engineering, 2011(10):189-195.
[6]Lu H S, Xu J, Zhang J W. Mission maintainability modeling of submarine power system [J]. Chinese ship research, 2008(04):14-18.
[7]Shen J, Li J, Bai M L. Mission maintainability evaluation of complex systems based on state assessment [J]. Fire and command and control, 2014(6).
[8]Han X F, Zhang Y H, Wang shaohua. Study on maintenance evaluation method of repairable unit combined task [J]. Journal of military engineering (11).
[9]Han X, Zhang Y H, Wang S H. assessment method for equipment mission success considering maintenance work [J]. Systems engineering and electronic technology (3).
[10] Sun Y C. Information theory method for reliability synthesis of exponential hybrid systems [J]. Acta aerodynamics sinica, 2005, 20(5):783-788.

Numerical simulation of the flow in a vertical vane type tank with micro downward acceleration

Jintao Liu[1,2], Yong Li[1*], Wen Li[1,2], Lei Chen[1] and Qiang Bi[1]

[1]Beijing Engineering Research Center of Efficient and Green Aerospace Propulsion Technology, Beijing Institute of Control Engineering, Beijing, 100094, China

[2]Tianjin Research Institute for Advanced Equipment, Tsinghua University, Beijing, 100083, China

Corresponding author: ljt86@zju.edu.cn

Abstract. Vane type propellant tank is widely used in propellant system of satellite at present for that it can be refuelled in space. The performance of a vertical vane type tank is determined by propellant management device (PMD). The VOF model with commercial code (Fluent) was used to simulate the gas-liquid distribution with micro downward acceleration. Two kinds of volume fractions of liquid were chosen for the simulation. The flow rate of the liquid can reach 1300 ml/s. The potential energy caused by gravity for the initial conditions would be released in three seconds. The liquid in the tank would keep steady state in 28 seconds. These results could provide a reference for high acceleration maneuvers of a satellite.

1. Introduction

Surface tension forces are obvious in space with micro acceleration and it can determine the liquid distribution in a tank. Traditional net type tank used in space is hardly to be refilled by liquid in PMD when there has gas inside PMD. Tank with vertical vane is a new type tank for the propellant which can be refuelled in space. In order to ensure gas free propellant delivery for a thruster, the location of gas and liquid in a vertical vane type tank must be evaluated after the tank geometry is designed.

Two methods were used to evaluate the control performance of vertical vane in micro gravity. One way was numerical simulation. Hu[1] studied the influence of parallel vane in a tank. The performance of the parallel vane was studied with different widths and angles. It was found that the flow rate increases with larger width of the vane, while the climbing height will decrease. Liu[2] numerical studied the flow distribution in a tank with parallel vane in different accelerations. Zhuang[3] studied the damping effects and the basic frequency of the propellant in a tank by numerical simulation. The other way is experiment. Rui [4] investigated the location of gas and liquid flow in a sphere vane type tank using different fractions of propellant as well as different acceleration vectors. Bolleddula [5] experimental studied the capillary flow in a vane type tank by drop towers. Li[6] investigated the capillary flow in a vane-type propellant tank. Zhuang[7] performed an drop tower test to study the flow distribution and refillable performance in a reservoir with micro gravity condition.

A new kind of tank with vertical vanes is established. Flow distributions with micro downward acceleration in space were investigated.

2. Tank Geometry

The model of the tank with vertical vanes is shown in Fig. 1. The tank is form by two hemisphere part and a cylinder part. The PMD has 12 small vanes, 4 large blades and 4 vanes. The diameter of the tank is 476 mm, and the height of the cylinder part is 396 mm.

Table 1. Parameters of the vane type tank

Diameter of the tank (mm)	476
Height in cylinder part (mm)	396
Number of blade	4
numbers of vanes	4
numbers of small vanes	12

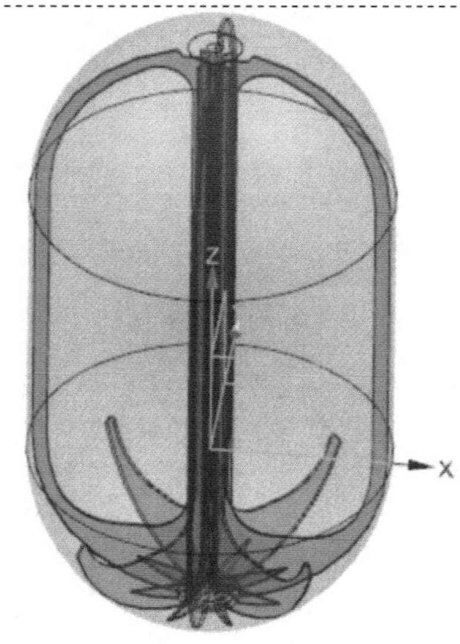

Figure 1. Vane type tank model

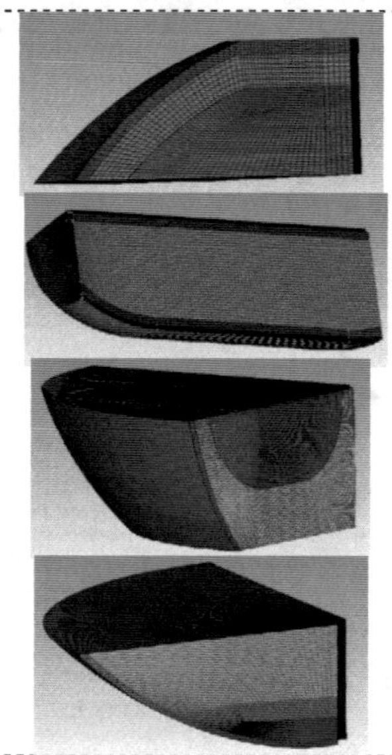

Figure 2. Mesh of each part of the tank

3. Computational Method

3.1 Volume of fluid model

The volume of fluid (VOF) method are usually used to simulate the gas and liquid separative flow and capture the gas-liquid interface based on a fractional volume of fluid.

The continuity equation for the gas and liquid flow is

$$\frac{\partial}{\partial t}(\rho\ )+\frac{\partial}{\partial x_i}(\rho u_i)=R \tag{1}$$

R is the source term. ρ is the mixture(gas and liquid) density. u_i is the averaged velocity for mass. For gas and liquid flow, for example, the density with different location is given by

$$\rho=\alpha_1\rho_1+\alpha_2\rho_2=(1-\alpha_2)\rho_1+\alpha_2\rho_2 \tag{2}$$

α is the volume fraction of the phase.

The volume fraction equation is shown as follows,

$$\alpha_1+\alpha_2=1 \tag{3}$$

3.2 Simulation conditions

FLUENT was used to simulate the gas and liquid distribution in the tank with VOF model. The mass conservation is performed by SIMPLEC algorithm. No slip boundary condition on the wall was used during the simulation. The gas in the tank is He and the liquid is N_2H_4. Properties in temperature 20 ℃ of He and N_2H_4 are shown in table 1. The surface tension coefficient between the gas and liquid if 74.76 dyn/cm. The time step for the unsteady flow was 0.0001 s. For all calculations, simulations should be convergence with residual error less than 0.0001.

Table 1. Property of the two phases

Fluid	Tempreture(℃)	Density(g/cm^3)	Viscosity (10^{-4}Pa.s)	Surface tension coefficient (dyn/cm)
N_2H_4	20	1.008	9.51	74.76
He	20	0.1625	0.199	/

Numerical simulations were performed to study the location of gas and liquid with different downward accelerations in the tank. The model's grids are divided into four part using unstructured hexahedron and tetrahedron.

4. Result and Discussion

The distribution of gas-liquid interface in the relocation process with 25% volume fraction is shown in Fig. 3. The propellant climbed to the top of the tank along the central column at six seconds. At 28 seconds, the force for the fluid in the tank is in equilibrium, and the distribution of gas-liquid interface is stable. When the propellant in the tank is within 12 seconds, the propellant on the guide plates on both sides merges with the propellant on the central column, which constrains the gas in the internal area of the tank. The distribution of gas-liquid interface in the relocation process with 50% volume fraction is shown in Fig. 4. It has the same rule with volume fraction is 25%. The distribution of gas-liquid interface is stable at 28 seconds.

The flow rate changes with time is shown in Fig. 5. The potential energy caused by gravity will be totally released when the flow rate of liquid reached the maximum value. It can be found that the time for the complete release of potential energy is shorter than three seconds. The maximum flow rate of liquid along the four vanes is about 1300 ml/s.

5. Conclusion

Two kinds of volume fractions of liquid in the vane type tank were evaluated. The flow rate of the liquid can reach 1300 ml/s. The potential energy caused by gravity for the initial conditions would be released in three seconds. The propellant climbed to the top of the tank along the central column at six seconds. The liquid in the tank would keep steady state in 28 seconds. These results could provide a reference for high acceleration maneuvers of a satellite.

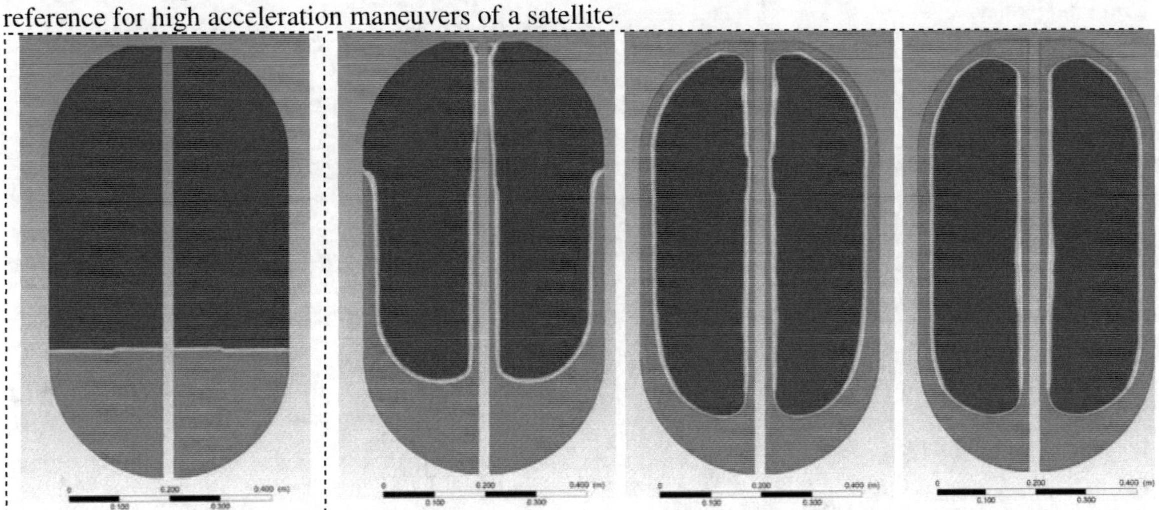

0 s	6 s	14 s	28 s

Figure 3. Relocation process of liquid in tank with volume fraction is 25%

IOP Conf. Series: Materials Science and Engineering **793** (2020) 012055 doi:10.1088/1757-899X/793/1/012055

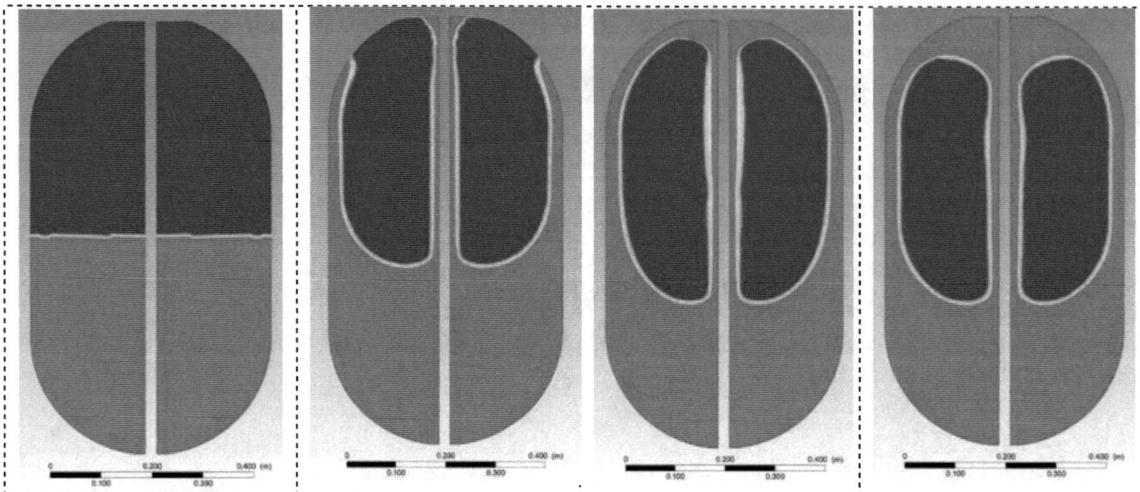

0 s	6 s	14 s	28 s

Figure 4. Relocation process of liquid in tank with volume fraction is 50%

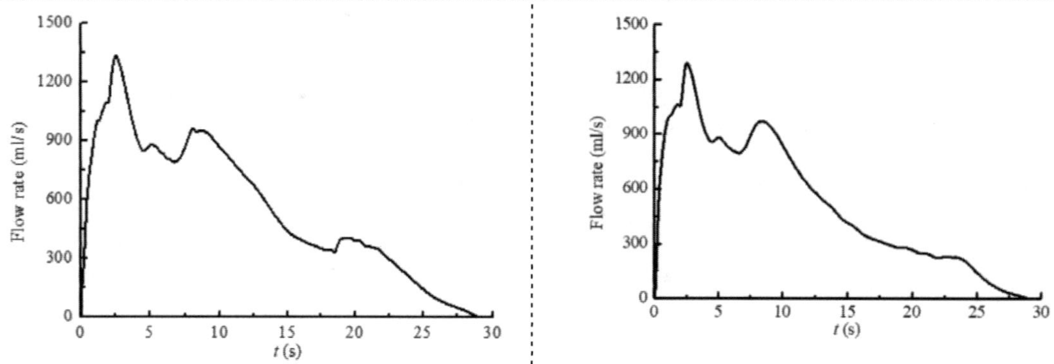

25%	50%

Figure 5. Liquid flow rate during relocation process with different volume fractions

References

[1] Hu Q, Li Y, Pan H, Liu J, Zhuang B, 2015, Numerical analysis and experiment research on fluid orbital performance of vane type propellant management device, IOP Conf. Series: Materials Science and Engineering 72 (2015) 042042.

[2] Liu J T, Zhou C, Wu Y L, Zhuang B T and Li Y, 2015, Numerical investigation of performance of vane-type propellant management device by VOF methods, IOP Conf. Series: Materials Science and Engineering 72 (2015) 042041.

[3] Zhuang B, Li Y, Pan H, Hu Q, Li Z 2012 Experiment Investigation on Transportation Velocity of the Fluid on Propellant Acquisition Vanes under Microgravity Environment, Aerospace Control and Application, 38(6) pp 1-5.

[4] Rui W., Numerical Simulation and Experimental Research of Orbital Liquid Transport and Interface Behavior in Vane Tank. Northeastern University,2014.

[5] Bolleddula D. A., Chen Y., Semerjian B., et al., Compound Capillary Flows in Complex Containers: Drop Tower Test Results. Microgravity Science & Technology, 2010, 22(4):475-485.

[6] Li J. H., Chen X. Q., Huang Y. Y., Capillary Flow in the Vane-Type Surface Tension Tank: Drop Tower Test Results. Applied Mechanics & Materials, 2012, 224:325-328.

IOP Conf. Series: Materials Science and Engineering **793** (2020) 012055 doi:10.1088/1757-899X/793/1/012055

[7] Zhuang B., Li Y., Pan H., et al., Experimental Investigation on Fluid-Storage Characteristic of Propellant Refillable Reservoir in Microgravity Environment. Aerospace Control & Application, 2014,2014(1):27-30,

IOP Conf. Series: Materials Science and Engineering **793** (2020) 012056 doi:10.1088/1757-899X/793/1/012056

Genetic analysis of remaining oil in thin and poor layers in Lamadian oilfield

Liu Wenchao

Production Sixth Plant of Daqing Oilfield,
Northeast Petroleum University
DaQing,Heilongjiang Province,163114,China
E-mail:20620054@qq.com

Abstract. Understanding the distribution of remaining oil and its genesis is the basic work of geological research. Lamadian oilfield has entered the late development stage of ultra-high water cut, the thick oil layer of water drive is decreasing gradually, and the thin and poor oil layer is becoming an important target of adjustment and tapping potential below 1.0m. In order to further improve the development effect of thin and poor reservoir, it is necessary to analyze the saturation field of the numerical model based on interpretation result, core analysis result of sealed coring well and water drive model. The remaining oil characteristics of thin and poor reservoir are studied by means of statistics, genetic analysis, dynamic analysis and static analysis.

1.Introduction

Taking the sedimentary facies as the research leader, this paper analyzes the sedimentary characteristics of the distributary plain subfacies and delta front subfacies, which are mainly developed in the thin and poor layers, and combines them with the permeability data of logging interpretation results. The reservoir with effective thickness less than 1.0 m is dominated by uniform water flooding, and the remaining oil in the thin and poor layers in Lamadian oilfield is mainly formed because of the influence of plane and interlayer heterogeneity, which leads to imperfect injection-production relationship. Through the analysis of remaining oil in thin and poor reservoirs under the sedimentary models of abandoned channels, interfluvial subfacies and branch-tuotransitional and sheet-sand subfacies in the delta front, the types and origin of remaining oil in the current facies-controlled model are clarified, in order to guide more effectively Lamadian oilfield extra-high water cut later on the remaining oil tapping program preparation.

2.Type of remaining oil in thin difference layer

After more than 40 years of development in Lamadian Oilfield, through fine adjustment of strata and infill well pattern, the control of well pattern on sand body is very high, and the proportion of effective sandstone producing thickness is 95.8%. Among them, the proportion of reservoir thickness below 1.0 m is only 10.1% , and there is some remaining oil.

Table 1. Statistical table of sand body producing degree

Effective thickness level (meter)	Perforation condition			Active proportion			Unused proportion		
	Number of layers (Number)	Sandstone thickness (meter)	Effective thickness (meter)	Number of layers (%)	Sandstone thickness (%)	Effective thickness (%)	Number of layers (%)	Sandstone thickness (%)	Effective thickness (%)

≥2.0	9	24.2	23.7	100.0	100.0	100.0			
1.0~2.0	203	292.7	273.2	97.5	97.8	97.8	2.5	2.2	2.2
0.5~1.0	814	690.7	606.6	96.0	96.2	96.1	4.1	3.8	3.9
<0.5	1189	623.0	428.3	93.4	93.1	93.8	6.6	6.9	6.2
Total	2215	1630.6	1331.8	94.8	95.4	95.8	5.2	4.7	4.2

The sedimentary characteristics of the distributary plain and delta front subfacies are combined with the permeability data from logging interpretation results, and the sedimentary facies is taken as the research leader. Based on the analysis of the sedimentary characteristics of the distributary plain and delta front subfacies, which are mainly developed in the thin and poor layers.And the permeability data in the logging interpretation results, the reservoirs with effective thickness less than 1.0m are mainly flooded by uniform water. The remaining oil in the thin and poor layers in Lamadian oilfield is mainly caused by plane and interlayer heterogeneity, which leads to imperfect injection-production relationship. Through the analysis of remaining oil in thin and poor reservoirs under the sedimentary models of abandoned channels, interfluvial subfacies and branch-tuotransitional and sheet-sand subfacies in the delta front, the types and origin of remaining oil in the current facies-controlled model are clarified, in order to guide more effectively Lamadian oilfield extra-high water cut later on the remaining oil tapping program preparation.

2.1Distribution characteristics of remaining oil in subfacies sedimentary environment of distributary plain

There are two types of remaining oil enrichment areas: one is that the remaining oil enrichment area is at the variation part of the river channel edge, the variation part of the crevasse sector edge, and the middle and upper part of the abandoned river channel. The second is the remaining oil enrichment area in the middle of two or more rivers in the marsh[1].

2.2 Distribution characteristics of remaining oil in delta front subfacies sedimentary environment

There are three situations in the remaining oil enrichment area: one is that the remaining oil enrichment area of the sheet sand in the delta is in the equivariant position at the side of the underwater channel[2]; The other is the remaining oil-rich area of delta-branch-tuotransitional subfacies in the sheet sand body between two or more channels, and the third is the isolated sheet sand body.

3. Static origin of remaining oil in thin difference layer

3.1 Analysis on the cause of formation of distributary plain subfacies control

The sub-facies of distributary plain is characterized by the deposition of each stage cutting down to the last stage. The middle part of the river has stronger ability to cut down than the side part. In the horizontal connecting section, the thin difference layers are mostly connected with the middle and upper parts of the channel body, and the thin difference layers formed by the variation of the channel side are of the same type as the contemporaneous deposits of the channel body.That is the thin difference layers sand bodies and the channel sand bodies are of the same origin. It is connected with the sand body in the middle of the channel. The remaining oil mainly exists in the middle and upper part of the channel sand body, and the thin and poor layers are formed after the physical properties of the sand body become worse.

3.2Analysis of the controlling causes of delta front subfacies

Intertidal zone exists in the sub-facies sedimentary environment of delta sheet sand, that is the sand body thickness decreases under the same conditions as that of fluvial sand body. Because it is washed and reformed by the lake tide. The lateral development range of sand bodies is enlarged, and the sand bodies extension distance along the river flow direction is reduced. In particular, injection-production wells are easily formed in the branch-tuo transitional subfacies in the underwater channel and sheet sand.Respectively the reservoir physical properties of underwater channel sand body and sheet sand body are quite different. The effective porosity and permeability of underwater channel sand body are much larger than that of sheet sand body.

4.Dynamic origin of remaining oil in thin differential layer

4.1Genetic analysis of plane contradiction remaining oil

According to the development data , the initial permeability of channel sand body is 0.5-0.8 darcy , and some reservoirs even reach 1.0 darcy. The thin differential layer is affected by the permeability range.,which results in low degree of production and almost no production of the off-balance sheet reservoir. Taking the abandoned channel subfacies as an example, the remaining oil is enriched in the middle and upper part of the reservoir sand body, especially in the meandering river sedimentary environment, the abandoned channel subfacies develops more, and the well logging curves show bell or pagoda type. The lithology and physical property difference in the upper part of the middle part is mainly the thin difference layer in the plane. Although it is connected with the river channel, but before the permeability difference is 3.3-6, and before the effective measures such as fracturing are taken to improve the permeability of the reservoir. Because of the phase change, the poor and poor layers interacted with the channel on the plane, and the control of the original well pattern on the sand body decreased.Finally the incomplete residual oil was formed in the sand body of the thin and poor layers.

4.2 Genetic analysis of remaining oil in interlayer contradiction

Because of different sedimentary environments and hydrodynamic characteristics, the lithology and physical properties of the sedimentary sand bodies are different vertically, and the physical properties are good when multiple reservoirs with different perforations are simultaneously developed. The reservoir with good porosity and permeability is used preferentially, and the reservoir with low permeability can not be used when the range of permeability between different reservoirs is more than 5[3]. In the fluvial sedimentary environment, the porosity and permeability of the sand bodies in the middle part of the channel is good, the initial permeability can reach 0.8-1.0 darcy, and the initial permeability of the thin difference layer is generally not more than 0.15 darcy. The permeability of off-balance sheet reservoirs is lower, almost all of them are less than 0.05 darcy, even less than 0.015 darcy. Individual layers can be up to 10 or more. According to the field experimental data , interlayer interference will be formed when the permeability range reaches more than 2.5, the interlayer interference is serious. Which leads to the low degree of production of thin and poor layers, and the reservoir outside the table is hardly used. Finally the incomplete residual oil is formed in the sand body of the thin and poor layer.

5.Conclusion

The remaining oil in distributary plain subfacies is mainly distributed in channel margin, crevasse fan margin, interchannel subfacies and abandoned channel subfacies. So there is no injection-production relationship between thin difference layer and channel sand body. The remaining oil in delta front subfacies is mainly distributed in sheet sand subfacies with different porosity and permeability characteristics, intermittent underwater channel sand body and isolated sand bar sand body.There is no perfect relationship between the thin difference layer and the underwater channel sand body or sand bar sand body.

Acknowledgments

I would like to express my gratitude to all those who have helped me during the writing of this theis.I gratefully acknowledge the help of my supervisor Professor Wang Yachun.I do appreciate her patience,encouragement, and professional instructions during my thesis writing.

IOP Conf. Series: Materials Science and Engineering **793** (2020) 012056 doi:10.1088/1757-899X/793/1/012056

References

[1] Zhao Wei, Zhao Yunfei, Wang Liming, et al. Reservoir production and comprehensive adjustment direction of Lamadian Oilfield [a]. Daqing Petroleum Geology and development, 2002, 21 (2): 26-28
[2] Han Weidong, Huang Fusheng, AI Ying, et al. Methods for tapping the remaining oil potential of Lamadian Oilfield [a]. Daqing Petroleum Geology and development, 2002,21 (3): 41-43
[3] Zhang Baosheng, Zhang Shujie, Zhong Ling, et al. Water washing conditions of various oil layers in LaSaXing oilfield [a]. Daqing Petroleum Geology and development, 2002,21 (6): 40-43

IOP Conf. Series: Materials Science and Engineering **793** (2020) 012057 doi:10.1088/1757-899X/793/1/012057

Parametric Energy Consumption Modeling for Cathode Coating Manufacturing of Lithium-Ion Batteries

Kai Zhou[1], Yongjian Su[1], Weiwei Liu[1] and Yelin DENG[2]*

[1]School of Mechanical Engineering, Dalian University of Technology, Dalian, Liaoning Province, 116024, China

[2]School of Rail Transit, Soochow University, Suzhou, Jiangsu Province, 215006, China

* Corresponding author: E-mail: yelin.deng@suda.edu.cn

Abstract. In the process of manufacturing the positive electrode coating for lithium batteries, the slow and energy-consuming drying process greatly restricts the shortening of the production cycle and the improvement of energy efficiency of power lithium batteries. The purpose of this paper is to calculate the evaporation rate of N-methylpyrrolidone (NMP) solvent under different drying parameters in the first stage of evaporation and to predict the main energy consumption changes in the drying process in this stage. This research shows that, in the first evaporation stage of lithium battery coating manufacturing, when the inlet wind speed changes at 2~12m s^{-1}, the evaporation rate of NMP solvent obtained in the coating will change in 1.18×10^{-3} ~2.88×10^{-3} g min^{-1} cm^{-2}. For every 1m s^{-1} increase in wind speed, the energy consumption of electric heating increases by 57.60kW on average.

1. Introduction

In recent years, the development of the electric vehicle industry has greatly promoted the development of power lithium battery [1-3]. However, there are many problems behind the explosive growth of power lithium battery industry, especially the high environmental impact of battery production. Among them, the drying process is the main time consuming and energy consuming link of positive electrode coating [2-4]. Compared with graphite negative electrode as a solvent [5, 6], the boiling point of positive electrode N-methylpyrrolidone (NMP) solvent is as high as 204℃, and its drying temperature is mostly in the range of 90-150℃ [7, 8]. The evaporation rate is relatively slow, leading to a long evaporation time and production cost. In addition, NMP is an explosive gas, and the volume ratio of the lower explosive limit in the air is 1.1% [9]. Therefore, evaporation requires large air volume to dilute the concentration of NMP in the air, which will cause high energy consumption [10].

Currently, Ahmed *et al* [9] measured the energy consumption of each process in the evaporation and recovery of NMP. Their results showed that producing a 10kWh battery pack required 420kWh of electricity to evaporate the NMP solvent. The actual heating capacity of the drying process is 45 times that of NMP evaporation. Therefore, the setting of inlet air temperature and air volume of coating drying has a great influence on the energy consumption of drying. Deng *et al* [2] used Primary Energy Demand as the indictor to analyze the "Cradle to Gate (CtG)" stage of the current lithium battery pack. The results show that the primary energy consumption in the CtG phase is about twice that of the battery material. The drying process of coating manufacturing is the main energy source of the single battery manufacturing process, accounting for 40% of the total energy consumption of the single battery manufacturing.

According to previous studies from Jaiser *et al* [11], the drying process of the electrode experiences a prolonged constant evaporation rate stage (first stage), next to a very short decreased rate stage. The reason from Jaiser *et al* is that capillary force within the porous structure drives the NMP solvent to the surface of the electrode, and thus, maintaining the evaporate rate of NMP. The dominant first stage evaporation also facilitates our calculation for the evaporation rate. Therefore, In this study, we will quantitatively calculate the evaporation rate in the first stage of evaporation under different drying parameters. Moreover, a parameterized process energy consumption model was established for the coated desiccants to calculate the change of main energy consumption in the drying system when the inlet wind speed was changed in the first stage of evaporation.

2. Method

2.1. The evaporation rate formula of the first stage of NMP solvent evaporation

According to the research of the same research group Su *et al*, we can get the evaporation rate formula of the first stage of the evaporation of NMP solvent [12].

First of all, we can use the type and the coating temperature (T_S):

$$q = h(T_{in} - T_s) \tag{1}$$

Where q is the heat flow density and h is the convective heat transfer coefficient.

Then we can also calculate the mainstream flow velocity of the coating surface (V_∞) through the following equation:

$$v_\infty = v_{in} \cdot k \tag{2}$$

Where V_{in} is the inlet wind speed of the drying box, k is the distribution wind coefficient, which is related to the vertical height (H) of the coating from the tuyere, the hydraulic radius (r) of the inlet and the space volume (V).

The evaporation rates of NMP can be calculated according to formula (3).

$$e_1 = \frac{c_{sa}(T_s) - c_\infty}{\gamma_{BL}} \tag{3}$$

Where e_1 is the evaporation rate at the first stage, the unit is (mol m^{-2} s^{-1}).

Finally, it can be concluded that the relation between T_s and v_∞ of e_1 in evaporation stage 1 is the formula (4):

$$\begin{aligned} e_1 = 5.9931\times10^{-6}\cdot v_\infty^{1/2}(2.67159\times10^{-7}\cdot T_S^{\ 5} - 3.7468\times10^{-4}\cdot T_S^{\ 4} + 0.1989\cdot T_S^{\ 3} \\ -47.29151\cdot T_S^{\ 2} + 4243.28404\cdot T_S) \end{aligned} \tag{4}$$

2.2. Energy consumption modeling of the drying system

In the dynamic drying system of lithium battery polar plate, the energy distribution includes: (1) temperature rise of polar plate substrate and coating solute; (2) temperature rise, vaporization and overheating of coating solvent; (3) electric heating energy consumption of dry medium (hot air); (4) energy consumption by the system for wind (fan); (5) energy consumption of wheel dehumidifier; (6) heat loss of drying equipment (negligible when small); (7) exhaust (waste air) sensible heat.

As shown in figure 1, the variable power of lithium battery positive plate desiccant comes from three parts: electric heater, fan and wheel dehumidifier. When the system begins to work, the fresh air is dried by the wheel dehumidifier. The dry air enters the heat exchanger and the return air of the drying box carries on the preliminary heat exchange in different pipelines. Then the dry air after preheating is sent to the heater for heating. After heating, it is sent to a drying oven to dry the wet pole. The return air after drying is returned to the heat exchanger and discharged as waste gas after heat exchange.

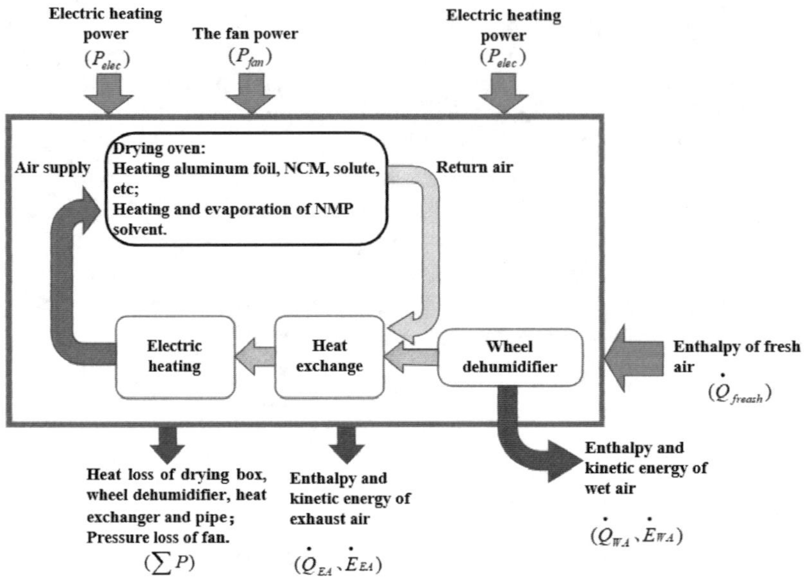

Figure 1. Variable power flow during coating drying

The drying system in figure 1 is a stable thermodynamic open system, and the energy entering the system and the energy leaving the system have the energy conservation relationship in steady-state. For the first stage of evaporation studied in this paper, we assume that a static wet electrode is dried in a drying box, and the heat balance equation is established as follows:

$$\sum P_{elec}+\sum P_{fan}+\sum P_{WD}+\dot{Q}_{fresh}=\dot{Q}_{WA}+\dot{Q}_{EA}+\dot{E}_{EA}+\dot{E}_{WA}+\sum P \quad (5)$$

Where: $\sum P_{elec}$, total electrical heating power, kW; $\sum P_{fan}$, total fan power, kW; $\sum P_{WD}$, total power of wheel dehumidifier, kW; $\sum P$, heat loss of drying box, wheel dehumidifier, heat exchanger and pipe and pressure loss of fan, kW; $\dot{Q}_{freash}$, enthalpy of fresh air, kW; $\dot{Q}_{WA}$, enthalpy of wet air, kW; $\dot{Q}_{EA}$, enthalpy of exhaust air, kW; $\dot{E}_{EA}$, the kinetic energy of wet air, kW; $\dot{E}_{WA}$, the kinetic energy of exhaust air, kW.

Main parameters in the heat balance equation are calculated according to the following formula:

$$\dot{Q}_{fresh}=q_m[1.005T_0+d(2501+1.86T_0)] \quad (6)$$

$$\dot{Q}_{WA}=q_m[1.005T_1+d(2501+1.86T_1)] \quad (7)$$

$$\dot{Q}_{EA}=q_m[1.005T_4+d_N(r_N+C_N T_4)] \quad (8)$$

Where: 1.005, average constant pressure specific heat capacity of dry air, kJ kg^{-1} ℃$^{-1}$; 2501, latent heat of water gasification, kJ kg^{-1}; 1.86, the average specific heat of water vapor at constant pressure, kJ kg^{-1} ℃$^{-1}$; q_m, mass flow of air kg s^{-1},T_0, initial air temperature, ℃; T_1 is the temperature of wet air, ℃; T_4, the temperature of exhaust air, ℃; d, moisture content of air, kg(water vapor) kg (dry air)$^{-1}$;d_N, NMP content of air, kg(NMP) kg(dry air)$^{-1}$; r_N, latent heat of vaporization of NMP, kJ kg^{-1}; C_N, average specific heat capacity of NMP at constant pressure, kJ kg^{-1} ℃$^{-1}$.

Because evaporation is an endothermic process. Heat is supplied either by convection, conduction, radiation, or a combination of methods. The solvent evaporates from the coating surface, while the latent heat absorbed by evaporation cools the coating surface. Therefore, heat and mass transfer are carried out at the coated surface.

Assume that heating is provided by convection of hot air only and that the substrate is impenetrable. If the internal resistance of the solvent to the coated surface is ignored, then the heat and mass balance when lumped parameter system. According to the solvent manual, the evaporation energy consumption formula can be obtained [13]:

$$\dot{Q} = Lb\left[h(T_{in} - T_s) - e_1 M_N r_N\right] \tag{9}$$

Where: L, length of wet electrode, m; b, width of the wet electrode, m; H, heat transfer coefficient; M_N, the molar mass of NMP.

The general heat transfer coefficient can be written as follows:

$$h = K v_\infty^n \tag{10}$$

Where, K, constant, are related to the physical properties of air and the collection size of the drying oven; n, 0.6~0.8 (related to nozzle shape)

The return air temperature T_2 can be obtained by formula (11), (12) and (13):

$$\dot{Q}_{in} - \dot{Q}_{RA} = \dot{Q} \tag{11}$$

$$\dot{Q}_{in} = 1.005 q_m T_{in} \tag{12}$$

$$\dot{Q}_{RA} = q_m[1.005T_2 + d_N(r_N + C_N T_2)] \tag{13}$$

Where: $\dot{Q}_{in}$, enthalpy of air at the inlet of the oven, kW; $\dot{Q}_{RA}$, Enthalpy of return air, kW; $\dot{Q}$, the evaporation energy consumption, kW.

The above is the analysis and calculation model of energy balance and energy consumption of the drying system. On the whole, the analysis model shows the balance between the energy input, energy utilization and energy loss.

3. Results

On the basis of lithium battery positive electrode coating manufacturing, the base material is aluminum foil, the coating solute is NCM material, and the solvent is NMP solvent. The composition of adhesives and conductive agents are ignored, and the typical dry conditions are selected. The relevant known data are listed in table 1.

Table 1. Relevant known data

Parameter	Data
Width of wet electrode (b)	0.38 m
Length of the wet electrode (L)	1 m
Initial air temperature (T_0)	25 °C
Air temperature at the inlet of the drying oven (T_{in})	115 °C
Coating temperature (T_s)	70 °C
Inlet wind speed of drying oven (V_{in})	2 m s^{-1}
Moisture content of air (d)	0.015 kg kg^{-1}
NMP content of air (d_N)	0.017 kg kg^{-1}
Average specific heat capacity of NMP at constant pressure (C_N)	6.44 kJ kg^{-1} °C^{-1}
Latent heat of vaporization of NMP (r_N)	439.5 kJ kg^{-1}
The molar mass of NMP (M_N)	98.70 g mol^{-1}
Ventilation coefficient (k)	0.7
K	6
n	0.6

According to the data in table 1, when the wind speed V_{in} is 2 m/s and the inlet wind temperature T_{in} is 115 ℃, the main drying results of the system are obtained as shown in table 2:

Table 2. Main drying results when V_{in}=2m/s and T_{in}=115°C

Parameter	Data
The first stage evaporation rate of NMP (e_1)	0.00199 mol m^{-2} s^{-1}
Mass flow of air (q_m)	1.72 kg s^{-1}
Enthalpy of air at the inlet of the oven ($\dot{Q}_{in}$)	198.79 kW
Enthalpy of return air ($\dot{Q}_{RA}$)	73.27 kW
The evaporation energy consumption ($\dot{Q}$)	125.52 kW
Electrical heating energy consumption ($\dot{Q}_{elec}$)	149.94 kW

The drying process of the system belongs to the energy exchange process between the electrode plate and air. The air enthalpy after heating is used for the latent heat of NMP solvent evaporation. In the energy consumption of the drying system in the first stage of evaporation, the heating air energy consumption is the main part of the energy consumption of the system, in which the electrical energy is converted into the enthalpy of the heated air without considering the impact of energy loss. In the first stage of evaporation, the evaporation of NMP consumes part of the energy, and the rest energy is discharged with the return air.

By changing the inlet wind speed, the evaporation rate and energy consumption of the system under different conditions are analyzed. On the basis of table 2, without changing the inlet air temperature T_{in}, the evaporation rate and energy consumption of V_{in} at 2-12 m/s are calculated, and the results are shown in figure 2 and 3.

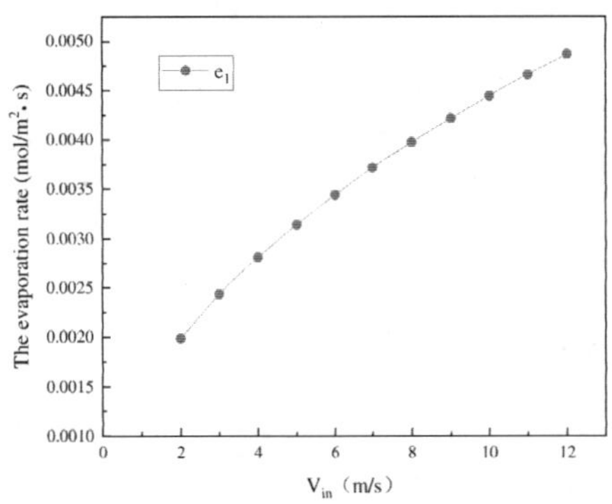

Figure 2. Evaporation rate variation of V_{in} at 2-12 m/s

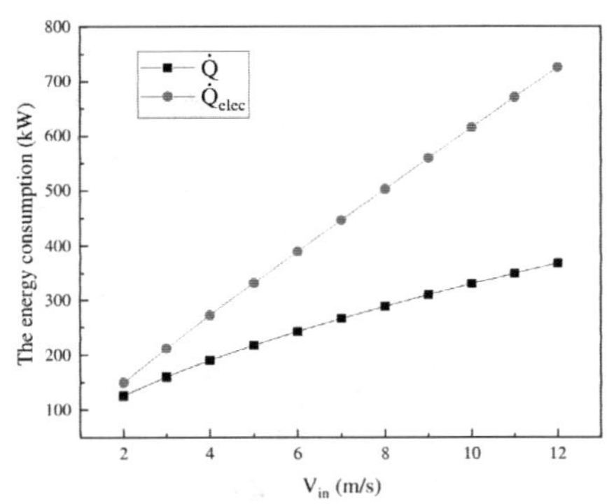

Figure 3. The energy consumption of V_{in} at 2-12 m/s

According to the figure 2 and figure 3 show that the inlet V_{in} during 2-12 m/s changes, the evaporation rate of e1 increases, however, as the wind speed increase gradually, slowly in the increase of evaporation rate is reduced, the scope of its change in 1.99×10^{-3} - 4.87×10^{-3} mol m^{-2} s^{-1}, or 1.18×10^{-3} - 2.88×10^{-3} g min^{-1} cm^{-2}. With the improvement of inlet, the inlet air enthalpy and return air enthalpy value is also in constant rise. Evaporation energy consumption and electric heating energy consumption are also increasing. Although the energy consumption of electric heating is also increasing, the speed of improvement is also decreasing very slowly. For every 1m/s increase of inlet wind speed, the energy consumption of electric heating increases by 56.70 kW on average.

4. Conclusion

In this paper, the evaporation rate of the first stage of evaporation is calculated. According to the principle of energy conservation, the energy balance of the drying system of lithium battery was established, and the calculation method of energy consumption was determined. By changing the inlet wind speed, the evaporation rate and energy consumption of the system under different inlet wind speeds are analyzed. When the inlet wind temperature is constant, the inlet wind speed is changed, and when V_{in} changes from 2-12 m s^{-1}, the evaporation rate e_1 in the first stage of evaporation increases with the increase of the inlet wind speed, and its variation range is 1.18×10^{-3} - 2.88×10^{-3} g min^{-1} cm^{-2}. At the same time, evaporation energy consumption and electric heating energy consumption are also constantly increasing. Whenever the inlet wind speed rises 1m/s, the energy consumption of electric heating increases by 56.70 kW on average.

Acknowledgments
We would like to acknowledge the supports from the National Science Foundation of China (Grant No.51905361), Fundamental Research Funds for the Central Universities (No. DUT18RC(3)008), Talent Project of Revitalizing Liaoning (No. XLYC1802106) and Dalian Innovation and Entrepreneurship Project for High-level Talents (No. 2018RD04) on this project.

References

[1] Antti Väyrynen,Justin Salminen. Lithium ion battery production[J]. The Journal of Chemical Thermodynamics,2011,46
[2] Yuan C, Deng Yelin, Li T, Yang F. Manufacturing energy analysis of lithium ion battery pack for electric vehicles. CIRP Annals - Manufacturing Technology 2017, 66(1)
[3] Deng Yelin, Li J, Li T, Gao X, Yuan C. Life cycle assessment of lithium sulfur battery for electric vehicles. Journal of Power Sources 2017, 343
[4] Matthias Thomitzek, Nicolas von Drachenfels, Felipe Cerdas, Christoph Herrmann, Sebastian Thiede. Simulation-based assessment of the energy demand in battery cell manufacturing[J]. Procedia CIRP,2019,80
[5] Morasch R, Landesfeind J, Suthar B, Gasteiger HA. Detection of Binder Gradients Using Impedance Spectroscopy and Their Influence on the Tortuosity of Li-Ion Battery Graphite Electrodes. Journal of the Electrochemical Society 2018, 165(14): A3459-A3467
[6] Li xujia, Zhao shilin. Energy consumption analysis of lithium battery pole plate drying system. Thermal science and technology 2015(6): 470-475 (in chinese)
[7] Baunach M, Jaiser S, Schmelzle S, Nirschl H, Scharfer P, Schabel W. Delamination behavior of lithium-ion battery anodes: Influence of drying temperature during electrode processing AU - Baunach, M. Drying Technology 2016, 34(4): 462-473
[8] Wood DL, Li J, Daniel C. Prospects for reducing the processing cost of lithium ion batteries. Journal of Power Sources 2015, 275: 234-242
[9] Ahmed S, Nelson PA, Gallagher KG, Dees DW. Energy impact of cathode drying and solvent recovery during lithium-ion battery manufacturing. Journal of Power Sources 2016, 322: 169-178
[10] Wood DL, Quass JD, Li J, Ahmed S, Ventola D, Daniel C. Technical and economic analysis of solvent-based lithium-ion electrode drying with water and NMP. Drying Technology 2018, 36(2): 234-244
[11] Jaiser Stefan, Funk Lena, Baunach Michael, Scharfer Philip, Schabel Wilhelm. Experimental investigation into battery electrode surfaces: The distribution of liquid at the surface and the emptying of pores during drying[J]. Journal of colloid and interface science, 2017, 494.
[12] Yongjian Su, Kai Zhou, Yelin Deng. Study on Predicting of Binder Distribution during Drying of the Coated Web of Positive Electrode for Lithium ion Battery, submitted to EPME 2019, International Conference on Energy, Power and Mechanical Engineering (Guangzhou, China)
[13] S. Kim, J. Hyun, Drying of coated film, in: G. Wypych (Ed.), Handbook of Solvents, ChemTec Publishing, William Andrew Publishing, Toronto, 2001, pp.386-417

IOP Conf. Series: Materials Science and Engineering **793** (2020) 012058 doi:10.1088/1757-899X/793/1/012058

Research on the Impact of upgrading of consumption structure on energy intensity under the background of urban and rural dual economy

Yan Li [1], Yuan Li[2], Tao Chen[1], Ying-ying Hu[1]
([1]State Grid Shanxi Economic research Institute,Taiyuan, Shanxi, 030002, China;
[2]State Grid Shanxi integrated energy service,Taiyuan, Shanxi, 030002, China)

[*]Corresponding author's e-mail: 2272176609@qq.com

Abstract: Based on the current situation of dual economic structure of urban and rural areas in China, this paper takes urban and rural areas as two independent research objects, and constructs the panel threshold regression model of consumption structure upgrading on energy consumption intensity. The results show that the upgrading of regional urban residents' consumption structure has double threshold effect on energy intensity, and the upgrading of urban residents' consumption structure has a restraining effect on energy intensity. With the upgrading of per capita income, it first decreases and then increases; the upgrading of regional rural residents' consumption structure has a double threshold effect on energy intensity,the upgrading of rural residents' consumption structure will significantly inhibit the energy intensity when the per capita income of residents is in the maximum range. In addition, under the threshold of per capita income, compared with rural areas, the upgrading of consumption structure of urban residents has a more prominent effect on the inhibition of energy intensity.

1. Introduction

Energy is not only the material basis of social production and people's life, but also the power source of economic development. With the development of economy, more and more energy needs will be generated, and the sustainable use of non renewable resources such as fossil resources will make energy shortage a common problem facing the international community. The extensive economic growth mode which depends on resource consumption aggravates the severity of energy shortage In China. China's energy consumption ranks first in the world, its energy intensity is higher than that of developed countries, and its energy efficiency needs to be further improved. In the critical period of economic development transformation and upgrading, it is of great significance to solve the contradiction between the limitation of energy resources and the sustainability of economic growth. Not only that, coal based energy consumption structure has produced a large number of greenhouse gases, and now China's carbon dioxide emissions have ranked second in the world. In order to solve these problems, general secretary Xi J P put forward the energy security strategy of "four revolutions and one cooperation" in 2014. He explicitly demanded that we must firmly control the total energy consumption, reduce the intensity of energy use, and put energy conservation in the whole process of economic and social development and various fields. However, to solve the energy problem, we should not only rely on the energy industry itself, but also combine the whole economic development mode and the structure of production and consumption.

The consumption of residents has become the main factor causing the increase of energy consumption[1][2].The existing research focuses on the relationship between the consumption level of residents and the total energy consumption and CO^2 emissions. Wei Y M say that the carbon emissions caused by the daily consumption of Chinese residents accounted for 30% of the total carbon emissions in China every year[3].The energy consumption caused by residents' daily consumption accounts for 45% - 50% of the total energy consumption[4].Moreover, the impact of residents' consumption on energy consumption is not a simple linear relationship, and there are obvious regional differences[5].Chen X H found that the expansion of urban-rural income gap is bad to the improvement of social energy efficiency[6].At present,the consumption expenditure scale of residents is increasing year by year, and the demand for consumption is becoming more and more diverse.there are few researches on the energy consumption intensity notice the changes in the consumption structure of urban and rural residents.Based on the background of the dual development of urban and rural areas in China,To discusses the relationship between the upgrading of consumption structure of urban and rural residents and the intensity of energy consumption respectively, providing a new idea for the correct understanding of China's energy problem .

2. research design

2.1 Variable selection and data source

energy consumption intensity is Interpreted variable , which is measured by the ratio of consumption data of ten thousand tons of standard coal and GDP at the end of each year. GDP is converted based on 1978.The upgrading index of consumption structure of urban residents and that of rural residents are selected as the main explanatory variables. Draw lessons from the existing research of domestic scholars[7],The upgrading index of consumption structure of urban residents is determined by the proportion of the sum of the consumption of health care, education, culture, transportation and communication of urban residents in each region to the total consumption of urban residents in each province. The upgrading index of consumption structure of rural residents is calculated by the same method.

The per capita income of urban residents and rural residents is the threshold variable of this paper. In order to eliminate the influence of price factors, the data are translated on the basis of 1978. With reference to the existing research, technological progress, population scale and industrial structure are included in the control variables. The percentage of regional invention patent authorization in the total number of patent authorization is used to measure the level of regional technological progress; the population at the end of each year is used to measure the population scale of each region, which is also treated logarithmically; the ratio of the output value of the secondary industry to the output value of the tertiary industry in each region is used to measure the industrial structure of the region.

Due to the lack of some data, in order to ensure the integrity of the research data and the accuracy of empirical analysis, this paper selects panel data from 29 provinces and municipalities in China, excluding Tibet and Xin jiang. The time span is from 2007 to 2017. The relevant data are from China energy statistical yearbook and China Statistical yearbook.

2.2 Model settings

Threshold model can be divided into single threshold regression model and multi threshold regression model. Multi threshold regression model is the extension of single threshold regression model. Single threshold regression model provides the basis for parameter estimation and hypothesis test of multi threshold regression model.

First, the regression equations of single threshold and double threshold are set as follows:

$$y_{it} = \beta_0 + \beta_1 Z_{it} + \theta_1 X_{it} I(q_{it} \le r_1) + \theta_2 X_{it} I(q_{it} > r_1) + u_{it} + \varepsilon_{it} \qquad (1)$$

$$y_{it} = \beta_0 + \beta_1 Z_{it} + \theta_1 X_{it} I(q_{it} \le r_1) + \theta_2 X_{it} I(r_1 < q_{it} \le r_2) + \theta_3 X_{it} I(q_{it} > r_2) + u_{it} + \varepsilon_{it} \qquad (2)$$

In the above formula ***y*** is the energy consumption intensity, Z is the set of control variables, and X is the upgrading index of consumption structure of urban and rural residents. q is the threshold variable expressed by the per capita income of urban and rural residents. β_0, β_1, θ_1, θ_2 are the output

IOP Conf. Series: Materials Science and Engineering **793** (2020) 012058 doi:10.1088/1757-899X/793/1/012058

elasticity coefficients of each variable, u_{it} are regional fixed effects and ε_{it} is the random interference term. r is the threshold value to be estimated, i and t represent provinces and years respectively. $I(\bullet)$ is an explicit function. When the condition of the function is satisfied, the value is 1. Otherwise, the value is 0.

3 Empirical analysis

3.1 The threshold effect of the upgrading of consumption structure on energy intensity in rural areas
Using stata14.0 software, taking the per capita income of rural residents in each region as the threshold variable, bootstrap (BS) method was used to conduct 500 self sampling, and then tested the hypothesis of single threshold, double threshold and triple threshold successively. It can be seen from table 1 that the upgrading index of rural residents' consumption structure has a threshold effect on energy intensity. The p-values of single threshold model and double threshold model are 0.000 and 0.036, respectively, indicating that single threshold and double threshold exist at the significance level of 1% and 5%, while the test results of triple threshold model are not significant, so this paper adopts the double threshold model for analysis. The two thresholds and their confidence intervals are further estimated. The results are shown in Table 2. The first threshold is 6.541 and the second threshold is 7.432. The LR figure is shown in Figure 1. X-axis represents threshold parameter, Y-axis represents LR value.The horizontal dotted line in Figure 1 represents the LR critical value of the likelihood ratio statistic at 95% confidence level. The interval between the two points where the horizontal dotted line intersects the LR test curve is the confidence interval of the threshold value. When the threshold value falls in this interval, the likelihood ratio statistic is less than the critical value, which means that the threshold value is valid.

Table 1 Results of the number of thresholds in rural area

Threshold number test	model	number of self sampling	F	P	critical value		
					1%	5%	10%
	Single threshold test	500	47.142	0.000	33.654	20.534	13.501
	Double threshold test	500	15.825	0.036	23.061	13.363	9.462
	Three threshold test	500	5.795	0.196	24.899	13.650	9.224

Table 2 Estimated threshold of rural area and its confidence interval

	critical value	95% confidence interval
The threshold effect of the upgrading of consumption structure on energy intensity in rural areas	7.432	[7.112, 7.945]
	6.541	[6.002, 6.557]

3.2 The threshold effect of the upgrading of consumption structure on energy intensity in urban areas
Referring to the above methods, taking the per capita income of urban residents in each region as the threshold variable, this part tests the threshold effect of upgrading the consumption structure of urban residents. Just as Table 3 says, The upgrading index of consumption structure of urban residents has a threshold effect on per capita energy intensity. The p-values of single threshold and double threshold

are 0.008 and 0.029, respectively, indicating that the single threshold and double threshold pass the test at the significance level of 1% and 5%, although the triple threshold model is also significant at the level of 1%, but because the third threshold is within the range of the first two thresholds, referring to the research of domestic scholars, it is decided to use the double threshold model for analysis, so as to reduce freedom. Table 4 shows the two thresholds and their corresponding confidence intervals under the double threshold model. The first threshold is 9.569 and the second threshold is 10.701. Similarly, LR charts show that threshold estimates are valid.

Table 3 Results of the number of thresholds in Urban area

Threshold number test	model	number of self -sampling	F	P	critical value		
					1%	5%	10%
	Single threshold test	500	28.371	0.008	25.825	17.722	14.210
	Double threshold test	500	18.414	0.029	39.804	21.796	15.165
	Three threshold test	500	15.659	0.004	11.768	5.612	4.514

Table 4 Estimated threshold of Urban area and its confidence interval

	critical value	95% confidence interval
The threshold effect of the upgrading of consumption structure on energy intensity in urban areas	9.569	[9.123, 9.998]
	10.701	[9.253, 10.778]
	9.188	[9.188, 10.288]

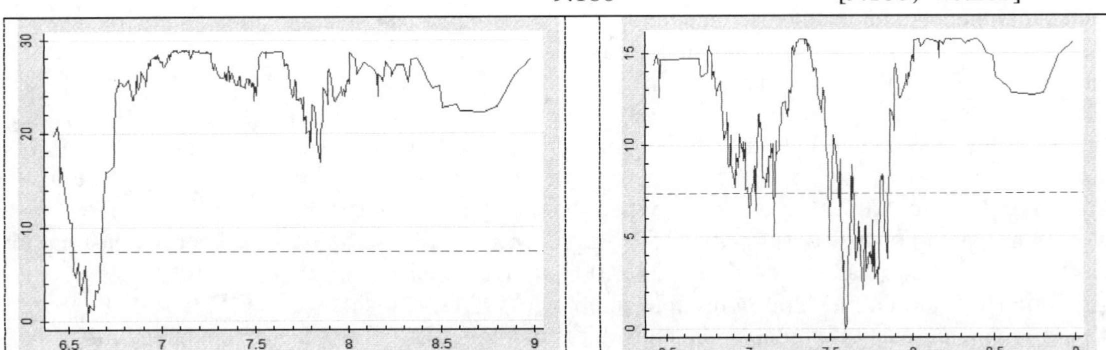

Figure 1 Two-threshold likelihood ratio test of per capita income of rural residents

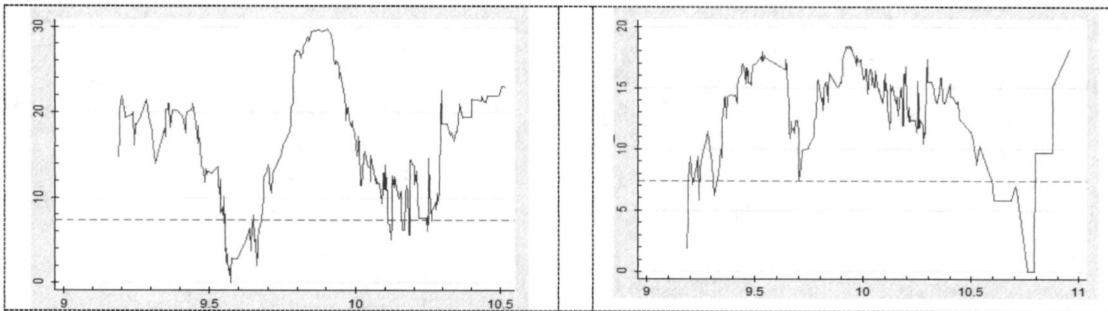

Figure 2 Two-threshold likelihood ratio test of per capita income of Urban residents

3.3 Regression results of threshold model

The threshold values obtained from the two tests are respectively substituted into formula 2, and the regression results are shown in Table 6. We can see the relationship between the upgrading of consumption structure and energy intensity of urban residents has different effects with the change of per capita income level of urban residents. In general, the upgrading of consumption structure of urban residents can significantly reduce the regional energy intensity. When the per capita income of urban residents is at the lowest level, the regression result of the upgrading of the consumption structure of urban residents to the energy intensity is -0.497 at the significance level of 5%. If the per capita income of urban residents increases to the middle range, the inhibition decreases to -0.251, which is because with the improvement of the income level, the housing, vehicles The increase of communication consumption has stimulated the development of high energy consuming industries such as urban industry and manufacturing industry. When the income of urban residents exceeds the maximum threshold, urban residents will tend to choose new energy-saving and environmental protection products with higher quality of life, and the consumption will turn to service industry and knowledge technology intensive industry with higher price level, which has high price and low energy consumption coefficient. In this range, the upgrading of consumption structure of urban residents shows the greatest inhibition on energy intensity, with a correlation coefficient of - 0.824, which is significant at the level of 1%.

The threshold regression results of upgrading the consumption structure of rural residents in the region are shown in column 2 of table 6. When the per capita income of rural residents is at the lowest level, the regression result of upgrading index of consumption structure of rural residents to energy intensity is not significant. This may be because the consumption of rural residents in the lowest income range is still at the level of basic necessities such as food, clothing and housing due to the limitation of economic development level. When the income of rural residents increases to the middle range, the upgrading of the consumption structure of rural residents promotes the energy consumption intensity, and the correlation coefficient is 0.007 at the significance level of 10%. because the consumption of automobiles and household appliances of rural residents increases and the energy consumption increases, but limited by the constraints of economic development, the energy intensity is reversed increase. When the per capita income of rural residents crosses the highest threshold, the consumption structure is further improved, and the inhibition coefficient is - 0.003, which is significant at the level of 10%. From table 6, we can also see that the upgrading of consumption structure of urban residents has a more prominent effect on the inhibition of energy intensity. Therefore, to solve the energy problem at this stage, we should not only adjust the consumption structure of urban residents, but also further stimulate the potential of rural residents in energy conservation and emission reduction.

IOP Conf. Series: Materials Science and Engineering **793** (2020) 012058 doi:10.1088/1757-899X/793/1/012058

Table 5 Threshold regression results

	The threshold effect of the upg-rading of consumption structure on energy intensity in urban areas			The threshold effect of the upg-rading of consumption structure on energy intensity in rural areas		
Threshold value	$r \le 9.569$	$9.569 < r \le 10.701$	$r > 10.701$	$r \le 6.541$	$6.541 < r \le 7.432$	$r > 7.432$
Main explanatory variables	-0.497** (-2.540)	-0.251* (-1.430)	-0.824*** (-2.880)	-0.176 (-1.84)	0.007* (0.95)	-0.003* (-0.16)
technological progress		0.010 (1.150)			0.004** (2.310)	
population scale (10000)		0.267** (2.140)			0.110* (0.820)	
industrial structure		0.168*** (1.150)			0.132*** (2.920)	
Constant term		3.634*** (3.770)			3.951*** (3.668)	

Note: * * * means significant at 1%, * * means significant at 5%, and * means significant at 10%.

The analysis results of the control variables are the same as those of the existing research. The rapid development of secondary industry and population scale also has a significant positive role in promoting energy intensity. Under the two regression models, technological progress has different effects on energy intensity.Because technological progress makes high energy consuming enterprises improve equipment and technology, the increase of enterprise productivity reduces energy intensity, but on the other hand, the increase of enterprise output will also lead to consumption growth.

4.Conclusions and suggestions

This paper draws the following conclusions: (1) the upgrading of consumption structure of urban and rural residents in each province has a double threshold effect on energy intensity. (2) the upgrading of consumption structure of urban residents has a restraining effect on energy intensity. With the improvement of per capita income, the upgrading of consumption structure of urban residents in the region will reduce the energy intensity of the region .(3)The upgrading of the consumption structure of rural residents improves the energy intensity when per capita income is distributed in the lowest range.But it turn to inhibit the energy intensity
when per capita income is distributed in the largst range. (4) compared with the rural areas, the upgrading of the consumption structure of urban residents has an effect on the energy intensity The inhibition was more prominent.

In this regard, this paper puts forward the following suggestions: (1) It is necessary to increase the publicity of low-carbon consumption concept, enhance the awareness of energy conservation of urban residents, especially promote the transformation of their residential consumption and transportation to a low-carbon lifestyle.we should give appropriate policy preference, encourage urban residents to transform into knowledge intensive industries and service industries, and then control energy intensity. (2) the energy consumption of rural residents is more to meet the basic living needs. We should give corresponding subsidies to eliminate high energy consumption household equipment in rural areas and promote energy-saving household appliances at low cost. In addition, we should popularize the knowledge of energy conservation and environmental protection to rural areas, encourage them to cooperate with the deepening implementation of the "coal to electricity" policy. (3)Adjust the industrial structure, guide the transformation and upgrading of the resource intensive industries with high pollution and emission to the technology intensive low-carbon industries as soon as possible. To increase regional R& D investment and cultivate innovative technical talents, promote the accelerated utilization of "energy saving and emission reduction" technologies.

Acknowledgments

It is grareful that The youth program of the National Natural Science Foundation of China（41401655）provides guidance and support in writing.

IOP Conf. Series: Materials Science and Engineering **793** (2020) 012058 doi:10.1088/1757-899X/793/1/012058

References

[1] Park H C, Heo E. The direct and indirect household energy requirements in the Republic of Korea from 1980 to 2000—An input–output analysis[J].Energy Policy.2007(35):2839-2851.
[2] Li Y. A test of the dynamic influence of household consumption on energy consumption[J]. Business economy research,2015(12):54-56.
[3] Wei Y M, Liu L C, Fan Y,et c.The Impact of Lifestyle on Energy Use and CO_2 Emission: An Empirical Analysis of China's Residents[J].Energy Policy,2007(35):247-257.
[4] Schipper L, Bartlett, et al.Linking life-styles and energy use:a matter of Time[M]. Annual Review of Energy,1989,14.
[5] Wang Y, ShiM J. The total energy consumption induced by the living consumption of urban residents in China[J]. Resource Science, 2009, 31(12):2093-2100.
[6] Chen X H, Li C Q, Zhang G R, et al. Analysis of the impact of income gap on energy efficiency in the process of Urbanization[J]. inquiry into economic issues, 2011(7):144-149.
[7] Huang J, Li J K. The characteristics, measurement and development of China's consumption upgrading[J]. China's circulation economy, 2018(4):94-101.

IOP Conf. Series: Materials Science and Engineering **793** (2020) 012059 doi:10.1088/1757-899X/793/1/012059

Research on Postal Dispatching System at Multiple Transfer Points of Postal Containers Based on Genetic Algorithm

Min Liu, Jie Meng , Guangping Li

College of Mechanical and Power Engineering

Chongqing University of Science and Technology

Chongqing, China

*Corresponding author's e-mail: 1liuyijunpp2007@163.com

Abstract. In this paper, a new comprehensive algorithm is proposed to solve the problem of postal dispatching at multiple transfer points of postal containers, and based on this, a postal dispatching system of postal transfers at multiple transfer points is established. In this system, Floyd algorithm combined with fuzzy judgment to find each key route, and to combine linear programming to set a small step size for the far three points on each key route, so as to realize the area planning and design of key points. Finally, the genetic algorithm is used to solve the postal planning of the remaining points in the area. To optimize the system, the compensation principle is used to carry out post and mail planning at nearly three points in the area and complete the post and mail planning for the entire area.

1. Introduction

Postal containers are conducive to improving the quality and efficiency of postal communications production, and enhancing the market competitiveness of postal communications. The use of postal containers and letter boxes to replace flexible containers such as pouches to encapsulate mail is the development tendency of postal communications. Practice in recent years has proved that there are many problems required to be solved in the promotion and application of postal containers, especially the problem of postal dispatching at multiple transfer points of postal container. Due to the complex and changeable interrelationships, it is particularly difficult to solve this problem[1]. Only by solving this problem, using postal containers on the vehicle mail road from central office to the city of each branch office and the secondary trunk line can be realized.

The postal dispatching system at multiple transfer points of postal containers is to provide the entire network of mail and post production scheduling information for each batch of intra-city transfers and secondary trunk lines. The information include the reference time for handover of the import and export containers (mail) at the branch office, the type of vehicle implementation, the delivery method, the arrangement of specific vehicle driving routes and specific tasks for the vehicles that can be controlled within this period. Under the premise of collecting the number of import and export containers of the central office and each junction within a certain time limit, combining all kinds of conditions and constraint conditions, according to the principle of minimum labor consumption, the best postal organization scheme of the day is calculated to determine the running schedule of the vehicle and the equipment of the vehicle.The number of containers in each batch of the central office and each junction is changing, so the choice of mail route and the dispatch of mail vehicles should be changed according to the specific situation, which is the key problem to be solved in the dispatching system[2].

2. The system description

The vehicle mail dispatching system is a system that is based on hundreds of contact points (branch offices), a variety of paths to each junction point and a number of import / export containers (mail) randomly generated by each batch, etc., which can not be planned and scheduled manually. Therefore, it is necessary to establish a postal dispatching system at multiple transfer points of postal containers, which requires a feasible mail scheduling for a city multi-point handover problem with more comprehensive historical data.

For the sake of analysis, note a central office as v_0 points, each branch is recorded as v_1, v_2, v_3,..., v_n (n is the number of branch offices). Assume that the time distance between each post office is l_{ij} (The time that takes for a vehicle to travel between two branch offices under normal traffic conditions, where i and j respectively represents the serial number of the corresponding branch office), the probability of traffic congestion on this road section is α_{ij}. In addition, the number of sent and received mail of any branch v_i is: F_i and J_i, respectively.

The problems to be solved by the system: firstly, according to the historical accumulation and the necessary supplementary investigation data, the basic operation route of the postal vehicle is put forward. Secondly, on the basis of the completion of the first step, the basic route is supplemented and the complete operation route of the mail vehicle is made. Thirdly, a mathematical model is developed to simulate the running process of each line (with one as the object). Fourthly, on the basis of solving the above three problems, propose a normal vehicle operation schedule for each line.

Specific tasks include two parts: designing driving routes and designing all-day mail dispatching vehicle schemes which are easy to operate and determine for working days. For the design of driving routes, taking into account the operating efficiency of the mail cars, there are always a number of routes that are more frequent in the shipping process of each batch, which are noted as the key routes. The determination of dispatching routes can be topological on the key routes. For the second task, it is required to determine departure schedule of the route, in which the total running time is the shortest. Therefore, the dispatching problem is abstracted into a clear mathematical model, and the specific method of solving the model is given.

3. System solving method

In order to solve the above problems, the solution process shown in Figure 1 can be used.

3.1. Use the improved FLOYD algorithm combined with fuzzy judgment to seek each key route

Set the central office as the starting point of the key route.

Determination of the end points of the key routes: the traffic map is traversed, and all points in the figure are divided. The central office is classified into the first stage. The points which are directly connected with the central office are classified into the second stage; the other points which are directly connected with the points of the second stage are in the third stage, and so on... all the point division phases are classified like this way. The point in the last stage, and all the points that are not directly connected to the point in the next stage from this stage are the end points for each key route.

Determination of the rest points for the key route: (take the v_i end route as an example.) With fuzzy judgment, the traffic condition of the route is converted into a time path, and the actual time path is the product of the general time path and the fuzzy function of the traffic condition. If there is an arc from v_d To v_i, then there is a path of length arcs [a] [i]. A huge number of trials are required since this path is not necessarily the shortest path. Whether the path (v_d, v_1, v_i) exists needed to be considered first. If it exists, the path length of (v_d, v_i)and(v_d, v_1, v_i) need to be compared. The shorter path is the key path where middle vertex sequence number from v_d to v_i is not greater than 1. Assume that one more vertex v_2 is added to the path, that is, if (v_d,..., v_1) and (v_1,..., v_i) are the key routes where the middle vertex sequence number from v_d to v_1 and from v_1 to v_i, respectively, not greater than 1, then (v_d,..., v_2,... v_i) may be the shortest path where the middle vertex sequence number from v_d to v_i is not greater than 2. Compared with the shortest path that has been obtained where middle vertex sequence number from v_d to v_i not greater than 1, the shorter path of them can be selected. And then add into one more vertex v_3, continue testing, and so on.

In general, if (v_d,..., v_k) and (v_k,..., v_i) are the key routes where the middle vertex sequence number from v_d to v_k and from v_k to v_i, respectively, not greater than k, then (v_d,..., v_k,... v_i) and the already obtained shortest path where the middle vertex number from v_d to v_i not greater than k have been compared, the shorter one is the key route where the middle vertex sequence number from v_d to v_i is not greater than k.

In this way, after n comparisons, the final result must be the key route from v_d to v_i. Based on this method, the key paths of each pair of determined end points can be obtained at the same time (fig.1).

A sequence of n-order square matrices is defined:

D (-1), D (0), D (1), ..., D (k), ..., D (n-1)

where

D(-1)[i][j]=arcs[i][j]

D(k)[i][j]=Min{D(k-1)[i][j]

D(k-1)[i][k]+D(k-1)[k][j]} 0<=k<=n-1

It can be seen from the above calculation formula that D (k) (i) (j) is the length of the key path where the middle sequence number from v_i to v_j is not greater than k. The Floyd algorithm is modified, and a path matrix P is added based on the original weight C and the minimum weight matrix A, initialized to 0. That is, suppose the shortest distance between any two points is a direct arc of two points, without passing through other points. When A [i] [j] is obtained by the FLOYD algorithm, let P [i] [j] be the point passing in the middle. In this way, the Floyd algorithm finally produces two matrices: the minimum weight matrix A and the path matrix P. For the path matrix P, the critical path of any two points can be obtained recursively.

3.2. Linear programming is used to set the step size of the far three points on each key route in a small range

By using the capacity principle of the car, a linear programming model is established based on the maximum useful box kilometers and the minimum useless box kilometers as follows:

1) In order to meet the capacity of the vehicle, consider the number of mail sent and received by the mail car at each transfer point and the quantity of the mail carried by the post-mail vehicle, the sum of them is not greater than the capacity of the vehicle.

2) The principle of early departure and late return is met, and the corresponding constraint relationship is established.

The so-called early departure and late return principle means that during a postal shipment, the export

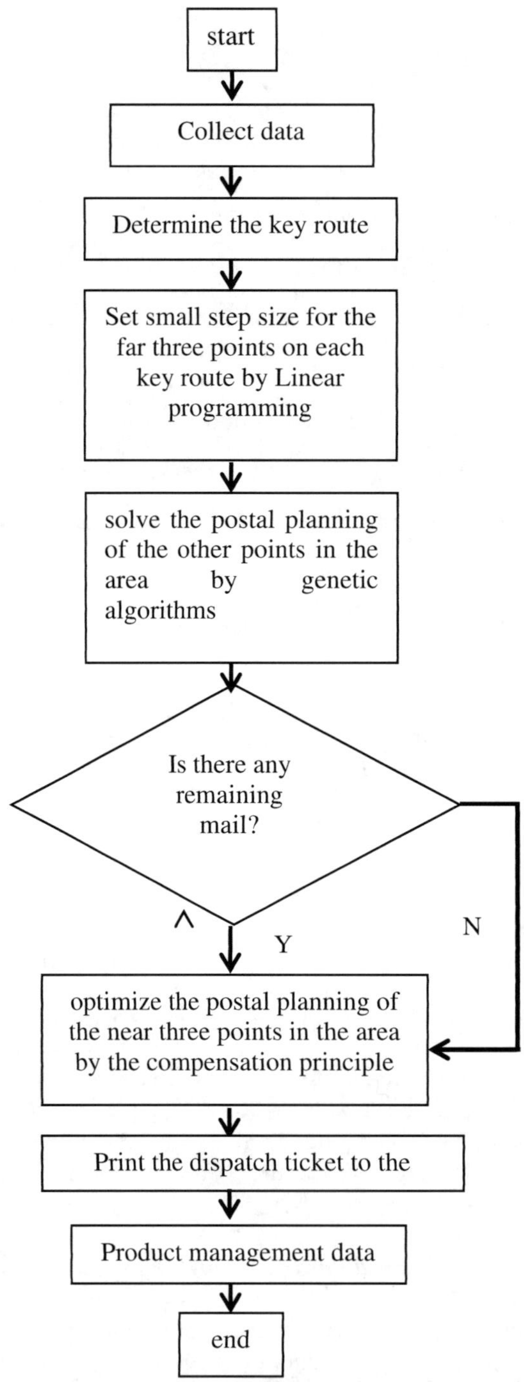

Fig. 1 Solution flowchart

container (mail) is required to be delivered to each branch office as soon as possible, and the imported container (mail) is required to recover the central office as late as possible within a certain period of time.

3) Establish the useless box kilometers discrimination function as the target for discrimination

An export or import container (mail) of a transfer point (branch office), the container (mail) of which is loaded into a mail car. If the same batch of mail is being sent, the post car must also pass through the transfer point (branch office), then during the course of the mail car passing through the transfer point twice, one kilometer of container (mail) received from the transfer point (branch office) is called a unit of useless container kilometers.

Solved this simple linear planning problem by computer, the postal planning of far three points on each key route can be got.

3.3 .Using genetic algorithm to solve the postal planning of other points in the area

The specific steps of the genetic algorithm are as follows:

1) Parameter encoding

The data of the traffic map are represented as the genetic structure data of the genetic space. It is planned to traverse the branches and use the sequence as the coding of the genetic algorithm.

2) Generation of initial population

Prepare an initial population for genetic algorithms (with historical data as the initial population) consisting of several initial solutions, that is, the first generation of evolution[3].

3) Fitness evaluation and detection: this test is used to measure the pros and cons of the offspring during the search process, that is, the objective function[4].

$$f = \begin{Bmatrix} a\left[\max[L(\varphi^{k}{}_{TSP})] - \min[L(\varphi_{TSP}^{k})]\right] + \\ (1-a)\sum_{k=1}^{m} L\left(\varphi_{TSP}^{k}\right) \end{Bmatrix}^{-1} \quad (1)$$

Where, φ_{TSP}^{k} represents the initial, and k represents the kth. $L(\varphi_{TSP}^{k})$ represents the total time course, a as the adjustment factor.

4) Design of genetic operations: selection, crossover, mutation.

5) Setting control parameters: The setting of control parameters can control the running time of the program and the accuracy of the results[5].

3.4 Use the compensation principle to plan the shipping of nearly three points in the area

In the postal planning process, in order to avoid a small number of transfer points (branch offices) that are not conducive to optimization of the transfer conflict, arrange the shipping of their containers (mails) to the remaining capacity on other paths passing this transfer point (branch offices).

4. Conclusion

Aiming at the difficulty of the multi-junction postal container transportation dispatching problem, it is proposed to use Floyd algorithm combined with fuzzy judgment to find each key route, and to combine linear programming to set a small step size for the far three points on each key route, so as to realize the area planning and design of key points. In addition, the genetic algorithms is used to solve the postal planning of the remaining points in the area. On this basis, a postal dispatching system of postal transfers at multiple transfer points is established. The test result show that this method is highly practical and can significantly improve the efficiency of postal transportation.

Acknowledgment

This research was funded by the National Natural Science Foundation of China under Grant No.51505049 and Chongqing Research Program of Basic Research and Frontier Technology under Grant No. cstc2018jcyjAX0690.

References

[1] The multi-depot vehicle routing problem with inter-depot routes. Crevier, Benoit,Cordeau, Jean-Fran?ois,Laporte, Gilbert. European Journal of Operational Research . 2007

[2] A scatter search for multi-depot vehicle routing problem with weight-related cost. Zhang,

IOP Conf. Series: Materials Science and Engineering **793** (2020) 012059 doi:10.1088/1757-899X/793/1/012059

Jun,Tang, Jiafu,Fung, Richard Y. K. Asia Pacific Journal of Operational Research . 2011
[3] Solution of a Large-Scale Traveling-Salesman Problem. G. Dantzig,R. Fulkerson,S. Johns. Operations Research . 1954
[4] Xinqi Hu, Operations Research, Beijing: Tsinghua University Press
[5] Xue Luzhong, Engineering Optimization Technology, Tianjin: Tianjin University Press

IOP Conf. Series: Materials Science and Engineering **793** (2020) 012060 doi:10.1088/1757-899X/793/1/012060

Research and Field Application of Oil-paper Insulation Material Aging State Based on FDS

LI Dandan[1], FAN Yongqiang[2], WANG Yaping[2], YUE Yonggang[2*], RENG Ming[3]

[1]Training Center of Inner Mongolia power (Group) Co., Ltd., Hohhot 010080, China

[2]Inner Mongolia EHV Power Supply Bureau, Hohhot, Inner Mongolia, 010080, China

[3]State Key Laboratory of Electrical Insulation and Power Equipment, Xi'an Jiaotong University, Xi'an, Shaanxi, 710049, China

[*]Corresponding author's e-mail: hvyue@163.com

Abstract: The effects of three factors, such as aging degree, water content and temperature, on the dielectric response characteristics of oil-paper insulating materials were studied by setting up a frequency-domain dielectric response experimental platform. In this paper, the frequency domain dielectric spectrum (FDS) test is carried out for the on-site transformer paper insulated casing, and it is found that the method is more abundant than the traditional power frequency dielectric loss method, which is less affected by the work frequency electromagnetic field, and can realize the nondestructive measurement of the aging state of the oil-grease insulation.

1. Introduction

For oil-immersed transformers and other power equipment, the insulation state directly determines the safe and stable operation of the power grid. With the increase of operating life, under the influence of many factors such as electricity, chemistry, mechanical force and temperature, the electrical performance of transformer oil-paper insulation will continuously decline. The electrical properties of transformer oil-paper insulation will continuously decline, which is manifested by breakdown voltage decline, dielectric loss and acid value increase, etc. The oil quality deterioration factors will produce a synergistic effect, and promote the accelerated development of aging.

According to the rules to determine the corresponding electrical parameters change state, the aging state of oil-paper insulation is evaluated in China by tests of different properties at present[1-3]. The main tests include dielectric loss at power frequency, insulation resistance, dissolved gas content in oil, furfural content, tensile strength (TS) and degree of polymerization (DP).The main reason for the deterioration of mechanical properties of insulating paper is the aging of cellulose insulating paper, the measured power frequency, pulse breakdown voltage strength and dielectric loss Angle of power frequency have little change in electrical properties. The dissolved contents of CO, CO2 and furfural in the oil are all related to the degree of polymerization[4-5], which can indirectly reflect the mechanical properties of the insulation.

Although the tensile strength and the degree of polymerization can best reflect the aging degree of oil-paper insulation, it is difficult to measure. Therefore, the dielectric loss, capacitance, insulation resistance, dc leakage, dissolved gas analysis in oil (DGA), partial discharge and other test methods in the current regulations of China for the insulation aging assessment of large power equipment are often not true and effective to reflect the oil paper insulation state of transformers.

IOP Conf. Series: Materials Science and Engineering **793** (2020) 012060 doi:10.1088/1757-899X/793/1/012060

2. Development of oil-paper insulation aging detection technology

In recent years, some new methods for evaluating the insulation state of oil-paper based on dielectric response have been widely studied. Dielectric response refers to the polarization of the dielectric under the action of an electric field. The insulation aging of transformer and the change of moisture and impurities in the insulation will cause the change of the polarization process of oil paper insulation, and then cause the change of dielectric response characteristics. On the contrary, the change of dielectric response characteristic of transformer oil paper insulation corresponds to the change of transformer insulation state and insulation life. Therefore, the insulation state and insulation life of the transformer can be evaluated by measuring the dielectric response of the oil-paper insulation. Compared with the traditional chemical testing methods, the dielectric response methods are all electrical measurements, without complicated and expensive chemical instruments and reagents, and without collecting insulation samples, the testing method is simple, the testing process will not cause damage to the transformer insulation. Compared with the traditional electrical measurement methods such as insulation resistance, power frequency dielectric loss and capacitance, the dielectric response method has a richer information content, which can more accurately reflect the insulation defects and insulation degradation that are difficult to be reflected by the traditional method, thus more reliably evaluate the transformer life. Therefore, the testing method of transformer life evaluation based on dielectric response is expected to supplement or even replace the traditional testing method, which has a broad application prospect[6-7].

3. Oil-paper insulation experiment platform based on frequency domain dielectric spectroscopy

3.1. Frequency domain dielectric spectroscopy

Frequency domain dielectric spectroscopy (FDS) is used to evaluate the aging state of insulating materials by measuring the complex capacitance, dielectric loss factor and complex dielectric constant of the medium varying with frequency in the low-voltage sinusoidal alternating electric field[8]. FDS method measures the complex capacitance and dielectric loss spectrum of insulating medium from low frequency to high frequency through sweeping frequency measurement, such as sweeping frequency measurement from 1mHz to 1kHz, and analyzes and studies the change law of the curve. It can be said that FDS method is an extension of the conventional dielectric loss test method, and its measurement principle is shown in 'figure 1'.

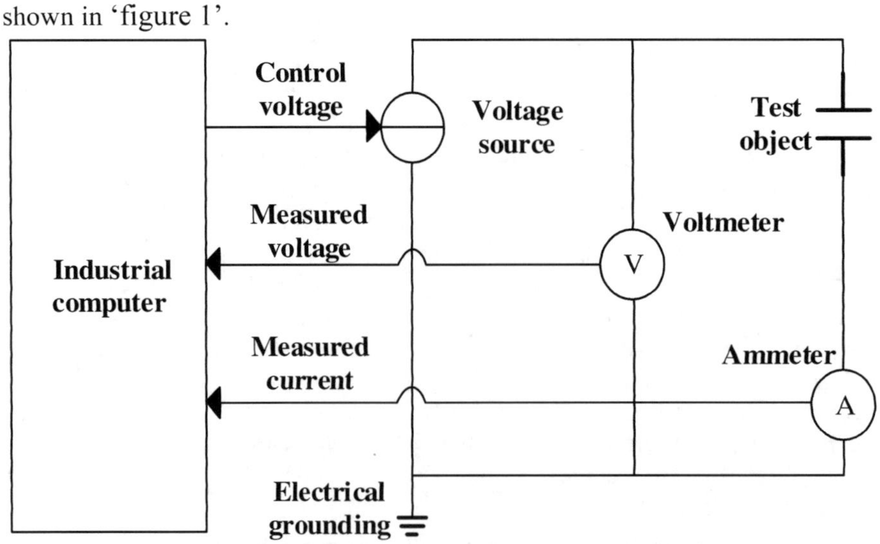

Figure 1. Schematic diagram of measuring principle.

The typical graph of FDS test is shown in 'figure 2'. The horizontal axis is the frequency and the vertical axis is the dielectric loss coefficient. The whole curve is s-shaped. Different factors affect different parts of the FDS curve, for example, water affects the low frequency region and the high frequency region. The transmissibility of oil affects the steeper middle part of the curve, while the geometry of the insulation affects the steeper "bulge" on the left.

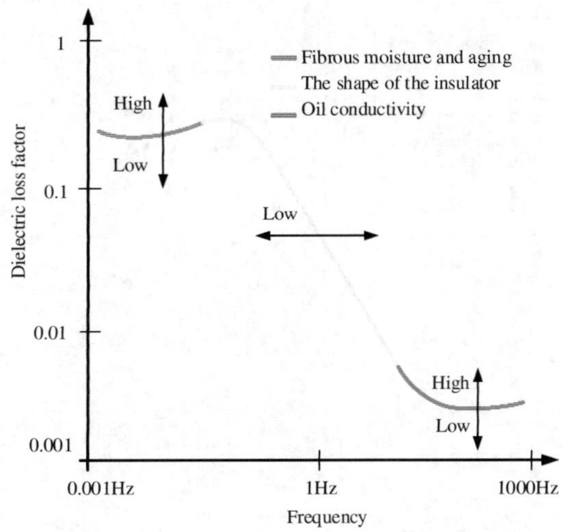

Figure 2. Relation curve between dielectric loss and frequency

3.2. Oil-paper insulation experiment platform

The oil-paper insulation test platform as ‘figure 3’. The system mainly consists of high-voltage copper electrode, measuring electrode and protection electrode. The position of the measuring electrode and the protection electrode was fixed, and the high voltage electrode was fixed through the card slot to ensure that the position of the cardboard was fixed during the long-term dielectric response experiment. There are air valves and oil valves on the tank body, which can guarantee the vacuum oil immersion or vacuum gas injection. The experimental temperature can be changed by coating the tank with an heating belt. The temperature sensor and barometer can monitor the experimental conditions in real time. The experimental tank is connected with the experimental device by the high-temperature resistant coaxial cable. The coaxial cable contains two wire cores for connecting the measuring electrode and the high-voltage electrode respectively, and the shielding layer of the cable is connected with the protection electrode.

IOP Conf. Series: Materials Science and Engineering **793** (2020) 012060 doi:10.1088/1757-899X/793/1/012060

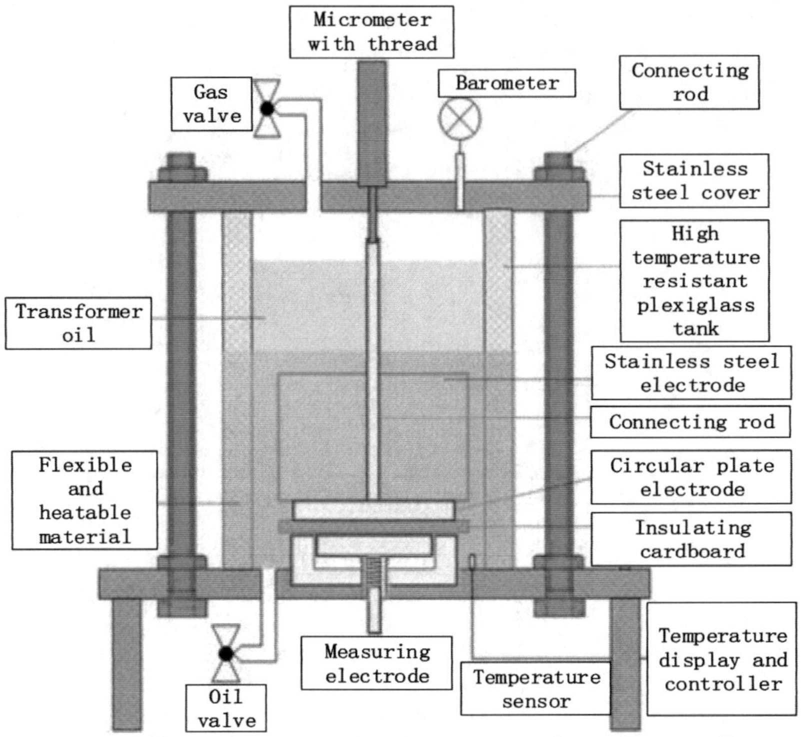

Figure 3. Experimental model of insulation aging of transformer oil-paper

4. Experimental study on aging of oil-paper insulation material

4.1. Influence of aging

The insulating board was heated and aged at 120°C, and 5 samples with different aging degrees (0 days, 10 days, 31 days, 50 days and 75 days) were obtained. The complex capacitance C* and tanδ frequency domain spectra of the samples were measured by FDS method, as shown in 'figure 4', 'figure 5' and 'figure 6'.

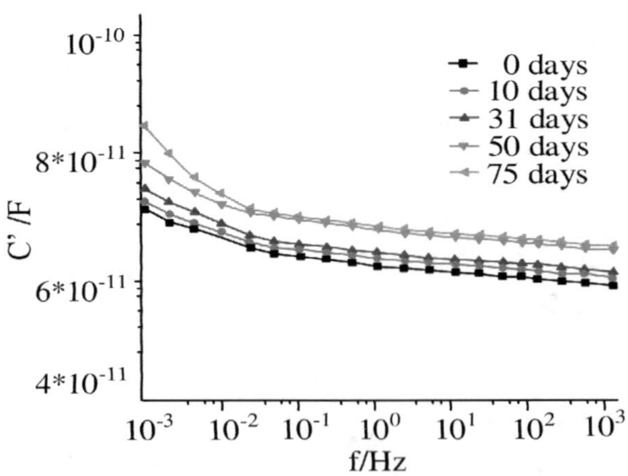

Figure 4. C′ frequency domain spectrum of samples with different aging degrees

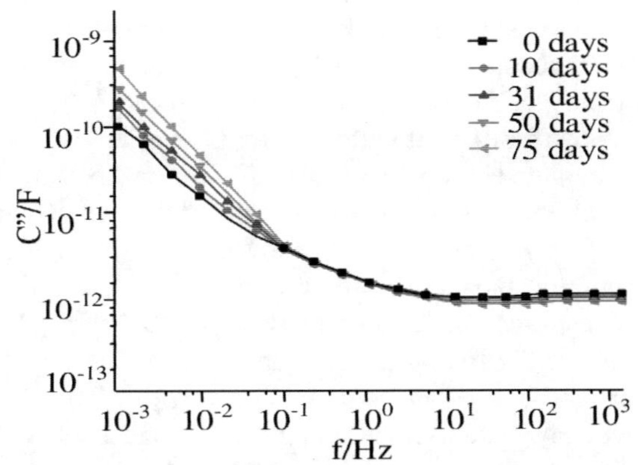

Figure 5. C″ frequency domain spectrum of samples with different aging degrees

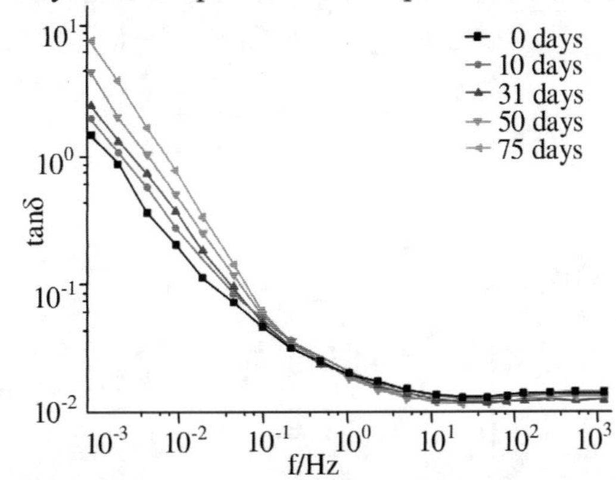

Figure 6. Tanδ frequency domain spectrum of samples with different aging degrees

As can be seen from ‘figure 4’, the real part C' of the complex capacitance of the samples with different aging degrees increases gradually with the aging degree in the whole frequency band range of 10^{-3}~10^{3}Hz, and for the samples with the same aging degree, the real part C' of the complex capacitance decreases continuously with the increase of frequency, and there is no coincidence in the samples with different aging degrees. As you can see from ‘figure 5’, for the different aging degree of the samples, after capacitance imaginary part C" within 10^{-3}~10^{0}Hz frequency range increases with aging degree is increasing gradually, and in the frequency range is greater than 10^{0}Hz is overlap phenomenon, for the same aging degree of the samples, after capacitance imaginary part C" within the frequency range is less than 10^{0}Hz decreases with frequency, the frequency range is more than 10^{0}Hz above change is not obvious, tend to be a straight and level. See from ‘figure 6’, the dielectric loss factor tanδ of oil-immersed insulation board within 10^{-3}~10^{0}Hz frequency range increases with aging degree of intensified, and in this range, the tanδ of the same sample decreases with the increase of frequency, in does not change significantly within 10^{0}~10^{3}Hz frequency range, the tanδ of samples with different aging degree overlap.

4.2. Influence of moisture

In addition to the FDS, which is affected by the aging degree of insulating paper, moisture is another important factor. The insulating paper with different moisture content was put into the insulating oil for a long time to stand still, and four test samples with different moisture content of 0.76%, 1.2%, 2.35% and 3.38% were obtained. The C* and tanδ frequency spectrum as shown in ‘figure7’, ‘figure 8’ and ‘figure 9’ were detected by FDS method.

As can be seen from ‘figure 7’, for samples with the same water content, the real part of the complex capacitance C' gradually decreases with the increase of frequency in the frequency band range of 10^{-3}~10^{-2}Hz, and the real part of the complex capacitance C' of samples with different water content in this range will continuously increase with the increase of water content, and the larger the water content of samples, the steeper the curve.When the frequency range is greater than 10Hz, the real part C' of the complex capacitance does not change much into a horizontal line, and the real part C' of the complex capacitance with different water contents overlaps. As you can see from 'figure 8', and after the change of capacitance imaginary part C" similar to the real component and within 10^{-3}~10^{0}Hz frequency range of the imaginary part of the same water sample after capacitance C" decreases with the increase of frequency, the frequency range of samples of different water content under the same frequency, complex capacitance imaginary part C "increases with the increase of water content, while in the frequency range is greater than 10^{0}Hz overlap phenomenon, the same water sample along with the increase of the frequency change is not big, straight and level.See from 'figure 9', the dielectric loss factor tanδ of the same water sample within 10^{-3}~10^{1}Hz frequency range along with the increase of the frequency decreases, and samples of different water content under the same frequency dielectric loss factor tanδ increases with the increase of water content, when more than 10^{1}Hz frequency range, The tanδ of the sample with the same water content did not change significantly with the frequency, while the tanδ of the sample with different water content also overlapped.

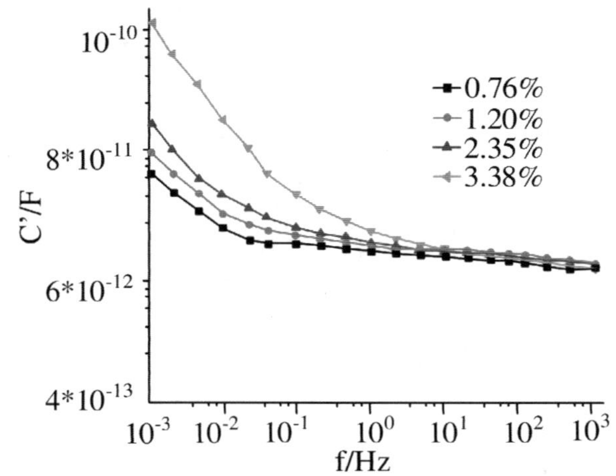

Figure 7. C′ frequency domain spectrum of samples with different water content

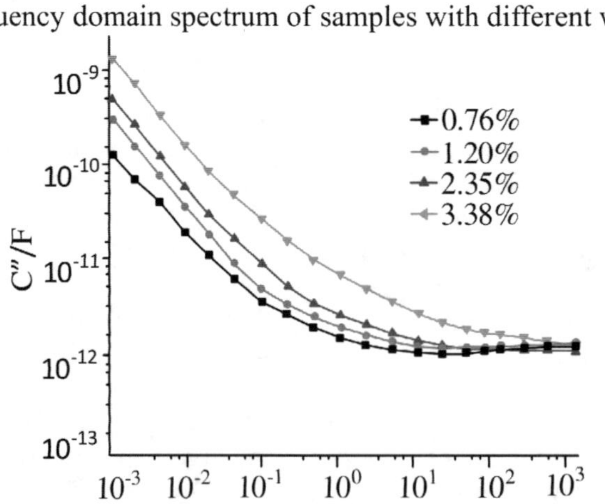

Figure 8. C″ frequency domain spectrum of samples with different water content

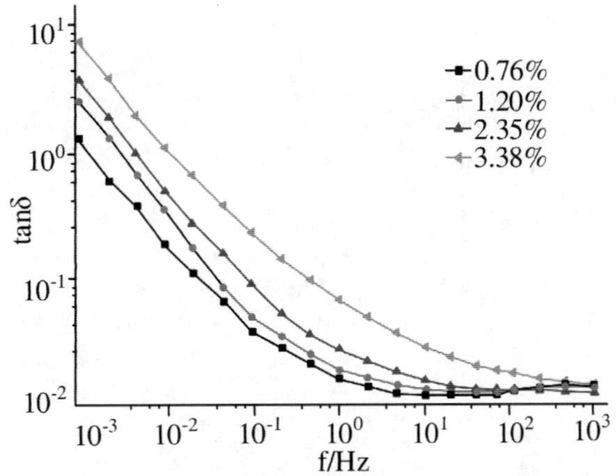

Figure 9. Tanδ frequency domain spectrum of samples with different water content

4.3. Influence of temperature

Temperature is another important factor that affects the spectrum of the dielectric. Especially in the northern region, the temperature difference between day and night and the harsh environment make the effect of temperature on the aging state of transformers more prominent. The complex capacitance and dielectric loss factor frequency domain spectrum of oil-paper insulating board with low water content without aging at 30℃, 50℃, 70℃ and 90℃in 'figure 10', 'figure 11' and 'figure 12'.

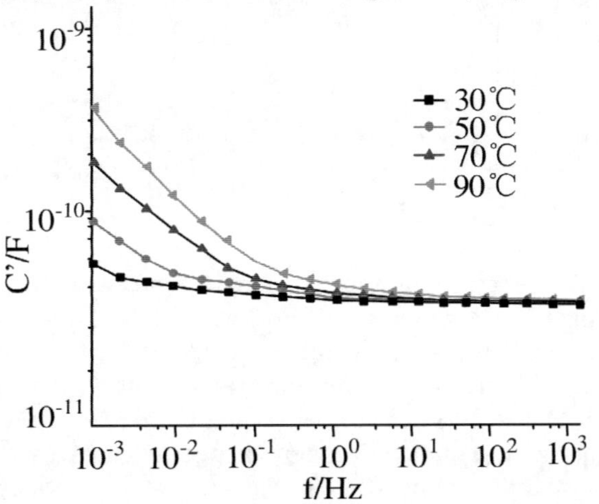

Figure 10. C′ frequency domain spectrum of samples with different temperature

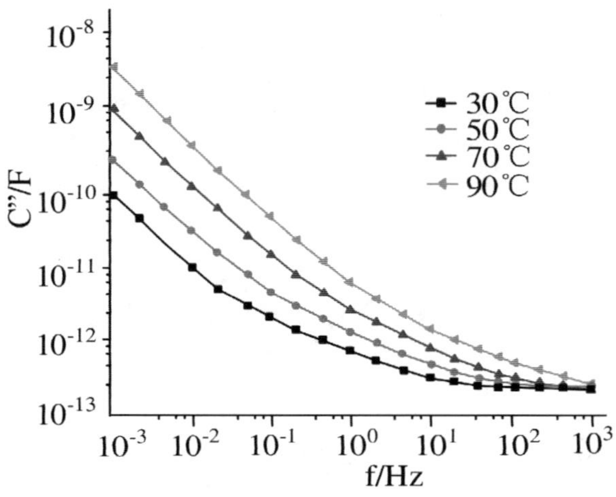

Figure 11. C″ frequency domain spectrum of samples with different temperature

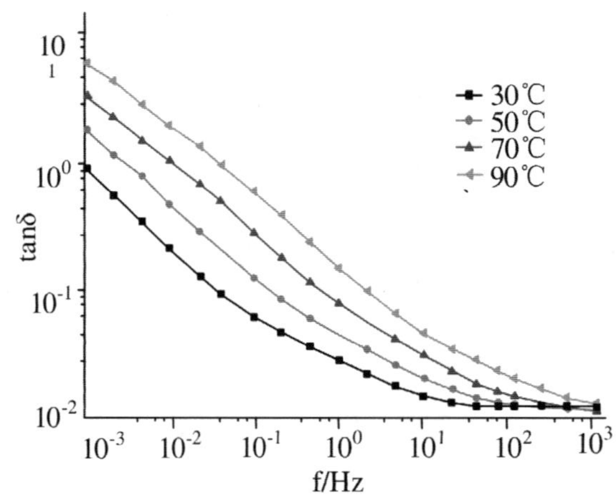

Figure 12. Tanδ frequency domain spectrum of samples with different temperature

As can be seen from 'figure 10', the real part C' of the complex capacitor gradually decreases with the increase of frequency in the frequency band range of 10^{-3}~10^{0}Hz, and the real part C' of the complex capacitor increases with the increase of temperature in the same frequency range. The higher the temperature of the sample, the curve is steeper. When the frequency range is greater than 10^{0}Hz, the real part C' of the complex capacitance does not change much into a horizontal line, and the real part C' of the complex capacitance with different water contents overlaps.As can be seen from the 'figure 11', the complex capacitance of imaginary part C" decreases continuously with the increase of frequency, and under the same frequency, complex capacitance imaginary part C" increases with the increase of temperature, see from 'figure 12', the dielectric loss factor tanδ of sample decreases with the increase of frequency, under the same frequency, the dielectric loss factor tanδ increases with the increase of water content, overlapping phenomenon occurring in the high frequency area.

5. Field test application of FDS

For the transformer bushing with good sealing, with the increase of operation life, the oil-paper composite insulation gradually ages under the action of thermal, electrical and chemical factors, resulting in the decline of insulation performance, thus affecting the safe and stable operation of the transformer. Therefore, the study on frequency domain dielectric property of oil-paper insulation of transformer bushing is of great significance to evaluate the insulation status of transformer as a whole. In order to further verify the effectiveness of the FDS method, 9 devices with different manufacturers, voltage levels and operating life were selected, as shown in table 1.At the same time, there is a positive correlation between operation life and equipment aging and water content.

Table 1.selection of transformer bushing on site

Number	Voltage Level (kV)	Run time (year)	Environment Temperature (°C)	The upper oil temperature (°C)	Casing temperature (°C)
A	110	4	32.0	42	38.6
A	110	5	35.0	50	44.9
A	110	8	34.0	42	39.3
A	110	9	29.4	46	40.3
A	110	10	30.0	44	39.3
B	110	4	38.0	49	45.3
B	110	6	33.9	40	38.0
B	110	7	29.5	36	33.8
B	110	8	28.0	35	32.5

The FDS test curves of 110kV bushing from different manufacturers are shown in 'figure 13' and 'figure 14'. Although the same manufacturer produces the same batch of casing with the same voltage level, the capacitance varies. The tanδ-f curve in FDS characteristic is mainly compared here. It can be seen that the tanδ-f curve increases with the increase of the operating life, only in the high-frequency region (1kHz~5kHz) is not affected by the operating life.The FDS characteristic law of field casing is consistent with experimental results.

As the use time of the equipment increases, the cellulose long chain fracture gradually in the insulating paper, the DP value constantly decreases, and solid insulation became uneven, interstitial, insulation aging makes the paper generated polarity material such as organic acid, compound, water, which is more easily immerse the amorphous cellulose area, form a more abundant oiled paper, acid paper and other interface.With the increase of carrier concentration and mobility, the establishment of interlayer interface polarization is promoted and the strength of polarization is strengthened, which finally leads to the increase of equipment loss, and the increase of temperature rise in the operating state further aggravates the aging.

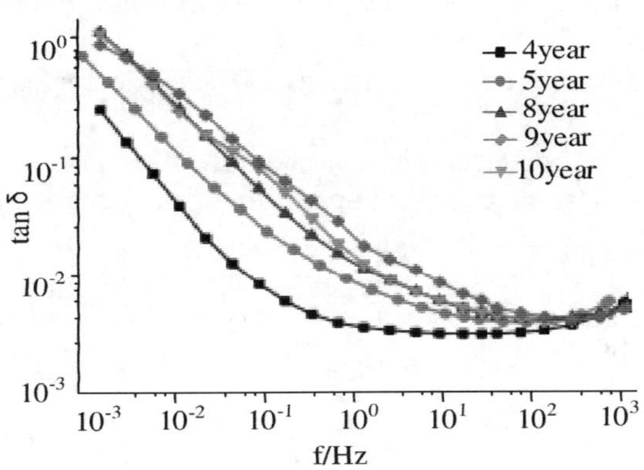

Figure 13. The casing in A factory

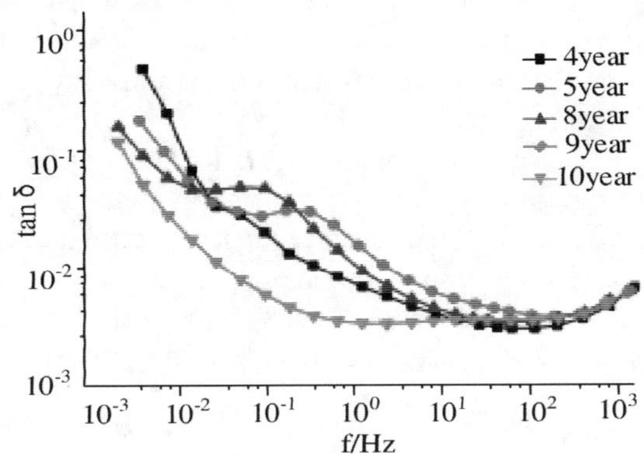

Figure 14. The casing in B factory

According to the FDS test, the dielectric loss factor tanδ and capacitance C of the above 110kV bushing operating frequency are calculated as shown in table 2 (A4 represents the bushing in A plant that has been running for 4 years, and so on).It is worth noting that the tanδ of each casing under power frequency are not more than 0.7%, The capacitance variation is also within ±5% of the nameplate capacitance, indicating that the casing in traditional preventive test performance is good, not by power frequency test parameters to diagnose the assessment, However, it is obvious from the reference curve that the running state of each casing is very different.The capacitance of these casings hardly changes with the operating life of the equipment, and no curve is shown.

Table 2. frequency data of 110kV casing pipe of different manufacturers

Number	Dielectric loss (%)	Capacitance (pF)
A4	0.226	364
A5	0.310	291
A8	0.360	338
A9	0.370	302
A10	0.484	304
B8	0.305	312
B4	0.311	284
B6	0.334	311
B7	0.451	314

It can be seen from table 2 that the power frequency dielectric loss gradually increases with the normal aging of the equipment as the operation life grows longer for the casing in plant A. For the casing in plant B, the power frequency dielectric loss is smaller when the casing has been running for 8 years and 4 years, and larger when the casing has been running for 6 years and 7 years.

6. conclusion

By setting up an experimental device for frequency domain dielectric response, the influence of aging degree, water content and temperature on the insulation state of oil paper insulation samples was analyzed, and the effectiveness of FDS in analyzing the insulation state of oil paper was verified. Through the analysis and research, the following conclusions are drawn:

（1）Temperature, water content and aging degree have great influence on the dielectric response characteristics in frequency domain.

（2）Compared with the traditional power frequency dielectric loss measurement method, the frequency-domain dielectric spectrum method contains more state information of the equipment and is less affected by the power frequency electromagnetic field.

（3）The results of laboratory research and field test are basically the same, which can be popularized and applied on a large scale in the field, and the spectrum library of FDS can be established to assist the general test for comprehensive diagnosis.

Reference

[1] Bao, Y., He, X.X., Zhang, M. (2017) Analysis of characteristic quantities in transformer life evaluation. J. Electroelectrics, 2:18-23.
[2] Yang, Q.P., Xue, W.D., Lan, Z.D. (2006) Research on insulation aging assessment of transformers.

J. Transformers, 5:1-5.
[3] Zhang, Q. (2015) Research on insulation state of high voltage bushing based on dielectric response method . D. Dalian university of technology.
[4] Jia, L.R., Liu, X. (2017) In the oil furfural analysis in the application of transformer life assessment. J. Journal of jiangxi electric power, 41 (2):9-12.
[5] Zhang, Y.X., Li, S.L. (2011) Variation law of polymerization degree of transformer insulating paper. J. High voltage technology, 37(10):2458-2463.
[6] Yang,L.J., Qi, C.L., Lu, Y.D. (2015) Characteristic parameters and evaluation methods of frequency domain dielectric spectrum for insulation state of transformer oil paper. J. Acta electrica technica sinica, 30(1):212-219.
[7] Chen, M.M., Wei, G., Liu, J. (2012) Effects of temperature and moisture on frequency-domain dielectric characteristics of oil-paper insulation system for transformers. J. Insulating materials, 45(1):57-61.
[8] Dong, M., Wang, L., Wu, X.Z. (2016) Research status and development of dielectric response detection technology for oil-paper insulation. J. High voltage technology, 42(4):1179-1189.

IOP Conf. Series: Materials Science and Engineering **793** (2020) 012061 doi:10.1088/1757-899X/793/1/012061

An insulation diagnosis method for battery pack based on battery model

Yanhui Zhang[1,2,3,a], Qingpeng Chen[1,2,3], Xiao Liu[1,2,3] , Zhile Yang[1,2,3,a], Shifeng Guo[1,2,3] , Wei Feng[1,2,3,a]

[1]Shenzhen Institutes of Advanced Technology, Chinese Academy of Sciences, Shenzhen 518055, China

[2]Guangdong Provincial Key Lab of Robotics and Intelligent System, Shenzhen Institutes of Advanced Technology, Chinese Academy of Sciences, Shenzhen 518055, China

[3]CAS Key Laboratory of Human-Machine Intelligence-Synergy Systems, Shenzhen Institutes of Advanced Technology, Chinese Academy of Sciences, Shenzhen 518055, China

[a]Corresponding author: Yanhui Zhang: zhangyh@siat.ac.cn Zhile Yang: zl.yang@siat.ac.cn Wei Feng: wei.feng@siat.ac.cn

Abstract: The collision and components aging of electric vehicles may lead to dramatic decrease of insulation performance in the complex application environment, which will reduce the insulation between battery pack and chassis, and will affect the performance of on-board equipment and electronic control unit, resulting in heat accumulation effect of leakage circuit and even vehicle fire. Therefore, it is critical to develop a real-time, accurate and reliable detection device to monitor the insulation resistance of the high-voltage system of electric vehicles, which is beneficial to the normal operation of electrical equipment, the safe operation of vehicles and the personal safety of passengers. In this work, an online estimation method of insulation resistance based on battery model is proposed.

1. Introduction

Electric vehicles are regarded as an effective way to solve energy and environmental problems, and have been strongly supported and promoted by various countries [1]. Unlike traditional fuel vehicle, batteries are the energy source of electric vehicles [2-3]. The battery system of electric vehicle has a high voltage characteristic with a range of 300-700V. Power battery packs are subjected to severe temperature, humidity, vibration, shock, corrosive gases and liquids as well as the reduced insulation performance. It has been reported that during the rapid acceleration process, the amplitude of the bus terminal voltage jitter changes very sharply, which directly affects the normal operation of the vehicle controller. Therefore it is required to be identified and eliminated promptly to avoid the traffic safety accidents [4].

With the constant upgrading of voltage specification, the requirements of insulation resistance between high-voltage electrical system and vehicle chassis also rises. The insulation performance of the electrical system will be reduced with the aging of cable insulation medium and the humid environment. Thus, it is necessary to detect the insulation resistance of battery system [5].

Various insulation resistance models of electric vehicle and detection methods have been developed, including voltmeter method, current sensing method, balanced bridge method, and unbalanced bridge method [6]. These methods determine the insulation performance of the battery system with the bus current and voltage parameters. The voltmeter method is only suitable for off-line measurements, which requires to disconnect the battery pack from external electrical equipment. Therefore, it is complicated and can be implemented online.

The conventional sensor methods are to detect DC fault in system with current sensor [7]. The balanced bridge method is to construct a bridge composing of equivalent resistance on the positive and negative bus bars. This method has advantage of high accuracy, however, it cannot identify the branch fault and can only detect the asymmetry fault. The unbalanced bridge method is based on the principle of circuit voltage division. The insulation resistance value is obtained based on a function relationship between the resistance and the voltage. However, the voltage across the battery pack varies greatly, especially during transient acceleration and deceleration of the vehicle. There is a strong signal interference, which affects the calculation accuracy of the insulation resistance.

In this paper, the equivalent circuit of battery pack insulation model is provided, and the voltage and state of charge are also analyzed in the process of battery charging and discharging.

2. System modelling

High-precision battery model is the basis for establishing fault diagnosis methods. This section mainly introduces the insulation fault diagnosis model. Considering the complexity and accuracy of the model, the Thevenin equivalent circuit model is used as the battery model to characterize the dynamic performance of lithium-ion batteries. And the battery pack equivalent circuit model for insulation fault diagnosis is established with the battery model. The Thevenin model [6] is the most representative equivalent circuit model, which reflects the polarization resistance and polarization capacitance in the internal chemical reaction. The simulation of the dynamic and static characteristics of the battery, and its circuit structure is shown in Figure. 1.

Where U_{oc} represents the open circuit voltage of the battery, R_r represents the ohmic resistance, V_B represents the terminal voltage of the lithium battery, R_p and C_p are the polarized resistance and capacitance, respectively.

$$E(t)=V(t)+i(t)R_r+V_{oc} \tag{1}$$

$$I=\frac{V_p}{R_p}+C_p\frac{dV_p}{dt} \tag{2}$$

The difference equation of the model can be obtained by arranging the above equations using transformation and z transformation, as shown in the equation (3).

$$E_k=V_k-a(V_{k-1}-E_{k-1})+b\times i_k-ci_{k-1} \tag{3}$$

If $\theta=\dfrac{R_1+R_2}{R_1R_2C}$, then the coefficients a, b, c can be described as follows.

$$\begin{cases} a=1-R_0(1-e^{-\theta T})/(R_0+R_p) \\ b=R_0 \\ c=R_0e^{-\theta T} \end{cases} \tag{4}$$

If we define $y_k=E_k-V_k$, the corresponding difference equation can be expressed as follows.

$$y_k=ay_{k-1}+bi_k+ci_{k-1} \tag{5}$$

Constructing observation vectors and Parametric vector observation vectors $\varphi_k=[y_{k-1},i_k,i_{k-1}]$, Parametric vector $\theta_k=[a,b,c]^T$. The basic observation model can be described as follows.

$$y_k=\varphi_k\theta_k \tag{6}$$

IOP Conf. Series: Materials Science and Engineering **793** (2020) 012061 doi:10.1088/1757-899X/793/1/012061

According to the relevant rules of the ampere-time integral method, the definition of battery charging state is described as follows.

$$SOC(\mathrm{t})=SOC(\mathrm{t}_0)-\frac{\eta}{Q_N}\int_{\mathrm{t}0}^{t} Idt \qquad (7)$$

Where η represents Coulomb efficiency coefficient, Q_N represents the rated capacity of the battery, I denotes the current,and SOC (t_0) is the state of charge at initial time.

Since the maximum leakage current occurs at the positive and negative electrodes of the battery pack, the positive and negative insulation resistances of the battery pack should be considered. The simplified model of the negative side insulation fault is shown in Figure. 1 (b) and the battery pack voltage V_B can be defined as:

$$V_{\mathrm{B}}=\sum_{i=1}^{n} \mathrm{V}_{i,t}+i(t)\mathrm{R}_{i,r}+\mathrm{V}_{i,\mathrm{oc}} \qquad (8)$$

3. On-board diagnosis of insulation fault

The equivalent circuit of battery pack insulation detection is shown in Figure. 2. For the battery pack model, the pack voltage can be written as:

$$V_{pack}=\sum_{i=1}^{n}(V_{ocv,i}-iR_{0,i}-V_{p,i}) \qquad (9)$$

where V_{pack} represents the pack voltage of the lithium-ion batteries, $V_{ocv,i}$ represents the cell open-circuit voltage of the i_{th} cell, $R_{o,i}$ represents the ohmic internal resistance of the i_{th} cell, i represents the current of the battery and $V_{p,i}$ represents the polarization voltage of the i_{th} cell. To calculate the negative virtual insulation resistance, the following equations are obtained:

$$\begin{cases}(i_3-i_1)(R_1+R_2)=V_B\\ i_1R_3+(i_1-i_2)R_4+(i_1-i_3)(R_1+R_2)=0\\ (i_2-i_1)R_4-V_s+i_2R_-+V_f=0\end{cases} \qquad (10)$$

Where $i_2 = V_f/R_f$. The negative virtual insulation resistance can therefore be calculated as:

$$R_-=-\frac{R_f}{V_f}(V_B+V_f-Vs-\frac{R_3V_B}{R_3+R_4})+\frac{R_3R_4}{R_3+R_4} \qquad (11)$$

Similarly, he positive virtual insulation resistance can also written:

$$R_+=-\frac{R_f}{V_f}(-V_B+V_f-Vs+\frac{R_3V_B}{R_3+R_4})+\frac{R_3R_4}{R_3+R_4} \qquad (12$$

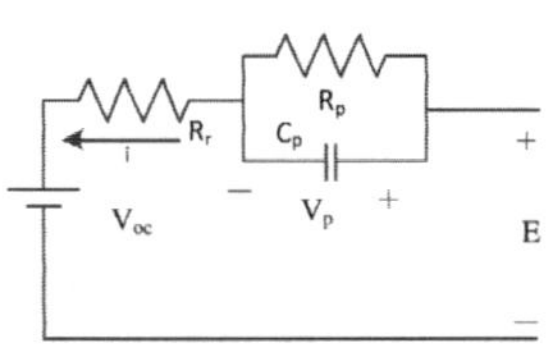

Figure.1 Equivalent circuit model

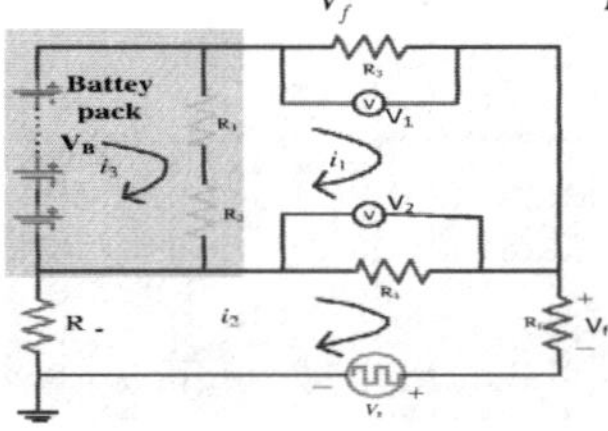

Figure 2. Battery pack insulation model

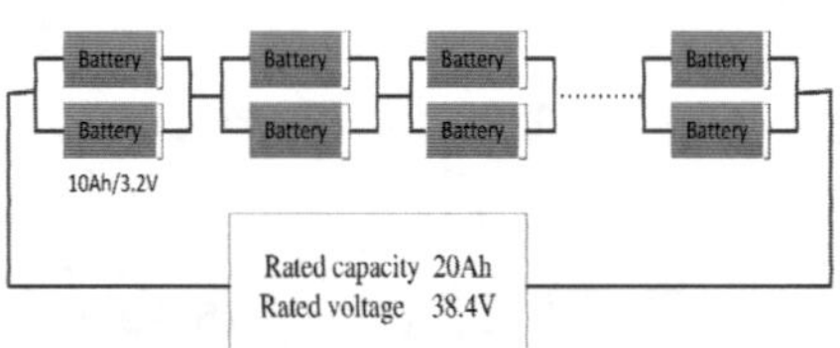

Figure 3. The topology of Battery pack

4. Experimental and discussion

This work intends to explore the dynamic response characteristics of battery pack, which consists of two batteries in parallel and twelve batteries in series, as shown in Figure 3. To further probe the dynamic characteristics of the battery during charging and discharging, the specific discharge curve of the battery is shown in Figure 4.

It can be observed from Figure 4 that the 12 batteries have the same current characteristics, however, the inconsistency of the internal resistance of the battery causes a certain deviation in the voltage. During the uniformly discharge experiment of the battery, the voltage difference is maintained at 0.015V when the state of charge is maintained between SOC 20%~90%. It indicates that the dynamic voltage has a great influence on the experimental results. When the battery voltage changes greatly, the estimation results of SOC will have a large estimation error, as shown in Figure 5.

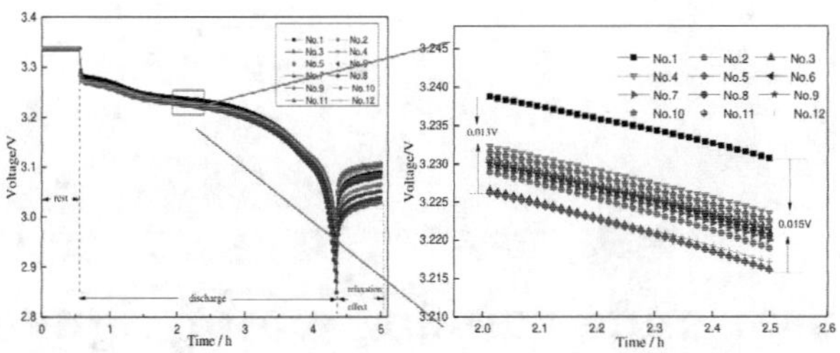

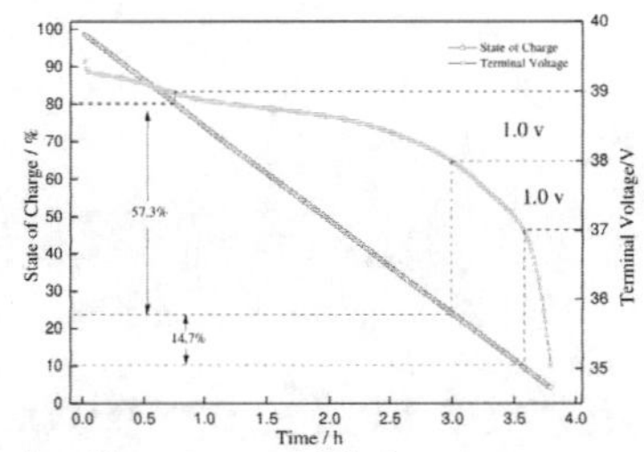

Figure 4. local voltages distribution Figure 5. Voltage and State of Charge during Charging and Discharging

5. Conclusions

In this paper, a novel method for insulation detection of lithium ion battery packs for electric vehicles based on thevnin battery model is proposed to improve the insulation detection performance of electric vehicle battery system. In addition, a more reliable insulation detection scheme for battery system is developed.

Acknowledgement

This project was supported by the National Natural Science Foundation of China (No.U1813222), Science and Technology Planning Project of Guangdong Province, China(No.2017B090910013), Science and Technology Project of Shenzhen（JSGG20170823140127645), and Knowledge Innovation Project of Shenzhen (JCYJ20170818153048647, JCYJ20180507182239617).

References

[1] S. Tong, M.P. Klein, J.W. Park. *J. Power Sources* 2015, **293**, 416–428.
[2] H. Rahimi-Eichi, U. Ojha, F. Baronti, et al.,2013, *IEEE Ind.Electron. Mag.* **7** ,4–16.
[3] Y. Wang, Z. Chen, C. Zhang. *Appl. Energy* 2017, **194,** 688–695.
[4] Y. Zheng, X. Han, L. Lu, J.Li, M.Ouyang. 2013, *J. Power Sources* **223**,136–146.
[5] H. Torkaman, F. Karimi.*Measurement,* 2015, **59,** 21–29.
[6] S. Grubic, J. Restrepo, T.G. Habetler.2013, *IEEE Trans. Ind. Appl.* **49** , 1358–1366.
[7] R.Gu, P. Malysz, H. Yang, A.Emadi. 2016, *IEEE T TRANSP ELECTR*, **2**, 417-431.

IOP Conf. Series: Materials Science and Engineering **793** (2020) 012062 doi:10.1088/1757-899X/793/1/012062

Design of self - diagnostic sensor and ordinary sensor signal combination device

Li Yuliang[1] Zhang Lihong[1] Zhou XiaoZhong[1] Lou Weidong[1]

[1]Hangzhou Cigarette Factory, ZJTI, Kehai Road 118, Hangzhou, China.

liyuliang@zjtobacco.com

Abstract. A device is designed to combine the signal of self-diagnosis sensor and ordinary sensor. The device comprises a high-speed optocoupler, a single-chip microcomputer and an output drive circuit. The high-speed optocoupler is responsible for collecting the level state of the signal of the self-diagnosis sensor and the signal of the ordinary sensor. Micro Controller Unit (MCU) is finally combined with the results of the operation and the self-diagnosis sensor signal state, real-time maintenance or the test results of the output drive circuit. The output drive circuit amplifies the control signal sent by the single-chip microcomputer to enhance the driving ability of the output signal; This device can intelligently identify the external input signal, and then perform the operation of and or merge with the external input signal, and output the control signal with diagnostic pulse according to the level state of the input channel.

1. Introduction

A large number of sensors are usually used in industrial control and industrial detection systems. Ordinary sensor output normal level, status of sustained high or low continuously. The diagnosis of self-diagnosis of sensor output signal is composed of a series of useful signal and its phase. Detecting the pulse in the presence of the sensor is known whether there is any power lost. When the abnormal situation is occurred, it is possible to remind workers check sensor or check the related circuit in a timely manner. These control signals may contain ordinary level signals or self-diagnosis signals. If the ordinary signal is directly combined with the self-diagnosis signal or the ordinary sensor is directly connected with the self-diagnosis sensor, the self-diagnosis function will be lost. When the control system does not detect the diagnostic pulse of the input signal, the system may drive the device to alarm, or even force shutdown, resulting in the whole system can not work. Therefore, it is necessary to design a device and a method to solve the above problems.[1-5]

2. Device design

A device for combining the signal of a self-diagnostic sensor with that of an ordinary sensor comprises a high-speed optocoupler, a MCU and an input device.[6] Out of the driving circuit, the high speed optocoupler is responsible for collecting the level of signals from the diagnosis sensor and ordinary sensor. MCU through its internal program intelligent detection of the state of the input signal, or operation, finally combined with the results of the operation and the self-diagnosis sensor signal state, real-time maintenance or reverse the signal, and output the test results to the output drive circuit; The output drive circuit amplifies the control signal sent by the single-chip microcomputer to enhance the driving ability of the output signal; One pin of MCU collects the input state of self-diagnosis signal every 90us, and another pin of MCU collects the input state of ordinary sensor signal every 90us. The level state is considered as the same as the input state of self-diagnosis signal. Both the general sensor

IOP Conf. Series: Materials Science and Engineering **793** (2020) 012062 doi:10.1088/1757-899X/793/1/012062

signal and the self-diagnosis sensor signal are high level effective for the control system. When the signal of self-diagnosis sensor is high current, no matter the signal of ordinary sensor is high level or low level, the signal of self-diagnosis sensor will keep the original waveform output. When the signal of self-diagnosis sensor is low level and the signal of ordinary sensor is high level, the signal of self-diagnosis sensor takes the reverse output. When the signal of self-diagnosis sensor is low level and the signal of ordinary sensor is low level, the signal of self-diagnosis sensor maintains the original waveform output. When the signal state of the self-diagnosis sensor and the ordinary sensor is determined in real time, the logic operation of the two kinds of signals and or is combined, and finally the related signals with diagnosis function are input to the control system according to the field control requirements. MCU hardware structure diagram is described as follows:

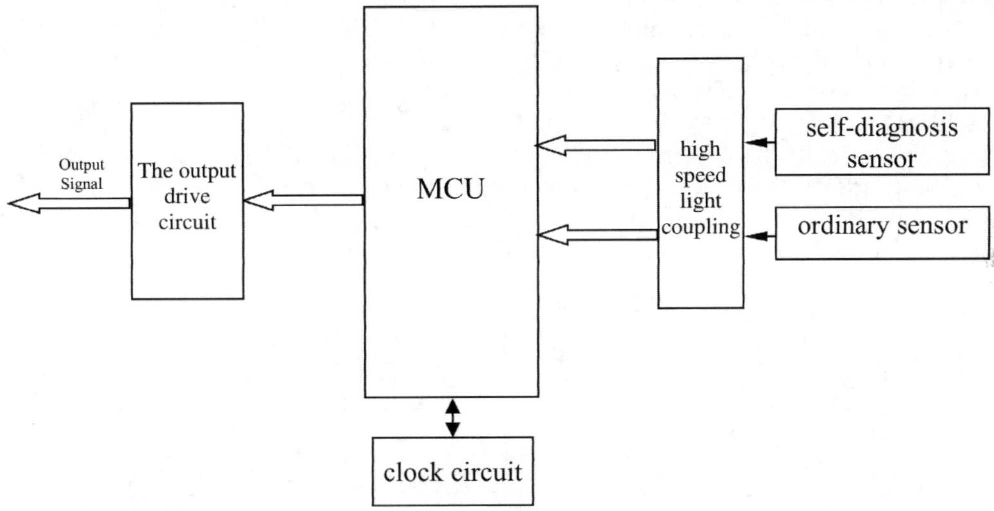

Figure 1. MCU hardware structure diagram

3. The realization principle of the device

The device used in this method includes high-speed optocoupler, single-chip microcomputer and output drive circuit. The high-speed optocoupler is responsible for collecting the level state of the signal of the self-diagnosis sensor and the signal of the ordinary sensor. MCU is finally combined with the results of the operation and the self-diagnosis sensor signal state, real-time maintenance or the test results of the output drive circuit. The output drive circuit amplifies the control signal sent by the single chip microcomputer to enhance the driving ability of the output signal.

3.1. Acquisition of signal level

A pin of MCU collects the input state of self-diagnosis signal every 90us and marks it with a variable LevelCount. If the current acquisition signal is high level, LevelCount is incremented by 1, and the upper limit value of LevelCount is 5. Otherwise, the execution of LevelCount is minus 1, and the lower limit value of LevelCount is 0. Real-time judgment of LevelCount value, when LevelCount$\geqslant$3, the level state is considered to be high; When LevelCount is less than or equal to 2, the level is considered to be low. The other pin of the MCU collects the input state of the ordinary sensor signal every 90us, and considers the level state as the same as the input state of the collected self-diagnosis signal.

3.2. Signal level determination

Both the general sensor signal and the self-diagnosis sensor signal are high level effective for the control system. When the signal of self-diagnosis sensor is high current, no matter the signal of ordinary sensor is high level or low level, the signal of self-diagnosis sensor will keep the original waveform output.

IOP Conf. Series: Materials Science and Engineering **793** (2020) 012062 doi:10.1088/1757-899X/793/1/012062

When the signal of self-diagnosis sensor is low level and the signal of ordinary sensor is high level, the signal of self-diagnosis sensor takes the reverse output. When the self-diagnosis sensor signal is low level and the ordinary sensor signal is also low level, the self-diagnosis sensor signal maintains the original waveform output.

3.3. Signal algorithm

When the signal state of the self-diagnosis sensor and the ordinary sensor is determined in real time, the logic operation of the two kinds of signals and or is combined, and finally the related signals with diagnosis function are input to the control system according to the field control requirements. The logic flow of the algorithm is as follows:

1) system initialization;
2) self-diagnostic sensor signals are collected;
3) determine the signal level state of self-diagnosis sensor;
4) collecting signals of ordinary sensors;
5) perform or operation;
6) determine whether it is consistent with the signal level of self-diagnosis sensor;
7) if yes, the self-diagnostic sensor signal will keep the output;
8) if not, take back the output of the self-diagnosis sensor signal.

The following figure shows the processed signal waveform:

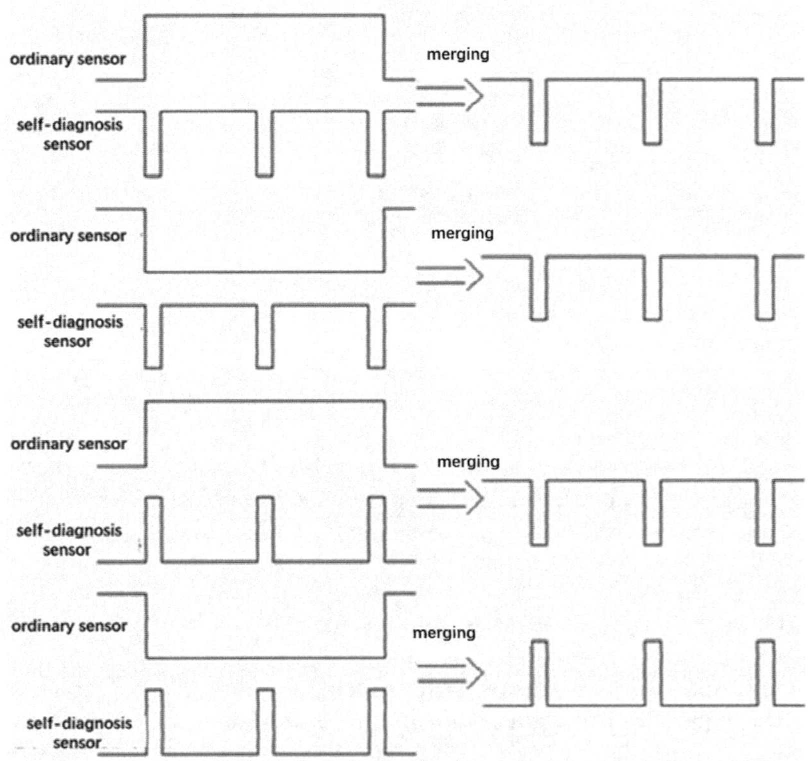

Figure 2. processed signal waveform chart

4. Conclusion

The device can intelligently identify the external input signal, perform the operation of merging with the external input signal and output of the control signal with diagnostic pulse according to the level state of the input channel. Since the self-diagnosis signal contains the self-diagnosis pulse with a width

IOP Conf. Series: Materials Science and Engineering **793** (2020) 012062 doi:10.1088/1757-899X/793/1/012062

of 200us, in order to improve the anti-interference ability of the system and ensure the correct collection of the self-diagnosis pulse, a high-speed optocoupler with a reaction speed of nanosecond is adopted in the input circuit for isolation, and a single chip microcomputer is selected to calculate and control the output of the whole system. The invention can not only realize the acquisition of self-diagnosis signal, but also the acquisition of ordinary level signal, and realize the combined output of the two kinds of different signals through related operation, which has a wide application space.

References

[1] Jia-Bin, Wen, et al. Sensor fault diagnosis and fault tolerant processing based on improved neural network [J]. Sensors and Microsystems,2019,38(10):132-134+138.
[2] Si-Cong Han,et al. Fault diagnosis and fault tolerance of a position sensor [J]. Power electronics technology, 2008,52(12):37-41.
[3] Tian-Tian Wei, et al. An online fault-tolerant and neighbor collaboration sensor node fault diagnosis algorithm [J]. Miniature microcomputer system, 2008,39(12):2669-2674.
[4] Rui Li, et al. Assessing Diagnosis Approaches for Wireless Sensor Networks:Concepts and Analysis[J].Journal of Computer Science & Technology,2014,29(05):887-900.
[5] Da-Qi Zhu, et al. Fast sensor fault diagnosis method based on ca-cmac [J]. Acta electronica sinica,2008(08):1646-1650.
[6] Guang Yang, et al. Sensor layout optimization based on fault detection and reliability constraints [J]. Acta electronica sinica,2006(02):348-351.

State of Health Estimation Framework of Li-on Battery Based on Improved Gaussian Process Regression for Real Car Data

Wei Xiao[1], Huachi Xu[1], Jun Jia[2*], Fei Feng[2], Wen Wang[3]

[1] Sichuan Energy Internet Research Institute, Tsinghua University, Chengdu 610213, China

[2] State Key Laboratory of Mechanical Transmissions, Chongqing University, Chongqing 400044, China

[3] State Grid Electric Vehicle Service Company, Beijing 100053, China

Junjia@cqu.edu.cn

Abstract. With the diversification of power batteries in the market, the estimation and management of State of Health (SOH) become more and more important. The building of the models based on experimental data needs much time and many resources, and the models do not have strong versatility. In this paper, a data-driven on-line SOH estimation framework is built by using the measured data of vehicle operating process. In case that the charging data is incomplete and with low precision, an indirect feature (IF) extraction and combination scheme based on Incremental Capacity Analysis (ICA) is built, so as to realize the mapping between IF and SOH based on Gaussian Process Regression (GPR), and Multi-Colony Particle Swarm Optimization (MPSO) is adopted to solve the local optimum of GPR hyper-parameters. Finally, after the algorithm is verified and evaluated by applying NASA (National Aeronautics and Space Administration) dataset and real vehicle data, and when the training amount of operation data is enough, the error of SOH estimation can be controlled within 2%, which shows that the algorithm has good versatility and estimation precision.

1. Introduction

Power batteries of multiple brands, types and different quality are existing in the market at the same time, and the management of the State of Health (SOH) of power batteries will become an important research issue. SOH represents the difference between an aging battery and a new battery [1]. In recent years, scholars have built many estimation methods of SOH, including model-based methods and data-driven methods.

The building of battery model requires much time and resources, which does not have strong versatility, and it is difficult to build a reliable model with actual vehicle operating data. Compared with model-based methods, data-driven methods are more flexible without considering external disturbances. Existing data-driven SOH estimation methods include Autoregressive Model [2], Neural Network [3], Support Vector Machine [4] and Gaussian Process Regression (GPR) [5], etc. GPR is a non-parametric modeling process based on Bayesian learning, which has Bayesian essence and good robustness, can provide uncertainty of prediction, and is more adaptable to the requirements of changeable actual operating conditions and low data precision than other data-driven methods. In recent years, many scholars have studied battery degradation through signal processing methods, such as Incremental Capacity Analysis (ICA) and Differential Voltage Method [6]. These methods convert

IOP Conf. Series: Materials Science and Engineering **793** (2020) 012063 doi:10.1088/1757-899X/793/1/012063

voltage platforms into observable peaks or valleys, and realize the analysis of aging degree and aging mechanism by observing the characteristic changes of curves during the aging process. Wang Z [7] realized the prediction of SOH by extracting the peak value of IC curve and peak voltage as indirect feature (IF) and training GPR model. However, most of the existing researches are based on high-precision and relatively complete experimental cycle, and the researches that are applicable to the actual operation data with poor accuracy and incomplete charging are still relatively rare.

Based on the above limitation, this paper designs a data-driven method for on-line estimation of SOH. In the following sections, the related algorithms are described briefly, the SOH Gaussian process regression model based on Multi-Colony Particle Swarm Optimization (MPSO) and ICA is built, the NASA (National Aeronautics and Space Administration) data set and the real vehicle data are analyzed respectively, and the feature extraction and combination method is constructed to realize the on-line estimation of SOH. Finally, the work of this paper is summarized and prospected.

2. Related Algorithms

2.1. Gaussian Process Regression

Gaussian Process Regression (GPR) is a statistical learning method based on Bayesian framework. The prior distribution is transformed into a posterior model by training of historical data, which gives confidence interval while outputting the mean value. The basic framework is shown in Figure 1.

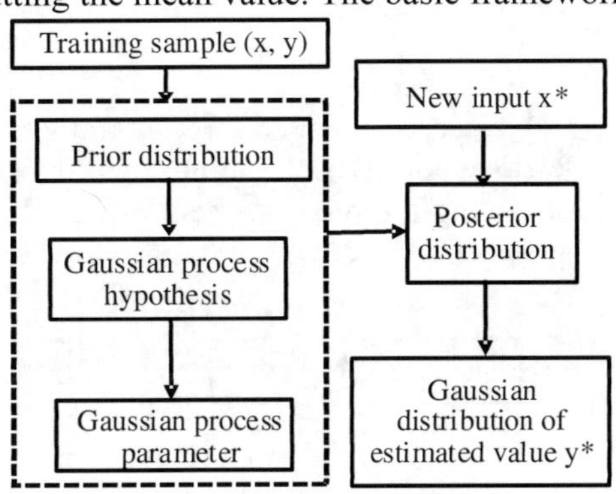

Figure 1. The basic framework of Gaussian process regression

GPR does not need to specify the specific form of the process $f(x)$, but only assumes that it obeys the joint Gaussian distribution, then y is the observed value disturbed by noise, i.e.:

$$y = f(x)+\varepsilon,\ \varepsilon\sim N(0,\sigma_n^2) \tag{1}$$

σ_n^2 is the noise variance, and the set of finite observed values of y constitutes a Gaussian process that obeys the Gaussian distribution:

$$y\sim N(0, Var(x) + \sigma_n^2 I) \tag{2}$$

$Var(x)$ is the covariance matrix of $\mathrm{n} \times \mathrm{n}$, the (i,j) element is the covariance function, i.e., the radial basis function (RBF):

$$Var(x_i, x_j) = {\sigma_f}^2 \exp\left(-\frac{1}{2\lambda^2}|x_i - x_j|^2\right) \tag{3}$$

Wherein, σ_f and λ are constants. When a new test set , also obeys Gaussian distribution, the predicted $\hat{y}^*$ and $Var(\hat{y})^*$ are:

$$\hat{y}^* = k^T(x^*)[Var(x) + \sigma_n^2 I]^{-1} \tag{4}$$

$$Var(\hat{y})^* = Var(x^*, x^*) - k^T(X^*)[Var(x) + \sigma_n^2 I]^{-1} k(x^*) \tag{5}$$

Wherein, $k(X^*) = [Var(x^*, x_1), \dots, Var(x^*, x_n)]^T$

The hyper-parameter $\theta = (\sigma_n, \sigma_f, \lambda)$ is obtained by maximum marginal likelihood according to the equations (6), and whether the selection of θ is accurate affects the prediction result. In this paper, the multi-colony particle swarm optimization is used to iteratively obtain the optimal hyper-parameter.

$$L(\theta) = \frac{1}{(2\pi)^{0.5\pi}|Var(x)+\sigma_n^2 I|^{0.5}} \times \exp\{-\frac{1}{2}y^T[Var(x)+\sigma_n^2 I]^{-1}y\} \tag{6}$$

2.2. Multi-Colony Particle Filter Algorithm

Many optimization algorithms are used for hyper parameter optimization of data-driven models, such as GA [6], linear programming [7], PSO[8], which greatly improves the accuracy of the model. In practical engineering applications, mature PSO is the most widely used. Particle swarm optimization (PSO) is a swarm intelligence algorithm, which has obvious premature convergence phenomenon and tends to fall into local optimum. In this paper, the particle swarm is divided into three sub-swarms evenly, which are evolved according to PSO, ω Adjustment Particle Swarm Optimization (ωAPSO) [9]and Cloud Adaptive Particle Swarm Optimization [10] (CAPSO). And then MPSO is used by updating the three sub-swarms iteratively. The particle richness is increased, so that the computation speed is effectively increased, and the probability of falling into local optimum is reduced.

3. Model Building and Verification of Model

3.1. SOH GPR Model Based on MPSO and ICA

Gaussian Process Regression (GPR) model is obtained through training by using the existing charging data to extract IF sequence and SOH sequence. In the future, the SOH value of the battery during the charging is obtained merely by extracting the relevant features during the charging, rather than by a complete testing to get the residual capacity. There are 3 steps in the basic framework.

Step1:

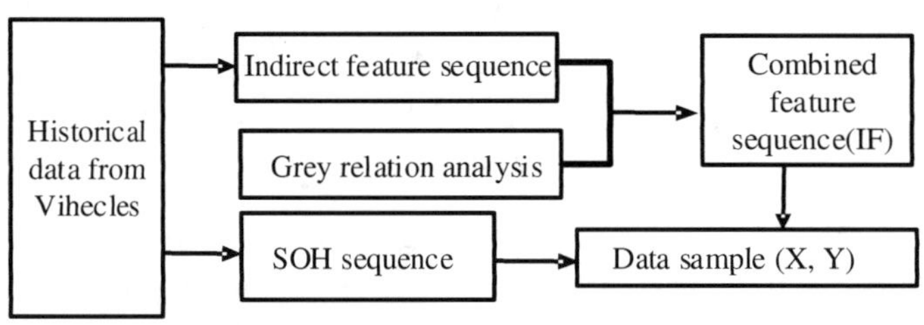

Figure 2. Get data set from historical records

As shown in Figure 2, extract the effective charging cycle of the experimental data or the real vehicle data, and obtain the IFs of the IC curve to form a combined feature, and calculate the SOH value of each cycle.

Step2:

The training set and the test set are divided, and the training data set is used to train the model. The hyper parameters of GPR are optimized by MPSO. Eventually, when the trained model can achieve satisfactory results on the test set, the training is over. As shown in Figure 3, the specific steps are as follows:

(1) Divide the sample data into training samples (x, y) and test samples (x*, y*) a.

(2)Calculate the marginal likelihood function and set the initial value of the hyper-parameter

(3)Divided the particle swarm into three groups equally after initializing, and calculate the current fitness value.

(4)Perform the iterative calculation according to the algorithm described in Section 2.2 to obtain the optimal hyper-parameter. If the fitness value meets a certain threshold twice in succession, it can be regarded as an iteration stop condition, but this may lead to premature convergence. Therefore, the number of iterations is selected as the stop condition.

IOP Conf. Series: Materials Science and Engineering **793** (2020) 012063 doi:10.1088/1757-899X/793/1/012063

(5)Take the test sample x* as the input and bring it into the trained GPR model, and output the SOH mean value and variance corresponding to the charging cycle, with the variance covering a 95% confidence interval.

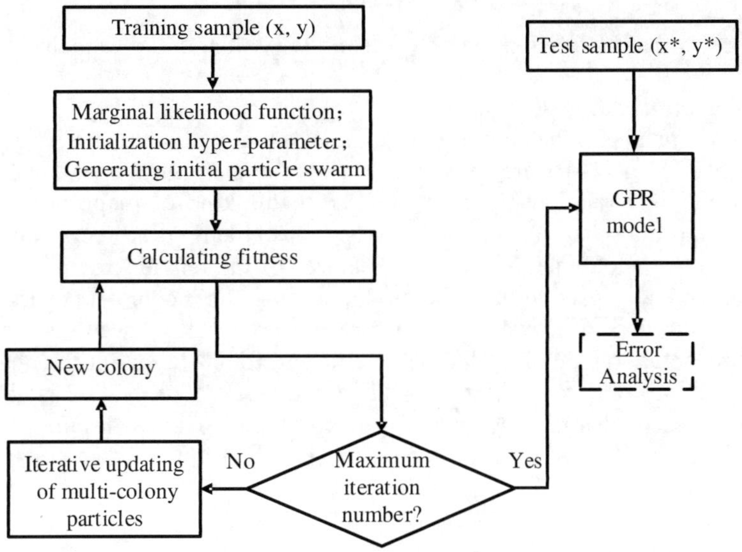

Figure 3. Training process of data-driven model

Step3:

Get the latest charging data and extract IC curve features. As long as one or more features are obtained, the feature combination method can be used to obtain IFs.

3.2. Verification Based on NASA Lithium-ion Battery Data Set

3.2.1. Data of NASA Lithium-ion Battery Aging Experiment

When performing the real vehicle data analysis, we did not find that multiple features exist in multiple cycles of vehicle charging data, so it is impossible to judge whether the combined method is correct. Firstly, the NASA experimental data set is used to verify the algorithm framework and optimize the feature combination method. Using 18650 lithium-ion batteries with a rated capacity of 2Ah, NASA Prognostics Center of Excellence conducted a large number of battery experiments under various conditions in Idaho National Laboratory [11]. In this paper, the experimental data of B5 and B7 battery is used for verification. A brief description of the battery test flow is as follows:

Table 1. Test Flow of NASA Lithium-ion Battery Data Set

Experiment Procedure	Stage	Test Item	Cut-off Condition	Remarks
Charging	Stage 1	Constant-current charging	Reach the upper cut-off voltage	The upper cut-off voltage is 4.2 V.
	Stage 2	Constant-voltage charging	The current drops to 0.01C	/
Discharging	/	Constant-current discharging	Reach the lower cut-off voltage	The lower cut-off voltage of B5 battery is 2.7 V; That of B7 battery is 2.5 V

Repeat the charging and discharging to make the battery age. Stop the experiment when the actual capacity of the battery drops to 70% of the rated capacity. The capacity-based SOH [1] is defined as:

$$SOH = (C_i/C_0) * 100\% \quad (7)$$

Where C_i is the capacity measured in the *i*th cycle, and C_0 is the initial capacity.

3.2.2. IC Curve Calculation and IF Extraction

Discharge voltage drop rate [12], the charging time under equal voltage rise [13], average voltage and temperature during charging [14] were extracted during the charging process as an indirect health factor and built its mapping relationship with SOH. But this kind of mapping relationship needs a relatively complete charge-discharge curve as the basis for extracting indirect health factors, which is not easy to achieve in the actual use of vehicles. Features of the IC curve, such as Peak Value, Peak Area, Peak Voltage, were also used to train the model because the incomplete charging curve can also be converted to an IC curve [15]. However, these methods are usually based on experimental data. In most cases during the actual operation of the vehicle, it is difficult to obtain the characteristics of the IC curve, or sometimes this feature is obtained, and sometimes another feature is obtained. Therefore, we make full use of each charging cycle and get combined feature training based on the grey correlation analysis, which effectively enriches the data set. The IC curve is calculated as:

$$\frac{dQ}{dV} = \frac{\delta Q}{\delta V} = \frac{Q_t - Q_{t-1}}{V_t - V_{t-1}} \quad (8)$$

Where Q is capacity, V is Voltage , and t is time point.

Here, the Savitzky-Golay [16] smoothing filter is used to process noise. The fitting result is shown in Figure 4. The point A corresponds to the Peak of the IC curve, the shaded area B is the Peak Area, and the point C corresponds to the V_peak.

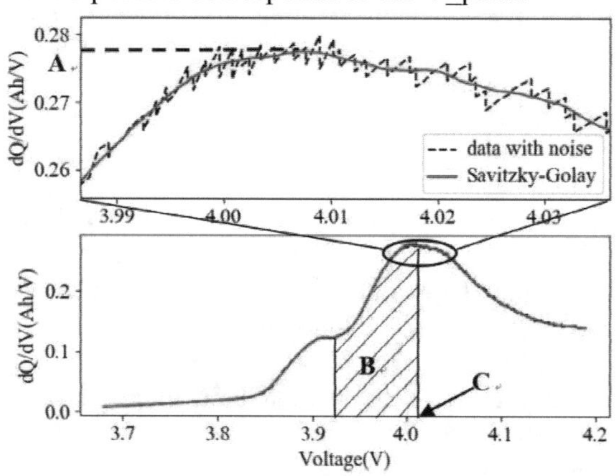

Figure 4. IC Curve Filtering Diagram of B5

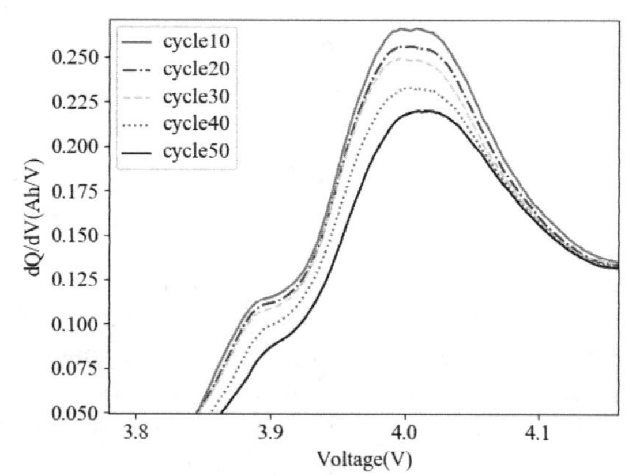

Figure 5. Change of IC Curves of B5

By selecting some cycles at equal intervals and plotting the IC curves on the same graph, Figure 5 shows that the I and II Peaks are significantly reduced, while the Areas are also reduced, and the V_peaks are increased. Peak, Area and 1/V_Peak of peak II are extracted as IF, and subject to normalization processing. The similarity is quantified by Grey Relation Analysis (GRA). GRA judges the correlation degree according to the similarity of geometric shapes between the sequence curves [16]. The grey correlation of SOH and Area and Peak are relatively high, as shown in Table 2. However, due to the error of the voltage sensor and the error introduced during the smooth filtering, it is difficult to obtain the accurate V_Peak, so the *r* of the 1/V_peak and SOH sequences is relatively low. In the real vehicle data analysis, the error is greater, so it is not used in subsequent calculations.

Table 2. Grey Correlation between Indirect Features and SOH

Battery No.	IF Sequence	grey correlation

B5	Area	0.9160
	Peak	0.8472
	1/V_peak	0.6662
B7	Area	0.8763
	Peak	0.8363
	1/V_peak	0.6042

3.2.3. Verification and Improvement of Algorithm

For the algorithm verification with single IF as input, taking the single IF test results of B5 battery as an example, the results are shown in **Figure 6**:

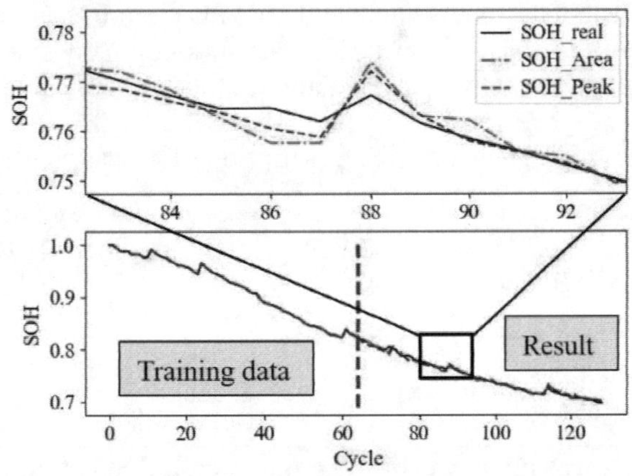

Figure 6. Test Results with a Single Indirect Feature as Input of B5

The left of the vertical dividing line is the training data, and the right is the results estimated by the trained GPR model. SOH_real is the SOH measured by experiment, SOH_Area is the SOH estimated with Area as IF, and SOH_Peak is the SOH estimated with Peak as IF. The root mean square error (RMSE) analysis of the two groups of batteries is shown in Table 3:

Table 3. Error Analysis of Estimated SOH by Single Indirect Feature

Battery No.	IF Sequence	RMSE
B5	Area	0.003759
	Peak	0.003307
B7	Area	0.002614
	Peak	0.001690

It cannot ensure that IFs are extracted from each circle due to the limitation of SOC range when the vehicle is actually running. In this paper, therefore, IFs are reasonably combined. The combined IFs are used as input to train the GPR model, realizing SOH estimation. The combination is $IF_{new} = a * Peak + b * Area$, where, $a + b = 1$. The estimation results under the different values of the coefficients a and b are analyzed below.

Table 4. Error Analysis of Estimated SOH by Combined IFs

Battery No.	a:b	RMSE
B5	1:1	0.003225
	$r_{Peak}:r_{Area}$	0.003160
B7	1:1	0.002200
	$r_{Peak}:r_{Area}$	0.002162

As is shown in Table 4, by analyzing the estimation results under multiple combinations, the IF estimation result obtained by linear combination using the ratio of grey correlation is the best. The linear combination equation is:

$$IF_{new} = \frac{r_{Peak}}{r_{Peak}+r_{Area}} * Peak + \frac{r_{Area}}{r_{Peak}+r_{Area}} * Area \tag{9}$$

3.3. Results Based on Real Vehicle Data and Discuss

3.3.1. Description of Actual Vehicle Operation Data

Currently, the monitoring center mostly collects the BMS data and the related data through vehicle-mounted terminals installed in the vehicle and uploads them to the cloud data center. The data used in the paper comes from the one-year operation data of a company's new energy minibus. Two vehicles (No. 2621 and 2672) are selected for data analysis in the paper. Table 5 gives a brief description of the actual vehicle data.

Table 5. Brief Description of Real Vehicle Data

Battery type	Lithium iron phosphate	
Series-parallel mode	3 cells in series per module/100 module in series	
Data upload frequency	Once for 10S	
Charging rate	1/3C	
Data acquisition accuracy	Voltage	0.01V
	Current	0.1A
	SOC	0.1%

At first, the pre-processing results of the original collected data were analyzed in Table 6. Although many charging cycles have been extracted, since the SOC range is not suitable for ICA, the number of cycles that can be effectively used for model training is relatively small.

Table 6. The pre-processing results of the original collected data

Vehicle No.	Total data points	Missing/error data points	Percentage of available data	Total charging cycles	Effective charging cycles
2621	2956608	2324190	78.61%	198	106
2672	2935872	2389800	81.40%	182	98

Figure 7 shows a statistical chart for SOC range of the vehicle during operation, with 88.8% of the SOC range not exceeding 50%. Figure 8 shows a statistical chart for SOC range when the vehicle starts charging, with 87.2% of the SOC range being greater than 50% at the starting.

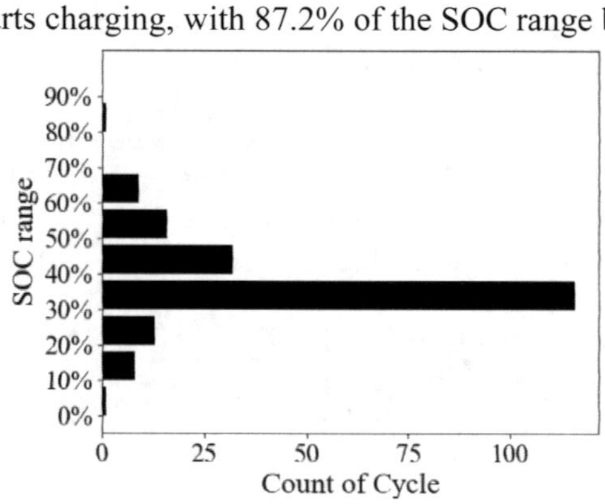

Figure 7. SOC Range at the Charging

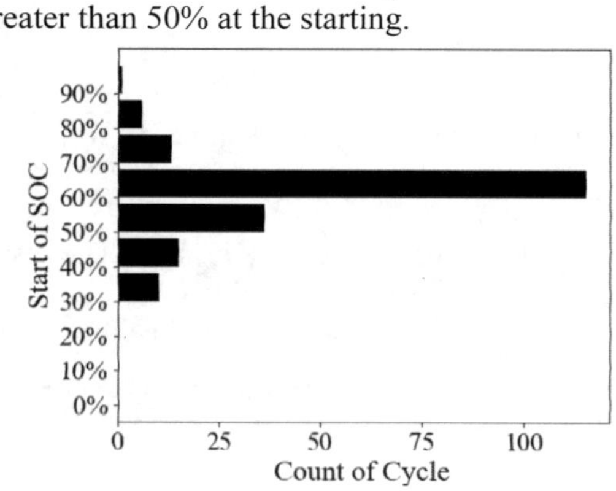

Figure 8. SOC When Start Charging

3.3.2. Data Preliminary Processing and Feature Extraction

Firstly, take the preliminary processing for the real vehicle data, including elimination of abnormal values, filling of missing values and so on. Secondly, extract all the charging cycles to identify the multiple segment data interrupted during the charging and connect them reasonably. And then, obtain the IC curve. As shown in Figure 9, A is the peak (Peak) and the shaded area B is the peak area (Area).

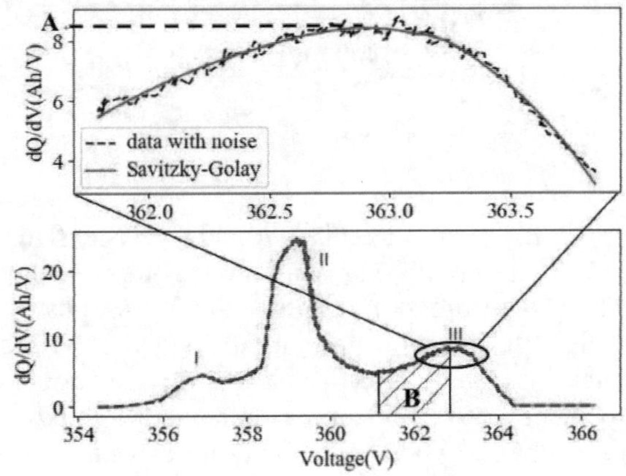

Figure 9. IC Curve Filtering Diagram for Real Vehicle Data

In actual vehicle operation data, Peak I and Peak II are rarely seen, so Peak III is selected. Table 7 shows the results of feature extraction for actual vehicles. It can be seen that the feature sequence is difficult to be completely obtained due to the data precision and under the impact of actual operation interval and the GRA between single feature and its corresponding SOH sequence is lower than the experimental data, so it is necessary to reconstruct the feature sequence. When only a single feature can be extracted, the feature is used as the IF of the current cycle. When both features can be extracted, IF is calculated according to equation (9).

Table 7. Feature Conditions Extracted from Real Vehicle Data

Vehicle No.	effective cycles	Feature type	Available Features	Grey correlation
2621	106	Area	73	0.8378
		Peak	85	0.6435
		Both	52	/
2672	98	Area	89	0.6274
		Peak	52	0.6106
		Both	44	/

3.3.3. Verification Results

IF sequences and SOH sequences are used as real vehicle data samples, and different proportions of cycles are selected as training samples to carry out algorithm verification. Here, we chose Support Vector Regression (SVR) to replace GPR as a model comparison. MPSO is adopted to obtain the optimal hyper-parameter of SVR. Because the predicted results are graphically similar, only three verification results are given for each model of vehicles, as shown in Figure 11 and Figure 12.

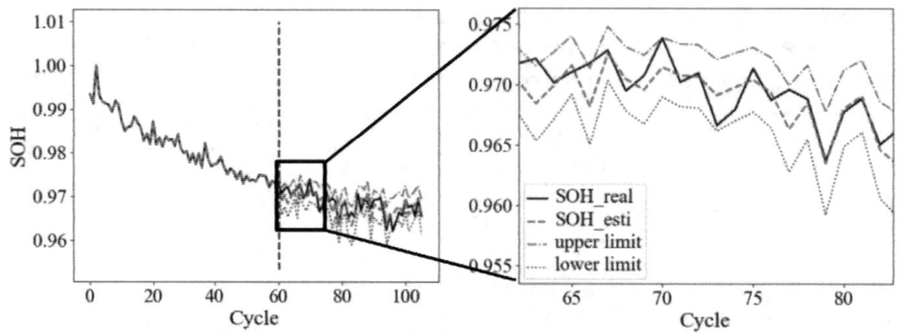

Figure 10. SOH Estimation Results with 60 Cycles as Training of No.2621

The GPR model gives confidence interval while giving the predicted mean value to calculate the upper and lower limits. The SOH estimation results with 60 cycles as training of No.2621 are taken as an example, as shown in Fig. 10. The solid lines are the actual SOH sequences. It can be seen that in actual operation, the vehicle has obvious capacity recovery phenomenon and capacity degradation curve fluctuations due to much battery idle time, but overall shows a slow downward trend. The vehicle operation mileage is about 50,000 miles, and the capacity drop is less than 4%. The dotted lines are the SOH estimated based on the IF extracted from the current cycle of the trained GPR model. The first curve and the fourth curve are the upper and lower limits of the 95% confidence interval of the estimated values, respectively. It can be seen that basically all the real values are within the upper and lower limits of the confidence interval.

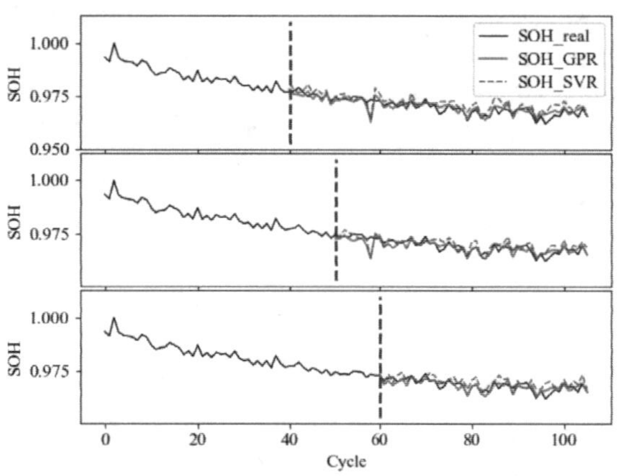

Figure 11. Estimated Results of Different Proportions of Training Data of No. 2621

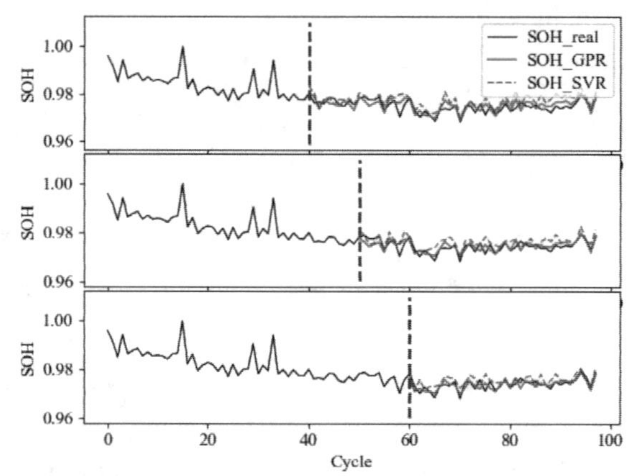

Figure 12. Estimated Results of Different Proportions of Training Data of No. 2672

It can be seen from the above figures that both the GPR and SVR model under different proportions of training data can achieve the estimation of SOH in the future, and the large the number of training is, the more accurate the model will be. Among them, GPR is better in most cases. Since GPR has a good confidence interval description ability, we conclude that GPR can show better performance in practical applications. However, the actual vehicle battery degradation is 3%-4% and the NASA experimental battery degradation is about 30%. When the degradation ratio difference of the two is not one order of magnitude, the RMSE is the same order of magnitude, as shown in Table 7. Although the maximum percentage error relative to the total capacity is small, the ratio of the maximum error value to the degradation at the cut-off time is above 10%, and the maximum error of No. 2621 is 30%. The computation accuracy of the actual data is far less than that of the experimental data.

Table 8. Error Analysis of Estimated SOH for Real Vehicle Data

Vehicle No.	Number of cycles used for training	Proportion (%) of training data to sample data	Data-driven model	RMSE	Maximum percentage error %
2621	40	37.74%	GPR	0.002245	1.09%
			SVR	0.002568	0.96%
	50	47.17%	GPR	0.002285	1.05%
			SVR	0.001952	1.05%
	60	56.60%	GPR	0.001823	0.38%
			SVR	0.001927	0.76%
2672	40	40.82%	GPR	0.002167	0.61%
			SVR	0.002648	0.98%
	50	51.02%	GPR	0.001938	0.52%
			SVR	0.002016	0.74%
	60	61.22%	GPR	0.001713	0.45%
			SVR	0.001716	0.72%

From the above results we can see that an accurate model has been trained for each vehicle. Now, the electric vehicle service platform can use the charging data uploaded by the vehicle to extract IFs. Regardless of the depth and range of charging cycle, as long as one or more IC features can be extracted, the platform can feed back the current SOH for supporting other applications.

3.3.4. Error Analysis

Temperature fluctuation

Although the BMS has a thermal management function, the public charging stations may be placed outdoors, which can cause severe temperature fluctuations due to change of seasons. The temperature has a certain influence on the shape of the IC curve [12]. As the seasons change, it spans a wide temperature range.

Module inconsistency

In this paper, the total current and total voltage data are used to analyze the SOH at the module level. Affected by the aging degree and temperature difference between cells, the module inconsistency will obviously increase. Even for the battery module with the same capacity, if the cells have large inconsistency, the IC curve shape will be obviously different [17], which is beyond the scope of this paper.

Data quality and calculation error

Due to the limitation of the current hardware facilities and network transmission capability, errors are introduced in the processes of measurement, recording and transmission of the actual operation data, which leads to the poor accuracy of the original data. In the follow-up analysis process, additional calculation errors are inevitably introduced in the preparatory processing, preliminary processing, secondary calculation, the filtering process, etc.

Under the cumulative effect of the above errors, the IF combination method and GPR estimation model constructed in this paper have good performance, and the error is within the acceptable range.

4. Conclusions

In this paper, an online SOH estimation framework based on GPR is designed and verified by real vehicle charging data. The accuracy of SOH estimation can be controlled within 2% with enough data training. The contributions of this paper include:

(1) For the incompleteness of charging curve in actual vehicle operation, the combination feature sequence is constructed by fully utilizing the extractable IC curve features, and the online estimation framework of SOH is realized based on GPR model, which can be used for partial charging curve data.

(2) The SOH estimation framework proposed in this paper can collect data synchronously to optimize the model in vehicle operation, which reduces the cost of experiment design and implementation.

(3) This method is suitable for all kinds of vehicles and batteries, and adaptable to the poor data accuracy condition to an extent.

Next, we will focus on the optimization of the algorithm framework in case of significant effects of temperature changes and inconsistencies in the battery modules.

Acknowledgements

This work was supported by the State Grid Corporation Science and Technology Project of China (Grant No. 52020118000G).

References

[1] Lu L , Han X , Li J , et al. A review on the key issues for lithium-ion battery management in electric vehicles. *Journal of Power Sources*, 2013, **226**:272-288.

[2] Liu D , Luo Y , Liu J , et al. Lithium-ion battery remaining useful life estimation based on fusion nonlinear degradation AR model and RPF algorithm. *Neural Computing and Applications*, 2014, **25(3-4)**:557-572.

[3] Bai G,Wang,and Hu C. A Self-cognizant Dynamic System Approach for Prognostics and Health Management. *Journal of Power Sources*,2015,**278**: 163-174.

[4] Berecibar M , Devriendt F , Dubarry M , et al. Online state of health estimation on NMC cells based on predictive analytics. *Journal of Power Sources,* 2016, **320**:239-250.

[5] Liu D , Pang J , Zhou J , et al. Prognostics for state of health estimation of lithium-ion batteries based on combination Gaussian process functional regression. *Microelectronics Reliability*, 2013, **53(6)**:832-839.

[6] Ecker M , Nieto N , KBitz S , et al. Calendar and cycle life study of Li(NiMnCo)O2-based 18650 lithium-ion batteries. *Journal of Power Sources*, 2014, **248**:839-851

[7] Wang Z , Ma J ,and Zhang L . State-of-Health Estimation for Lithium-Ion Batteries Based on the Multi-Island Genetic Algorithm and the Gaussian Process Regression. *IEEE Access*, 2017, **5**:21286-21295.

[8] Qin T , Zeng S , and Guo J . Robust prognostics for state of health estimation of lithium-ion batteries based on an improved PSO-SVR model. *Microelectronics Reliability,* 2015, **55(s9 - 10)**:S0026271415300792.

[9] Eberhart R,Kennedy J. A new optimizer using particle swarm theory. *Proc of the 6th Int Symposium on Micro Machine and Human Science*,Vagoya,**1995**: 39-43.

[10] Marinakis Y , Marinaki M . A Hybrid Multi-Swarm Particle Swarm Optimization algorithm for the Probabilistic Traveling Salesman Problem. *Computers & Operations Research*, 2010, **37(3)**:432-442.

[11] Saha B, Goebel K, Poll S, et al. Prognostics methods for battery health monitoring using a Bayesian framework. *Instrumentation and Measurement, IEEE Transactions on 2009*, **58(2)**: 291-296.

[12] Ye J , Zhang S , and Yang W . Gravitational search algorithm with Gaussian process, for lithium-ion batteries state of health (SOH) estimation. *International Conference on Information Technology & Management Innovation*. **2015**.

[13] Weng C , Cui Y , Sun J , et al. On-board state of health monitoring of lithium-ion batteries using incremental capacity analysis with support vector regression. *Journal of Power Sources*, 2013, **235**:36-44.

[14] Hao L , Mingxin H , Yiheng Z , et al. Prediction for state of health of lithium-ion batteries by genetic algorithm and support vector regression. *Journal of Nanjing University of Science and Technology*, 2018.

[15] Peikun S , Zhenpo W . Research of the Relationship between Li-ion Battery Charge

IOP Conf. Series: Materials Science and Engineering **793** (2020) 012063 doi:10.1088/1757-899X/793/1/012063

Performance and SOH based on MIGA-Gpr Method. *Energy Procedia*, 2016, **88**:608-613.

[16] Chen D, Chen Y Q, and Xue D. Digital Fractional Order Savitzky-Golay Differentiator. *IEEE Transactions on Circuits & Systems II Express Briefs*, 2011, **58(11)**:758-762.

[17] Weng C , Feng X , Sun J , et al. State-of-health monitoring of lithium-ion battery modules and packs via incremental capacity peak tracking. *Applied Energy, 2016*, **180**:360-368.

IOP Conf. Series: Materials Science and Engineering **793** (2020) 012064 doi:10.1088/1757-899X/793/1/012064

Research and application on rapid recognition of CNC machine tools' running state

Mengmei Li[1] , Yuanmeng Xia[1, 2] and Xuezhen Chen[1, *]

[1] CNC processing plant, AVIC Chengdu Aircraft Industrial (Group) Co., Ltd., Chengdu, Sichuan, 610091, China

[2]University of Electronic Science and Technology of China， Chengdu，Sichuan, 611731， China

[*]Corresponding author's e-mail:532015901@qq.com

Abstract: A rapid recognition of CNC (Computer Numerical Control) machine tools' running state based on time domain characteristic parameters and frequency domain correlation coefficient was proposed. The experiment proved the accuracy and efficiency of the mentioned method，which made the states recognition rate reach 93.33% and the identification efficiency improve 66.67%. In this scenario， a theoretical foundation is established for the condition-based maintenance of CNC machine tools.

1. Introduction

As China gradually develops from a large manufacturing country to a powerful manufacturing country, CNC machine tools are becoming more and more large, high-speed, high-precision and intelligent, which dominate an important position in the manufacturing industries. On the one hand, the CNC machine tools improve production efficiency and product quality; on the other hand, the sudden failure of CNC machine tools seriously affects the on-time parts delivery, which extremely wastes all kinds of production resources and causes huge losses. Therefore, modern advanced manufacturing companies are not satisfied with the breakdown maintenance. It is urgent to improve the reliable operation of CNC machine tools by identifying the running states of the CNC machine tools and implementing the condition-based maintenance [1].

At present, the identification methods of CNC machine tools' running states include traditional methods and intelligent methods. The traditional method is the viewing method, which is checked by well-experienced workers applying visual inspection, hand touch, etc, but it's time-consuming, and the inspection quality is limited by the labor quality and cannot be standardized. The intelligent method collects information by using various advanced instruments, and uses knowledge-based methods such as artificial neural networks, fuzzy theory, and expert systems to identify the running state. The intelligent method is a research hotspot in the field of identification, and also a development tendency, but there are still some shortcomings. The artificial neural network converges slowly and needs a large number of samples for training; the conclusion of the fuzzy theory is not accurate enough; the expert system shows weak recognition ability and poor self-learning ability when it is applied to the CNC machine tool with complex structures. Nowadays, intelligent methods research is focused on top-level design and has not been extensively engineering applied [2-3].

In order to improve the accuracy and efficiency of CNC machine tools' running state recognition, this paper proposes a method based on time domain characteristic parameters and frequency domain correlation coefficient. Firstly, collecting vibration signals under normal, attention and abnormal conditions of CNC machine tools by the overall sensor installation method. Then, through the sensitivity analysis of the time domain characteristic parameters and the frequency domain correlation coefficient, four time domain characteristic parameters and the frequency domain correlation coefficient are selected as rapid identification indicators. Finally, the judgment standards of state recognition is established to complete the rapid recognition model of the CNC machine tool, which can rapidly and efficiently identify the running state of CNC machine tools.

2. Theoretical basis

2.1. Time domain characteristic parameters

The time domain analysis of the vibration signal is carried out by amplitude analysis in the time domain waveform, and the dimensional characteristic parameters (including peak value, peak-peak value, average, absolute average, valid value, variance, skewness and kurtosis) and dimensionless characteristic parameters (including peak factor, pulse factor, margin factor, waveform factor, kurtosis factor) of the signal are calculated by the amplitude probability density function [4].

The amplitude probability density function p(t) of the signal x(t) is:

$$p(x) = \lim_{\Delta x \to 0} \frac{P[x<x(t)\leq x+\Delta x]}{\Delta x} = \lim_{\Delta x \to 0} \frac{1}{\Delta x} [\lim_{T \to \infty} \frac{T_x}{T}] \tag{1}$$

Where T_x is the total time that the signal x(t) is in the $(x, x + \Delta x)$ interval.

The amplitude probability density function reflects the distribution pattern of the signal in amplitude, and the probability density functions of different signals are different. The generation and development of the fault will cause a change in the amplitude distribution of the signal, that is, a change in the amplitude probability density function. Generally, if the device is in normal running state, its probability density function curve is close to the normal distribution. When the running state is abnormal, it will be doped into the extra periodic signal to change the original amplitude distribution.

The selection principles of the time domain characteristic parameters : 1 the stability of the time domain characteristic parameters is high under the same states; 2 the sensitivity of the time domain characteristic parameters is high under different states. The dimensional characteristic parameters, such as RMS, are gradually increasing with the occurrence and development of faults, with high stability and low sensitivity. The dimensionless characteristic parameters, such as margin factor, increase significantly in early failures, but if the fault develops at a certain extent, it decreases, with high sensitivity and low stability.Therefore, it is necessary to select both the dimensional characteristic parameters and dimensionless characteristic parameters considering sensitivity and flexibility.

2.2. Frequency domain correlation coefficient

When the state deteriorates or the fault occurs and develops, it often causes a change in the signal frequency structure. In engineering applications, the time domain signal is converted to a frequency domain signal by Fast Fourier Transform (FFT) to obtain a relationship between frequency structure and frequency.

Correlation indicates the extent of similarity between two kinds of signals. The larger the absolute value of the correlation indicates, the stronger the correlation, and vice versa.

The Pearson product-moment correlation coefficient of X and Y is the quotient of the covariance of the two variables and the standard deviation product of the two, i.e.

$$\rho_{xy} = \frac{cov(X,Y)}{\delta_X \delta_Y} = \frac{E(X-\mu_x)(Y-\mu_y)}{\delta_X \delta_Y} \tag{2}$$

IOP Conf. Series: Materials Science and Engineering **793** (2020) 012064 doi:10.1088/1757-899X/793/1/012064

The above is the overall correlation coefficient. Generally, the covariance and standard deviation of the sample calculation are used instead of the covariance and standard deviation of the population. The correlation coefficient formula of the sample is:

$$r = \frac{\sum_{i=1}^{n}(x_i-\bar{x})(y_i-\bar{y})}{\sqrt{\sum_{i=1}^{n}(x_i-\bar{x})^2}\sqrt{\sum_{i=1}^{n}(y_i-\bar{y})^2}} \tag{3}$$

The frequency domain correlation coefficient of the same states is high, and the frequency domain correlation coefficient of different states is low. The correlation analysis between the test signal and the normal state signal can be used as the basis for judging whether the state changes [5-6].

2.3. Rapid recognition model of CNC machine tools' running state

The time domain and frequency domain indicators respectively characterize the running states of the CNC machine tool from different aspects. Therefore, this paper comprehensively selects the time domain characteristic parameters and the frequency domain correlation coefficient to make a rapid identification model of the CNC machine tool' running states. The specific implementation steps are as follows:

- For different monitoring parts, the overall sensor installation method is adopted to collect the vibration signals of the CNC machine tool under normal, attention and abnormal states;
- According to the selection principle of the characteristic indicators, appropriate time domain characteristic parameters and frequency domain correlation coefficient are selected as rapid recognition indicators through the sensitivity analysis.
- Establish the judgment standard of state recognition and complete the rapid recognition model of the CNC machine tools' running state.

3. Engineering application

In order to verify the feasibility of the proposed method, the Y-axis running states of a large horizontal CNC machine tool is rapidly recognition.

3.1. Rapid recognition model establishment

Considering the actual demand of production and processing, the sensor installation method of directly mounting the sensor to the transmission table or column is different from the traditional sensor installation method, disassembling the CNC machine cover and mounting to specific components such as bearings, rolling elements, nut holders, etc, collecting the vibration signals of the CNC machine tool under normal, attention, and abnormal states [7].

The sensitivity analysis of the time domain characteristic parameters is carried out when the Y-axis of the CNC machine tool is under normal, attention and abnormal states, as shown in Fig. 1. Due to limited space, only 3 tests are selected for each state.

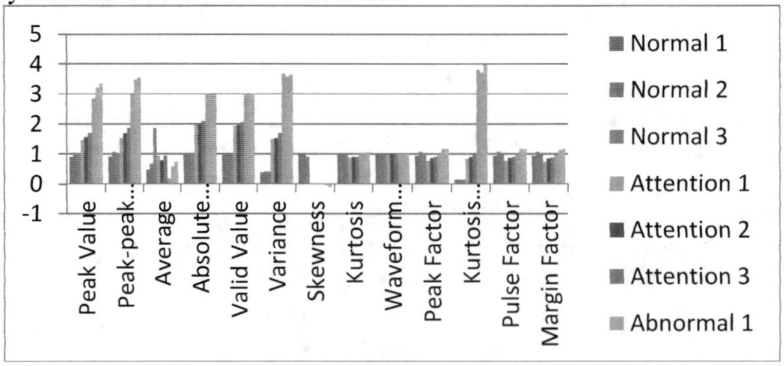

Figure 1. The time domain characteristic parameters of the Y-axis under normal, attention, and abnormal states (normalized)

It can be seen from Fig. 1 that the average value, skewness, and waveform factor cannot accurately identify the running states; peak, peak-to-peak, kurtosis, crest factor, pulse factor, and margin factor identify the running states normally; absolute average, root mean square, variance and kurtosis factor can accurately identify the running states and are selected as rapid identification indicators.

Calculate the frequency domain correlation coefficient of the Y-axis under normal, attention, and abnormal states, as shown in Table 1.

Table 1. The frequency domain correlation coefficient of the Y-axis under normal, attention, and abnormal states

States	Normal	Attention	Abnormal
Normal	0.874	0.584	0.153
Attention	/	0.891	0.511
Abnormal	/	/	0.929

It can be seen from Table 1 that the correlation is very high if the running state does not change, both greater than 0.8, and the correlation coefficient is reduced to less than 0.6 if the running state changes.

According to sensitivity analysis of the time domain characteristic parameters and the frequency domain correlation coefficient, a rapid recognition model of the Y-axis running state is established, and the judgment standard of the state recognition is determined, as shown in Table 2.

Table 2. The judgment standard of the Y-axis state recognition

Indicators / States	Root mean square	Absolute average	Variance	Kurtosis factor	Frequency domain correlation coefficient
Normal	<1.5times	<1.5times	<2times	<4times	>0.6
Attention	1.5-2.5times	1.5-2.5times	2-7times	4-20times	0.2-0.6
Abnormal	>2.5times	>2.5times	>7times	>20times	<0.2

3.2. Application verification

After establishing the rapid recognition model of the Y-axis running states, a total of 30 tests from April 2019 to June 2019 are carried out to verify the method. The correct recognition rate is 93.33% with 28 correct tests. 2 incorrect tests are caused by the interference of sensors' signal lines when the data is collected. The axis state recognition takes an average of 0.5 hours, compared with the previous state recognition of taking 1.5 hours, the recognition efficiency is increases by 66.67%.

The root mean square of the vibration signal collected from the Y-axis on May 15th is 1.5 times of the normal state (April 14th), the absolute value 1.5 times, the variance 2.3 times, the kurtosis factor 4 times, and the frequency domain correlation coefficient between the test state and the normal state is 0.286, so the state is rapidly recognized as the attention state. After dismantling and troubleshooting, it was diagnosed that there was a crack in the Y-axis balance cylinder seal ring and immediately replaced it. After the replacement was done on May 19th, the rapid state recognition was performed again, and the four time domain characteristic parameters and the frequency domain correlation coefficient all returned to values of the normal state.

The rapid identification analysis of the Y-axis running states on April 14th, May 15th and May 19th is shown in Figure 2 and Table 3.

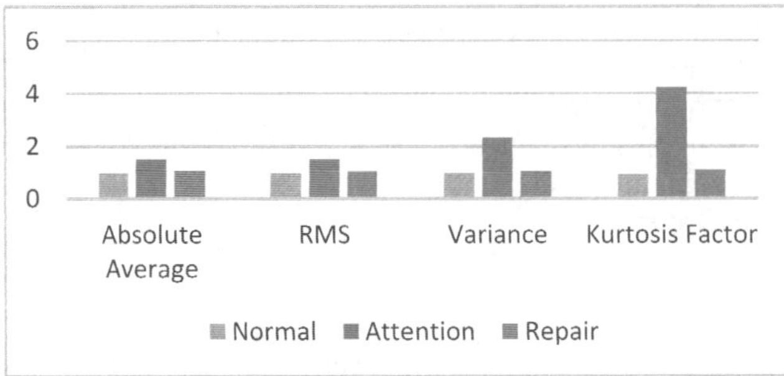

Figure 2. The time domain characteristic parameters (normalized) of the Y-axis under normal, attention, and repair states

Table 3. The frequency domain correlation coefficient of the Y-axis under normal, attention, and repair states

The frequency correlation coefficient	Normal	Attention	Repair
Normal	0.953	0.286	0.938
Attention	/	0.948	0.284
Repair	/	/	0.959

Through the application of the above theoretical methods and technical specifications, the rapid recognition of CNC machine tools' running state is proved. In engineering applications, the established rapid recognition model effectively detects the shaking of the Y-axis caused by the crack in the balance cylinder seal ring. The rapid identification of the states guides workers to implement condition-based maintenance, saving maintenance cost and improving production efficiency and quality.

4. Conclusions

- Considering the actual demand of production and processing, a rapid recognition model of the CNC machine tool' running state is established, which overcomes the shortcomings of low accuracy and efficiency of the traditional identification methods.
- The effectiveness and feasibility of the rapid recognition method is verified by the real engineering application. It can identify the running states of CNC machine tools and guide the condition-based maintenance, which has some value in engineering application.

Funding information: This work is supported by the Major Project of National Science and Technology(Grant No.2017ZX04002001).

References

[1] Deepam Goyal, B. S. Pabla. Condition based maintenance of machine tools-A review [J].CIRP Journal of Manufacturing Science and Technology.2015, 10: 24-35.
[2] Venkat Venkatasubramanian, Raghunathan Rengaswamy, Kewen Yin, et al. A review of process fault detection and diagnosis: Part I: Quantitative model-based methods [J]. Computers &Chemical Engineering. 2003, 27(3): 293-311.
[3] REICHARD KM. Integrating self-health awareness in autonomous systems[J]. Robotics and Autonomous Systems, 2004, 49(1 /2) : 105-112.
[4] Choi S, Pazouki E, Baek J, et al. Iterative Condition Monitoring and Fault Diagnosis Scheme of Electric Motor for Harsh Industrial Application[J]. IEEE Transactions on Industrial Electronics, 2015, 62(3): 1760-1769.
[5] R. L. Florom, J. E. Hiatt, and R.L. Smith. 1987. Wayside Acoustic Detection of Railroad Roller Bearing Defects[C]. Paper No. 87-WART 11.
[6] LIU G, YU J. Gray correlation analysis and prediction models of living refuse generation in

Shanghai city [J]. Waste Management, 2007, 27（3）：345-351.
[7] Chen Xuezhen, Li Chunlei, Li Lianyu, Xia yuanmeng. Research on State Monitoring of CNC Machine Tool based on Dual Dimension Feature [J]. APISAT2018.

IOP Conf. Series: Materials Science and Engineering **793** (2020) 012065 doi:10.1088/1757-899X/793/1/012065

Design and achievement of hardware-in-the-loop simulation system for strapdown semi-active laser seeker

Wenbin Zhou[1*], Chunxi Zhang[1], Siyuan Wang[1], Longjun Ran[1]

[1] School of Instrumentation Science and Opto-electronics Engineering, Beihang University, Beijing, 100191, China

[*]Corresponding author's e-mail: zhwbing@outlook.com

Abstract. According to the work characteristics of semi-active laser-guided weapon system, at first, the optical characteristics of the laser received by seeker were analyzed; and mathematical model of the optical signal power received by seeker was established. Secondly, a hardware-in-the-loop simulation(HILS) system was designed for strapdown seeker. In order to simulate energy and laser spot's continuous variations, a high-precision continuous dynamic laser energy attenuation system and a digital micromirror device(DMD) spot projection system were designed respectively. In the end, the HILS system was used to evaluate the performances of strapdown seeker and the seeker was tested for response coherence among the four quadrants, field of view(FOV), linear field of view, minimum received optical power, and angle measurement accuracy.

1. Introduction

Semi-active laser guided weapons are widely used because of high precision and strong anti-interference ability[1]. As the core device of precision guided weapons, it is an urgent problem for research and development department to evaluate the performance of laser seeker. Live-fire target shooting is one of the best methods for weapon performance measurement, but the measurement cost is expensive and it cannot obtain large amount of data resource. Hardware-in-the-loop simulation(HILS) introduces test components into the simulation loop, and corresponding performances are tested through simulating real working environment[2,3]. HILS has become an important performance evaluation method of new weapons. At present, HILS technology for guidance weapons is studied constantly by major military powers[4]. Therefore, it has important scientific and strategic significance to establish HILS platforms for the terminal guidance weapons.

Semi-active laser guidance is completed with the help of target designator and laser guided missile; the designator and the missile are mutually independent. The designator keeps laser spot irradiating target surface. The seeker can receive the laser echo signals diffusely reflected by target in real time, within an effective range and calculate the angle of sight which can generate guidance information to guide a missile to the target[5]. According to the similarity principle of simulation technology, the HILS system should implement the following functions[6]:

- simulate the real attitude of projectile.
- provide a laser irradiation environment, and simulate spot size.
- simulate laser energy received by seeker.
- receive instructions from missile-borne computer and complete 6-DOF trajectory simulation.

IOP Conf. Series: Materials Science and Engineering **793** (2020) 012065 doi:10.1088/1757-899X/793/1/012065

2. Optical characteristics of laser received by seeker

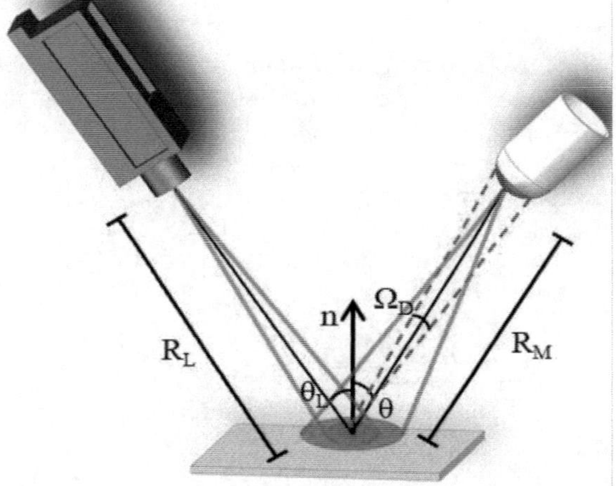

Figure 1. model of semi-active laser guidance system

Semi-active laser guidance process can be modeled as shown in figure 1. The relationship between the emitted laser power, P_L, of target designator and the received optical power, P_S, of seeker can be expressed by[7]:

$$P_S = \frac{1}{\pi} P_L \rho_T \gamma \Omega_D T_R T_F e^{-\sigma(R_L+R_M)} \cos\theta \tag{1}$$

where ρ_T is the target average reflectance; γ is the ratio of the target's scattering area to the illuminated spot area; Ω_D is the target scattering solid angle; T_R is the transmittance coefficient of the seeker optical system; T_F is the transmittance coefficient of the seeker filter; R_L is the distance between the designator and the target; σ is the attenuation coefficient of the atmosphere; R_M is the distance between the seeker and the target; θ is the angle between the target surface normal and the optical axis of the seeker; the target scattering solid angle Ω_D can be given by:

$$\Omega_D \approx \frac{\pi D^2}{4R^2_M} \tag{2}$$

where D is the clear aperture of the seeker optical system.

Assuming that the laser spot emitted by the designator is illuminated on the target totally, the target scattering area and the illuminated spot area are the same approximately (γ=1). In this case, equation(1) is simplified to:

$$P_S = \frac{D^2}{4R^2_M} P_L \rho_T T_R T_F e^{-\sigma(R_L+R_M)} \cos\theta \tag{3}$$

Fixed objects or slow-moving objects are usually attacked by semi-active laser-guided weapons. The distance, R_L,between designator and target keeps basically unchanged during the strike. It can be seen from equation (2) that the solid angle of the spot to the seeker is increasing as the distance between seeker and target decreases constantly, which is equivalent that the spot size captured by the seeker is increasing constantly in the process. It means that HILS system need to change the spot size constantly on the diffuse reflection screen. Assuming that the target reflectance, ρ_T,is known, other variables can be regarded as constants except R_M in the right side of equation (3), and the received optical power, P_S,of the seeker is a continuous function of R_M. This requires that the laser energy can be updated in real time according to the distance between the seeker and the target during the simulation.

3. Design of HILS system

The strapdown semi-active laser seeker consists of a quadrant PIN detector (QD), an optical system, and a shell. The QD and the optical system are fixed in the shell. In order to evaluate the strapdown seeker comprehensively, a HILS system designed in this paper is composed of a laser, an energy attenuation system, a digital micromirror device(DMD) spot projection system, a diffuse reflective screen, a three-axis turntable, simulation computer, etc. The diagram of the HILS system is shown in figure 2.

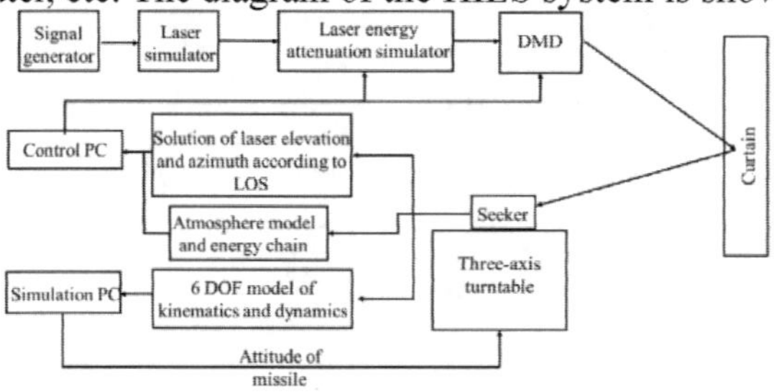

Figure 2. diagram of HILS system

3.1 General equipment

A 50mW, 1064nm pulse laser is used in the HILS system, with a maximum repetition frequency of 4 kHz. Laser frequency and laser code can be controlled by external trigger. The maximum rotation speeds of the three-axis turntable in the three axes of pitch, yaw and roll are 80 °/s、80 °/s、3600 °/s; and the minimum step angle is 0.01°. The three-axis turntable is connected to a control computer adjusting the seeker attitudes according to the angle information calculated by the seeker, which can test the guidance and control system of the seeker.

3.2 Energy attenuation systems

In order to meet continuous dynamic laser energy attenuation, a half-wave plate-polarizer combination attenuation system is designed to control the laser energy. The system includes two linear polarizers, a half-wave plate, and a high-precision stepping motor. The optical components are place as show in figure 3.

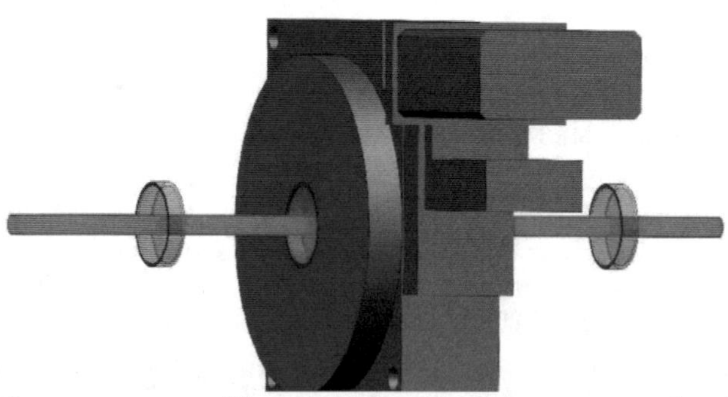

Figure 3. components of laser energy continuous attenuation device

The two linear polarizers are placed in parallel and the direction of the transmission axes is coincident. One is polarizer, and the other one is an analyzer. A half-wave plate fixed on a high-precision stepping motor is inserted between the two polarizers. The fast axis of the half-wave plate is consistent with the transmission axis of the two polarizers. The steeping motor drives the half-wave plate to rotate around the optical axis. The laser energy attenuation rate can be adjusted continuously between 0 and 1, by changing the angle between the fast axis of the half-wave plate and the transmission axis of the two polarizers. The transmittance, T , of the energy attenuation system satisfies the following express[8]:

$$T = \cos^2(2\theta) \tag{4}$$

where θ is the angle between the fast axis of the half-wave plate and the transmission axis of the two polarizers.

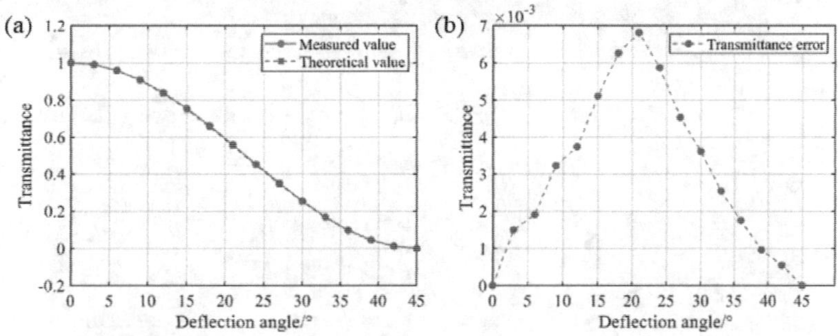

Figure 4. (a) Fitting curve of transmittance (b) error curve of attenuation system

In order to attenuate laser energy precisely, the relationship between rotating angle of the half-wave plate and the transmittance should be calibrated. The calibration can eliminate the effects of inconsistent optical axes, the misalignment transmission axis of the linear polarizers, and the misalignment zero position. The half-wave plate's fast axis is rotated from 0 ° to 45 ° with 3 ° as a step. The transmittance is measured at each point and the motor execution curve is obtained by fitting curve for measured data (figure 4). The maximum transmittance error is only 7×10^{-3} . The calibration result shows that the attenuation system can meet continuous and high-precision laser energy attenuation in HILS system.

3.3 DMD spot projection system

DMD spot projection system is designed to meet continuous laser spot size adjustment. The system consists of a beam expander, a DMD (German, ViALUX V-7001; refresh rate 22KHz, 7.6μm x 7.6μm/1microns, micromirror array 1064 pixel x 768 pixel) and a Nikon lens whose focal length is 80 mm. The laser beam is irradiated onto the DMD after being expanded by 5times. The reflected beam of the DMD is imaged on a diffuse reflection screen by a Nikon lens.

Each aluminum mirror of the DMD is a light switch. When the mirror is "ON" state, the mirror reflects incident laser into the pupil of the projection lens. When the mirror is "OFF" state, the mirror reflects incident laser out of the pupil of the projection lens[9]. When the DMD projection system and the diffuse reflection screen are fixed, the spot size and the spot shape on the diffuse reflection screen are achieved by adjusting the switch of DMD micromirror array. DMD can realize arbitrary shape output by controlling each pixel switch. When the system works, the spot shape, size and position of the projection graphics should be preloaded into the DMD.

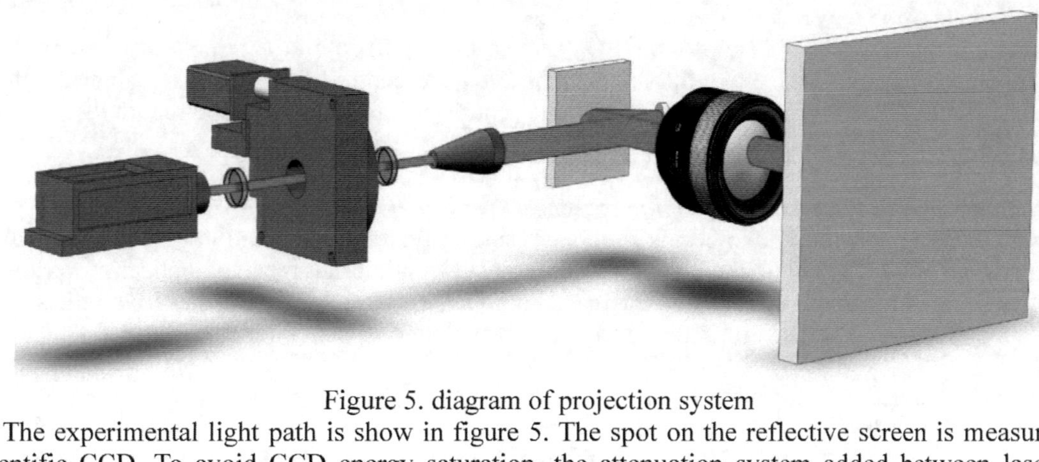

Figure 5. diagram of projection system

The experimental light path is show in figure 5. The spot on the reflective screen is measured by scientific CCD. To avoid CCD energy saturation, the attenuation system added between laser and

IOP Conf. Series: Materials Science and Engineering **793** (2020) 012065 doi:10.1088/1757-899X/793/1/012065

projection system adjusts the incident laser intensity.

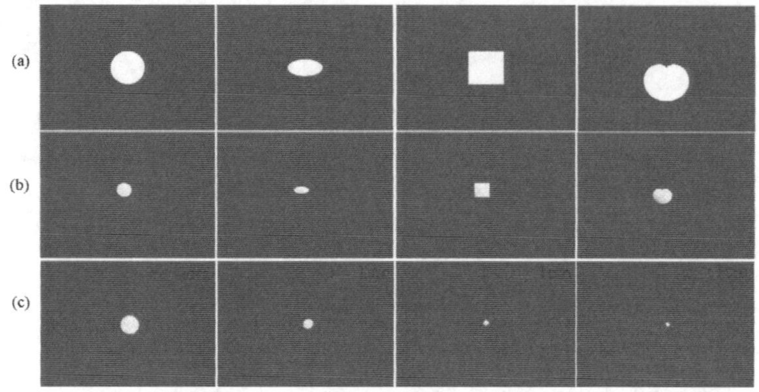

Figure 6. the results of projection test

The figures in figure 6(a) are preloaded into DMD, the spot on the reflection screen is shown in figure 6(b) taken by CCD camera. By comparison, the experimental results show that the projection system can meet different spot shapes simulation.

Assume that designator is 1 km away from target, the laser beam divergence angle is 0.4 mrad, and the laser is vertically irradiated on the target, the diameter of the spot is 0.4 m on the target. When the seeker is 5 m away from the diffuse reflection screen, and the distance between the missile and the target are 200m, 400m, 800m, 1000m respectively, the spot corresponds to 10 mm, 5 mm, 2.5 mm, 2 mm on the reflection screen. Figure 6(c) shows the simulated spot size by the projection system. The results show that the DMD spot projection system can simulate the spot changes well during the flight of the seeker.

4. Test results of the seeker

4.1 Response coherence among the four quadrants test

A large-area, uniform-intensity spot is needed to illuminate the whole detector for calibrating the response coherence among the four quadrants. The four-quadrant detector (QD) is fixed on a three-axis turntable and rotated around its optical axis. The position of the QD is adjusted until the output voltage values are not changed due to rotation. Record the output voltage of the four channels and calculate the response coherence among the four quadrants. The uncalibrated result is 0.946; after calibration, the response coherence of the QD is 0.984, with an increase of 3.8 %. The response coherence meets the requirement for use after calibration.

4.2 Field of view and linear field of view test

When measuring the angle of field of view(FOV), the seeker is fixed on the three-axis turntable first. Then, adjust the position of seeker so that the spot is on the center of the QD. Last, make angular movement to the seeker in both directions around the centrally aligned position and record its response. The experimental results show that the FOV is 30.28 ° and the linear FOV is 5.96 °.

4.3 Minimum received optical power of the seeker test

The laser beam is divided into two beams by a beam splitter prism, one is used to irradiate the seeker, and the other is used to irradiate the optical power meter. The minimum received optical power of the seeker is measured by gradually reducing the received optical power until the SNR is 1 with the help of energy attenuation system. The minimum received optical power of the seeker is 1.769nW.

4.4 angle measurement accuracy test

The seeker is about 8 m away from the diffuse reflection screen. The spot is 10 mm on the diffuse reflection screen, and the field angle of the spot to the seeker is 1.25mrad, which means that the missile

IOP Conf. Series: Materials Science and Engineering **793** (2020) 012065 doi:10.1088/1757-899X/793/1/012065

is 0.2 km away from the target. The angle measurement accuracy of the seeker can be measured by the absolute error, which refers to the difference between the deflection angle controlled by turntable and the angle processed by the seeker. When the seeker works in the linear FOV, the pitch angle, ε ,and the yaw angle, θ , of the spot to the seeker can be calculated by the following equations:

$$\begin{cases} x = \dfrac{R}{2}\dfrac{(V_1+V_4)-(V_2+V_3)}{V_1+V_2+V_3+V_4} \\ y = \dfrac{R}{2}\dfrac{(V_1+V_2)-(V_3+V_4)}{V_1+V_2+V_3+V_4} \\ \varepsilon = \arctan\left(\dfrac{y}{f}\right) \\ \theta = \arctan\left(\dfrac{x}{f}\right) \end{cases} \tag{5}$$

where V_1 , V_2 , V_3 , V_4 are the output voltage of the photosensitive sectors; R is the radius of the photosensitive surface; (x, y) is the coordinates of the spot center on the photosensitive surface; f is the focal length of the seeker's optical system.

Table 1. error of angle measurement

deflection angle (°)	processed angle(°)	error (°)
0	0	0
0.22	0.24	0.02
0.44	0.45	0.01
0.66	0.70	0.04
0.89	0.88	0.01
1.13	1.09	0.04
1.38	1.44	0.06
1.64	1.72	0.08
1.93	2.02	0.09
2.26	2.28	0.02
2.81	2.73	0.08

Here, we only list part results of the yaw because the yaw presents a symmetric distribution. The angle measurement accuracy in the linear FOV is an important index of the seeker. The result in table 1 shows that the angle measurement accuracy of the strapdown seeker is smaller than 0.1° within the linear FOV, which meets the work requirements of the system.

5. Conclusion

In this paper, to meet the demand of actual simulation tasks and functional requirements, we designed a HILS platform for strapdown semi-active laser seeker. In order to simulate the real flight environment better, the irradiation characteristics were studied from received energy of seeker and spot size on the diffuse reflection screen. Concrete schemes for laser energy continuous attenuation system and spot size continuous control system were put forward. It made the simulation environment closer to the real flight environment of the seeker. Some performances of the strapdown seeker were tested repeatedly. The results indicates that the angle measurement error of the seeker is smaller than 0.1° within the linear FOV, which means the seeker has the ability to hit the target precisely.

IOP Conf. Series: Materials Science and Engineering **793** (2020) 012065 doi:10.1088/1757-899X/793/1/012065

Reference

[1] Yang, SQ. (1999). Precision guided weapon and technology. Infrared and laser engineering, 28:3-5.
[2] Shen, YF, Deng, FL, Ke, XZ. (2002). Scheme design of semi-physical simulation system for a laser-guided bomb seeker. Infrared and laser engineering,31:74-77.
[3] Su, JG, Fu, MY.(2007). The application of simulation to the test and evaluation of guided weapons. Fire control and command control,32:6-9.
[4] Zhang, W, Zhang H, Zhang, XJ. (2016). Simulation on target acquisition of strapdown laser guided munitions in whole terminal trajectory. Infrared and laser engineering,45:23-27.
[5] Gao, SY, Liu H, Zhang HW, Zhang X, Chen J. (2019). Improve the detection range of semi-active laser guidance system by temperature compensation of four-quadrant pin detector. Sensors,19:2284-2294.
[6] Fan, SP, Lin, DF, Lu YL, Zong R. (2014). Design and achievement of hardware-in-the-loop simulation system for guided shell based on laser semi-active seeker. Infrared and laser engineering,43:394-397.
[7] Manojlovic, L. M. ,Barbaric, Z. P. . (2009). Optimization of optical receiver parameters for pulsed laser-tracking systems. IEEE,58:681-690.
[8] Yin, BY, Li, GY, Li, XC, Fan W. (2014). High power laser facility front-end energy control system. Acta photonica sinica,43:414-420.
[9] Hu, Y, Gao JB, Wang J. (2007). Study of dynamic infrared scene projection technology based on digital micromirror device(DMD). In: international society for optics and photonics. Beijing. 124-128.

AUTHOR INDEX

Institute of Physics
Dirac House, Temple Back
Bristol BS1 6BE UK

ISSN: 1757-8981
ISBN 978-1-7138-3505-9
9 781713 835059

CURRAN ASSOCIATES INC.
proceedings.com